12th Process Plant Safety Symposium 2010

Topical Conference at the 2010 AIChE Spring Meeting and 6th Global Congress on Process Safety

San Antonio, Texas, USA
21 – 25 March 2010

ISBN: 978-1-61738-434-9

Printed from e-media with permission by:

Curran Associates, Inc.
57 Morehouse Lane
Red Hook, NY 12571

Printed by Curran Associates, Inc. (2010)

For permission requests, please contact AIChE
at the address below.

AIChE
3 Park Avenue
New York, NY 10016-5991

Phone: (203) 702-7660
Fax: (203) 775-5177

www.aiche.org

Additional copies of this publication are available from:

Curran Associates, Inc.
57 Morehouse Lane
Red Hook, NY 12571 USA
Phone: 845-758-0400
Fax: 845-758-2634
Email: curran@proceedings.com
Web: www.proceedings.com

TABLE OF CONTENTS

Managing Management of Change

Jack Chosnek, Ph.D., P.E.
Knowledge*One*
PO Box 580465
Houston, TX 77258
jc@knowledge1.net

Prepared for Presentation at
American Institute of Chemical Engineers
2010 Spring Meeting
6th Global Congress on Process Safety
San Antonio, Texas
March 22-24, 2010

UNPUBLISHED

Managing Management of Change

Jack Chosnek, Ph.D., P.E.
Knowledge*One*
PO Box 580465
Houston, TX 77258
jc@knowledge1.net

Keywords: process safety, management of change, MOC, metrics

Abstract

The Center for Chemical Process Safety (CCPS) has designated MOC as one of the areas where metrics need to be developed. One of the issues that CCPS is concerned about is the quality of the MOCs as much as the quantity of those that have not been completed. One of the reasons is that people tend to concentrate on the mechanics of MOC rather than the content because of limited resources and the need to be in compliance. So how do we ensure quality as well as prompt closure of changes? By recognizing the elements that are crucial to make a good MOC and using a database to accomplish the mechanics, thus freeing personnel to concentrate on the important requirements.

It will be proposed that a good MOC requires three main elements to be very well executed: the technical description of the change, the safety analysis (which includes its review) and the Pre-Startup Safety Review. Having achieved quality it is important to manage the process efficiently and this can be done with the help of a database that automatically provides the necessary metrics. The requirements of high-quality MOCs, the necessary workflow to have an efficient process, and the metrics that are necessary to do a good job in managing the MOC process will be discussed.

1. INTRODUCTION

A watershed event in process safety, the 1974 Flixborough explosion in which 28 people perished and many more were injured, was due to a mismanaged change as its root cause. The change was to replace a malfunctioning reactor in the middle of two other reactors with a piece of 20" pipe. The reactor was part of a series of five low-conversion reactors used for the high-temperature oxidation of cyclohexane, in which the product of one reactor flowed into the next reactor. The change was done to minimize loss of production since repairing the reactor could have taken up to three months. It is obvious from the results, a pipe break caused by the pipe expansion and failure of the support (temporary scaffolding), that no safety analysis had been conducted. The review of the change had been done while engineers and chemists stood next to the equipment and decided that the change was technically feasible and would indeed minimize production losses. They then proceeded to implement the change.

This incident eventually led to the formulation of practices for managing changes that was codified in the U.S. by OSHA under 29 CFR 1910.119 in 1992. But, after all this time incidents caused by mismanaged change still happen. In the Citgo Refinery incident of July of 2009, in which there was a release of large amounts of HF resulting in severe burns to one employee, OSHA citations indicate deficiencies in the MOCs for changes in the equipment that failed and caused or contributed to the incident [1].

The Center for Chemical Process Safety (CCPS) considers MOC one of the main elements (out of five) that can be precursors to incidents and thus should be part of the leading indicators of process safety [2].

2. CAUSES FOR MOC FAILURE

We won't address here *recognizing* the difference between a change and a replacement in kind as this would be a training flaw in the general process safety system. Avoiding the MOC and calling it a replacement in kind on purpose is addressed here as one of the problems of the MOC system.

We have determined the need for an MOC *system* that works. What are the most common problems of MOC that lead to incidents? The most common problem is the lack of completion of an MOC, either lack of completion of a permanent change or timely removal of a temporary change. When looking deeper, the significance and safety impact of the change was not well understood and the proper priority was not allocated leading to the unfinished implementation. The second problem is poor performance of the safety analysis resulting in added risk to the process, and the third problem is poor technical description of the change resulting in a different change than the one intended. CCPS in Appendix G of their book on MOC [3] mentions some other common MOC problems and proposed solutions. The recurring theme of the problems mentioned seems to be complexity and inefficiency of the MOC system.

3. EFFICACY VS. EFFICIENCY

But inefficiency is not the main problem, as quality of the MOC is the most important part of the change process. If the safety analysis is incorrect or the change was not done as indicated, serious problems can result that could lead to an incident. It seems that the efficacy of the MOC, i.e. ensuring that the desired change is the one implemented and that the change does not impact safety, should be the principal consideration. Yet most MOC systems concentrate on the process, i.e. the workflow, rather than on the content. Nevertheless, it is our contention that an inefficient system leads to an ineffective one. The reason is that with limited time and resources the emphasis lies with compliance rather than quality. As the complexity of the MOC process increases less time is available for working on the content. Furthermore, a new element is introduced: as the time interval between the start and finish of a change increases, the risk to the facility increases. This is due to the chain of incomplete changes that are part of the MOC that can easily lead to errors (e.g. unimplemented P&ID changes that can accidentally be used by other projects, tie-ins that can become dead legs, etc.).

We will discuss the elements that make a good MOC and show how to make the process efficient.

4. ELEMENTS OF A GOOD MOC

There are three elements that are crucial for having a successful change: the technical description of the change, the safety analysis, and the Pre-Startup Safety Review (PSSR). Let us see why.

1. The **technical description** is crucial for two reasons: it allows us to conduct a proper safety review and it allows us to ensure that the change that will be implemented is the desired change and not a similar one. We can do a safety analysis only of what is described and do it correctly if we understand the technical basis and the assumptions that have been made. Similarly, with a good technical description we avoid creeping changes within the change. Without a detailed technical description, the final product may be different from the conceptual one.
2. A good **safety analysis** is the crux of the MOC. The purpose of doing an MOC is to ensure that the change doesn't unknowingly increase risk or compromise the safety of the existing process. If the change is a complex change then a PHA may be necessary for a good safety analysis. If a PHA is not considered necessary, then peer review of the analysis should be undertaken to help ascertain that it was done correctly. Ensuring that a review of the safety analysis was performed by knowledgeable personnel is part of managing the MOC process and a part that usually is not undertaken. Often, the review is perfunctory and done by a supervisor who is only seeking to move the process forward and complete the MOC. A good MOC system needs to spell out how the review of the safety analysis is to be undertaken.
3. The **Pre-Startup Safety Analysis (PSSR)** makes certain that the desired change was the change that was implemented before it is activated. Again, to accomplish this we need to have a good technical description. The PSSR will also check that the personnel who are going to operate the new or modified equipment have been *properly* trained on it and that the operating procedures have incorporated the change. Lastly, it will check that a safety review was performed and that if there were any recommendations coming out of the review, they were correctly implemented. But, the PSSR will not check the quality of the safety review. If a quality review was not done immediately following the safety analysis, the opportunity to discover any potential hazards is lost.

So we see that if we prepare a good technical description, do a good safety analysis (which includes a review) and verify that the change is technically correct and we have the tools and training to operate the change, we will have implemented a safe change. The rest is ancillary and the MOC process is built to make certain that these three elements will be in place when making a change.

5. THE MOC PROCESS

The MOC process is a work flow that regulates the way an MOC will be conducted. As such, it ought to be a company approved procedure that guarantees the MOC process will contain the necessary elements and be properly executed. This procedure should allow management to estimate the resources necessary for carrying out all phases of the MOC, including the resources for the safety analysis, its review, training and PSSR elements, as well as the work and resources necessary to build or modify the equipment involved.

The work flow ought to be relatively simple: propose the change, obtain preliminary agreement on its need and availability of resources, prepare a detailed description of the change, conduct a safety analysis and have it reviewed, obtain approval to proceed as described and analyzed, build the change (preferably through the plant's work order system already in place but referencing the MOC), and verify it. This simple work flow is shown in Figure 1.

Required steps of the workflow should include (responsible person in parenthesis):

1. Appoint an MOC Coordinator to coordinate the MOC from start to finish (Line Management)
2. Develop a detailed technical description of the change (MOC Coordinator)
3. Determine the completion date of the change or date of removal of a temporary change (MOC Coordinator)
4. Develop a checklist of items and responsible persons that will ensure that the technical specifications and the safety requirements will be fulfilled (MOC Coordinator)
5. Perform a safety analysis of the change (MOC Coordinator; if a PHA is required, the MOC Coordinator will arrange for it)
6. Request approval of items 2 through 5 (MOC Coordinator)
7. Appoint a safety analysis reviewer (Line Management)
8. Review the safety analysis (MOC Safety Analysis Reviewer)
9. Make an assessment that all safety and technical requirements can be implemented, that sufficient resources are available to do so in the time required, and ratify the safety analysis by approving the MOC (Line Management)
10. Perform the work (MOC Coordinator will monitor progress with the persons in the checklist assigned to do the work)
11. Determine that the work is complete and ready for activation (startup) of the change (MOC Coordinator)
12. Ensure that all PSSR requirements are met:
    a. Technical specifications are met (MOC Coordinator)
    b. Safety recommendations have been incorporated (MOC Coordinator)
    c. Personnel have been trained and SOPs have been updated (Operations)
13. Coordinate startup date (MOC Coordinator)
14. Complete documentation and enter date when this step is completed signaling completion of MOC (MOC Coordinator)

6. MANAGING THE MOC PROCESS

As can be seen, there are quite a few steps in completing an MOC; and if a paper system is used to verify every step, there could be a lot of paper being carried around by, and to many people. This cumbersome process is a barrier to timely completion of the MOCs and we need to streamline it if we want to have an efficient system. As mentioned above, in the absence of an efficient system people will spend their time in verifying that other persons have done their part and won't have time to dedicate to the important issues. Many facilities require a signature for each step (and sometimes each sub-step) in order to have a record of who did what when. It becomes difficult to determine the status and progress of an MOC. Managing the process then evolves into a reactive system where corrective action is needed increasing the amount of work needed and a dislike for doing it.

A database where everyone can log in to indicate completion of an MOC or their assigned part to it or to view its status is needed in order to eliminate the bothersome mechanics of the system. The database should be able to automatically provide a status report of each MOC, collect the metrics that will help us manage the facility's (or the company's) MOC process and give us a measure of how well we are doing in this area.

The necessary features of a database in order to manage the process efficiently are:

1. Automatic assignment of a protected (cannot be changed), unique number to the MOC when it is created.
2. Have protected creation (automatically set) and modification dates of the MOC (the modification date serves as an audit tool).
3. Secure login that identifies and records the person doing the assigned functions of MOC coordination, safety review, approval, and confirmation of the various elements of the PSSR, as well as the privileges of each one of those persons in terms of what they can enter or modify in the MOC. The logins will serve as electronic signatures.
4. Required classification of the MOC as either permanent or temporary.
5. Required target date for completion of a permanent MOC or a firm date of removal of a temporary change. This date will be used to monitor whether MOCs are being closed out and to measure how many are open (in progress). It should allow a supervisor with approval privileges to extend the time of a temporary MOC after recording the justification for the extension.
6. It should have practically unlimited space to enter descriptions, safety analysis, checklist items, justification, etc. and have the capability of attaching relevant documents.
7. Requires following the preset work flow, that is, no data can be entered in one section unless the previous section has been completed. It automatically logs in the dates of completion of every section.
8. It has the capability of creating one-button e-mails that automatically include the relevant MOC information to request approval, ask for status, request input, warn of the approach of the target date, etc.
9. It should declare the MOC closed out only after confirmation that the documentation has been updated.
10. The database should be easily and fully searchable and reports should be able to be sorted by each heading.

With all the capabilities described above, it's possible to greatly reduce the time wasted in bureaucratic activities and have more time to concentrate on the important issues. These capabilities should also give management a picture (through automatic database reports) of:

1. The current number of open MOCs, past-due MOCs, and those that whose target date is approaching (let's say within 30 days)

2. The status of each MOC with dates for each stage of the workflow and the responsible person for the MOC (see Figure 2)
3. A one-button capability to send individual e-mails to the group of persons involved in one of reports (open, past due, due in 30 days) containing the relevant information.

The facility's process safety coordinator should be charged with monitoring the database to see the status of the MOC system and be able to communicate with management when the system starts to falter. For example, if there too many open MOCs it is a signal that the risk to the facility has increased (the cumulative potential for error is large) and an indication that too many changes are being made at one time. If there are too many past-due MOCs is an indication that not enough resources or priority have been allocated to these changes. If past-due MOCs are temporary MOCs, then it is an indication that the safety of some processes is being compromised. The MOCs that will become due in 30 days will serve as a warning of the amount of work and resources that will be needed in the short term in order not to decrease the safety of the facility. The e-mail feature of the database should be an excellent aide for this communications not only to management but also proactively to the people involved in the MOC process.

7. METRICS

Although we have seen how to easily manage the MOC system on a day-to-day basis, we need to have a measure of how well the system is doing long term. For this type of evaluation we need metrics. We will discuss first the CCPS metrics which are an industry-wide effort and then proposed use of the database to gather company metrics that are actual and directly related to the management of the MOC system.

7.1 CCPS Metrics

The metrics proposed by CCPS [2] are leading metrics for incidents and would have to be collected over a few years and be correlated with incidents and near misses with the participation of enough companies to be statistically significant. For companies participating in this effort, the database proposed in this paper would help collect these data and may help to avoid incurring the deviations described. The CCPS metrics for MOC are:

1. Percentage of audited MOCs that satisfied all aspects of the site's MOC procedure
2. Percentage of audited changes that used the site's MOC procedure prior to making the change
3. Percentage of start-ups following plant changes where no safety problems related to the changes were encountered during re-commissioning or start-up

We can see that by using the database for the MOC system we can easily satisfy the first and second metrics since the database is continuously audited and it forces the user to comply with all its aspects and all the changes have to follow the procedure. The only changes that may possibly not follow the procedure are changes classified as replacement-in-kind that bypassed the system because of a failure to recognize the change or because it was purposefully ignored. The former failure is a training problem and as for the latter, hopefully by having an easy-to-comply system there will be much less of an incentive to risk the wrath of management. A periodic audit of the *work order* system should uncover the culprits in both cases. In order to capture the third metric a note can be added to the specific MOC where safety problems occurred. Hopefully, they will be few and far between; and by having a procedure that concentrates on quality rather than compliance, they will be rare.

7.2 Management Metrics

In order to understand how the system is behaving over time, we need to collect data on the process as we go and analyze it at certain time intervals. As mentioned before, we want to increase safety and minimize risk by having a well performing system. Safety will be increased by performing good safety analyses, and risk will be minimized by minimizing the number of MOCs that are open at a given time and avoiding extending temporary MOCs past the time in which they were assessed to be safe.

By creating a graph of the number of open and past-due MOCs, both temporary and permanent, over a time period we can see if we are allocating the necessary resources to the system. Tracking the time element may help personnel understand why resources were not available at the needed time - perhaps because of a turnaround or a hiring freeze - and improve future planning. If, in addition, we compare these numbers with the number of MOC starts and completions over the same time period, we may realize that too many changes are undertaken given the facility's resources or perhaps not enough priority is given to their completion. The database should provide these graphs automatically (see Figure 2). By inspection of the database we can easily see if there is a specific stage of the MOCs that is being the bottleneck and address the specific problem (Figure 3).

8. CONCLUSIONS

We have shown that an MOC system that results in safe changes can be developed. The key to developing that system is to put emphasis on the important parts of an MOC such as the technical description, the safety analysis and the PSSR, and having tools that automatically take care of the mechanics and give us the time to focus on those emphasis elements. In addition, these tools should also easily provide us the information and feedback that will allow us to manage the system well.

Figure 1. Simple MOC Workflow

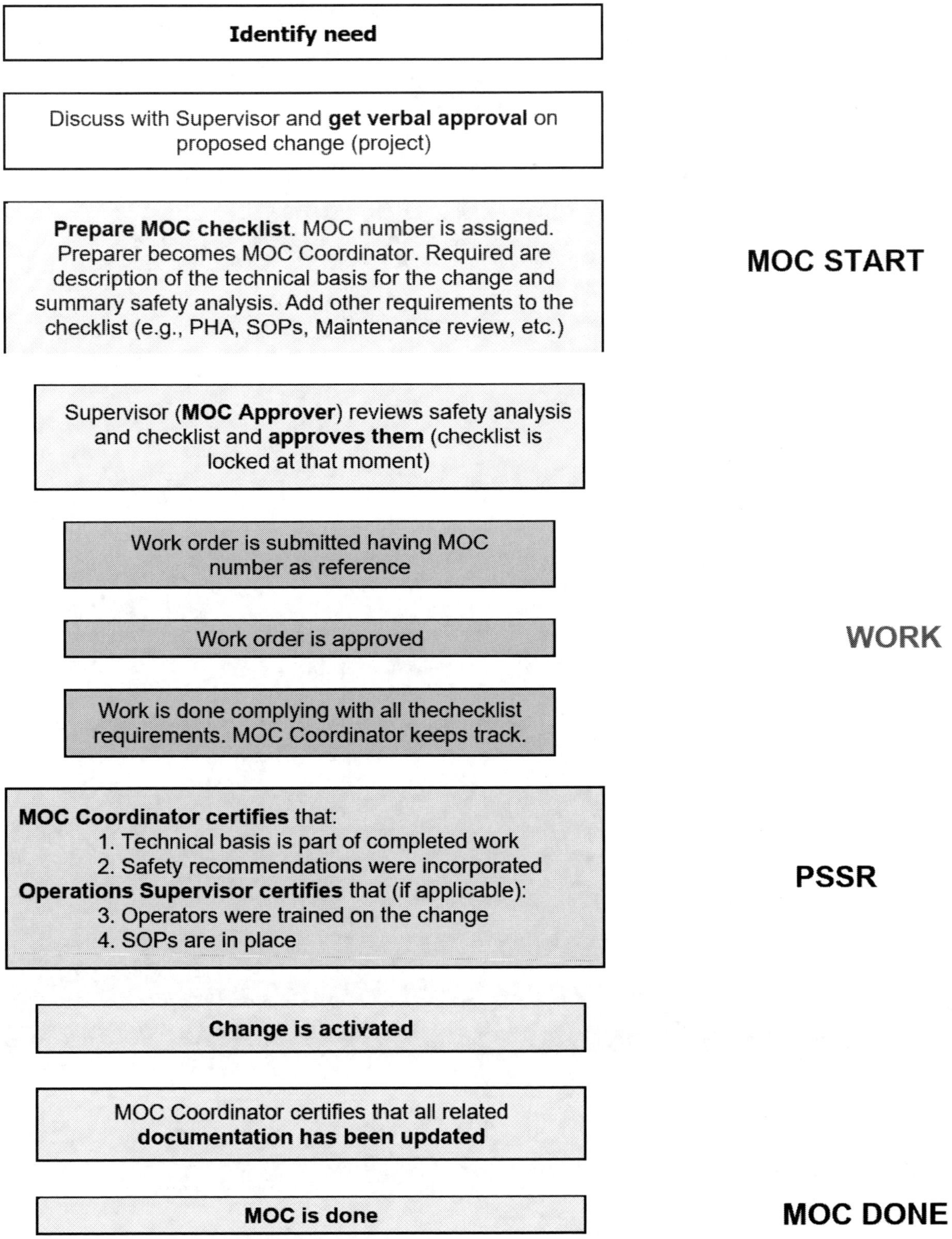

Figure 2. Open and Past-Due MOCs over a period of time compared to starts and completions over the same time period.

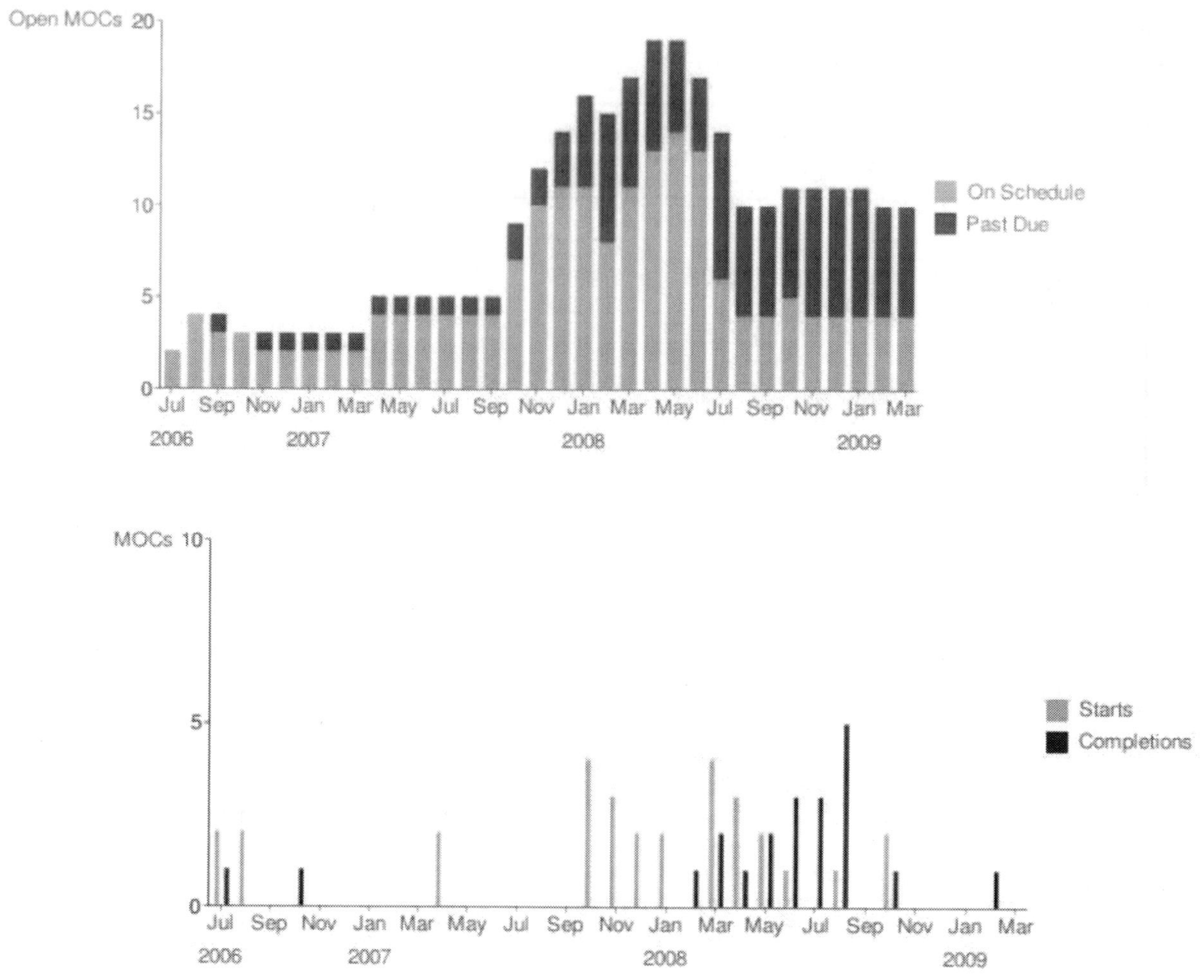

Figure 3. Database screen showing status of all MOCs

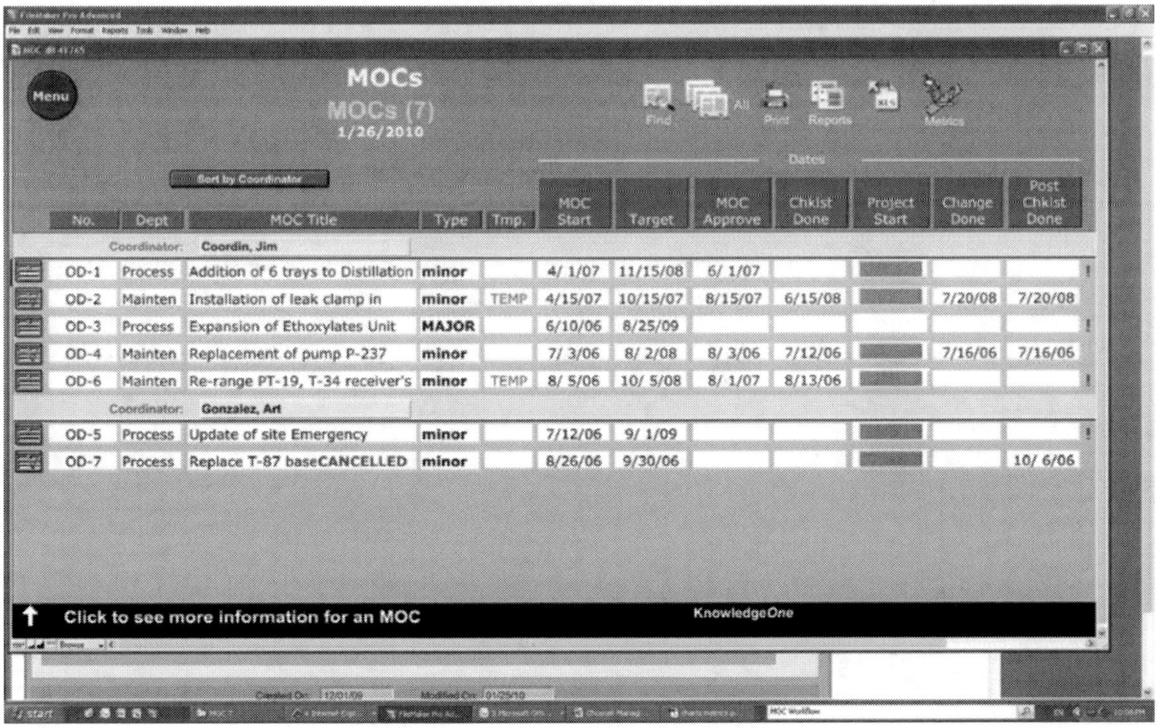

9. REFERENCES

1. U.S. Dept. of Labor, OSHA, Citation and Notification of Penalty to CITGO Refining & Chemicals Company LP, January 15, 2010.
2. Center for Chemical Process Safety (CCPS), "Process Safety Leading and Lagging Metrics – You Don't Improve What You Don't Measure", CCPS Website Publication, 2008.
3. Center for Chemical Process Safety (CCPS), "Guidelines for Management of Change for Process Safety", Wiley-Interscience, 2008.

Are You Tired of Playing "Whack a Mole" with Your Process Safety Results - If so, it's Time for a Change

David E. Cummings, MS, CIH

Global Process Safety Management Competency Leader
Safety, Health, Environment and Sustainable Growth Center
E. I. du Pont de Nemours and Company
Wilmington, DE, USA 19898
(302) 774-9558
david.e.cummings-1@usa.dupont.com

Prepared for Presentation at
American Institute of Chemical Engineers
2010 Spring Meeting
6th Global Congress on Process Safety
San Antonio, Texas
March 22-24, 2010

UNPUBLISHED

Are You Tired of Playing "Whack a Mole" with Your Process Safety Results - If so, it's Time for a Change

David E. Cummings, MS, CIH

Global Process Safety Management Competency Leader
Safety, Health, Environment and Sustainable Growth Center
E. I. du Pont de Nemours and Company
Wilmington, DE, USA 19898
(302) 774-9558
david.e.cummings-1@usa.dupont.com

Keywords: Management of Change – Personnel, Training

Abstract

Recurring or chronic performance problems over time within any process safety management system at a site or within a corporate structure can introduce additional unintended risks, generate internal and external compliance issues, and impede continuous improvement. Ineffective management of changes to technology, facilities, people, internal procedures, and organizational structures all contribute to potential process safety performance issues. This paper will review and briefly summarize several unique process safety incidents which were attributed to inadequate management of change implementation. Additionally, all the elements of a process safety system, including MOC procedures, may be in place, but without knowledgeable people to apply them, the likelihood of achieving and maintaining process safety excellence is greatly diminished. While not always obvious, many process safety performance issues can be traced back to a central aspect of a management of change program – effective control of personnel changes throughout the manufacturing organization and across many key roles that impact process safety. This includes personnel transfers, retirements, re-assignments, and other related people changes that must be anticipated and managed. A detailed discussion on the key elements of a "management of change – personnel" process is described, including practical tools and solutions that can assist organizations to maintain and strengthen this element of PSM.

1. Introduction and Background on DuPont

Development and sustainability of a comprehensive global process safety management program is critical within a large, multi-national company such as DuPont in order to manage a diverse set of process risks and positively contribute to business success (1). The program includes detailed and specific mandatory technical elements applied to all 180 global manufacturing locations to help ensure consistency. DuPont's overarching PSM philosophy is to develop and maintain PSM technical expertise within site operations to the greatest extent possible in order to address routine needs and issues, and rely on regional and corporate teams and resources for strategic planning and to assist sites with complex or unusual PSM technical issues. Based on this approach, a detailed Management of Change – Personnel (MOC- P) program is critical to maintain individual and organizational knowledge and capabilities over time within sites, regions, and corporate functions to prevent incidents and sustain PSM program performance and leading indicators.(2)

About DuPont

DuPont is a global science based product and services company that operationally is organized around thirteen (13) main businesses across four global regions, which serve over 200,000 customers worldwide with annual revenues of $30 billion. Founded in 1802, the company operates at more than 180 manufacturing sites in 70 countries involving 64,000 employees.

DuPont operates many different manufacturing technologies involving both higher hazard processes and lower hazard operations. These technologies, products, and associated process safety hazards (fires, toxics, explosions, reactives, and mechanical energies) vary widely across each business and site.

2. PSM Incidents Related to Management of Change Personnel

Inadequate management of personnel changes can result in serious process safety incidents and near misses. The involvement and deficiencies associated with the MOC-Personnel element as part of an incident can be subtle and may not be readily obvious unless the investigation team conducts a detailed review of the facts and specifically evaluates the relevance of personnel who may be new to their roles and how their knowledge and expertise (or lack thereof) may have contributed to the event. MOC-P factors will be identified only as part of a thorough root cause failure analysis focused on underlying management systems where lack of knowledge or human error (both omission and commission) could have played a role. The following past incidents within DuPont are described briefly as examples where MOC-Personnel was involved.

Incident A - Near Miss Involving Inadequate Design of Electrical Grounding on Rail Car Spot

An electrical arcing incident occurred at a newly installed rail car spot designated to unload flammable liquids and wastes. The project engineer and site operations contact were newly assigned and were not familiar with internal DuPont design standards (Figure 1), project procedures, and requirements for this type of equipment and operation. Further, a Prestart Up Safety Review was not fully completed per site policy using tools and checklists which should have identified these defects. The incident concluded that, among other factors, the new technical and management personnel did not receive adequate training and develop competency to address all the PSM systems and procedural requirements within their assigned roles for this project.

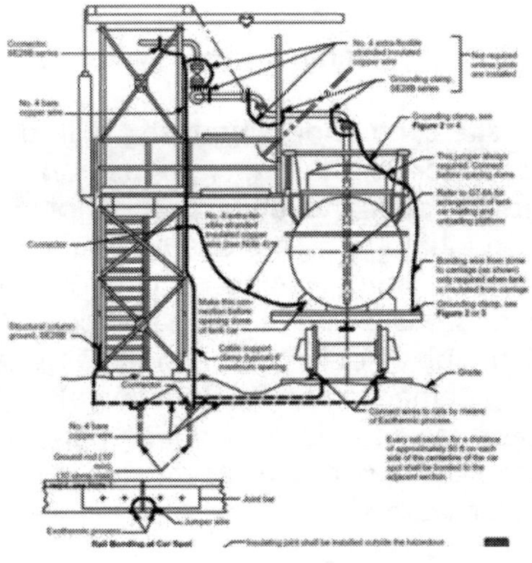

Figure 1: Electrical Grounding and Bonding Diagram for T/C Unloading Spot

Incident B – Polymer Line Plug in Line Break Job with Gas Release and Fire

A large hardened polymer plug developed in the high pressure separator (30,000 psig) and associated piping which handles flammable gas (Figure 2) as a result of a process upset and troubleshooting that had extended outside of standard operating conditions. Upon process shutdown, a line break permit and maintenance plan was developed and conducted in order to clean out the equipment. After completing a first break, the polymer plug released and a substantial pocket of pressurized gas which was trapped behind the plug discharged and was ignited. The incident concluded that, among other factors, several people in key operations, technical, and maintenance roles that were involved in developing the line break plans (and were relatively new to the operating area) did not fully understand the process technology for polymer manufacturing (including equipment design, polymer operations outside of SOC's, and consequences of deviation). Therefore, they did not recognize the potential for the hazards of trapped high pressure gas buildup due to system plugs.

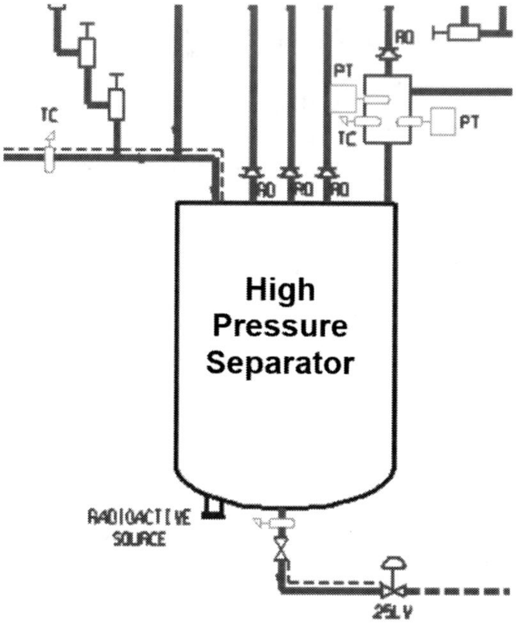

***Figure 2*:** High Pressure Separator and Associated Process Piping

Incident C - Contractor Deviation during High Pressure Water Jetting on Process Equipment

An incident occurred involving a site contractor conducting high pressure water jetting (20,000 psi) on plant equipment used for specialty chemical manufacturing. The site contractor field coordinator assigned to manage and audit the jetting contractor was new in the assignment and had not completed the internal (mandatory) high pressure water jetting training at the time of the incident. The contractor's jetting equipment was not fully functional for line moling and emergency shutdown per site standards which resulted in a serious jetting injury. It was later identified that the contractor should also not have been jetting certain equipment components due to the potential for equipment damage. The incident investigation concluded that the contractor coordinator did not possess adequate knowledge on the jetting procedures and equipment design requirements to manage and audit the contract work with a specific focus on preventing operating failures and incidents.

Figure 3: High Pressure Water Jetting of Process Equipment

3. Management of Change – Personnel Internal Requirements

Site Based Minimum Staffing and Change Management Criteria

Each DuPont site is required to develop and apply minimum criteria involving changes over time for key roles that are essential to develop and maintain PSM program elements and performance. Key site PSM roles are outlined below in Table 1. These minimum criteria are intended to proactively manage the number of personnel changes over time across the organization to maintain knowledge and experience within each of the identified roles. The following examples are provided to illustrate these change criteria:

- The Site Leadership Team (SLT) consists of the site manager, five unit operations leaders, and three functional leaders (maintenance/engineering, human relations, and EHS). The site's MOC-P criteria allows a maximum of two personnel changes of the total nine leadership roles within a 12 month period.
- The Manufacturing technical group within an operating unit consists of six technology resources, including three chemical engineers, two MT Chemists, and one process control/DCS resource. The unit MOC-P criteria allows a maximum of two personnel changes within a 12 month period
- A large site PSM Committee consists of a PSM Leader and fourteen members (each of which leads one element of PSM). The site MOC-P criteria allows a maximum of four personnel changes on the PSM committee within a 12 month period

Sites typically maintain a staffing matrix of all exempt roles (technical, maintenance, line management , PSM/EHS, and support roles) and operating personnel (operating technicians and mechanics) to track the number of changes per unit time and verify compliance with change criteria.

These change criteria should be managed when personnel changes occur involving planned transfers, reassignments, promotions and other typical HR strategies however, little or no control can be applied for other "reactive" personnel changes such as resignations, and major restructuring, business shutdowns, or reductions in forces. With these types of unusual conditions, sites are required to develop a specific intervention plan to manage uncontrollable personnel changes.

Table 1: Key Site Roles Related to PSM MOC- Personnel

Site Role	Description
Manufacturing Technical	Chemical Engineers and similar technical roles within operations who manage process safety information, new products, and changes to technology
Reliability Engineering	Mechanical engineers and similar roles within operations who manage Mechanical Integrity and Quality Assurance
R&D Personnel in support of Operations	Chemists, technicians, and other research personnel who work with operations on product formulations, trials and new technologies
Line Management	First Line Supervision, Area/Unit Management, Site Managers and all associated operations leadership
Process Control Engineering /E&I	DCS and process control engineers, E&I resources associated with SIS design and operations
Contractor Administration	Field personnel who develop, administer, and oversee contactor work scope and routinely audit performance
Equipment Examiners/NDE Resources	API Inspectors, QA personnel, and other employees who schedule and conduct MI tests and inspections for all types of PSM critical equipment components
Operators and mechanics	Operations personnel who conduct day to day work on process equipment
PSM and Fire Safety Coordinators and Committee members	Site PSM Leaders, PSM committee members, Fire Safety resources, HTM resources, leveraged technical resources
Emergency Response Leaders	Fire Chiefs/Brigade Leaders, Incident Commanders, Security, off -site emergency response Managers
Non DuPont Personnel in Key PSM Roles	Contractors in key PSM roles such as Supervisors of resident contractors, engineering design, PSM consultants or retirees

Off Site Roles and Management of Change - Personnel

In addition to site based MOC- Personnel systems, key off site PSM roles should also be included in a management of change process. Table 2 lists typical roles within DuPont where an MOC-personnel approach is applied, including regional and corporate PSM resources, Operations and Business Leaders, and special focus teams. These roles typically do not impact operating sites on a daily basis, however, these leadership roles may directly involve PSM decisions on strategic objectives, resource staffing, finances, project approvals, and leadership performance reviews which flow through Operations and Businesses leadership

Table 2: Key Off Site Roles Related to PSM MOC- Personnel

Key Off Site PSM Roles	Description
Regional Operations Directors	Senior Operations Leader within each region with manufacturing oversight for 40+ sites
Regional EHS Leaders	Health and Safety Manager within each Region including leadership for Regional PSM resources
Regional PSM Leaders	PSM Technical Leader for each region PSM program
Corporate Engineering PSM Consultants	Senior PSM technical experts in DuPont Engineering who complete complex PSM technical work/training
Corporate PSM Competency Personnel	Senior PSM resources in corporate EHS who lead global PSM elements and initiatives
Business EHS Leaders	Health and Safety leaders with strategic planning roles in support of all 13 businesses

Business Supply Chain Leaders	Senior Operations leaders in each business
Business Engineering Managers	Engineering managers involved with new projects, new plants, and infrastructure renewal for businesses
Business Technology Managers	Senior Technology Managers with oversight and development of site technology and R&D personnel
Global Operations Vice Presidents	Executive leaders for regions and businesses
Sister Plant Technology Leaders	Leveraged personnel leading DuPont Sister Plant Technology (SPT) networks with PSM focus
Highly Toxic Material Guardian Team Leads	Leveraged personnel leading DuPont Highly Toxic Materials teams (e.g., chlorine, HF, HCL, S03, etc)
Global Sourcing Commodity Team Leads	Sourcing/Purchasing personnel who lead commodity team for procurement of PSM critical equipment

4. Departing Incumbents

Personnel moves involving key site roles listed in Table 1 relative to process safety management occur within any organization frequently due to a number of reasons, including transfers, promotions, terminations or downsizing, resignations, and long term disability management. Loss of experienced personnel can dramatically impact any organization's capability and process safety performance over time if not managed effectively.

Identification of PSM Roles, Knowledge and Skills Associated with Departing Personnel

A key step in the MOC-Personnel process is to formally catalog a list of specialized PSM and other skills possessed by the departing incumbent before the person exits the site. Depending on the reason for departure, this review typically is conducted by the line manager in coordination with PSM/EHS resources several months or weeks prior to the last day on site. The intent is to document specialized skills, training, tools, knowledge or roles that should be either reassigned within the organization, re-developed in new or other personnel, contracted out to external resources, or other means provided to complete the work. Sites typically use a standardized checklist or document to review any special skills and knowledge, and document those relevant items that require tracking and follow-up. One typical document to accomplish this is a "Skills Knowledge Abilities and Functions (SKAF) Checklist" which contains a detailed set of criteria for review. A partial list of content examples for an SKAF checklist are provided in Table 3

Table 3 - List of Skills, Knowledge, Abilities and Functions - Departing Incumbents

Site PSM Role or Action Item	Purpose and Outcome
Specialized skills or qualifications	Reassignment of specialized roles and skills related to plant operations, maintenance, technical, process control , or similar subjects
PSM Software and Information Technology	Retention of PSM Software, licensing, electronic team rooms, or information data management systems
Certifications and Licenses	Reassignment of key certification roles involving site PSM engineering, operations, and maintenance elements
PHA Resource or Facilitation	Reassignment or development of key roles involving Process Hazards Analysis facilitation, resourcing, training, quantitative consequence analysis for fire/explosion modeling
PSM Auditing (First or Second Party)	Reassignment or development of key roles involving PSM auditing program elements such as qualified auditors, trend analysis

RCFA Resource or Facilitation	Reassignment or development of key roles involving Incident Investigation facilitation, training, trend analysis
Key PSM Leadership Roles	Reassignment and transition involving key site PSM roles such as Element leaders, PSM committee members, off site PSM team participation or leveraged activities
PSM Related Projects and Improvements	Transfer of longer term PSM improvement projects, annual objectives,
PSM Related Records	Transfer of key PSM related hard copy and electronic records within individual custody (not already retained in centralized data management systems)
PSM Related Action Items	Reassignment of all open or unresolved PSM action items to alternate personnel as part of turnover
PSM Regulatory Roles and Responsibilities	Reassignment, document updates, external notifications involving regulatory permits, roles, contacts
PSM related contract or vendor oversight	Transfer of specific contracts or oversight roles of third parties who are involved in support of PSM (engineering designers, T&I resources, equipment vendors/suppliers)
Participation in external PSM activities	Reassignment of key roles involving participation with external PSM teams and activities (LEPC, CAP, state, federal , industry groups or technical associations)

5. Newly Assigned Personnel in Key PSM Roles

Similar to departing incumbents, newly assigned personnel on site may be due to a number of typical personnel actions including new hires, transfers, promotions, or reassignments due to organizational changes. In all cases, its critically important to develop and apply a consistent plan for new personnel in order to instill short term PSM knowledge and demonstrate a fundamental competency on PSM management systems and specific PSM information related to the assigned role and operating unit. The following section outlines standard elements of this change process for new personnel at all DuPont global sites

Short term PSM Knowledge and Competency Demonstration

Each newly assigned person in a site role identified in Table 1 is required to complete a PSM Knowledge and Competency Demonstration within 90 days in order to ensure the individual has a fundamental and consistent understanding of key PSM information. The key areas of information that are involved with this competency include the following:

Understanding of DuPont PSM elements, essential features, and technical requirements

Understanding of how the local site PSM program is structured, including roles and responsibilities, policies and procedures, regulatory coverage, and documentation management , including expectations for line leaders who review and approve PSM related documents such as PHA reports, incident investigations, MOC's, and verification of completed action items.

The process technology and process safety information associated with the operation(s) and areas where the individual is assigned

Knowledge of hazard information, consequence analysis and key administrative And engineering layers of protection based on existing Process Hazards Analyses

A formal means to complete this competency demonstration exists at each site, however the format and detailed methodology may vary based on local practices and organizational structure. This can include a formal panel review over several hours where the individual must respond to questions and discussion points from experienced peers, a written test and field practicum on process equipment and operations, or a combination of computer based evaluation and interviews with key PSM resources and line managers.

Long Term PSM Training and Development Plan

In addition to a short term competency demonstration, newly assigned personnel and line management are also required to develop a longer term PSM training and development plan, The training content for this plan will vary based on the role of the individual, and this plan typically includes more formal classroom technical instruction from either internal or external sources. The plan and any actions must be tracked within the site action tracking database and is reviewed annually for updates or modifications. A list of "typical" training courses and a brief description and training objective currently offered internally within DuPont is provided in Table 4 (external training courses may also be utilized to satisfy individual developmental needs).

Table 4: Examples of Internal DuPont PSM Technical Training

PSM Technical Training	Length	Description
Process Safety and Risk Management	2 days	Technical or maintenance engineers, operations line management, PSM leads
Process Safety Management Auditing (initial and refresher)	2 days	1st and 2nd party PSM auditors
Process Technology	1 day	Manufacturing technical resources
Process Hazards Analysis Facilitation (initial and refresher)	5 days/ 2 days	Site and region PHA resources
Consequence Analysis for Fires and Explosions	5 days	PHA resources
Consequence Analysis for Toxic Releases	5 days	PHA resources
Chemical Reactivity Evaluation and Management	2 days	Manufacturing technical resources
Incident Investigation	1 day	All personnel
Layer of Protection Analysis and PFD Calculations	2.5 days	PHA, E&I, process control and manufacturing technical resources
Inherently Safer Technologies	1 day	R&D , manufacturing technical
Contractor Safety Management	1 day	Site contractor coordinators
Operational Discipline Workshop	1 day	All personnel
Safety Instrumented Systems Overview and PFD Calculations	3 days	PHA, E&I, process control and manufacturing technical resources
Pressure Relief System Design	3 days	Manufacturing technical and design
Combustible Dust Hazards	1 day	Manufacturing technical resources
Furnace and Fuel Safety	1.5 days	Site Operations and technical personnel operating burner management systems
Fire Protection Systems Management and Fire Technology courses	Variable	Site Operations, Fire Safety , and PSM resources

6. Corollary PSM Elements Connected to MOC-Personnel

PSM Auditing

Compliance with specific site and corporate MOC-personnel requirements is evaluated during all first and second party PSM audits which are conducted at higher hazard sites every three years. The audits evaluate personnel changes within the previous period based on organizational and staffing charts, and will evaluate specific personnel changes (and associated knowledge transfer and training) involving both departing incumbents and newly assigned personnel through interviews and documentation reviews.

Incident Reporting and Trend Analysis

DuPont's current practice involves corporate reporting of all PSM incidents within a common incident electronic database. As part of each formal incident investigation, the applicable elements of PSM are tagged when they involve either a key aspect or contributing factor related to the incident based on the applicable root cause failure analysis. Management of Change – Personnel is one of fourteen PSM elements which may be identified as an area which needs to be strengthened as part of incident follow-up, therefore, learnings and trend analyses may be generated periodically where MOC-P is related to incidents entered in the database.

Annual PSM Business Reviews

Each of the thirteen DuPont businesses conducts an annual PSM program review to evaluate systems, performance, technical strategies, and future leadership focus areas. Management of Change –Personnel involving changes in key Operations Leadership and business roles is assessed as part of these annual reviews to help ensure compliance with corporate practices. Where businesses sponsor key global PSM teams on highly toxic materials, chemical reactivity, and sister plant technology networks (3), personnel changes in leadership or membership on these teams is also evaluated as part of the annual review.

7. Summary

The fundamental principle of DuPont's Management of Change - Personnel PSM element states the following:

> "Recognizing that people are the essential ingredient that is interwoven through all the elements of PSM, it is important to maintain a minimum level of (1) specific direct process experience and (2) knowledge and skill in managing process safety within the site line and support organizations. Loss of minimum levels of experience and knowledge through personnel movements and organizational change, like changes to technology or facilities, potentially invalidates prior assessments and risk based assumptions. Accordingly, it is essential that sites understand the required skills, knowledge, and capabilities for personnel in key PSM roles and that personnel changes (at all levels) are analyzed and controlled with pre-established criteria and training so that minimum levels of experience and knowledge are fully maintained. All other PSM elements and management systems may be in place but without trained and capable people to make decisions and administer the program in an ongoing basis, the likelihood of continued and sustainable safe operations are diminished over time."

It is not unusual to experience a 20 - 30 % turnover rate for key technical, operations, and leadership roles within a site over a three to four year period given historical trends involving transfers, new hires, retirements and other traditional personnel moves. An integrated and consistent MOC-Personnel process within each site, region and PSM corporate functions is critical to maintain organizational capabilities and individual employee knowledge in order to maintain and improve performance, as well as implement PSM continuous improvement initiatives.

Sites, businesses and companies which experience recurring issues involving PSM performance (for example; incidents with repetitive or similar causes, repeat audit findings, increasing trends in overdue PSM action items, or failure to meet strategic objectives) typically will evaluate their management systems and leadership commitment to understand why these undesirable outcomes exist. As part of a complete PSM management system, changes in personnel across all roles that can impact site operations should be controlled in order to maintain individual and organizational knowledge and capabilities over time. The DuPont model provided in Figure 4 and associated MOC-Personnel tools, checklists, and training materials are one means to accomplish this objective.

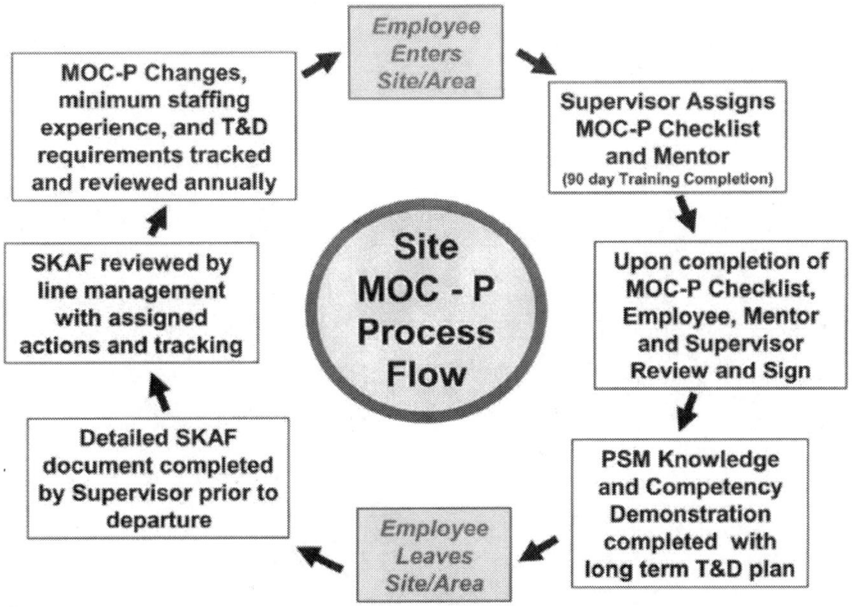

Figure 4: Overview of Site MOC-Personnel Process

Consistent tools and methods to manage personnel changes are also a key part of a reliable MOC- Personnel system. This may include checklists, training materials, and tracking systems such as the following:

- Skill Knowledge Abilities and Functions Checklists for departing personnel
- Initial Training and Skills training packages, Checklists and tracking mechanisms for newly assigned personnel
- Specialized checklists for key Roles – Site PSM Leaders, PSM Committee members, Special focus PSM teams, Operations and Regional Managers
- Standardized Skill and Competency Demonstration methods and evaluations
- Standardized PSM audit criteria to evaluate MOC-P performance
- Databases to monitor personnel changes across the site organization opposite maximum change criteria

Comments or questions on this paper or DuPont Process Safety Management program elements may be directed to the author or through DuPont Safety Resources at the following web address:
http://www2.dupont.com/Safety_Consulting_and_Products/en_US/index.html

References

1. Guidelines for Implementing Process Safety Management Systems, Center for Chemical Process Safety, American Institute of Chemical Engineers, 1994, ISBN 0-8169-0590-8

2. Guidelines for Management of Change for Process Safety, Center for Chemical Process Safety, American Institute of Chemical Engineers, 2008, ISBN 978-0-470-04309-7

3. Cummings, David E., *PSM Knowledge Management and Applications across Common or Sister Plant Technologies*, 10th Process Plant Safety Symposium – New Orleans, LA April 8, 2008

Training Management of Change Approvers

Karen Study
Process Safety Technology Leader
The Dow Chemical Company
6519 State Highway 225
Deer Park, TX 77536
281-228-8679
kstudy@rohmhaas.com

Prepared for Presentation at
American Institute of Chemical Engineers
2010 Spring Meeting
6th Global Congress on Process Safety
San Antonio, Texas
March 22-24, 2010

Training Management of Change Approvers

Karen Study
Process Safety Technology Leader
The Dow Chemical Company
6519 State Highway 225
Deer Park, TX 77536
281-228-8679
kstudy@rohmhaas.com

Keywords: Management of Change, Production, Leader, Review, MOC, Reviewer, Authorization, Manager, Training

Abstract

Many companies have made great strides in change management systems since OSHA's Process Safety Management Standard 29 CFR 1910.119 was first promulgated in 1992. This standard required "implementation of written procedures to manage changes… that affect a covered process"[i]. Although significant improvements have been made, there is one area of change management that is often overlooked. Changes in personnel such as staffing levels, work hours and reassigned job responsibilities are frequently either not covered or are covered inadequately by existing change management systems. This paper addresses one element of personnel change by reviewing a system in use at Dow Chemical Company for conducting critical MOC approver endorsement reviews for new production managers. Key questions answered will include: who should attend the review, what are the minimum expectations of a new production manager and why is this review so critical to process safety.

I Summary

To properly manage change, process safety hazards and lines of defense must be well understood. Therefore, it is of utmost importance for personnel who approve critical Management of Change (MOC) reviews to be knowledgeable in the hazards of a process and the associated layers of protection. At the Dow Chemical Company, a process has been in place for decades to verify that critical MOC approvers are properly trained and knowledgeable. The process is called the New Leader Review. This paper will describe the New Leader Review process, its importance, and how it is implemented at Dow.

II The New Leader Review Process at Dow

The goal of the New Leader Review (NLR) is to ensure that those who are responsible for approving critical MOC reviews understand the desired and undesired chemistry, the key process safety hazards for the facility and have an understanding for how these hazards are managed. New Leaders Reviews also explore operational topics and administrative programs. This review enables the New Leader to have a command of the primary issues and to effectively function as the critical MOC approver.

Critical MOC reviews are defined as reviews for changes which have a potentially significant impact. Examples of critical MOC reviews include:

- Changes in a process that change toxicity information, permissible exposure limits, physical data, reactivity data, corrosivity data, thermal and chemical stability data, or maximum intended inventory,
- Changes that impact normal operating limits or safety instrumented systems,
- Changes that require or impact hazard assessments (e.g., Hazard and Operability studies, Fault Tree analyses or other internal methodologies),
- Changes that impact a layer of protection as defined by Layer of Protection Analysis (LOPA),
- Changes to electrical or hot-work area classifications,
- Changes that require deviation from design codes or standards,
- Changes that impact relief system designs or
- Changes that impact a regulatory permit.

The new leader review is focused on the facility MOC program owner which is typically filled by the plant production manager. After assuming a production manager role at Dow, the new manager has 90 days to complete the New Leader Review (NLR) process. After successfully completing the review, the new manager will be authorized to approve critical MOC reviews. In the interim, an alternate approver must be used. That role is typically filled by the previous incumbent or a technical expert for the facility.

Preparation for the New Leader Review involves development of a NLR package. Many members of the production and technical staff are usually involved in the development of this package. Once a NLR package is developed for a facility, updating the package for the next NLR is a much simpler effort. The finished package will contain valuable information which can be further leveraged for process hazard analyses and staff training. Major elements of the NLR package include:

1) <u>Plant Overview</u>
   a) High level overview of operations which includes what is done, scale of operations, and the age of the facility,
   b) A facility layout drawing,
   c) Plant organization and staff levels,
   d) Level of experience at the site,
   e) The business and technology alignment,
   f) How process control is implemented at the site,

g) Recent key operating changes,

h) A description of the MOC process in use at the plant and

i) Key mitigation equipment used in the plant

2) Process Overview and Major Hazards

a) A simplified block flow diagram for the process,

b) Main reactions, key side reactions and impurities,

c) Major unit operations and

d) Key reactive chemicals and process hazards for each unit operation.

3) Material & Physical Properties Overview

a) A high level overview which includes details about how raw materials are received and stored, material stability, etc..

4) Compatibility Chart

5) Key Lines of Defense

a) A summary of potential hazards and the associated layers of protection and

b) Plant policies for bypassing safety systems and any administrative controls used to manage isolation valves,

6) Chemical Exposure Index (CEI) and Fire and Explosion Index (F&EI) Calculations Summary

a) The CEI and F&EI are Dow specific calculations used to establish relative degrees of hazard for a process.

7) Current variances and major plant initiatives

a) A summary of the gaps in compliance with company and/or regulatory standards and

b) A summary of major initiatives the plant is working on.

8) Previous process hazard assessments

a) A summary of major action points from the latest hazard assessments and their status.

b) A review of key sections from the previous hazard assessment.

9) Previous incidents

a) Key incidents in the technology as well as the specific plant incident history

Preparation of the NLR package is an extensive effort which typically includes contributions from many members of the production and technical staff. The time required to develop the NLR package will vary significantly based on the existing documentation, the status of previous hazard assessments and LOPAs, the availability of production and technical staff and the level of familiarity with the NLR process. Likewise, the time the production manager requires to learn the material before the panel presentation will vary significantly based on the manager's knowledge of the facility, the NLR process and basic chemistry and production issues.

The NLR meeting itself typically takes 3 to 8 hours to complete. The factors which most impact the length of the meeting are the new leader's presentation style, the number of questions posed by the panel and the complexity of the plant being reviewed. The NLR panel will typically include representatives from process safety, research and development, manufacturing, process engineering, site Environmental, Health and Safety, operations and may also include business technical personnel. The NLR package is distributed to panel members in advance of the actual meeting to allow members time to review the material and prepare any questions in advance. The review format includes a presentation of the NLR package to the panel by the production manager. Although the NLR package may be prepared by contributions from numerous people, it is the new production manager's responsibility to present this package to the NLR panel. One important aspect of the NLR is that questions raised during the review are addressed to the production manager first. If the manager does not have the knowledge to answer a specific question posed, others may offer information. If, during the course of the review, minor gaps in required knowledge are found, action points are made to address those gaps. If significant gaps are found, the production manager may be asked to reconvene the panel at a later date to obtain approval after addressing issues that came up during the review. At the end of the NLR, a consensus is reached by the review panel to determine if critical MOC approver status will be approved, conditionally approved, or declined for the production manager.

There are several expectations the review panel requires to be fulfilled before the production manager becomes a critical MOC approver. Most importantly, the production manager must demonstrate understanding of the key hazards and issues for the facility for which they are ultimately responsible to manage change. The following list includes the areas where the leader must demonstrate understanding:

1) Key chemistry practiced and major unit operations in the plant,
2) Hazards associated with the unit operations including fire and explosion potentials as well as community exposure issues,
3) Main raw materials in use and the hazards associated with each,
4) Potential consequences associated with worst case scenarios and how they might occur,
5) Key mitigation equipment used in the plant such as dikes, sprinkler systems, ventilation systems, etc.,
6) Key historical safety related incidents in their plant as well as in similar plants,
7) General knowledge of reactive chemicals testing which includes existing data and resources available,
8) The potential for reactive chemical incidents at the plant,
9) Plant training programs and communication methods in use,
10) Plant emergency plans,
11) Plant policies for bypassing safety systems (e.g. interlocks) and any administrative controls used to manage isolation valves,

12) Status of compliance with regulations and company standards and any associated variances,
13) Status of LOPA implementation at the plant,
14) Status of burner management reviews and mechanical integrity programs, and
15) Status of process hazard analysis action points.

As would be expected, this list includes the major elements in the NLR package. With a well prepared NLR package and a commitment by the production manager to learn the material, the reviews typically go well and have few action points requiring additional training.

III Conclusion

Although the New Leader Review process in use at Dow Chemical Company is time consuming for the new production manager as well as other staff, it is a valuable tool to prepare the manager to serve as an MOC approver. There are more traditional means for the leader to gain the necessary knowledge; however, this particular review process has proven to be a valuable "crash course" for new leaders at Dow. It forces the manager to quickly learn the major strengths, weaknesses, key hazards, and the key lines of defense for the plant. Without this process, there is a much larger potential that the production manager would not know what to focus on during MOC reviews. In the words of Trevor Kletz, "We don't know what we need to know until we know what there is to know."[ii]. If the manager doesn't know something is a hazard, how can he or she possibly be expected to protect against it?

IV Acknowledgements

Much of the information presented in this paper was summarized from internal Dow guidance and training documents written by Anne Wallin, Kevin Shaughnessy, Dave Frurip, Rhonda Zowarka and others. Without the work of these individuals, the New Leader Review process would not exist at Dow in its present form and this paper would not have been possible. Additionally, the author would like to thank Kevin Shaughnessy for reviewing this paper for content and clarity.

V Disclaimer

This information was developed for Dow Chemical Company's internal use. This information is available to you without any guarantee that it is appropriate for your purposes as conditions and methods of use of the information is beyond the author's control. Dow Chemical Company disclaims any liability for use of this information by persons outside of the company.

VI References

[i] Process Safety of Highly Hazardous Chemicals (29 CFR 1910.119), U.S. Occupational Safety and Health Administration, May 1992, *available at* www.osha.gov.

[ii] Kletz, Trevor A., "Don't just pass the parcel: Accidents that would not have occurred if those involved had talked together," *Journal of Loss Prevention in the Process Industries*, Volume 22, Issue 6, November 2009, Pages 667-671.

LOPA and Human Reliability – Human Errors and Human IPLs

By: William Bridges
Process Improvement Institute, Inc. (PII)
1938 Riversound Drive
Knoxville, TN 37922 USA
+1-865-675-3458
wbridges@piii.com

Prepared for Presentation at
American Institute of Chemical Engineers
2010 Spring Meeting
6th Global Congress on Process Safety
San Antonio, Texas
March 22-24, 2010

UNPUBLISHED

LOPA and Human Reliability – Human Errors and Human IPLs

By: William Bridges
Process Improvement Institute, Inc. (PII)
wbridges@piii.com

Keywords: human factors, Layer of Protection Analysis, LOPA, process safety, PSM, human error, independent protection layer, IPL, SIS, SIL, safety instrumented system

Abstract

Layer of Protection Analysis (LOPA) is a simplified risk assessment method that provides an order of magnitude estimate of the risk of a potential accident scenario. Humans can be the cause on an accident scenario (the Initiating Event [IE]) or humans can serve or participate as an independent protection layer (IPL). In either case, estimating the likelihood of the human error and measuring the human error rate at a site are troublesome tasks within the LOPA framework, which is one of the reasons some companies do not give any credit for a human IPL.

Identifying and sustaining independent protection layers (IPLs) is the heart of LOPA. Each IPL must be:

- independent of the initiating event (IE) and the other IPLs
- capable (big enough, fast enough, strong enough, etc.)
- maintained or kept in practice/service
- validated/proven to provide the probability of failure (PFD) chosen

and all of the above must be documented and audited periodically to ensure compliance with these definitions.

There are many types of IPLs, and some are more trustworthy than others, hence the difference in the PFD of IPLs. As just mentioned, one possible type of IPL is a Human IPL. These include preventative steps that may stop a scenario from progressing once it is initiated, but more typically the human IPLs are responses to alerts or alarms or troubling readings and sample results.

This paper discusses the data needed for adequately counting the human in a LOPA (and other risk assessments), and includes discussion of the theory of human factors. One key focus of the paper is on practical means for collecting raw data in a plant setting for substantiating the error rates for the site, and especially for crediting a human IPL. The method for data collection covers the training requirements that should be met, proof drills for response to alarms, simulations and tests, and frequency of proofs, and of course the effect of human factors on human error rates. *Actual plant data and tests are included in the paper to provide the reader with some examples of how a simple data collection and validation method can be set up within their companies.*

Human Error Fundamentals

Human errors are sometimes mistakenly called procedural errors. This is not true anymore than saying all equipment errors are due to design errors. Over the past 5 decades of research and observation in the workplace on human error, we have come to know that human error probability depends on many factors. These factors (described in more detail in *Human Factors Missing from PSM,* LPS/GCPS 2010), include:

- Procedure accuracy and procedure clarity (the number one most sited root cause of accidents):
    - A procedure typically needs to be 95% or better accuracy to help reduce human error; humans tend to compensate for the remaining 5% of errors in a written procedure.
    - A procedure must clearly convey the information (there are about 25 rules for structuring procedures to accomplish this) and the procedure must be convenient to use.
    - Checklist features – These should be used and enforced either in the procedure or in a supplemental document

- Training, knowledge, and skills
    - Employees must be selected with the necessary skills before being hired or assigned to a department.
    - Initial Training – There must be effective, demonstration based training on each proactive task and each reactive (e.g., response to alarm) task.
    - Ongoing validation of human action is required and usually must be repeated (in either actual performance or in drills/practice) at least once per year *(as discussed later in this paper).* For human IPLs, the action must be demonstrated to be "fast enough" as well.
    - Documentation – the performance of the humans must be documented and retained to demonstrate the error rates chosen are valid.

- Fitness for Duty – Includes control of many sub-factors such as fatigue (a factor in a great many accidents), stress, illness and medications, and substance abuse.

- Workload management – Too little workload and the mind becomes bored and looks for distraction; too many tasks per hour can increase human error as well.

- Communication – Miscommunication (of an instruction or set of instructions or of the status of a process) is the second or third most common cause of human error in the workplace. There are proven management systems for controlling communication errors.

- Work environment – Factors to optimize include lighting, noise, temperature, humidity, ventilation, and distractions.

- Human System Interface – Factors to control include layout of equipment, displays, controls and their integration to displays, alarm nature and control of alarm overload, labeling, color-coding, fool-proofing measures, etc.

- Task complexity – Complexity of a task or job is proportional to the (1) number of choices available for making a wrong selection of similar items (such as number of similar switches, number of similar valves, number of similar size and shaped cans), (2) number of parallel tasks that may distract the worker from the task at hand (leading to either an initiating event or failure of a protection layer), (3) number of individuals involved in the task, and (4) judgment or calculation/interpolation, if required. For most chemical process environments, the complexity of the task is relatively low (one action per step), but for response actions (human IPLs) there are almost always other tasks underway when the out-of-bounds reading occurs or the alarm is activated.

In addition to the human factors listed, other considerations for use of a human as an IPL include (1) time available to perform the action and (2) physical capability to perform the action.

When using human error data for IEs and IPLs, the site must ensure that the factors above are consistently controlled over the long-term and that they are controlled to the same degree during the mode of operation that the LOPA covers. For instance, if the workers are fatigued following many extra hours of work in a two week period leading up to restart of a process, then the human error rates can increase by a factor of 10 times or 20 times during startup.

Revealed versus Unrevealed Errors for Human. As with equipment failures, human errors can lead to a revealed fault in the system (the flow does not start, for instance) or to an unrevealed fault (the block valve downstream of a control valve is left closed, but the failure is not revealed until the control is needed/used). If the error is revealed, then the error can be corrected or compensated for. If the restoration/correction time is sufficiently short, then the probability of being in the failed state is much less than for an unrevealed failure that is only discovered upon testing or inspection.

General Relationship of Human Factors Consideration to LOPA

Every risk assessment must consider the likelihood and effect of human factors. For LOPA, poor human factors can lead to higher human error rates that increase IE frequencies and that increase the PFD of a human IPL. The table on the next page summarizes the human factor issues that relate directly to LOPA. The table contrasts the impact of good and poor human factors on initiating event frequency (IEF) and on the PFD of independent protection layers (IPLs). If all of the factors were optimized, one would expect the IEF to be about 0.01 per task for IEs and about 0.05 for response to alarms or call for actions (human IPLs), but such an analysis must be made on a case-by-case basis, especially for the PFD of IPLs.

Considerations for Getting Low Human Error Rates

Issue	Requirement for claiming a low error rate for humans causing (initiating event) an accident (usually skipping a step or doing one wrong)	Requirement for using a low probability of failure for a human as a protection layer against an accident (usually a response to alarm)	Comments
Time to perform action	NA	Time to detect deviation, diagnose the problem, decide on proper action, and take action is recommended to be less than 1/2 the time required to reach the consequence of interest.	The reliability of the indication (sample & analysis in lab, field reading, etc.) or annunciation (alarm of the deviation) may limit the value for the human action IPL. Operator must be reliably available.
Capable	NA	It is physically possible to perform the control action required to forestall the consequence of interest, and the capability has been demonstrated and is not the limiting factor in the human action (e.g., the operator is strong enough to do the required task)	If the final control requires closing a manual block valve, the action and strength required must be verified in the field (some valves take several operators more than 1 hour to close, even with a valve wrench or cheater). This usually requires preventive maintenance or equipment use as part of a 'critical action.'
Procedure	Step must be in procedure. Not completing the step as stated and in that sequence (most aspects of errors of omission and commission) is the IE. The step is a critical step and is indicated as critical in the procedure or with a warning or caution. The procedure must follow "best practices" for control of human error; best practices relate to content accuracy, page format, and step presentation. See rules for procedures presented elsewhere.	Step must be in a trouble-shooting guide or similar contingency procedure (including emergency shutdown procedure), that is in paper form or reliably available on demand electronically (including via linkage to trouble-shooting guides within a distributed control system [DCS] for an alarm point) that describes how to respond to a process deviation. The response procedure follows best practices as well; see rules for procedures development and writing.	Note that we assume there is a 100% chance of skipping the required step or doing it substantially wrong if the step is not in the written procedure or if the step written is wrong or in the wrong sequence.
Checklist	Checklist in place and its use is enforced.	NA	Without a checklist, the error rate goes up by a factor of 3 to 5 times.

Issue	Requirement for claiming a low error rate for humans causing (initiating event) an accident (usually skipping a step or doing one wrong)	Requirement for using a low probability of failure for a human as a protection layer against an accident (usually a response to alarm)	Comments
Initial Training	There is initial training focused on this step as a critical step	There are initial training and drills on the response to a deviation; and the deviation is annunciated.	Note that we assume there is a 100% chance of skipping the required step or doing it substantially wrong if the step is not emphasized in initial training.
Ongoing validation (or practice) of human action	Steps must be practiced, but for seldom-used procedures, this practice (normally in the form of a talk-through for shutdown or startup of a continuous unit) is not until just prior to use. Practice rate increases reliability per action (due to increase in skill) but Actual Use Rate also increases the number of opportunities for failure. These factors can offset each other completely. See limitations on error rates that account for "practice" versus "error rate per task demand."	Steps must be practiced routinely enough to assure the reliability of the action under the increased stress caused by the alarm/indication. The time it takes to complete the action must be measured during each practice/drill, to ensure response time limits are not exceeded. Each person who is expected to implement the action "on demand" must perform the practice/drill per the assigned frequency. The minimum amount of practice or drills and type of practice should be at least one per year. (Note: One refinery practiced one scenario for a human IPL, selected at random, every shift and found no issues on shift workload and actual had improvement in general worker performance.) It is possible to group similar IPLs along the classification of similar types of actions and similar available response times. If such grouping is possible, then practicing one action will in effect be practicing all similar actions. For determination of sample size for tests/drills based on number of human IPLs and number of human responders, refer to standard methods for sample size determination.	Note that we assume there is a 100% chance of skipping the required step or doing it substantially wrong if the step is not practiced as often as specified here.
Documentation of validation of human action	Document each practice (i.e., each startup and each shutdown) and document failures rates for 'critical steps'. Keep the file of the validation associated with the task and also with each worker.	Document each practice (i.e., each response) and document failures rates for 'critical response action.' Keep the file of the validation associated with the 'action' and also with each worker. If a sample plan method is used, the same data as above is still required, and further, the data should be checked each quarter to ensure against inadvertently using the same IPL and same operator too often.	Note that we assume there is a 100% chance of skipping the required step or doing it substantially wrong if the step is not practiced as often as specified here or if there is No documentation of the completed refresher training or just-in-time training for an IE, or of drills for an IPL.

<table>
<tr><th>Issue</th><th>Requirement for claiming a low error rate for humans causing (initiating event) an accident (usually skipping a step or doing one wrong)</th><th>Requirement for using a low probability of failure for a human as a protection layer against an accident (usually a response to alarm)</th><th>Comments</th></tr>
<tr><td>Control of related physiological and psychological stressors</td><td colspan="2">• Fatigue – Company policy and enforcement to limit fatigue, including limit of hours worked per day and per week, and including restrictions on hours worked outside of work. Can affect error rates by factor up to 20X.
• Other fitness of duty issues – Company policy and enforcement to limit effect of alcohol & drug abuse, illness, prescription drug effects, and personal life stress effects. Can affect error rates by factor up to 50X.
• Workload – Workload needs to be managed to optimize worker performance. Enough stress (time/task load-based), but not overstress. (Workload is also related to the complexity of the work. Complexity increases mental workload, even if the number of tasks per hour is not affected.) Workload evaluation may be necessary. Can affect error rates by a factor up to 10X.
• Communication –The second most common cause of human error, the site must have a policy and rules and follow-through for controlling verbal and visual communication. Rules include repeat-back on instructions received and use of common jargon. Poor control of communication rules can increase human error rates by 10X.
• Work environment – Temperature, lighting, noise, distractions, ventilation, etc., has been optimized to improvement worker performance. Can affect error rates by factor up to 5X.
• Human-System Interface – Includes control-display-information integration, labeling, error-proofing designs, color-coding, alarm management, etc. Poor human-system interface control can increase human error rates by 10X or higher (especially if "norms" are violated, where a negative effect of 20-50X is assumed).
• Complexity – Complexity of a task or job is proportional to the (1) number of choices available for making a wrong selection of similar items (such as number of similar switches, number of similar valves, number of similar size and shaped cans), (2) number of parallel tasks that may distract the worker from the task at hand (leading to either an initiating event or failure of a protection layer, (3) number of individuals involved, and (4) judgment or calculation/interpolation, if required. For most chemical process environments the complexity of the task is relatively low (one action per step), but for response actions there is almost always other tasks underway when the out-of-bounds reading or the alarm is activated. Complexity is difficult to predict (since it is not known when a human action will be needed), but higher complexity can increase error rates by 2X to 10X.</td><td>All of the listed human factors (beyond the procedure and training and skills listed above) have a large, independent influence on human error rates. A large negative swing in any one of the factors can increase human error rate by a factor of 3 to 50 times.</td></tr>
</table>

Issue	Requirement for claiming a low error rate for humans causing (initiating event) an accident (usually skipping a step or doing one wrong)	Requirement for using a low probability of failure for a human as a protection layer against an accident (usually a response to alarm)	Comments
Qualitative hazard evaluation	There has been a hazard evaluation of the scenario for skipping the step and doing it wrong and a hazard evaluation team has evaluated the consequences and safeguards.	The human action in this case is a safeguard that is called into play once a sequence of events is initiated. The human action (following the trouble-shooting guide or the emergency shutdown procedure) is to make a change of state in the system to prevent propagation of the event to the stated consequence. The hazard evaluation has evaluated the human action (and any annunciation device unavailability) in judging that this "critical responsive action" is capable, reliable, and audited.	It is essential that a company finds accident scenarios that can arise during all modes of operation, including startup, shutdown, emergency shutdown, and online maintenance.
Use of multiple successful human actions within a scenario	Making multiple mistakes (including shortcutting multiple steps) is the same probability as making a single mistake, if the "safeguard" steps amount to ensuring the same state in the system. The human who makes the errors sees this as performing the task by method B instead of method A. Method B could have different steps or less steps that method A.	If one person in a workgroup is assumed to make the mistake that is the initiating event, then it is assumed that no one in the workgroup can be part of the protection layers. Further, supervisors are also part of the workgroup, due to the trust/relationships that build quickly. Seemingly unrelated workgroups, such as maintenance and operations, usually build enough trust to rule out counting as separate groups. For a group/individual to be counted as a safeguard against the errors by another human error, the detection and response group must not share staff, must have an incentive to find mistakes and report/respond, and near miss reporting and auditing both indicate that independent measurement and action is occurring. An exception that is normally allowed is to re-use a workgroup or individual if the second action/response is separated by distance (location) and/or separated by enough time and/or there is a reminder alarm for the same alarm (the alarm that was missed earlier) that recurs often enough in the allowed time to still avoid the consequence.	It is nearly impossible to use a human response safeguard, if a cause of the predicted accident sequence is a human error.

TYPES of HUMAN ERRORS of IMPORTANCE IN LOPA

Human error comes in many forms, but the primary ones of interest in LOPA are:

- Errors in following proactive procedures, such as startup of a unit, that results in an initiating event of a LOPA accident scenario.
- Errors in responding to a call for action; if performed correctly such actions would interrupt the scenario and prevent the consequence.

The examples below illustrate both types of human error as they relate to LOPA.

Example of IE caused by Human error

In NUREG/CR-1278 – The Human Reliability Handbook (Dr. Alan Swain was the primary author), error rates were tabulated from various sources, including some empirical data. For errors that could be considered initiating events, these were summarized in Table 15-3 of that document, which is extracted and included in slightly amended form below:

Omission of Item	Human Error Probability
When procedures with check-off provisions are correctly used:	
(1) Short list, less than or equal to 10 items	.001
(2) Long list, greater than 10 items	.003
When procedures without check-off provisions are used, or when available check-off provisions are incorrectly used:	
(3) Short list, less than or equal to 10 items	.003
(4) Long list, greater than 10 items	.01
(5) When written procedures are available and should be used but are not used:	.05

The table above shows the estimated probabilities of errors of omission per item of instruction when use of a written procedure for a task is required. The probabilities in the tables above assumes that human factors are optimized as well, which includes providing an accurate and easy to follow procedure, good control of fitness for duty factors (including controls against fatigue), good controls against miscommunication, training/practice (or actual performance) at least once per year, etc., as described earlier.

Since startup is when most accidents occur due to human error (when human error is the initiating event), it is appropriate to use that as an example. Startup procedures normally have more than 10 steps, indeed more than 50 steps is more typical. So, if all human factors are optimized, the probability of human error taken should be 0.003 per step. If practice with the procedure is only done once per year or less often, then the IEF would be 0.005, which we would round up for Basic LOPA to 0.01. It may seem conservative to round up, but consider that there are few (if any) chemical plant sites that have optimized most of the human factors. Poor control of certain human factors can have a multiplying effect on the human error of 3 to 50 times. For this reason alone, it is dangerous to use the data above as representative of the error rates at a given site. Some analyst believe that in the absence of site data, these values can be applied, but perhaps it is best to scale the values upward (use higher base error rates) if the site determines that the human factors described earlier in this paper and in other resources are Not optimized.

Example human error related to human-based IPLs

Again, from NUREG/CR-1278 – The Human Reliability Handbook, estimates were provided for the probability of human error to respond correctly and in time following an alarm (annunciation). The Figure below is extracted from the handbook for response by control room personnel to an abnormal event. This represents the time it takes to diagnose the problem given there are No other alarms and given that all human factors are optimized, including practicing the response once per year or more often. The error probabilities given are the joint probability of failing to notice the alarm and failure to diagnose the situation correctly.

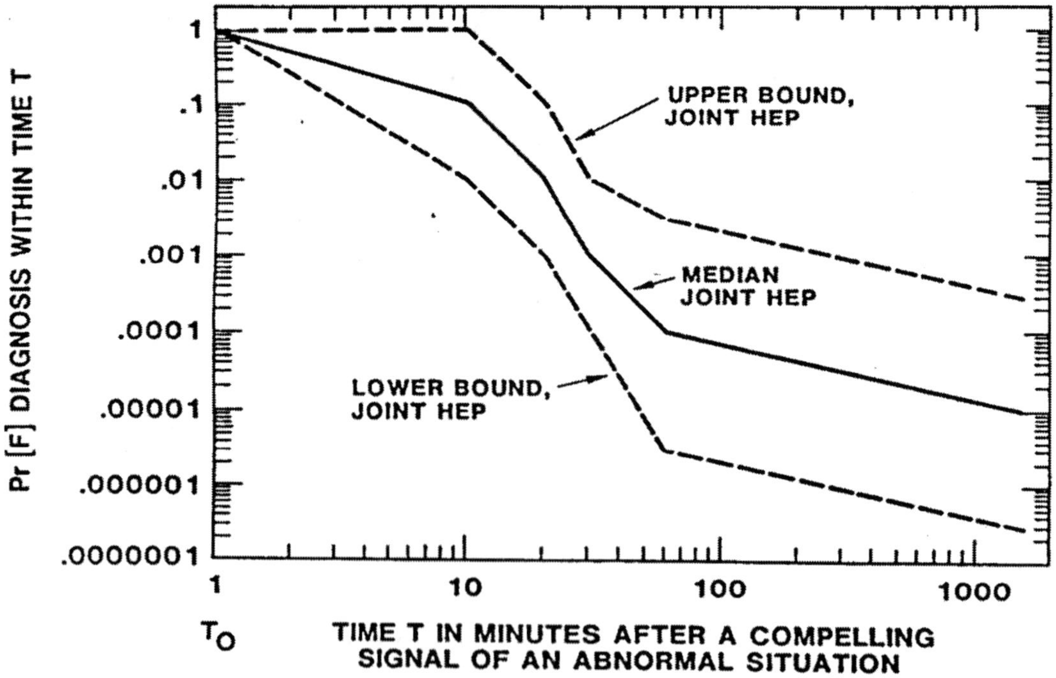

Note from the figure above that the more time to detect and diagnose the problem, the lower the chance of error. (HEP refers to Human Error Probability. Pr refers to the probability of making a diagnosis error.)

This figure only tells part of the story. After diagnosis, the operator, perhaps with help of others, must then respond appropriately to prevent the scenario from propagating to the next protection layer or to the ultimate consequence. The sum of the time to detect, diagnose, and complete the proper response must happen within the available "process safety time" (PST). The process safety time is the interval from when initiation of the event begins until the time when no further action can prevent the consequence. The PST must be known for a site to judge if a specific human response can qualify as an IPL. Some organizations require that the operator's overall response must be accomplished in much less (normally half) of the PST, for the human response to be considered an IPL (given other criteria of an IPL is met; see LOPA 2001 for guidelines on qualifying an IPL). It is generally assumed that if the problem is detected (such as by hearing an audible alarm) and if the diagnosis is correct, then the action is likely to be "more" accurate than the detection/diagnosis (though this conclusion does not account for the speed of the response action). The norm therefore is to use a PFD of 0.1 if the response can be performed within the PST (or within ½ the PST for some organizations) and the PST is equal to or greater than 20 minutes (so the complete detect-diagnose-response steps are accomplished in 10 to 20 minutes, depending on the organization's rules for LOPA). A value of better than this (lower error rate) is normally not allowed in Basic LOPA. From an advanced topic perspective, however, Human Reliability Analysis (HRA) will often use a PFD of 0.01 for response times greater than 40 minutes (PST greater than 80 minutes). Performance of HRA and documenting the basis for lower probability values requires a level of effort and a degree of education/experience far beyond the required learning to perform Basic LOPA.

HUMAN ERRORS as related to IEs

Human errors that result in IEs are usually because of operator errors during startup (roughly 65% of all accidents in chemical-related process plants occur during startup) or errors during normal shutdown or emergency shutdown (together these represent 10% of the IEs caused by human errors). The remaining errors usually occur during normal operation and include such issues as performing a step wrong during operation in manual mode (such as when a process controller is bypassed) or by turning the wrong switch or opening the wrong valve.

To find these errors, a very good system for reporting and investigating near misses is needed (but that topic has been covered in other research and papers, see Bridges 1997 and 2008). The US Nuclear Regulatory Agency (NRC) in 10 CFR 55.45 and 55.59 requires that all power plant operators be tested once per year on abnormal procedures. This testing is mostly related to humans involved as IPLs. 10 CFR 55.45 and 55.59 also allude to the possibility of using a test of a representative sample of human responses, but we address this option later in the paper.

To help minimize human errors that could be IEs, many companies will review a normal shutdown or normal startup just before its next use. This has proven effective in controlling human error in chemical-related industries. Nuclear power plants also perform such "just-in-time (JIT)" refresher training on normal shutdowns and normal startup procedures.

If a site records the errors that are made during such tasks, then the company can validate the average IEF for that task and so the site will have a better estimate of the error probability. Further, if the measured rate is too high, they can take proactive steps to lower the error rate.

Errors can be measured either in actual practice in the field, or via simulators, or roughly by analysis of near misses reported (if the number of near misses reported is at least 20 times higher than the number of loss events or accident; Bridges 1997 and 2008; CCPS 2003).

Below is an example of data from one site that was collected to estimate the IEF for mistakes made during normal startup:

> ***EXAMPLE - Initiating event frequency data collection and use:*** In one company, the operating group was able to provide the data on the following page for errors made during startup of various units. Based on this data, the average probability of human error was 0.0071 and the number of errors per year that could lead to a known accident sequence was 0.17 (but note that if all errors are counted, the average error rate would be 0.29/yr). For basic LOPA, this value is rounded up to an IEF of 1/yr, for initiating events arising from these and similar procedures at the site. ***See the Table on the next page for the raw data.***

Chemical-related process plants should collect such data to understand and validate the initiating event frequency caused by human error. This data would then be used to validate the values for IEF used in LOPA and other risk assessments.

Site data on Human Errors related to Initiating Event Frequency (IEF)

Task for which a Potential Initiating Event exists	***Number of steps in the procedure***	***Number of Uses of the procedure***	***Number of Human Errors Noted***	***Number of Years of Data***	***Number of Errors That Initiated a Potential Accident Sequence***	***Human Error Probability per Step***	***Error Rate (probability of error per year)***
1. Startup of Process A	46	4	2	6	1	0.0109	0.17
2. Startup of Process B	71	5	3	6	1	0.0085	0.17
3. Startup of Process C	55	4	2	6	2	0.0091	0.33
4. Startup of Process D	56	4	0	6	0	0.0000	0.00
					Average:	0.0071	0.17
						IEF for LOPA for site:	**1/yr**

HUMAN ERRORS as related to HUMAN RESPONSE IPLs

A 10^{-1} PFD value for a human response indicates that the correct response occurs at least 9 out of 10 times (or no more than 1 wrong response for every 10 attempts). Most organizations will have identified many human responses involving a number of personnel, as part of their LOPA studies. Some organizations believe that if they have a procedure and a training program in place, that they can claim the PFD value of 0.1 for a Human IPL. This is no truer for a human IPL than it is for an active component IPL. A Human IPL must be tested and validated the same as a component IPL. On the next few pages are the options for validating human IPLs; these (1) include 100% testing for each human responder and for each action to be taken and (2) a sample plan testing of random combinations of human responders and actions.

Using a 100% Individual Test Plan Approach for Validation of Human IPLs.

One method to prove the operators will reliably respond for each Human IPL trigger is to have each operator demonstrate they can individually respond to each alarm (or other trigger). This response can be demonstrated by walkthroughs in the field or using simulators of the process unit. The nuclear power industry uses this approach for validating response by control room operators, but many in the chemical industry perceive this will take "too much time." As an example of the effort required, one nuclear power plant allocates about 200 hours per year per operator for refresher training activities, including 60 hours per year per operator for demonstration of skill in responding to critical alarms (from Tennessee Valley Authority [TVA] internal requirements to meet the performance-based requirements of NRC regulations 10 CFR 55.45 and 55.59). But is this time "too much"? We'll explore that answer in the rest of the paper.

Consider the following as an example of the application of this method of testing for 100% of the responses by 100% of the operators at an actual chemical plant:

> ***Background:*** The operating area being evaluated has 25 operators and 130 identified (from a hazards analysis and LOPA) human response IPLs.
>
> ***Validation Test***: The "tests" are documented on a set of index cards that call out various alarm conditions; these are the events (triggers) for a scenario identified in the hazards analysis (and LOPA) to be protected by the given human response IPL.
>
> ***Demonstrated Result:*** A correct answer (success of the IPL) is the desired response to the alarm scenario. A wrong answer (failure of the IPL) is any other response than the one desired or a response that takes too long.
>
> ***Estimation of Resources Requirements for 100% Testing Scheme:*** How much test effort is needed to ensure that training and retraining programs are sufficient to validate a 10^{-1} value for the PFD of all of these identified human response IPLs? An example estimate is as follows:

1. **Determine the number of test to be performed.** This would be the number of human response IPLs multiplied by the number of people who are expected to respond (perform as the human IPL) at some point in the future during their own shift. This example would yield 2750 (25 x 130 = 2750) discrete tests for one test of each combination of trigger and human responder (which makes up the 2750 IPLs).

2. **Determine the test frequency**. It is difficult to get consensus on this value. One documented example is from the nuclear industry. The US NRC regulation for control room operators (10 CFR 55.45 and 55.59) requires annual demonstration of proficiency in response to critical alarms and signals. This in fact would likely Not be enough testing to give a 90% chance of proper response to every alarm, except that response to one alarm is normally similar to response to other alarms, so the operator is in essence getting more **practice on similar alarms** as they perform each demonstration. Operators under this regimen of recertification have shown a 90% chance or better of proper response (detection, diagnosis, and action) within 10 minutes of an annunciation of an event of interest (from internal power plant records and also inferred from, Swain, 1983 and NUREG 1278 data, since the basic control room regimen of testing each operator with each action each year was essentially the same in the 1970s as they are now). For this example, a frequency of 1 validation per year is chosen.

3. **Determine the time required to perform each test.** Assuming 10 minutes of allowed response time for success (per alarm/trigger), then the test time for the organization would be 27,500 minutes or about 460 staff-hours (22 hours per operator per period). With a frequency (test period) of once per year per alarm, this equates to about 1% of the normal staff-hours for a worker in the USA. (Note: An equal amount of time would also be required to record/document the training record for the test [once the tracking system for such demonstrations is set up], but this is likely not a load the operator will have to bear.) An investment of 1% of each operator's time seems relatively inexpensive for the relative risk-reduction afforded by a Human IPL. (Note that testing only a sample of these responses, discussed later, would reduce the load considerably.)

EXPERIMENTAL DATA related to Human Response IPL.

Many companies have suggested that collecting data on human responses in order to validate that a human IPL is justified will be very difficult. However, the actual effort to collect such data is low and the benefits are great, as demonstrated by several chemical plants and refineries and nearly all nuclear power plants that have tried this.

To help demonstrate this, PII designed a simple test for use in measuring the response to an alarm condition. The test was not meant to measure the probability of detection of the alarm, but rather was meant to measure the time and success in determining and accomplishing the proper response to critical alarms as part of human IPLs. Two chemical plants belonging to large organizations (one in Malaysia and one in the USA) performed the test.

Test setup: The test involved having multiple operators in one unit of one plant/site perform responses to critical process alarms. These alarms were related to human IPLs. The actual response and time of response was measured, but essentially the tests were setup as a "pass or fail" – in other words, the test were to determine if the operators were able to respond as desired/expected.

Each test took 10-15 minutes to administer and less than 1 minute to record the data. The data was administered to multiple operators on multiple shifts. The tests were administered by a shift supervisor, a shift engineer, or in some cases, a process safety coordinator; though in the future, it is likely fine for a lead operator to administer the test. The overall effort to administer the test was deemed minimal, though there was resistance to such performance/validation test at first (as would be expected).

A human response IPL "failed" if the operator could not perform the required action to prevent the hypothetical outcome within the process safety time (definer earlier).

To run each test, the plants printed a data card (the size of an index card) and handed it to an operator chosen at random. Below is an example of such an index card.

IPL Verification Test			
Response Task:	Process Safety Time:		Response Time:
LAH for Tank 173	15 minutes		5:20 minutes
Date of Test:	Time/Shift:		Employee Number:
7/23/2008	07:35/A		23122
	Pass/Fail:	Pass	

Note that the card contains an estimate of the PST – as defined earlier in this paper and elsewhere, this is the time an operator has to perform the task once the alarm is received until it is too late to take any further action. The time it took to print and hand out the index card was minimal. The person administering the test then timed the operator response and recorded the results. Any time less than the PST was a successful validation. (Again note that these tests did not validate the probability of an operator failing to detect an alarm.) The total time for the test varied, but the two sites that performed the test considered the administration time to be minimal; the largest time effort is simply for someone other than the operator to be there to "independently" measure the operator's response time (i.e., time to administer the test). Perhaps in the future, another operator could likely administer most proof tests and then the site management could audit some percentage of the tests to help ensure against bias. If the operators test each other, then the time to administer a test is likely not significant enough to measure since they have to be "there" regardless for their other duties.

Aside: Currently the Safe Automation guidelines by CCPS and the upcoming guideline for IPLs and IEs (pending, 2010) both stipulate that the action must be completed within ½ of the PST. But, the ½ PST rule was arbitrarily chosen and has no statistical basis that the author is aware of.

For the most part, the tests came with little warning and occurred on all shifts. Several critical alarms were tested using various operators, all randomly selected. (It is anticipated that unless a sample plan is used, each operator will perform roughly one response related to each human IPL each year.) The time to respond was recorded on the index card.

Based on such raw data, the site was able to (1) evaluate the degree to which they were controlling human factors for the units, (2) identify which human responses qualify as IPLs, and (3) validate that the response is accurate enough and quick enough to qualify for the PFD used as the credit for the IPL (which for basic LOPA, the PFD is limited to a value of 0.1).

The table on the next page provides a sample of the site-specific data for several similar Human IPLs from the two sites (one in Malaysia and one in Texas, USA). For the Malaysia site, the data was from the same operating area consisting of multiple operators across 4 rotating shifts of 8 hours per shift and for the Texas site, the shift was 12 hours.

The sites believed that it is probably reasonable to use a sampling of human error for similarly qualified humans doing similar response or proactive tasks. A sample plan of perhaps only 5% to 10% of the number of human-task pairs may be necessary to have a valid statistic for human error for a "type of action." Sampling is valid for human actions because of how the mind processes information and actions for similar situations. Obviously, sampling can greatly reduce the measurement and documentation load for validation of human error rates. Human performance sampling is discussed in more detail next in this paper.

Human Error Data from Two Sites (USA and Malaysia) related to Independent Protection Layers (IPLs)

IPL No.	Response Task	Number of Test Performed	Average Response Time (minutes)	Process Safety Time (minutes)	Number Failures	PFD (average)	LOPA PFD
Company A (USA) Site 1							
IPL 1	ABP: High Temp in Generator	6	2.25	10	0	0	0.1
IPL 2	ABP: Loss of Acid Flow to Generator	12	2.166666667	10	0	0	0.1
Company B (Malaysia), Site 1							
IPL 1	LICA-HL-05007	10	5.054	15	0	0	0.1
IPL 2	TI 09010 low alarm on 09-K001 discharge	10	4.943	15	0	0	0.1
IPL 3	LAL-21004	10	6.962	15	0	0	0.1
IPL 4	LAH 22002	10	6.058	15	0	0	0.1
IPL 5	PIAH 21006	10	5.036	15	0	0	0.1
IPL 6	PIC-29014	10	5.352	15	0	0	0.1

Below are responses from the sites to questions related to the measurements:

- ***How much labor did each performance test take (on average)?***
    - The majority of the labor was with the actual operator going out to the field to 'take action' on the 'alarm'. Prep time was 1 hour for the initial recording of 10 data points for each of 2 to 6 IPLs, the prep time for each test was minimal as the cards were prepared beforehand and the recording takes less than 10 seconds.
    - The test time was about 15 minutes per operator per test.
    - Recording time was negligible.

- ***How much work do you think this would be longer-term (after getting in the habit and working any bugs out)?***
    - Not much difference in time as the testing process was picked up after the 4th to 5th data point in the unit, so it did not take long to get to "minimal" effort. One benefit of the test was the re-enforcement of "what to do" with all parties involved (the person doing the test, the person recording the test, and the other operators who noticed the test in progress).
    - The hard part is getting that up-front commitment; after doing a few tests, doing more was no issue.

- ***Do you feel such data is necessary for each human response IPL? (In other words, do you think a 5-10% sample of related type of responses would suffice?)***
    - The response for all the IPLs were the same when the same operator acts on different alarms or when different operators act on the same alarms. If this is the case, we probably can use a sample to test a "general" response to a type of alarm.
    - The site should first screen which response can be grouped together into general types of response IPLs. Then a lesser percentage will need individual validation drills.
    - Perhaps do just 1 or 2 drills per shift per group per year for the simpler ones on some periodic basis; that gives you a chance to test feasibility (make sure valves not locked, make sure valve wrench or other tools are available, etc).

- ***What improvements should be made to the tests/validation scheme?***
    - If we make the simulation a little bit sophisticated, we may be able to induce stress and surprise (maybe create some disorientation or carry out the tests in not-so-smooth situations – such as when Gene Hackman commenced a missile drill at the end of a minor crisis in the movie *Crimson Tide*) so that we can get closer to the actual critical situations. This would obviously need to be under very controlled conditions.
    - Perhaps with simulators or other methods, we could also include "detection and diagnosis" of the alarm in the same test.

- ***Who administered the test?***
    - Company 1. Plant production engineer
    - Company 2. As agreed upfront, about 80% was by the Shift Supervisor; 20% managed to be administered by the PSM Engineer or Shift Executive (Engineer) although more in an observer capacity.

- ***Who do you think should admin the tests in future?***
  - Same or a Lead Operator, which is basically an expert from the operator ranks.
  - Shift Supervisor level. In the Malaysia company, this is the highest level for non-degree holders in operations (or it can also be a newly graduated engineer).

- ***What unit was this in?***
  - This was a small batch plant in Texas
  - Ammonia Plant and Urea Plant in Malaysia

- ***Overall feeling of the exercise?***
  - The staff said it was easy to do and didn't have much difficulty in setting it up.
  - Initially, probably due to not fully seeing the whole picture, it was seen as a burden to do. Convincing was required to explain the potential benefits. But the plant staff rose to the occasion and the actual implementation was not so bad judging by what was able to be achieved even with the unplanned shutdowns that happen to occur in the month when the validation tests were run.

Using a Statistical Sample Plan Approach in the Future for Validation of Human IPLs.

It is important to ensure that a 10^{-1} value is indeed valid before using a human IPL. Testing every involved person on every identified response is not that burdensome, as shown in the test cases presented earlier that was run at two chemical plant sites. But, this burden could also be wasteful if testing the response to one alarm by one individual is essentially the same as testing the response of similarly trained and drilled individuals to the same or similar alarms.

Rather than testing 100% of the responses by 100% of the operators, it may be valid to use a sampling plan. This may be especially true for groups of responses that are similar in action and response time. US NRC alluded to a "sampling" of human response in 10 CFR 55.59, indicating that this may be acceptable for validating human response to triggers (i.e., for validating Human IPLs).

Statistical techniques developed decades ago are used to establish a basis of acceptance of all kinds of products, raw materials, components, etc. These methods can also be used to validate Human IPLs.

The sample plan approach must group similar type of actions and similar response time requirements. For a sample plan approach, choice of responder and trigger must be chosen at random. The lot size is the product of the number of responders multiplied by the number of similar response actions in the group.

As a rough rule of thumb, the sample should be about 10% of your total population of data, but not smaller than 30 and not greater than 350 to 500. The sample size and number of failures before the PFD is invalid is related to the confidence level and margin of error that is acceptable to the organization. A confidence level of 95% and an error margin of 5% indicate that the result of your testing (validation of a Human IPL) will be within +/-5% of the true PFD 95% of the time the validation testing is performed.

The correct sample size is a function of those three elements – your universe (how many people multiplied by the number of actions; each pairing makes up a Human IPL), the desired error margin, and the preferred confidence level. For IPL validation purposes, it is likely reasonable to use a 5% error margin at 95% confidence. Below are typical sample sizes (the first at a 10% error margin, the second at 5%):

50 in the population, sample 33 or 44
100 in the population, sample 49 or 80
200 in the population, sample 65 or 132
500 in the population, sample 81 or 217
1000 in the population, sample 88 or 278

The trend above approaches a limit for a 10% error margin that hardly moves above 350 in the sample size no matter how large the population (and approaches a limit of 500 for a 5% error margin). The sample size can also be approximated using the equation below:

Sample Size, ss:

$$\text{ss (for infinite population)} = \frac{Z^2 * (p) * (1-p)}{c^2}$$

Where:

- Z = Z value (e.g. 1.96 for 95% confidence level)
- p = percentage picking a choice, expressed as decimal (0.9 used for Human IPL sample size determination)
- c = confidence interval, expressed as decimal (e.g., .05= ±5%)

This calculation then must be corrected to account for a finite population (validation SS):

$$\text{Validation SS} = \frac{\text{Ss}}{1 + \dfrac{\text{ss} - 1}{\text{population}}}$$

Meeting acceptance criteria for a human response means that the procedures, training, retraining, communication control, and all other human factors are achieving the desired result of no more than 1 wrong response in 10 demands. Rejection means that the PFD of 10^{-1} is not valid (or that the control of human error needs improvement for the PFD to remain valid). When applying this sampling plan to a validation of administrative IPLs, the "acceptance" criteria (the desired response) is the response that prevents the consequence that is being considered in the LOPA, e.g., *the operator response to the alarm prevents the overflow.*

Sampling plan schemes rely on developing a valid definition of a group to be the sample. For Human IPLs, the group is a combination of similar operators (likely all operators are treated equally for this purpose) and similar responses to similar triggers. The latter grouping takes careful consideration by a multiple-disciplinary team (with heavy emphasis on the operators on the team composition), to ensure the grouping of IPLs makes sense (i.e., use expert judgment). Although it is likely that all operators can be lumped into the same statistical group (if they are selected at random for validation drills of IPLs), the triggers and responses will need to be grouped to ensure that validation of one trigger/response is essentially the same as validating other triggers/response within the same group.

The upcoming guideline for IPLs and IEs (CCPS, 2010, pending) provides guidance on choice of sample size and on grouping of alarms and workers who respond to them. Included in the discussion is a method derived from ANSI Z1.4 (now incorporated by reference with US MIL standards for lot sampling). Examples calculations and the resulting savings achievable by use of sample plans are also provided in the guideline.

The table on the next pages provides an example of the relative savings possible by using a sampling of human response IPLs instead of testing each operator for each IPL.

Sample Size and Acceptance Criteria Estimation for Human IPL Validation

Employee	Alarms	pop = actual size of population	ss = sample size based on infinite population	Z value	Confidence Level	p = anticipated PFD	c = confidence interval	Validation SS = sample size adjusted for actual population	% of actual population sampled	Expected average number of failures per test period using the sample size	Maximum allowed failures to have XX% confidence (95% if Z= 1.96) that the entire population has less than 1-p error rate (at the chosen confidence interval)	From Hypothesis Test, for 95% confidence level and approximate PFD of 0.1 for entire population
	Double-tailed Z test						"+/-"					
25	130	3250	138	1.96	95%	0.9	0.05	133	4	13.3	12	8
25	13	325	138	1.96	95%	0.9	0.05	97	30	9.7	9	6
25	130	3250	195	2.327	97.5%	0.9	0.05	184	6	18.4	17	11
25	13	325	195	2.327	97.5%	0.9	0.05	122	38	12.2	11	7
	Double-tailed Z test						"+/-"					
25	130	3250	54	1.96	95%	0.9	0.08	53	2	5.3	5	2
25	13	325	54	1.96	95%	0.9	0.08	46	14	4.6	4	2
25	130	3250	76	2.327	97.5%	0.9	0.08	74	2	7.4	7	3
25	13	325	76	2.327	97.5%	0.9	0.08	62	19	6.2	6	2

HUMAN ERRORS as related to Impact non-human IEs and IPLs (such as SIS and Relief Systems)

In addition to having a direct impact as an IE or as failure of a human IPL, human error can of course impact all other IEs and IPLs as well. This is because ALL safeguards are ultimately controlled and maintained by humans. Typically, a base human error rate of 1 error in 50 steps is reasonable, or for excellent control of human factors, an error rate of 1 error in 100 steps may be achieved. Below are some examples of how human factors control the reliability of all IPLs and IEs:

> ***Example related to IE – wrong materials of construction received:*** If the humans fail to control or detect errors during selection and use of materials of construction, then perhaps the wrong grade of steel or other material will be used in a highly corrosive environment and then the IEF will be many times higher than expected. Industry data indicates that the materials delivered are different that what we specify about 3 to 7% of the time. Positive material identification (PMI) detect and correct such failures, theoretically, but plant data suggests that using 100% PMI at receiving will lower the composite error rate to 1% and that another 100% PMI of "as-welded in the field" lowers the combined error rate to about 0.3%. This still leaves a large number of components that may be of the wrong materials.
>
> ***Example related to IPL – high integrity SIS:*** Safety Instrumented Systems (SIS), can be designed to provide risk reductions of 1, 2, or 3 orders of magnitude. However, even the best designed SIS can be negated if (1) the root valves for the sensors are left closed or (2) a bypass valve around an emergency isolation valve is left open (other errors are also possible of course). The current versions of SIS standards do not explicitly require validation of the control of human factors and the resulting human errors that can leave the valves in the wrong position. (In addition, the humans are always capable of revaluing trip points, etc., in the safety-rated PLC, given enough time to learn these shortcuts and given enough spurious trips to make them want to use such a shortcut.) Although an SIL 4 SIS is allowed by IEC 61508, with a PFD of $<10^{-4}$ to $\geq 10^{-5}$, looking at it pragmatically, we do not believe it is possible to control the human errors associated with in situ testing low enough to maintain a SIL 4 rating. Similar, due to the complexity of the analysis of human errors associated with in situ testing, basic LOPA only allows use of SIL 1. SIL 2 and SIL 3 rating is difficult to validate statistically, due to the consideration of the human factor, as a result, SIL of 2 and 3 require human reliability event tree or similar methods to quantify, and so is considered beyond basic LOPA.
>
> ***Example related to IPL – pressure relief systems:*** Many relief system IPLs can have a PFD of 0.01 or better, assuming the rest of the process does not interfere or restrict the flow. However, if you have a block valve upstream or downstream of the a relief device, then leaving it closed will negate the entire value of the relief system. A human reliability analysis of existing plant practices is needed to establish the PFD is better than 0.1, and excellent sustained control of human factors is necessary to maintain a PFD of .01 or better for a relief system; because of the relatively high probability that a

human will leave a block valve closed upstream or downstream of a relief valve. Common industry data for human error rates indicates that the error of leaving such a block valve closed will likely be the dominating failure mode for the relief system, so including block valve upstream or downstream of a relief device (though allowed by ASME with administrative controls) will need to be evaluated by a site before assigning a PFD for the combination system (comprised of the relief system and block valves). The conservative PFD of 0.1 is therefore used as the default value for relief devices that have blocks valves upstream or downstream of relief devices, unless a human reliability for the plant proves otherwise.

Conclusion

Control of human factors is key to achieving high reliability of processes. Poor human factors lead to higher initiating event rates and less or no value for IPLs. Proper control of human factors can be achieved by following industry best practices (not minimal regulations and codes/standards).

Failing to control human factors will result in the time invested in LOPA to be wasted, and more importantly will lead to more accidents. Simply installing or claiming IPLs will not change the critical dependency all plants face on controlling risk – risk control always depends on human factor control.

References

1. Tew, R. and Bridges, *W., Human Factors Missing from PSM,* LPS/GCPS, 2010.

2. US Nuclear Regulatory Commission, *Training Requirements for Nuclear Power Plant Operators,* 10 CFR 55.45 and 55.59

3. NUREG/CR-1278 – *The Human Reliability Handbook*; guidelines from the US NRC on Human Reliability Analysis.

4. *Layer of Protection Analysis (LOPA) Guideline*, AIChE/CCPS, 2001.

5. *Guidelines for Independent Protection Layers and Initiating Events*, AIChE/CCPS, 2010 [pending].

6. *Guidelines for Investigating Chemical Process Incidents, 2nd Edition,* CCPS/AICHE, 2000.

7. Bridges, W., *Gains in Getting Near Misses Reported,* 8th Conference, ASSE-MEC, Bahrain, 2008.

Use and Misuse of Enabling Conditions and Conditional Modifiers in Layers of Protection Analysis (LOPA)

J. Wayne Chastain
Engineering Associate, Plant Protection Technical Services
Eastman Chemical Company
P.O. Box 511
Kingsport, TN 37662
chastain@eastman.com

Prepared for Presentation at
American Institute of Chemical Engineers
2010 Spring Meeting
6th Global Congress on Process Safety
San Antonio, Texas
March 22-24, 2010

UNPUBLISHED

Use and Misuse of Enabling Conditions and Conditional Modifiers in Layers of Protection Analysis (LOPA)

J. Wayne Chastain
Engineering Associate, Plant Protection Technical Services
Eastman Chemical Company
P.O. Box 511
Kingsport, TN 37662
chastain@eastman.com

Keywords: layer of protection analysis, LOPA, conditional modifiers, enabling events

Abstract

Enabling conditions and conditional modifiers are some of the most contentious factors used when performing a Layer of Protection Analysis (LOPA). Enabling conditions are situations which must occur simultaneously with a given initiating event to allow the specific cause for a scenario to propagate to a consequence of interest. Some of the most common enabling conditions are related to modes of operation such as start up and shut down of a plant or unit operation. Often an analyst will use an enabling condition when a particular scenario requires the simultaneous occurrence of two initiating event failures.

Conditional modifiers are normally defined as the three independent factors of probability of ignition, probability of occupancy, and probability of injury. The use of each of these conditional modifiers is only applicable when LOPA is used to evaluate the potential for injury from a given initiating event. If a company's risk criteria are based on release of material from primary containment and not the subsequent impact to personnel, then conditional modifiers are not used in the LOPA method.

Inclusion of enabling conditions and conditional modifiers in the LOPA method allow more accurate modeling of the risk of a given scenario from a life safety perspective. However, use of these factors is subject to potential pitfalls and misuse. Avoiding improper use of these important factors can help to prevent gross misestimates of the risk of events and assist in the safe design and operation of facilities for which LOPA is used for the risk analysis.

1. Introduction

LOPA has become one of the most popular forms of risk assessment in the chemical processing industry since the publication of **Layer of Protection Analysis: Simplified Risk Assessment** in 2001 [1]. This risk analysis method combined the traditional single scenario risk evaluation common to qualitative analysis methods to actual values for failure rates found in traditionally much more detailed quantitative risk analysis methodologies such as fault tree and event tree analysis.

As with other forms of risk analysis, careful consideration has to be given to the rules used in the application of LOPA in order for the values generated by the analysis to be meaningful. LOPA has rules which allow the method to be consistently applied to give a reasonably conservative estimate of the risk of an event defined by a single cause – consequence pair. Application of LOPA to the simplest of these situations is a straight forward task. However, there are many situations that are encountered in risk assessment when the simplest application of the LOPA method will not suffice and will either give unreasonable results from the analysis or no results at all can determined. In some of these cases, the use of enabling events can allow the analyst to reach a conclusion using the LOPA method as opposed to being forced to use a more detailed quantitative methodology. While the use of enabling events can extend the LOPA method, this extension holds the potential danger of introducing additional error into the analysis, indicating that risks are adequately controlled, when in fact they are not.

In most QRAs the endpoint of the analysis is an evaluation of the harm to individuals either inside or outside the plant perimeter. Many companies use loss of primary containment as the endpoint of the analysis for LOPA. However, other companies have extended the LOPA method to the same endpoint used in QRA. In this case, the concept of conditional modifiers comes into play. Three traditional conditional modifiers are used in quantitative risk analysis and can be applied in LOPA;

1. Probability of ignition of a flammable release,
2. Probability of occupancy, and
3. Probability of injury.

While the use of these conditional modifiers has the potential to increase the accuracy of the LOPA method in predicting harm to individuals, they can be misused in ways which will underestimate the risk of the event.

2. Layer of Protection Analysis Overview

For a number of years prior to 2001 several companies had been using risk analysis methods focused on predicting the likelihood of single cause – consequence pairs. With the development of the CCPS text **Layer of Protection Analysis: Simplified Process Risk Assessment**, the concepts around these methods which had been in development by several companies and in the literature were standardized around three simple rules for the use of the technique. These rules, which have received a great deal of subsequent attention in the literature, are that safeguards which prevent an event from occurring should be a) independent, b) effective, and c) auditable in order to be included in LOPA [1].

In the CCPS text, additional concepts which were not required for the application of the simplest incarnation of the LOPA method were introduced. Some of these concepts which were not fully developed in the text included enabling conditions and conditional modifiers.

3. Enabling Conditions

Enabling conditions are typically used in the bounds of a LOPA which allow the analyst to address more complex scenarios than could be evaluated using a more basic application of the method. Proper use of enabling conditions allows the LOPA technique to be applied to a broader range of evaluations. Enabling conditions must be used with care since misuse of this aspect of the LOPA methodology can lead to under prediction of the frequency of an event.

The published guidance on the use of enabling conditions in **Layers of Protection Analysis: Simplified Process Risk Assessment** is very limited. The CCPS LOPA text indicates that a scenario evaluated by LOPA may include "enabling conditions or events that have to occur or be present before the initiating event can result in a consequence." From a fault tree analysis or logic perspective, the initiating event and the enabling condition pass through an AND gate; both must be present for the event to occur [1].

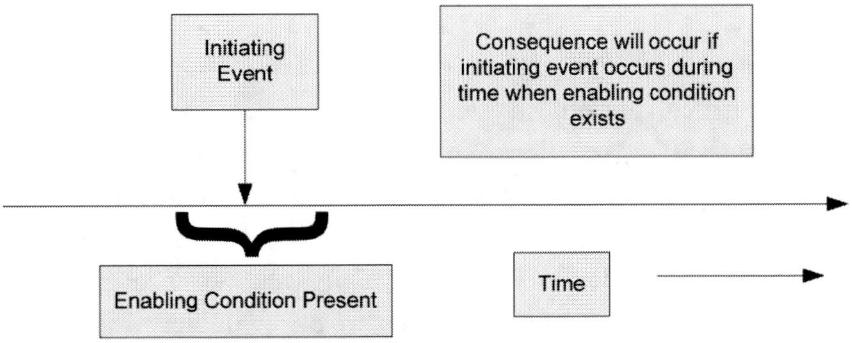

Figure 1: Coincident initiating event and enabling condition

3.1 Enabling Conditions for Concurrent Independent Events

LOPA evaluates a single cause – consequence pair to evaluate its frequency of occurrence. Most events evaluated in LOPA involve a single failure acting as an initiating event and the reliability of the independent protection layers which can serve to prevent or mitigate the event of interest. There are scenarios which are caused by two independent failures, both of which must occur to lead to the initiation of an event. The initiating failure in a LOPA is expressed as a frequency. Due to the linear nature of a LOPA, all other failures in the LOPA must be expressed as probabilities (i.e. dimensionless numbers).

One use of enabling conditions in LOPA is to evaluate the probability that an additional failure has occurred either prior to or concurrent with the initiating failure. In this case, the enabling condition in the LOPA is treated like an independent protection layer from a computational standpoint; however, the probability of the enabling condition being present is determined from the perspective that it is another initiating event.

As an example, isolated running of a certain pump can result in a hazardous event. The initiating failure for the event is an operator starting a pump that is isolated and filled with liquid in the field. By procedure, when the pumps in the field are switched, the pump being shut down is to either be left 1) valved in and ready to run or 2) isolated and drained of process fluid. Leaving the pump in the isolated and liquid filled condition is itself an error and must occur prior to the operator error in starting the pump in the unsafe condition.

In this particular application the valving of the pump is changed 10 times per year. The operator is judged to fail to follow the procedure and leave the pump in an unsafe condition one in one hundred times that the pump valving is changed. From an initiating event perspective this event will occur once in 10 years. However, since this is going to be treated as an enabling condition, an assumption is made that once the pump is set up incorrectly, it is left that way until another valving change is made. In this case that would be 1/10 of the year. Since the pump is in this state once per 10 year period, the average probability that the pump is in the state of being isolated improperly is 0.01 or 1%.

3.2 Time-at-Risk

Enabling conditions are also often used to evaluate the likelihood of scenarios which can only occur during certain critical periods in the operation, generally referred to as "time-at-risk." Startup, shut down, activation, reaction exotherm, chemical addition, and many other system states can involve unique hazard scenarios which are not present during other portions of the operation. Some of these operational states are only applicable to batch processes, which offer a great variety of operational modes to be evaluated for unique hazards. Continuous processes typically have fewer of these operational modes to be evaluated, but all processes can pose hazards that are exclusive to start up and shut down.

When used properly, time-at-risk as an enabling condition can greatly increase the precision of a LOPA for scenarios restricted to certain operating modes. Without its application, the risk of such a scenario may be significantly overstated in a LOPA.

An example of the improved accuracy of a LOPA using time-at-risk can be shown in its application to a multipurpose batch processing facility. In a particular reactor, 100 different products can be manufactured. One of these products uses components and chemistry which poses a hazard to the operators. The other 99 are low temperature; low toxicity aqueous blends which do not pose a hazard to the operators or other personnel. The processes all have equal time in the batch equipment. The instruments which can fail and result in a hazardous event are active and used during the manufacture of the innocuous processes which will reveal such a failure when it occurs. In this case, the time at risk is only 1% and the LOPAs conducted on the hazardous batch process could include a value of 0.01 as a probability of running the hazardous process as an enabling condition.

The above example points out several key factors involved in using time at risk appropriately in a LOPA. The first of these issues is avoiding dilution of the risk value through the use of operational modes. In order to understand how this can happen with improper use of time at risk it is important to understand the basis on which risk criteria are set for use in LOPA.

Most company's application of LOPA involves evaluating single cause – consequence scenarios and comparing the calculated frequency to a risk criteria used for scenario evaluation. The recently released CCPS text **Guidelines for Developing Quantitative Safety Risk Criteria** indicates that most risk criteria set by governmental bodies are either individual risk or societal risk criteria [2]. In both cases, these risk criteria are cumulative in nature, and would involve summing the risk from all of the scenarios which could impact a certain individual or population. Single scenario risk criteria serve as a proxy for individual risk and societal risk criteria in most applications of LOPA. However, when dealing with time-at-risk factors in a study, this relationship must be accounted for to prevent an under evaluation of the overall risk.

In the example above, only one process out of the 100 run in the equipment had the potential for harm. In this case, time at risk is not improper to use since the population of operators is only exposed to a comparable risk 1% of the time. Compare this with a situation in which 100 processes are run in a similar unit, but all of them pose similar hazards. In evaluating the risk from one process being manufactured in the equipment, time-at-risk would not be an appropriate enabling condition to use in the evaluation. To do so would expose the operator to 100 times greater risk than the original derivation of the risk criteria would indicate.

Another way of accounting for this type of situation is to use the time-at-risk factor as an enabling condition, but also modify the risk target used in the LOPA by the same factor. However, making this modification is an unnecessary complication. If both the risk and the risk criteria are modified by the same factor, the same reliability of the controls will be indicated by the results of the LOPA.

The second issue which must be addressed to ensure that time-at-risk is an appropriate enabling condition in a LOPA is the time of the failure that initiates the event in relationship to the process state of interest and whether the failure will be revealed. In the example above, the failure which initiates the scenario of interest was indicated as being revealed immediately. However, if the failure could occur at any time, but only be revealed when the process state of interest occurs, then the time-at-risk should not be taken into account in the analysis. Assume that a level measurement on the reactor can fail resulting in overfilling of the unit. This failure would be revealed during any of the recipes run in that unit and time-at-risk could be properly taken into account. Conversely, assume that a weigh tank on the reactor system is only used for the one in one hundred hazardous batch. In this case, without taking special precautions, the level measurement on the weigh tank could have failed at any point in the operation of the equipment, but this failure would not reveal itself until the hazardous chemistry was being conducted in the equipment. This would eliminate time-at-risk from consideration.

One way to address the issue of unrevealed failures eliminating time-at-risk from consideration is through testing. Normally testing of instrumentation and equipment is conducted on a regularly scheduled frequency. In this case, if the equipment prone to unrevealed failures is tested immediately before starting the highly hazardous process in the equipment, then the time-at-risk factor can be taken into account as an enabling condition in the LOPA and will reduce the risk from the scenario. Full functional testing of the equipment prone to unrevealed failures ensures that only failures which occur in the limited duration event will result in the scenario of concern.

These same considerations also apply to other operational modes when considering time-at-risk in a LOPA. The same thought process should be applied when evaluating the risk from start-up, shut down, turn around, activation, or specific steps within a batch process such as addition or exotherm.

4. Conditional Modifiers

Conditional modifier is the term used to describe several factors that come into play in a LOPA when the end point being evaluated is human harm, harm to businesses, or the environment. Of these potential end points, harm to humans is the most likely end point which will use conditional modifiers. For those analysts whose LOPA methodology uses a risk criteria based solely on loss of containment of a specific amount of material, conditional modifiers are an improper addition to the LOPA technique and can result in an under evaluation of the risk of scenarios being evaluated.

The most common conditional modifiers which are used in LOPA where the end points are related to human harm are:

- Probability of ignition
- Probability of occupancy
- Probability of injury

Each of these conditional modifiers will be addressed in detail below. Similar to the use of enabling conditions above, proper use of conditional modifiers can result in greater accuracy in evaluation of the risk from scenarios. Misuse of these conditional modifiers can result in under estimation of the risk of a scenario.

4.1 Probability of Ignition

Releases of flammables or combustibles are used by many companies as the end point of concern for LOPAs. However, when human harm is the end point of concern, ignition of a release of flammables or combustibles is normally required to reach the end point.

The CCPS LOPA text gives limited guidance on the use of probability of ignition as a conditional modifier. The text indicates conservative values which may be used in LOPA for this modifier in the absence of a more detailed or accurate model. The limited guidance provides the following values for probability of ignition:

- 1.0 for releases caused by collisions,
- 1.0 for large releases close to fired equipment,
- 0.5 for releases in general process areas,
- 0.1 for releases in remote process areas, like a tank farm [1].

It should be noted that for organizations that restrict values in their LOPAs to order of magnitude factors only, releases in general process areas would be rounded to a 100% chance of ignition.

The values for likelihood of ignition presented from the LOPA text include several critical assumptions. The most important of these is that the area where the release occurs is properly electrically classified for the material released. Any known violations of the electrical classification for an area should increase the likelihood of ignition to 100% for any release.

4.1.1 Survey of Emergency Response Data

Emergency response data is collected for the Eastman Chemical Company site at Kingsport, Tennessee. Each run made by the emergency response personnel on site is recorded in a database. Events are classified into several categories, including fire calls, fire standbys, and accidental discharges. A fire call on the site typically involves an emergency run where flames or smoke have actually been observed, either by the individual calling emergency services or the responders themselves. A fire standby may involve a release of material or a potential release of material but does indicate that fire engines were sent to the scene in case a fire occurred. An accidental discharge indicates that material has been lost from its primary containment but there was no fire or concern of a fire from personnel making the call to emergency services or by the emergency responders.

These three categories of emergency events were analyzed for 2000 – 2009 for the Kingsport, Tennessee site of Eastman Chemical Company. Adequate data was available for approximate analysis to yield an indication of the number of emergency calls which involved a fire either due to the presence of flames or smoke and the total number of events which involved a release of material from primary containment. In analyzing the data, several types of events had to be excluded from the datasets to obtain the appropriate subset of events. Each item in the fire call data set was evaluated to eliminate from the dataset events such as electrical fires, ballast failures in lights, and automobile engine fires. Each item in the accidental release data set was evaluated to eliminate spills of nonflammable materials including spills related to overfilling of sewer lines, water spills, and caustic releases. Each item in the fire standby data set was evaluated to eliminate from consideration events involving overheating of equipment, smoke coming from equipment due to rubbing of belts, and other items unrelated to a release of material from equipment.

Once all of the modifications were made to the data sets, it was possible to evaluate the fraction of events at the Eastman Chemical Company Kingsport site over the decade in which a fire resulted from a release of flammable or combustible material. 20% of releases resulted in a fire. This value has been rounded to the nearest tenth so as not to over represent the accuracy of the data set that it was drawn from. The majority of these releases and fires on the plant site are in the operating area as opposed to being located in remote tank farm areas. This data supports the conservative value recommended in the LOPA text of 50% for the likelihood of ignition in general processing areas.

4.1.2 Immediate vs. Delayed Ignition

In certain cases, particularly those involving releases of flammable gases or vapors or releases of combustible liquids above their atmospheric boiling point, delayed ignition of a release can result in a greater consequence than immediate ignition. Immediate ignition of such a release will typically result in a jet fire. Jet fires can have extremely negative consequences; however, except for knock-on effects, these are typically local in nature. Compared with the local consequences associated with jet fires, delayed ignition of such a release can result in a vapor cloud explosion which can have much greater effects impacting large scale areas both inside and outside a facility.

In such a case, even for high energy releases or releases near fired equipment, delayed ignition should be considered as a possible outcome. Failure to consider delayed ignition of the flammable cloud could significantly under estimate the potential risk of such a loss of containment.

4.1.3 More Detailed Probability of Ignition Models

Guidelines for Chemical Process Quantitative Risk Analysis, 2nd Edition, describes a more detailed model for evaluating the likelihood of ignition of a release [3]. The use of this model involves identification of specific potential ignition sources. These sources may include:

- Flares
- Boilers
- Fired Heaters
- Static Electricity

- Vehicle Traffic
- Electrical Motors
- Hot work (welding and cutting)
- Lightning
- Overhead high voltage lines
- Mechanical sources such as sparks, friction, impact, vibration
- Chemical reactions

This model involves two components, a presence factor and a strength factor. The presence factor is the probability that an ignition source will be present to ignite the flammable release. For example, flares may not be burning 100% of the time and vehicle traffic in a restricted area may be significantly less than continuous.

The strength factor is a measure of the likelihood that the ignition source will ignite the released mixture if the ignition source is present when the release occurs. The CPQRA text can be consulted for representative ranges of values for the strength factor for various types of ignition sources. It is important to note that the strength factor may have to be adjusted for releases of different chemicals due to the properties of the chemicals released. The autoignition temperature and the minimum ignition energy of the material released influence the probability of a given ignition source to ignite the mixture.

The potential impact that the fuel properties has on the likelihood of ignition of the release, using this more complex model indicates that a third factor may be appropriate for extending the method. By setting a strength factor for a standard material, a material factor could applied to the overall likelihood of ignition to adjust the value for the minimum ignition energy and / or autoignition temperature of a given fuel.

The detailed model presented in CPQRA was developed for quantitative risk analysis and may be too complex for application in all but the most detailed LOPA analyses. This probability of ignition model is more applicable to cases where detailed dispersion modeling is conducted and the resulting fires and explosions are modeled with complex effects models on personnel and installations.

The CPQRA text does present a simpler, alternative methodology for ignition probability developed for the Canvey Island study by the U.K. HSE and presented in "Canvey – A Second Report" [4]. This model differs from that presented in **Layer of Protection Analysis: Simplified Process Risk Assessment** but is of a similar level of detail and complexity and therefore more applicable to LOPA implementation.

The HSE model presented in the reference provides the following probabilities of ignition based on the sources of ignition.

Sources of Ignition	Ignition Probability
None (no ignition sources readily identifiable)	0.1
Very Few (release in a remote area)	0.2
Few (release near noncontinuous operations such as road / rail facilities)	0.5
Many (release near a plant or resulting from a nearby fire or explosion)	0.9

Table 1: Probability of Ignition Model from HSE "Canvey – A Second Report"

This model uses higher probabilities of ignition than that presented in the LOPA book. For a release in a general processing area, a value of 0.5 was suggested in the LOPA text. In this table, a value is 0.9, which would typically be rounded to 1.0 in a LOPA, is suggested. Based on the data set analyzed for the site noted above, a value of 0.9 for the probability of ignition in general processing areas is high with appropriate electrical classification and control of ignition sources. One adjustment which could reasonably be made to the LOPA probabilities of ignition suggested in the original text from the HSE report would be to increase the likelihood to 0.5 for releases in remote areas which are near non-continuous operations which may serve as ignition sources.

4.1.4 Probabilities of Ignition inside Process Equipment

One likelihood of ignition scenario not addressed by any of the above resources is an appropriate value to use for the probability of ignition when a flammable mixture is formed inside process equipment. Unlike releases outside the process, ignition sources are typically limited inside process equipment. Several factors have to be considered when determining this probability.

- Properties of the material
- Position in the flammability envelope
- Grounding and bonding of equipment
- Electrical classification

Normally processes are not designed so that a flammable mixture is present in the equipment under normal operation. A process designed to operate in the flammable range under normal conditions typically poses an undue risk. To prevent the formation of a flammable atmosphere in the process equipment, a common practice is to exclude oxygen or maintain the oxygen level in the equipment below the minimum oxygen concentration for the material present in the process.

Alternatively, the composition in the vapor space of process equipment can be maintained below the lower flammability limit by control of the temperature of the process. Certain upset conditions can result in the formation of flammable atmospheres inside equipment which is processing flammable or combustible material above its flash point.

The formation of a flammable vapor space in equipment usually requires an ignition source in the process to result in a deflagration. Cases where an ignition source is not required are experienced when the temperature in the process equipment is higher than the autoignition temperature of the flammable materials in the process. Even when the material is well below its autoignition temperature, if catalysts are present in the system, then these can act to promote combustion reactions and result in ignition. Non-polar materials in processing equipment can cause a spark and subsequent ignition of a flammable vapor due to the buildup of a static charge on the fluid. In this case, bonding and grounding will not necessarily act as a control to prevent a spark. Where possible, the use of an antistatic agent in non-polar materials can help to alleviate the risk of such materials from building up a static charge.

Given the formation of a flammable mixture in processing equipment, the amount of energy required to initiate a deflagration in the equipment is related to ratio of fuel to oxygen and the amount of inerts in the system. The lowest minimum ignition energy for a given fuel is typically a mixture of the fuel and oxygen near the stoichiometric ratio. The addition of inerts to the system increases the energy required for ignition. Movement away from the stoichiometric ratio of fuel and oxygen either by increasing the fuel or oxygen will also serve to increase the energy required to ignite the mixture. Reported minimum ignition energies in literature sources are typically measured in air at values near the stoichiometric ratio. Because of these factors, the position in the flammability envelope can be considered when determining the likelihood of ignition.

Process equipment is typically grounded, bonded, and electrically classified such that any instruments in contact with the process are intrinsically safe or protected with appropriate barriers to prevent ignition of flammable atmospheres. Any deviation from appropriate bonding and grounding or issues with inappropriate electrical classification can greatly increase the probability of ignition of flammable vapor spaces in process equipment.

4.1.5 Areas of Possible Misuse of Probability of Ignition

Probability of ignition can be misused by underestimating the value. When conservative but reasonable values are used for likelihood of ignition, a LOPA reflects actual risk from an event more accurately than always assuming a 100% likelihood of ignition. However, there are several cases when a likelihood of ignition should be assumed to be 100%.

High energy releases such as those from overpressure of equipment rated for high pressure, boiling liquid, expanding vapor explosions (BLEVEs), and runaway reactions, should be assumed to provide an ignition source for a release. In these cases, the failure of the vessel itself or the impact of shrapnel from the vessel against nearby equipment is very likely to result in ignition of released material.

Special attention should also be given to materials with particularly low minimum ignition energies or autoignition temperatures. These materials have the potential for finding a source of ignition much easier than typical materials. Acetaldehyde, as an example, has an autoignition temperature of 130 C. A release of this material will, in almost any process setting, encounter surfaces above its autoignition temperature.

4.2 Probability of Presence (Occupancy)

The occupancy conditional modifier is a simpler factor to evaluate than the previously discussed likelihood of ignition. Conceptually, if the end point of interest is human harm, then individuals must be present in the location of the event in order to be impacted.

Very little guidance is given in **Layer of Protection Analysis: Simplified Process Risk Assessment** regarding the likelihood of presence of individuals. One reason for this may be the simplicity of this factor. In general, the average occupancy of the area can be used as an indication of occupancy during any release. When determining the occupancy in the area, ensure that the scope of personnel included in the evaluation not only encompasses operations personnel, but also includes maintenance personnel, insulators, painters, engineers, and any other individuals that may be in the area.

In certain cases, corporations will use risk targets for end points involving multiple individuals for large scale events. In this case, normally occupancy is determined for buildings which may be impacted by the event. The buildings will likely be occupied continuously resulting in 100% occupancy, only on day shift resulting in approximately 30% occupancy, or have very limited occupancy which might result in a value of 10% or lower.

4.2.1 Events Leading to High Occupancy

Two types of events result in occupancy higher than the average value indicated above. In the first case, personnel presence is required to initiate the event. In the second case, the response to the initiation of the event is to send personnel to the site of the developing scenario. In both of these cases, occupancy should be assumed to be 100% in the LOPA.

In certain scenarios operators or other personnel are required to be in the location of an incident because an action on their part becomes the initiating failure for the scenario being analyzed. This type of scenario occurs most frequently in processes which are less automated and require greater physical interaction between the operator and the process. As an example, batch reaction systems which require manual charging of components to a unit increase the occupancy for the process during the time when mischarges could result in an unexpected and undesired reaction and an event. Some plants still operate using local panel control of chemical processes as opposed to remote control from a control room. In these cases operators are almost always present in the operating area and subject to the consequences of any event if the process is running. This type of situation can also present itself if the scenario initiation is due to starting a pump, agitator, or other motor and remote start of the motor is not available or contraindicated by the operational practice or procedures of the plant. In all of these cases, the occupancy for the event should be assumed to be 100%.

Operations response to an ongoing event can serve to increase the occupancy. Evaluating this change in the occupancy value for a scenario is more difficult than the previous situation. To determine the impact of operator response to an event, procedures must be evaluated and typically one or more operators must be interviewed to ascertain what the response of operators will be during an event. If the common practice in the operating area is to send a field operator to the location of an event to investigate the situation "eyes on" then this will serve to increase the occupancy to 100% for most events.

As an example, consider a situation in which a level control valve in a feed stream has failed open and a vessel is quickly overfilling. If the response of operations personnel monitoring the operation from the control room is to ask a field operator to go out to the vessel and inspect the equipment to troubleshoot the situation, then the field operator will likely be in the area if the vessel overfills and a release occurs. The occupancy in this case is not the average value assuming the random likelihood of vessel overfill and occupancy in the area occurring simultaneously. However, if the operating procedures and practices in the area indicate that the response of the control operator to this situation would be to close another remotely actuated valve or to stop the pump remotely that is overfilling the vessel and only when the operator has confirmed that the level in the vessel is stabilized at a safe value is a field operator sent to the location, then the average occupancy for the event may be assumed. This mode of operation may be counter intuitive for many facilities, so the LOPA analyst should ensure that appropriate operational discipline is in place before using a low value for occupancy.

4.3 Probability of Injury

The third common conditional modifier used for studies where human harm is the end point of the analysis is the probability of injury of personnel. There is little guidance in **Layer of Protection Analysis: Simplified Process Risk Assessment** on the use of probability of injury as a conditional modifier other than a small amount of qualitative guidance [1]. This qualitative guidance is reproduced in the table below:

Type of Event	Probability of Injury (Fatality)
Pool fire	Moderate to low
Flash fire	High
Toxic vapor exposure	Dependent on vapor concentration, duration of exposure, and ability to move out of the cloud which is impacted by the ability to detect the vapor, how quickly the person is incapacitated by the vapor, and the availability of escape routes

Table 2: Qualitative Likelihoods of Injury for LOPA Analysis [1]

Layer of Protection Analysis: Simplified Process Risk Assessment indicates that several analysts use a default value of 0.5 for the probability of injury in most situations. This value is increased to 1.0 for toxic exposures that are difficult to detect, overcome the person exposed quickly, or when inadequate escape routes are available. In the absence of more detailed modeling which will require greater expertise and time and effort spent in the analysis, this level of detail is likely appropriate. If additional detail is needed resources are available. The practitioner is referred to the **Guidelines for Chemical Process Quantitative Risk Analysis, 2nd Edition** as a starting point for delving into this topic [3].

A variety of models are available for several different types of events which can result in injury. Toxicity models are available to determine the impact to personnel by exposure to chemicals. Heat flux models can be used to determine the impact to personnel from a pool fire, jet fire, or flash fire. Overpressure models are used to determine the possible effect from blast waves on personnel and on structures which may house personnel. Other types of effect models exist, but these are the models which are most likely to be used in a LOPA.

4.3.1 Toxicity Models

Application of toxicity models require some degree of dispersion modeling in order for an evaluation to be conducted. ALOHA is a free dispersion modeling software that is part of the CAMEO program developed by the U.S. Environmental Protection Agency and the National Oceanic and Atmospheric Administration [5]. This program or any of several commercially available programs can be used to conduct modeling which will give concentrations of chemicals at varying distances based on the application of standard dispersion models. Modeling of releases is a specialty which requires detailed knowledge. In general, evaluating releases using dispersion models should be left to experts in the field.

A variety of toxicity models exist for modeling the impact of personnel exposure to different concentrations or dosages of chemicals. Point models are very commonly used in the industry and provide easily applied endpoints for evaluation. Common point values are Emergency Response Planning Guidelines (ERPGs) developed by the American Industrial Hygiene Association, Threshold Limit Values developed by the American Conference of Governmental Industrial Hygienists, in particular the ceiling value (TLV-C) and the short term exposure limit values (TLV-STEL), Immediately Dangerous to Life and Health (IDLH) levels developed by the National Institute for Occupational Safety and Health, and several others [3].

Use of the point values for evaluating the impact of toxic materials is the easiest application of toxicity models. In this case several of the point values mentioned above are possible endpoints for evaluating the impact of various concentrations on personnel. IDLH values can be used to evaluate a concentration which could be assumed to cause a 50% chance of fatalities. The use of a 50% chance of fatality in this case is a reasonable approach, since the assumption with IDLH concentrations is that personnel can and should flee the area if this concentration is exceeded and the response to the IDLH concentration is based on a maximum of 30 minutes of exposure. The caveats in Table 2 above would apply to this situation and vapors which are hard to detect, are fast acting, or in situations where personnel cannot easily flee the exposure, a value of the likelihood of injury of 100% should be assumed at IDLH concentrations. IDLH values are derived for a healthy population of workers. For this reason, application of this value for more sensitive populations of individuals which would be encountered for releases outside the facility would not necessarily be a conservative approach.

Probit models offer a more detailed level of analysis of personnel exposure response to specific dosages of chemicals. The impact to personnel exposed to chemicals is more properly evaluated as a dose – response relationship than an effect based solely on the concentration to which personnel are exposed [3]. The number of Probit models developed for chemical species are very limited and although they are the most detailed effect model available for most chemical exposures, their application within a LOPA, in general, implies a greater degree of accuracy than other aspects of the method such as the frequency determination.

The U.K. Health and Safety Executive has endorsed an alternative approach to determining values for use in a dose – response model that are more easily developed and applied within a simplified method such as LOPA. This method uses available animal test data to determine a Dangerous Toxic Load (DTL) for a Specified Level of Toxicity (SLOT) and a Significant Likelihood of Death (SLOD). Specifically, the SLOD is directly applicable to LOPA since it can be applied based on data obtained from ALOHA or other dispersion modeling software and is meant to predict the dose that will yield a fatality rate of 50% [6]. For an example of how this method can be applied please see Appendix A.

4.3.2 Thermal Effects Models

As indicated above, and shown in the examples in the LOPA text, a value of 0.5 was used as a default for probability of personnel injury for events where exposure to fire or high levels of heat occur. This approach gives a reasonable but conservative adjustment for the risk of an event related to fires and exposure to heat from fires when the end point of interest is human harm. An exception to this default value is the probability of injury for personnel present in a flash fire. Flash fires are short duration events which will not expose personnel who are present to very high thermal flux, but do have a high likelihood of injury. Because of this, personnel within the area where a flash fire will propagate should be assumed to have a probability of injury of 100%.

More detailed models are available for determining the likelihood of injury on exposure to fire or high levels of heat. Once again, the reader is referred to **Guidelines for Chemical Process Quantitative Risk Assessment, 2nd Edition** as a reference which reviews many of the approaches which could be applied within a LOPA. Thermal effects on personnel is a function of the thermal flux the person experiences and the time the person is exposed to the thermal radiation, which should be recognized as being very similar to a thermal dose – response effect. The CPQRA text presents a figure from Mudan which summarizes the data of Eisenberg and Mixter for a range of injury levels for different thermal fluxes and times of exposure [3, 7, 8, 9]. This figure is reproduced below.

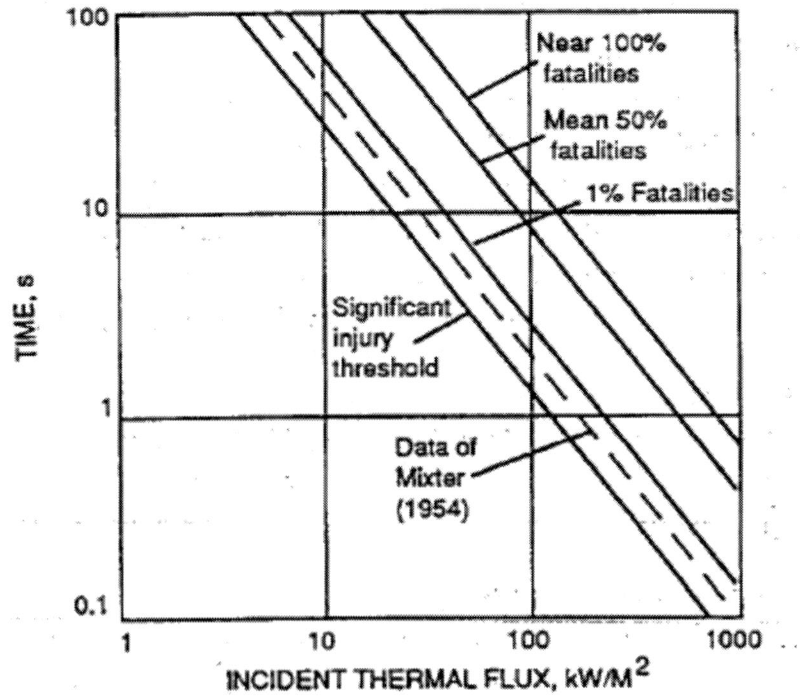

Figure 2: Serious Injury / Fatality Levels for Thermal Radiation

In order to use Figure 2, the thermal flux level, duration of exposure, and distance from the source are required inputs. Thermal flux can be calculated by one of several models presented in the CPQRA book. This is a detailed topic and any practitioners are referred to the CPQRA text for use of these detailed methods. Durations are provided by the consequence model used or an estimate of the time to extinguish the fire. It should be noted, that for long duration events, personnel are likely to take evasive action. Exposure times will be the minimum of the duration of the fire or the time to leave the area exposed to the thermal radiation [3].

4.3.3 Explosion Models

Explosions can occur as the consequence of a scenario being evaluated by LOPA due to deflagration, runaway reaction, failure of a relief system, or other means that causes overpressure of a vessel. Explosions can also occur after a release of flammable materials if the material disperses into a confined area or structure and is subsequently ignited. Harm to people can be caused by the overpressure from these events or from projectiles launched by the energy of the events.

The guidance in the LOPA text, as summarized above, would indicate that barring any additional information, personnel exposed to an explosion would have a 0.5 probability of injury. More detailed evaluation of explosions and their effects is another area that requires detailed knowledge and expertise. Explosion effects are normally evaluated for their impact to structures and to people. Similar to the evaluation of thermal effects, evaluation of the effect of an explosion requires the explosion to be characterized in terms of its overpressure and impulse. The methods in **Guidelines for Chemical Process Quantitative Risk Assessment, 2nd Edition** can be used to characterize the explosion or commercially available computer programs can be used for this purpose. Once an explosion has been characterized, the text provides Probit models for the effects of explosion overpressure on structures as well as people outside of buildings or structures [3].

5. Future Work

Enabling events are an integral part of LOPAs while conditional modifiers are means of obtaining greater accuracy when the end point of the analysis is human injury. Based on recent projects approved by the Technical Steering Committee of CCPS, conditional modifiers for use in LOPA are going to see additional development. A project to survey available literature on probability of ignition and document the best available technology is ongoing. This project currently includes a screening method which should be applicable to the level of detail appropriate for use in LOPA and a detailed method which is better suited for application in QRA. A second project specifically on conditional modifiers in LOPA has also been approved. While this project will likely provide a brief overview of likelihood of ignition, given the probability of ignition project, it should address in detail the likelihood of injury and occupancy factors used as conditional modifiers in LOPA.

6. Conclusions

Enabling events and conditional modifiers are an important aspect of LOPA, particularly for companies that use human harm as the end point for consideration. Proper application of these factors can allow a LOPA analysis to have greater accuracy than excluding these values. However, misuse of these factors can result in underestimation of the risk of a scenario and insufficient safeguards being in place to prevent or mitigate an event. For this reason, care must be used when applying these factors to a LOPA.

7. References

1. Center for Chemical Process Safety (CCPS), **Layer of Protection Analysis: Simplified Process Risk Assessment**, American Institute of Chemical Engineers, New York, NY, 2001.
2. Center for Chemical Process Safety (CCPS), **Guidelines for Developing Quantitative Safety Risk Criteria**, American Institute of Chemical Engineers, New York, NY, 2009.
3. Center for Chemical Process Safety (CCPS), **Guidelines for Chemical Process Quantitative Risk Analysis, 2nd Edition,** American Institute of Chemical Engineers, New York, NY, 2000.
4. U.K. Health and Safety Executive, **Canvey – A Second Report**, Her Majesty's Stationary Office, London, U.K., 1981.
5. U.S. Environmental Protection Agency, **Computer-Aided Management of Emergency Operations**, Washington, D.C., 2010, http://www.epa.gov/oem/content/cameo/index.htm.
6. U.K. Health and Safety Executive, **Assessment of the Dangerous Toxic Load (DTL) for Specified Level of Toxicity (SLOT) and Significant Likelihood of Death (SLOD)**, Bootle, Merseyside, U.K, 2010, http://www.hse.gov.uk/hid/haztox.htm.
7. Mudan, K. S., "Thermal Radiation Hazards from Hydrocarbon Pool Fires." *Proc Energy Combust Sci*, Vol. 10, No. 1, pp. 59-80.
8. Eisenberg , N.A., C. J. Lynch, and R. J. Breeding, **CG-D-136-75 and NTIS AD-015-245: Vulnerability Model: A Simulation System for Assessing Damage Resulting From Marine Spills**, U. S. Coast Guard, 1975.
9. Mixter, G., **Report UR-316: The Empirical Relation Between Time and Intensity of Applied Thermal Energy Production of 2+ Burns in Pigs**, Rochester, N.Y., Univerity of Rochester, 1954.

Appendix A

Use of Dispersion Modeling for Consequence Determination

Equipment processing acetic acid has the potential for a relief event. The relief calculations indicate that the relief flow of acetic acid from the equipment will be 10,000 lb/hr released at a height of 50 feet. A building containing personnel on site is located 120 feet away from the release point.

ALOHA 5.4.1 is used to model the dispersion of the acetic acid. The input data for the dispersion case is shown below.

SITE DATA:
Location: KNOXVILLE, TENNESSEE
Building Air Exchanges Per Hour: 0.50 (enclosed office)
Time: January 28, 2010 1048 hours EST (using computer's clock)

CHEMICAL DATA:
Chemical Name: ACETIC ACID, GLACIAL Molecular Weight: 60.05 g/mol
TEEL-1: 5 ppm TEEL-2: 35 ppm TEEL-3: 250 ppm
IDLH: 50 ppm LEL: 54000 ppm UEL: 160000 ppm
Ambient Boiling Point: 242.5° F
Vapor Pressure at Ambient Temperature: 0.016 atm
Ambient Saturation Concentration: 16,948 ppm or 1.69%

ATMOSPHERIC DATA: (MANUAL INPUT OF DATA)
Wind: 3 miles/hour from N at 3 meters
Ground Roughness: open country Cloud Cover: 5 tenths
Air Temperature: 70° F Stability Class: B
No Inversion Height Relative Humidity: 50%

SOURCE STRENGTH:
Direct Source: 10000 pounds/hr Source Height: 50 feet
Release Duration: 60 minutes
Release Rate: 167 pounds/min
Total Amount Released: 10,000 pounds

Figure A.1 Input Data for Example Acetic Acid Release

The dispersion of this release predicted by ALOHA is shown below in Figure A.2.

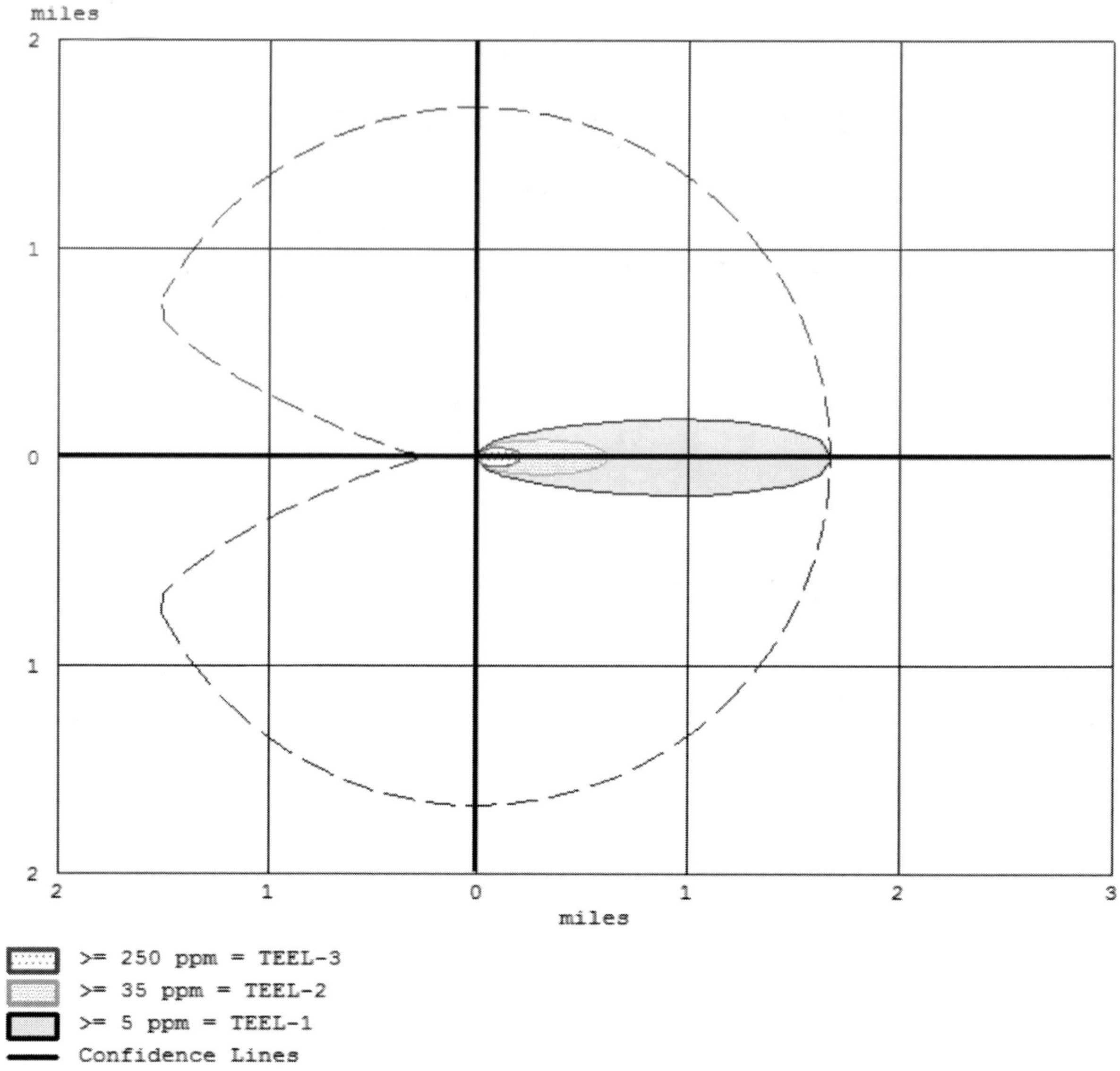

Figure A.2 Dispersion Results for Acetic Acid Example Release

The results from the ALOHA model show areas exposed to Temporary Emergency Exposure Limits (TEELs) for acetic acid. TEELs are used in lieu of ERPG concentrations when ERPGs have not been set for a given chemical. The method for established TEELs were set by the Department of Energy's Subcommittee on Consequence Assessment and Protective Action. [3].

ALOHA can also evaluate the concentrations at a certain distance downwind from the release and at a given crosswind distance. The use of this functionality in ALOHA or other dispersion modeling software can provide the concentration experienced by a population of interest. In this example, an office building is located 120 feet away from the release point. The model can predict the concentrations at this location and using the building type can also predict the concentrations present in the structure. For the example case, this is shown in Figure A.3.

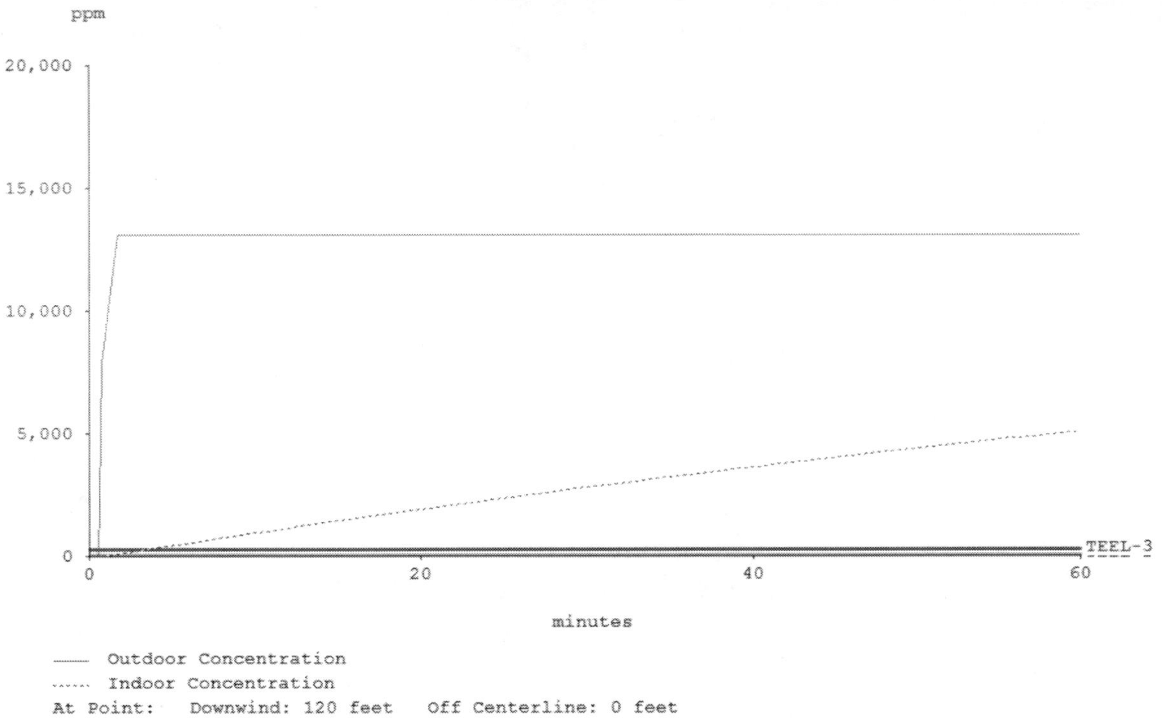

Figure A.3: Results of Dispersion Model for Acetic Acid Release Example 120' Downwind

The Specific Level of Toxicity (SLOT) Dangerous Toxic Load (DTL) for acetic acid is $7.5x10^4$ ppm*min. The dosage resulting in a Significant Likelihood of Death (SLOD) is $3x10^5$ ppm*min. The SLOD dosage predicts the value that will result in a 50% mortality rate which is a useful value for LOPA evaluation.

From the figure above, the concentration external to the building at 120 feet from the release point can be seen to rise quickly to a concentration of approximately 12,500 ppm. At this concentration, the SLOD dosage would be reached after approximately 24 minutes. At these concentrations personnel would be expected to try to reach the shelter of the building. If the release continued for one hour, the concentration in the building is predicted to rise to a final value approaching 5000 ppm. For conditions outside, once the release is terminated, the concentration will quickly drop to zero. However, inside the building, the concentration will decrease at a rate similar to the rate at which it increased. Because of the shape of this concentration curve the peak value multiplied by the release time will give an indication of the dosage. In this case, 5000 ppm times 60 min gives a dosage of $3x10^5$ ppm*min. This is the SLOD dosage for acetic acid and a 50% mortality rate would be predicted for anyone located in the building for the duration of the release and the hour after the conclusion of the release. Use of the SLOD dosage in this case will not indicate a 50% chance of fatality, but instead in a LOPA would indicate a 100% chance of fatality for 50% of the personnel exposed to this dosage in the structure.

Is It Really an Independent Protection Layer?

A. M. (Art) Dowell, III, PE
A M Dowell III PE
16430 Locke Haven Dr, Houston, TX 77059
adowell3@comcast.net

Prepared for Presentation at
American Institute of Chemical Engineers
2010 Spring Meeting
6th Global Congress on Process Safety
San Antonio, Texas
March 22-24, 2010

UNPUBLISHED

Is It Really an Independent Protection Layer?

A. M. (Art) Dowell, III, PE
A M Dowell III PE
16430 Locke Haven Dr, Houston, TX 77059
Adowell3@comcast.net

Keywords: LOPA, Layer of Protection Analysis, IPL, independent protection layer, risk assessment, SIF, safety instrumented function, SRS, safety requirements specification

Abstract

Layer of Protection Analysis (LOPA) is a simplified chemical process risk assessment tool that is used to determine if sufficient independent protection layers (IPLs) are in place to meet an organization's risk tolerance criteria for a particular hazardous scenario. LOPA can determine how much additional risk reduction is needed for the scenario and can help in specifying the additional protection layers.

The CCPS books, *Layer of Protection Analysis – Simplified Process Risk Assessment* (2001) [1], and *Guidelines for Safe and Reliable Instrumented Protective Systems Center for Chemical Process Safety* (2007) [2] identify required attributes of independent protective layers. In more than 15 years experience working with LOPA, we have frequently observed confusion about these attributes among LOPA analysts, protection layer implementers, and operations personnel.

Therefore, this paper discusses the characteristics of independent protection layers, gives examples to avoid confusion and misapplication of the characteristics, gives an easily remembered phrase to screen candidate IPLs, and provides sample tools to confirm IPL compliance with the characteristics. The paper discusses the importance of a Safety Requirements Specification for SIFs (Safety Instrumented Functions) and proposes analogous documentation for other IPLs. The paper gives guidance to determine when protection layers or scenarios require more detailed analysis such as event tree analysis or fault tree analysis.

1. Introduction

For a given scenario, Layer of Protection Analysis evaluates the severity of the consequence, the frequency or likelihood of the initiating cause, and the risk reduction of protection layers that can prevent or stop the scenario from proceeding to the consequence [1]. The mitigated frequency of the consequence is given by [Eq. 1].

$$f_c = f_i \times PFD_1 \times PFD_2 \times \ldots PFD_n \qquad \text{[Eq. 1]}$$

where f_i is the frequency of the initiating cause,
PFD_n is the probability of failure on demand of the nth independent protection layer,
f_c is the mitigated frequency of the consequence.

LOPA then compares the mitigated frequency for the consequence to the organization's risk tolerance criteria for consequences of that severity to determine if there are sufficient independent protection layers or if additional IPLs should be added.

Probably the most difficult aspect of LOPA is determining which safeguards are really effective independent protection layers. That is, which safeguards can actually prevent the consequence reliably? First we will examine the characteristics of independent protection layers, and then we will illustrate how to apply the characteristics.

2. Characteristics of independent protection layers

In 2001, CCPS [1] required that independent protection layers (IPLs) should be unique and applicable, independent, dependable, and auditable.

In 2007, CCPS [2] extended the requirements to Seven Characteristics of IPLs. These core attributes are

- Independence
- Functionality
- Integrity
- Reliability
- Auditability
- Access Security
- Management of Change

Independence

The performance of a protection layer is not affected by the initiating cause of a hazardous event or by the failure of other protection layers.

This characteristic probably presents the most difficult challenge in LOPA. For example, if the consequence is loss of containment by overflow of a tank, and the initiating cause is failure of the level control loop (sensor, controller, and valve), the protection layer to prevent the overflow should not rely on the level control loop. A failure of the level control loop element (sensor, controller, and valve) that causes the scenario would also disable the protection layer.

Functionality

The required operation of the protection layer in response to a hazardous event.

Functionality means that the protection layer by its design can detect the beginning of the scenario and provide the appropriate response to the process in time to prevent the consequence from occurring. Functionality includes the "unique" and "applicable" characteristics from CCPS [1]. For example, a pressure relief valve is designed to open at a pressure that is low enough, and the relief valve and its piping are large enough, to prevent the pressure in the vessel from exceeding the design pressure. In LOPA, we do not depend on risk reduction from something that is broken, such as a man way that just happened to be open during the high-pressure event in the vessel.

Integrity

Related to the risk reduction that can reasonably be expected given the protection layer's design and management.

Integrity is quantified as the probability of failure on demand (PFD) for the protection layer. For example, on average, at one end of the range, a safety integrity level (SIL) 1 safety instrumented function would be expected to operate successfully 9 times out of 10 demands and to fail 1 time out of 10 demands. (At the other end of the range, the SIL 1 SIF would be expected to operate successfully 99 times out of 100 demands and to fail 1 time out of 100 demands.)

Reliability

The probability that a protection layer will operate as intended under stated conditions for a specified time period.

For example, if the protection layer is a nitrogen purge of a vessel for five minutes, reliability is the probability that, once the purge starts, it continues for the specified five minutes.

Auditability

Ability to inspect information, documents and procedures, which demonstrate the adequacy of and adherence to the design, inspection, maintenance, testing, and operation practices used to achieve the other core attributes.

Everything breaks, rusts, or runs out of gas. Auditing, inspecting, testing, and repairing are critical to ensure that the protection layer will still function.

At one plant, personnel insisted that it was not possible for a dike to fail. During the walk through the tank farm, we observed a hole through the dike at about 50% of the dike height, but there was no grout to seal the annular space between the pipe and the hole. The dike protection layer was compromised until we found the hole and repairs could be completed.

Access Security

Use of administrative controls and physical means to reduce the potential for unintentional or unauthorized changes.

Access security examples include locking the room housing the safety system logic solver and password protection for the safety system logic solver configuration.

Management of Change

Formal process used to review, document, and approve modifications to equipment, procedures, raw materials, processing conditions, etc., other than "replacement in kind," prior to implementation.

For example, when a new product is introduced to a reactor, the management of change process should confirm that the reactor pressure relief system is sufficient for the runaway scenarios.

3. Applying the characteristics of independent protection layers

During LOPA, the analyst must determine the independence, functionality, and integrity of the existing protection layers in the plant or the design. The analyst must identify new or modified protection layers to reduce the frequency of the scenario to meet the organization's risk tolerance criteria. For the new or modified protection layers, the independence, functionality, integrity, and reliability must be specified. The other three characteristics of IPLs should be addressed during the design, implementation, operation, and maintenance of the protection layer.

The SRS (Safety Requirements Specification) [3] is used to document the independence, functionality, integrity, and reliability requirements developed in LOPA. During the design of the IPLs, the specific design details to achieve the independence, functionality, integrity, and reliability requirements of the IPL are added to SRS. The SRS also includes the requirements for auditability, access security, and management of change.

3.1 Keywords for Independence, Functionality

The authors of [CCPS 1] recommend an easily remembered set of keywords to screen candidate IPLs (Table 1).

Table 1: Keywords for IPLs

"3 Ds"	**"4 Enoughs"**	**The Big "I" – Independent**
• Detect • Decide • Deflect	• Big Enough • Fast Enough • Strong Enough • Smart Enough	• Of Initiating Cause • Of Other IPLs

The "3Ds" and the "4 Enoughs" address the characteristic of functionality. The Big "I", of course, addresses the characteristic of independence.

Many protection layers have the ability to **detect** or sense that an initiating cause has occurred and the scenario is moving toward an undesired consequence. Many protection layers have the ability to **decide** to take action that can **deflect** or prevent the consequence from occurring. For example, the spring in a pressure relief valve detects the increase in pressure, the spring "decides" to open the valve, and a properly designed relief valve is large enough to prevent (or deflect) the vessel pressure from exceeding the design pressure.

(One exception to the "3Ds" is a dike or bund. The dike does not need to detect or decide to take action, but it must be large enough to hold the contents released to it.)

The protection layer must be big enough, fast enough, strong enough, and smart enough to accomplish its purpose. For example, the pressure relief valve must be **big enough** and must open **fast enough** to prevent the overpressure. An isolation valve to prevent an overpressure surge must be **strong enough** to withstand the incoming pressure. For complex systems, the protection layers must be **smart enough** to execute the functions in the right sequence and for the specified periods of time. For example, a purge system may be required to flow nitrogen just long enough to purge a specified vessel and then the system is required to isolate the vessel.

Note that the **smart enough** characteristic was identified after the CCPS LOPA book [1] went to press.

Finally, the Big "I" requires that the protection layer be independent of the initiating cause and independent of other protection layers. Frequent conflicts with the Big "I" include

- Using the initiating cause sensor for the protection layer sensor.
- Using the same sensor for two protection layers.
- Using the initiating cause valve for the protection layer final element.
- Using the same final element for two protection layers.
- Using the initiating cause controller for the protection layer logic solver.
- Using the same logic solver for two protection layers (it may be possible to use the same safety-certified PLC as the logic solver for two protection layers if its probability of failure on demand is low enough).

LOPA requires independence between a protection layer and the initiating cause, and between other protection layers, because LOPA is a simplified risk assessment tool. The assumption of independence simplifies the calculations in the analysis. If independence is violated, the result can be a significant increase in risk.

For example, consider a scenario with two apparent protection layers.

1. Operator response to alarm, PFD of 0.1
2. SIF, PFD of 0.1

For independent protection layers, the total risk reduction, q, is given by the product of the PFDs for each protection layer, [Eq. 2].

$$q = PFD_1 \times PFD_2 = 0.1 \times 0.1 = 0.01 \qquad \text{[Eq. 2]}$$

However, if the same sensor is used for the alarm and the SIF, a problem with the sensor will prevent both protection layers from working. Using LOPA rules, one protection layer will be used for risk reduction, but not both, for a total PFD of 0.1, [Eq. 3].

$$q = PFD_1 = 0.1 \qquad \text{[Eq. 3]}$$

A fault tree analysis could be done for the two layers to account for the common mode of the sensor. The result of the fault tree analysis can be entered into the LOPA for the combination of the two protection layers.

Over the years, we have found the keywords in Table 1 to be very effective to determine quickly if a candidate IPL is truly an IPL.

3.2 Example scenario using the keywords for independence and functionality

Consider a scenario where the consequence is overflow of a flammable material from a storage tank and the initiating cause is failure of the tank level control loop to stop the flow of material into the tank as the level increases. The dialogue in the LOPA might go like this:

"Oh, we are protected against overflow. We have a high-level alarm."

"Let's apply the "3Ds". Detect, decide, deflect. What detects?"

"We have a high-level alarm on the sensor for the level control loop."

"Okay, that can detect. We'll come back to look at independence in a moment. The basic control system can decide to sound the alarm. What about deflect?"

"Hmm. ... Well, I guess the operator will go out and close the manual inlet valve to the tank."

"Is the operator response included in the troubleshooting guide, the basic operator training program, and the recurrent training program?"

"No, I guess we should add it to those programs."

"Okay, now let's apply the "4 Enoughs". Is the operator response big enough to stop the overflow? Yes. Is the operator response fast enough to stop the overflow? If the tank is filling at the maximum rate, how long will it take from the time the alarm sounds until the tank overflows? Note that we need 10 minutes time for operator response to an alarm where the alarm condition is clearly understood and the action is always the same."

"We will take an action to ensure there is at least 10 minutes time for the operator to respond to the alarm."

"It's clear that the operator response to alarm is strong enough. Is the operator response to alarm smart enough? What if the density of the fluid in the tank is lighter than the design specification for the level sensor?"

"Well, I guess the tank might overflow before the alarm goes off. We will have to look at that problem."

"Now let's look at the Big "I". Is the operator response to the high-level alarm independent of the initiating cause? What if the cause is the level sensor reading lower than actual level? Then the alarm will not go off in time."

"So we need an independent sensor. We can add a fixed high-level probe to detect high level. That will also address the smart enough (low-density) issue."

"We need to confirm that we have appropriate alarm management for this unit so that the high level alarm won't be lost in an alarm shower. Finally, if we need more than one IPL to meet the risk tolerance criteria, then we need to be sure that any other IPLs are independent of the operator response to alarm IPL."

This dialogue is a composite from several actual conversations between a LOPA analyst and project team members. The keywords in Table 1 provide simple tests for the functional and independence characteristics of an IPL and provide guidance to clarify the thought process to modify candidate protection layers into true IPLs. The IPL requirements identified in this dialogue would be incorporated into the SRS.

3.3 Example scenario using a spreadsheet test for independence and functionality

There are sometimes scenarios where additional detailed documentation might be necessary to identify clearly whether candidate protection layers meet the requirements. For these scenarios, it might be helpful to construct a spreadsheet with the characteristics of IPLs for each candidate protection layer (safeguard) listed across the top and the scenarios listed down the side. A sample for the tank overflow dialogue is shown in Table 2. The "comment" feature of the spreadsheet can be used to explain the choice of "yes" or "no" (shown as notes in Table 2). In particular, such a table can be useful for analysis, discussion, and documentation in situations where there are a number of existing safeguards and the historical culture has assumed that the safeguards are sufficient. An analysis guided by the Characteristics of IPLs spreadsheet can determine which safeguards are really IPLs, and provide a sound basis for calculating the risk for the LOPA scenarios.

3.4 Integrity

The integrity (or probability of failure on demand) for an IPL is a function of the failure rate of the components of the IPL, the architecture of the IPL (number of channels, diagnostics), and the test frequency of the IPL. The LOPA will specify the required PFD of each IPL as part of the SRS. The IPL designers will configure the IPL to deliver the required integrity and document that configuration in the SRS.

3.5 Auditability

"If you never test your IPL, you will find out that it is broken when your plant blows up."

Thus, it is critical that the IPL designers provide the ability to test the IPL periodically to ensure that all its components are working correctly and that the designers specify the appropriate test frequency. It is critical that operations provide timely opportunities to test the IPL and that any deficiencies are repaired in a timely manner. The auditing/testing procedures and frequency will be also documented in the SRS.

Now consider again the operator response to the high level alarm to prevent the tank overflow. Many facilities do a good job to test the high level sensor and the alarm annunciation. However, do all facilities test the knowledge and ability of the operator to make the correct response when the alarm sounds? Do they test the appropriate operator on every shift? Do they test the department relief operator or the "step up" vacation relief operator? Does anyone test that the manual inlet valve to the tank still has a handle and can still be operated?

We must recognize that it is extremely difficult to maintain procedures. Routine auditing of the existence and knowledge of procedures for IPLs is absolutely critical!

In one facility, an engineering specialist asked the control operator, "What would you do if the high temperature alarm sounded on this self-reactive chemical storage tank?" The operator responded, "I guess I would call the foreman." At about the same time, a process safety specialist asked the facility for a copy of the emergency response procedures for high temperature in the tank; he wanted to share the procedures with a sister facility. The facility said, "We don't have any procedures." The process safety specialist replied, "Hey, I helped you write those procedures three years ago! Here's a copy of the n-1 draft." It took the facility a month to find a paper copy of the final draft of the emergency response procedures. When these procedures were written, such procedures should have been subject to document control and should have been available in electronic form.

Moral: audit procedural IPLs frequently. Use a system comparable to the SIF SRS to document procedural IPLs.

A facility had two level sensors for a critical tank. When the tank overflowed, the investigation found that one of the sensors had been broken for months, but was not repaired because "We have a spare." Operations were not aware that two functioning sensors were needed for the required integrity level for the IPL.

Moral: timely repair of deficiencies is needed.

3.6 Access Security

Who has access to the configuration of the logic solver that annunciates the high level alarm? What are the procedures to ensure that the set point for an analogue high level alarm is not changed? What are the procedures to ensure that the high level alarm is not disabled or inhibited? The facility must ensure that there is appropriate access security to the physical components and to the software of the IPL to prevent deactivation. The security requirements are part of the SRS.

3.7 Management of Change

In the competitive climate, facilities are always making changes to improve capacity, quality, and environmental performance, and to reduce cost. The organization must have a management of change system that maintains the capability of existing IPLs when a change is made, and that improves IPLs when additional risk reduction is needed for a change. The management of change system must ensure that IPLs are not disabled by a change, either advertently or inadvertently.

Again, considering the operator response to the high-level alarm in the tank IPL, in the future, a process improvement specialist may ask, "Why is the set point for this alarm so low? We could be using a lot more of the volume of this tank." The management of change system must be able to retrieve the specifications for the operator response to high-level alarm IPL, must be able to recognize that the alarm set point provides sufficient time for the operator to make the response, and must be able to retain the required setting for a manual response or to automate the response for a higher set point. The SRS is a key document to provide the correct information to the management of change system.

4. When to go beyond LOPA

In the examples above, there are systems where two or more protection layers share common components. LOPA can take risk reduction credit for one of the layers but not for the others.

Fault tree analysis can be used to determine the risk reduction that can be achieved with layers that share common components.

Another example is a system where several safeguards can detect the beginning of a scenario but there is only one safeguard that can prevent the scenario from proceeding to the consequence. In a reactor, potential for a runaway reaction might be detected by a temperature alarm and by an analyzer. Regardless of which way the scenario is detected, the only way to prevent the scenario from proceeding is a troubleshooting procedure with specific actions depending on diagnosis. The LOPA might be tempted to claim both the temperature alarm and the analyzer as IPLs, but they both depend on the procedure to stop the scenario. It would be better to use an event tree to analyze the combination of the two detection mechanisms and the single prevention procedure.

5. Conclusion

This paper discusses the core attributes of independent protection layers used in LOPA. The key words in Table 1 provide an easily remembered way to evaluate the functionality and independence of candidate protection layers. The example spreadsheet in Table 2 gives a more detailed tool to document functionality and independence. The information to be added to the Safety Requirements Specification for protection layers is illustrated. Examples are given for all seven core attributes. It is suggested that an analogous document to the SRS be used for other protection layers to document the core attributes.

6. Disclaimer

The examples given in this paper are for illustrative purposes only. They are composites drawn from real-world observations, but the examples are not intended to reference any existing facilities.

This information was developed by the author solely for his own use. The author is making this information available in this paper without any guarantee that it is appropriate for your purposes as conditions and methods of use of the information is beyond his control. The author disclaims any liability for use of this information by other persons.

7. References

[1] Center for Chemical Process Safety, *Layer of Protection Analysis: Simplified Process Risk Assessment*, American Institute of Chemical Engineers, New York, NY, 2001.

[2] Center for Chemical Process Safety, *Guidelines for Safe and Reliable Instrumented Protective Systems Center for Chemical Process Safety,* American Institute of Chemical Engineers, New York, NY, 2007.

[3] *ANSI/ISA 84.00.01-2004, Functional Safety: Safety Instrumented Systems for the Process Industry Sector - Part 1: Framework, Definitions, System, Hardware and Software Requirements*, ISA, Research Triangle Park, North Carolina, 2004

Table 2: Sample IPL Characteristics Test

			Safeguard 1										Safeguard 2									
	Scenarios	# of Safeguards that meet IPL criteria	Description	Detect?	Decide?	Deflect?	Big Enough?	Fast Enough?	Strong Enough?	Smart Enough?	Independent of the Initiating Cause?	Independent of other IPLs?	Description	Detect?	Decide?	Deflect?	Big Enough?	Fast Enough?	Strong Enough?	Smart Enough?	Independent of the Initiating Cause?	Independent of other IPLs?
1	LIC loop failure overfills tank with release	1	Operator response to alarm from level control loop sensor	N[1]	Y	Y	Y	Y	Y	Y	N	N[2]	Operator response to alarm from fixed point level probe	Y	Y	Y	Y	Y[3]	Y	Y	Y	Y
2	Fluid in tank has lighter density than specified in level transmitter design, overfills tank with release	1		N	Y	Y	Y	Y	Y	N	N	N[2]		Y	Y	Y	Y	Y[3]	Y	Y	Y	Y

Notes:

1. If the LIC loop sensor fails, it cannot detect. If the LIC controller fails, it may not be able to annunciate the alarm.
2. Not independent of the operator response to alarm from the fixed point level probe.
3. Ensure there is at least 10 minutes for the operator to respond to the high level alarm at the maximum fill rate.

Consistent Consequence Severity Estimation

Angela Summers, PhD
SIS-TECH Solutions
12621 Featherwood Dr Suite 120 Houston, TX 77034
asummers@sis-tech.com

William Vogtmann, Steven Smolen
SIS-TECH Solutions
12621 Featherwood Dr Suite 120 Houston, TX 77034
bvogtmann@sis-tech.com, ssmolen@sis-tech.com

Prepared for Presentation at
American Institute of Chemical Engineers
2010 Spring Meeting
6th Global Congress on Process Safety
San Antonio, Texas
March 22-24, 2010

UNPUBLISHED

Consistent Consequence Severity Estimation

Angela Summers, PhD, PE
SIS-TECH Solutions
12621 Featherwood Dr Suite 120 Houston, TX 77034
asummers@sis-tech.com

William Vogtmann and Steven Smolen
SIS-TECH Solutions

Keywords: LOPA, Risk Analysis, Consequence, Quantitative

Abstract

Most risk analysis methods rely on a qualitative judgment of consequence severity, regardless of the analysis rigor applied to the estimation of hazardous event frequency. Since the risk analysis is dependent on the estimated frequency and consequence severity of the hazardous event, the error associated with the consequence severity estimate directly impacts the estimated risk and ultimately the risk reduction requirements. Overstatement of the consequence severity creates excessive risk reduction requirements. Understatement results in inadequate risk reduction.

Consistency in the Layers of Protection Analysis (LOPA) can be substantially improved by implementing consequence estimation tools that assist team members in understanding the flammability, explosivity, or toxicity of process chemical releases. This paper provides justification for developing semi-quantitative look-up tables to support the team assessment of consequence severity. Just as the frequency and risk reduction tables have greatly improved consistency in the estimate of the hazardous event frequency, consequence severity tables can significantly increase confidence in the severity estimate.

1. Introduction

Various types of process hazards analyses (PHA) are now in widespread use throughout the process industry [1]. These analyses evaluate process deviations that potentially lead to hazardous events and the safeguards that are implemented to reduce the likelihood of each event. Many PHAs incorporate risk analysis to determine the residual risk of each identified event. This is accomplished by qualitatively estimating the likelihood of the event given its causes and safeguards. The PHA process can be supplemented with Layers of Protection Analysis (LOPA) to provide an order of magnitude estimate of the hazardous event frequency by assessing the frequency of the initiating events that lead to the hazardous event and the probability that the safeguards fail [2].

One of the authors has often stated "A big risk is not addressed by a big list: it is addressed with the right list of independent protection layers." A big risk generally involves an event judged to have a consequence severity of a potential significant injury or fatality. PHA teams often deal with the big risks by making a big list of safeguards, thus making the event seem much less likely when the team is asked to qualitatively assess the potential event likelihood considering the causes and safeguards. In most PHAs, safeguards are recommended if they reduce the event likelihood a little or a lot. Yet, the expectation is that everything on the list is safety-related and is managed with sufficient rigor. Instead of a big list, an owner/operator needs the right list of protection layers that are known to demonstrate the claimed risk reduction.

The reduction of the big list to the right list is one of the primary benefits of LOPA. During the PHA, the team identifies process deviations that potentially lead to hazardous events and assesses the hazardous event frequency given the identified safeguards. LOPA seeks to analyze the root causes (or initiating causes) and enabling conditions that result in process deviations (or initiating events). Understanding the likelihood of the initiating causes and conditions enables the team to estimate the initiating event frequency.

In LOPA, the team recommends independent protection layers (IPL) rather than safeguards. IPL are designed and managed according to good engineering practices that are known to provide the claimed risk reduction. The characterization of the initiating event frequencies and IPL risk reduction in order of magnitude terms allows teams with little mathematical inclination to estimate the hazardous event frequency based on information and data that correspond to key performance indicators [3]. In most cases, the team is provided with look-up tables of typical initiating cause frequencies and IPL risk reduction values that can be adjusted based on specific site experience, so the estimated hazardous event frequency is generally very consistent from team to team. As advanced LOPA procedures make more precise estimates of the frequency of the events and probability of IPL failure, the hazardous event frequency estimate becomes more quantitative, but the method remains inherently semi-quantitative.

The logical next step in improving LOPA is to introduce guidance for estimating the consequence severity. The following discussion presents the authors' proposal for semi-quantitative selection of consequence severity associated with flammable releases. Simple look-up tables are presented that rely on equipment design specifications and process operating conditions – information that should already be maintained by the site and provided to the team that is conducting the analysis. Dispersion modeling based on conservative assumptions was used to correlate estimated release rates to "zones of damaging overpressure." Then, these zones were tied to the consequence severity rankings that typically support risk analyses during the PHA and LOPA. The look-up tables have been demonstrated to yield a consistent basis for estimating the consequence severity, leading to greater consistency in the risk estimate regardless of the team experience with harmful events.

2. LOPA is a Great Tool!

LOPA is an excellent tool for assessing a wide variety of process hazard scenarios and then crediting or applying protection layers of appropriately robust design. LOPA, as a semi-quantitative analysis, allows for efficient evaluation of process hazards by a multi-disciplined team of representatives that are experienced with the equipment under study. The results are based on a better estimate of the hazardous event frequency and therefore provide a stronger basis for recommending safety functions or layers.

LOPA supports performance-based process safety because it ties various quality assurance metrics or key performance indicators to the hazardous event frequency. As owner/operators track the challenges on their IPLs and failures of their IPLs, LOPA provides a means to tie order of magnitude estimates of historical performance to industry benchmarked values.

Fortunately, most teams have little experience with hazardous events, since a properly designed and operated chemical process will have adequate IPLs to reduce the event frequency to a sufficiently low level. On the other hand, they do have experience with the frequency of root causes (or initiating causes), the frequency of process deviations (or initiating events), and the frequency of work orders (or reported failures) of IPLs. The estimated hazardous event frequency can be understood by all disciplines and those disciplines that impact the estimate can better understand their role in hazardous event propagation. Done right, it gives the organization a firm and consistent basis for investment in protective systems.

3. LOPA is Inconsistent!

As organizations progress with LOPA throughout the entire organization, invariably the results are compared and questions asked. When LOPA results for similar process units are compared from company to company, facility to facility, and team to team, it is not unusual to see variation in the risk estimate for similar hazardous events. Since most LOPA procedures have well defined methods for estimating the hazardous event frequency, the inconsistency in the risk estimate is generally due to variation in the estimated consequence severity.

Most LOPA methods rely on a qualitative estimate of the consequence severity. Table 1 provides an example of typical consequence severity descriptions. These descriptions are essentially statements of harm which are not easily estimated by the team, which is expected to judge what happens when a process deviation leads to a hazardous event and what harm could potentially be posed by the hazardous event. While they may have experience in process deviations, hopefully, they have little experience with hazardous events and the resulting harm.

Table 1. Qualitative Consequence Severity Ranking

RANKING	SAFETY	ENVIRONMENTAL	ASSET
5	Multiple fatalities across a facility and/or Injuries or fatalities to the public	Catastrophic off-site environmental damage with long-term containment and clean-up	Expectant loss greater than $10,000,000 and/or substantial damage to buildings located off-site
4	Hospitalization of three or more personnel (e.g., serious burns, broken bones) and/or one or more fatalities within a unit or local area and/or Injuries to the public	Significant off-site environmental damage (e.g., substantial harm to wildlife) with prolonged containment and clean-up	Expectant loss between $1,000,000 and $10,000,000 and/or extended downtime with significant impact to the facility operation and/or minor damage (e.g., broken windows) to buildings located off-site
3	Hospitalization injury (e.g., serious burns, broken bones) and/or multiple lost work day injuries and/or Injury to the public	On-site release requiring containment and clean-up and/or off-site release causing environmental damage with quick clean-up	Expectant loss between $100,000 And $1,000,000 and/or downtime of several days severely impacting the facility operation
2	Lost work day injury and/or recordable injuries (e.g., skin rashes, cuts, burns) and/or minor impact to public	On-site release requiring containment and clean-up by emergency personnel and/or off-site release (e.g., odor) but no environmental damage	Expectant loss between $10,000 and $100,000 and/or downtime of more than day causing impact to facility operation and/or reportable quantity event
1	Recordable injury and/or no impact to the public	On-site release requiring containment and clean-up by on-site personnel.	Expectant loss of less than $10,000 and/or downtime of less than a day with minor impact to the facility operation

The severity estimate is often bounded with qualitative guidance that the team should estimate the "worst case consequence" or "worst credible consequence." For example, the worst credible scenario is "the most severe incident of all identified outcomes and their consequences that is considered plausible or reasonably believable" [4]. The interpretation of these terms and their definitions has proven to be highly variable. The perceived error in this severity estimate is one of the chief criticisms of PHA and LOPA. And, unfortunately, perception often drives investment, so if it is perceived that the team was excessively conservative in the consequence severity estimate, it is easier for management to reject PHA and LOPA recommendations as unwarranted.

There are many legitimate reasons for variation in the consequence severity estimate across an organization, including differences in the specific process quantities, layout, location, equipment pressure ratings, etc. These differences are recognized by the respective teams and they may result in different consequence severity rankings. In many cases, these specific differences are not fully described in the PHA or LOPA documentation, so it is difficult for people who were not part of the team to understand why the differences exist. So, it is important to remember that not all apparent variation in results is bad; indeed it often reflects the very purpose that PHA and LOPA are performed on specific equipment and revalidated periodically to capture lessons learned in operating and maintaining the equipment.

A significant amount of variation is introduced due to perceptual differences of the hazardous situation created by the hazardous event. Bias may also be introduced into teams by the presence of risk-taking or risk-adverse people. Direct experience with the hazardous event is a strong influence on the estimate. For example, where a site or team member has experienced a fire and no one was hurt, the assumed outcome for any fire may become "no one gets hurt". If there is experience where people were hurt or killed by the fire, even due to extremely unusual conditions, all events resulting in a potential fire may be regarded as a fatality event. While no one should take any fire lightly, the concept of risk analysis certainly incorporates the idea that large fires should receive more attention and effort in prevention than small ones, other factors being equal.

These qualitative biases may be great enough to push the severity estimate to a more or less conservative result than is appropriate. Overstatement of severity creates excessive risk reduction requirements and costs. Understatement results in inadequate risk reduction and a higher potential for loss events. Giving teams a consistent basis for the severity assessment can significantly reduce variability.

4. Conditional Modifiers Can (not) Fix it!

Basic LOPA analyzes the risk for each hazardous event resulting from one or more process deviations. Basic LOPA focuses the risk analysis on preventing hazardous events. This agrees with OSHA 29 CFR 1910.119 [5], which is dedicated to not just "minimizing the consequences of catastrophic releases of toxic, reactive, flammable, or explosive chemicals" but also to "preventing" such releases. The typical LOPA risk criteria are that the maximum hazardous event frequency should not exceed $1x10^{-4}$/yr for events potentially leading for a worker fatality or $1x10^{-5}$/yr for events potentially leading to a public fatality [6]. The overall frequency of all deviations leading to the same hazardous event should achieve these risk criteria. Many risk matrices have similar risk criteria built into the relationship between frequency and consequence severity as illustrated in Table 2 [4].

Table 2. Example Risk Matrix

CONSEQUENCE SEVERITY	REQUIRED RISK REDUCTION FACTOR				
5	100,000	10,000	1,000	100	10
4	10,000	1,000	100	10	TR
3	1,000	100	10	TR	TR
2	100	10	TR	TR	TR
1	10	TR	TR	TR	TR
FREQUENCY (1 in x years)	1	10	100	1,000	10,000

Hazardous Event Frequency = $1x10^{-4}$/yr for impact to worker or $1x10^{-5}$/yr for impact to public

Some owner/operators use LOPA to support risk criteria based on the harmful event frequency. Some regulatory authorities require that the maximum individual risk or societal risk be determined and reported as part of permitting or safety cases. Once the hazardous event risk is known, the team assesses site specific conditions (e.g., probability of ignition and probability of occupancy) to determine the harmful event frequency. The probability values assumed for conditional modifiers whether used in LOPA or QRA (quantitative risk analysis) should be justified through application-specific analysis. Typical risk criteria are a maximum individual risk of $1x10^{-5}$/yr for a worker or $1x10^{-6}$/yr for the public [6]. Assessment against these risk criteria requires that the overall frequency of all deviations leading to the same hazardous event be determined.

Conditional modifiers are often proposed as a means to deal with overstated consequence severity. Since the conditional modifiers reduce the frequency, the "risk" is likewise lowered. However, the use of conditional modifiers also depends on the basis for the risk analysis. It is not appropriate to use conditional modifiers if the risk criteria are based on the hazardous event frequency (e.g., loss of containment) rather than on direct harm frequency (e.g., fatality). Regardless of the frequency basis, the use of conditional modifiers does not address or minimize the error associated with the consequence severity estimate. It does provide a means to rationalize that the likelihood of such harm is lower; however, the consequence is still overstated.

5. Templates Can (not) Fix it

Some owner/operators have tried to get consistency in the LOPA by providing teams with standard LOPA scenarios. This is especially attractive in organizations with virtually identical units or highly standardized installations in multiple locations. These template scenarios can be very useful in giving specific guidance to a team. In many cases, the templates list IPLs that are considered good engineering practice, so teams have guidance on not only the risk but also the means for risk reduction. The use of templates does not restrict the team's ability to analyze the risk of a specific installation, since the team is still expected to identify significant differences between their installation and the template and to make adjustments to the template as necessary. However, such templates have proven to be difficult to develop, approve, and implement, predominantly because consensus on an organization-wide consequence severity ranking is difficult to achieve.

6. Developing Consequence Severity Guidance

The authors developed a set of consequence severity look-up tables to support PHA and LOPA using dispersion modeling. The **K**eep **I**t **S**imple principle was applied to the modeling effort, since the model was intended to be conservative and support order of magnitude consequence severity estimation. It was also desirable to have tables that based the severity on something that the team could assess themselves using information typically provided to the PHA or LOPA team. The results also needed to be presented in a format that made the tool practical for the team to use. Ultimately, a set of look-up tables were developed which yielded a severity ranking or range of severities, depending on the equipment type and release conditions.

Since many events involve the release of flammable materials, the first look-up tables addressed flammable hydrocarbon releases. Alkanes and alkenes, including C1-C10, are widely used in the process industry, so these were selected for modeling using ALOHA®. It was determined that this set of hydrocarbons yielded relatively similar zones of damaging overpressure for a given hole diameter and release rate. Consequently, the tables were simplified by treating these compounds as a single class of hydrocarbons.

Guidance was also needed to relate the initiating event conditions to an expected hole diameter or damage estimate. This led to the development of tables relating pump size (shaft size or HP) to an equivalent seal hole diameter, % overpressure to an equivalent hole diameter, or firebox type to equivalent damage. This paper will only present the pump size correlations for the loss of a mechanical seal.

6.1 Disclaimer

This guidance is intended to produce consistent results for similar hazardous events, but no severity estimate is a perfect forecast of actual events. Where plant conditions are significantly different from the assumptions, additional studies and/or consequence modeling should be considered. The PHA or LOPA team is advised to reach a consensus on consequence severity, considering site specific factors, the consequence severity tables, specific modeling results, and other available information. Ultimately, the guidance is intended to release teams from concern that they are not treating safety seriously if they don't consider all releases as high severity and the opposite concern that they are being too conservative about the process if they predict that many scenarios pose severe consequences.

6.2 Dispersion Modeling

Keeping everything simple was probably the highest hurdle to overcome. It was easy to become overwhelmed with the different modeling tools and the large number of parameters that could be specified. Fortunately, analysis paralysis was ended by accepting the standard assumptions in consequence analyses performed to support the EPA Risk Management Plan [7]. The simplifying assumptions made were:

- Weather/topographical Conditions
  - Wind: 3 knots from South at 3 meters
  - Ground Roughness: open country
  - Air Temperature: 95° F
  - Stability Class: B – software determined

    NOTE: ALOHA provides the stability class based on information about the time of day, wind speed, and cloud cover

  - No Inversion Height
  - Cloud Cover: 5 tenths
  - Relative Humidity: 50%
- Congestion
  - Congested – not open space, difficult to walk through

The dispersion modeling software chosen was ALOHA which was developed by the EPA's Office of Emergency Management and NOAA's Emergency Response Division [8]. ALOHA is an atmospheric dispersion model used for evaluating releases of hazardous chemical vapors, including toxic gas clouds, fires, and explosions. Using input about the release ALOHA generates a threat zone estimate. A threat zone is the area where a hazard (such as toxicity, flammability, thermal radiation, or damaging overpressure) is predicted to exceed a user-specified level of concern.

The major reasons for selecting this software were that it was:

- developed by EPA/NOAA for use in a site's Risk Management Plan (RMP) and Emergency Response Plans (ERP)
- familiar to Local Emergency Planning Committees (LEPC) since many companies used the software for their RMP case(s)
- user friendly
  - often provided suggested/recommended values for parameters making it more suitable for the what-if type of modeling required
  - includes guidance on severity based threat zone
  - availability of large database of hazardous materials
- available at no cost to the user as well as the general public (LEPC)

6.3 Covered Chemicals

The consequence severity look-up table addresses the hazards associated with the release of flammable liquids and vapors in the range of C1-C10 (alkanes and alkenes). The consequence severity of the hazardous event was defined by the zone of damaging overpressure generated when the flammable hydrocarbons were released from a defined hole diameter and a specified rate. This model did not consider jet fires that might impinge process equipment and cause additional failures.

6.4 Hole Diameter Estimate

Hole diameter look-up tables were developed for vessel and piping overpressure and seal leaks. This paper only presents the results for estimating the hole diameter for single mechanical seals. As shown in Table 3, the hole diameter can be estimated based on the pump shaft size at the seal or on horsepower. The shaft size (Table 3A) may be a more accurate predictor of hole diameter, but experience indicates that horsepower (Table 3B) is more commonly available to the team. The information may be included on unit P&IDs or in mechanical data sheets.

Table 3. Estimated equivalent hole diameter for single mechanical seal failure.
A. Shaft Size versus Equivalent Hole Diameter

Shaft Size vs Equivalent Hole Diameter							
Shaft	0.5	1	2	3	4	6	8
Hole	1/8	3/16	1/4	3/8	1/2	5/8	3/4

Note: The equivalent hole diameter estimate assumes that the carbon faces and their support within the seals provide no sealing. The remaining leak path is limited by the clearance between the shaft and the seal housing/support. An annular radius of 0.01 was used for small shaft sizes (up to 3 inch) and 0.016 was used for larger shafts.

B. Horsepower versus Equivalent Hole Diameter

Horsepower vs Equivalent Hole Diameter							
HP	<5	5-10	10-50	50-150	150-300	300-600	>600
Hole	1/8	3/16	1/4	3/8	1/2	5/8	3/4+

Note: The equivalent hole diameter estimate was developed by reviewing the output shaft size of typical industrial motors rated for various HP. The review indicated that the pumps could be grouped for simplification of the look-up table. The stated equivalent hole diameter is based on the shaft sizes for the upper boundary of each group.

6.5 Release Rate Estimate

For the flammable hydrocarbons, separate release rate tables were developed for liquid and vapor releases. The team determines whether the release will be liquid or vapor considering process conditions when the hazardous event occurs. Table 4 is for liquid releases through holes of 1/16 inch to 6 inches in diameter at pressures of 10 to 50,000 psig. The presence of an "X" in the table indicates a release that exceeds the threshold of 10,000 lbs/min, which was determined to correlate with a category 5 consequence severity.

Table 4. Liquid release rate (lb/min) as a function of pressure (psig) and hole diameter (inch)

Liquid Release Rate (lb/min) Pressure (psig) vs. Hole diameter (inch).										
Psig / inch	1/16	1/8	1/4	1/2	1	1.5	2	3	4	6
10	2	8	31	120	500	1100	2000	4500	7900	17900
50	4	17	69	280	1100	2500	4400	9980	17700	X
100	6	25	98	390	1600	3500	6300	14100	X	X
150	8	30	120	480	1900	4300	7700	X	X	X
300	11	42	170	680	2700	6100	10900	X	X	X
500	14	55	220	880	3500	7900	X	X	X	X
1000	19	77	310	1200	5000	11200	X	X	X	X
2000	27	110	440	1800	7000	X	X	X	X	X
6000	47	190	760	3000	12100	X				
10000	61	250	980	3900	X	X				
30000	110	420	1700	6800	X	X	X	X	X	X
50000	137	550	2200	8800	X	X	X	X	X	X

"X" = no details are provided, since the release is equivalent to category 5.

6.6 Zones of Damaging Overpressure

Modeling showed that the majority of releases reach steady state with clearly identified zones of hazardous conditions within 5 minutes, which is consistent with the EPA's RMP guidance to use an endpoint of 10 minutes for scenario modeling. Figure 1 is a screen capture showing a typical output from ALOHA® [8]. The ALOHA output depicts three predefined "zones of damaging overpressure." The yellow zone (outer ellipse) is an overpressure greater than 1 psi. This pressure typically causes minor damage to structures and minimal direct harm to humans. The ALOHA output includes a reference to shattered glass to indicate that 1 psi is a minor impact. The orange zone (inner ellipse) is an overpressure greater than 3.5 psi. This pressure typically causes structural damage to buildings and direct harm to humans. The ALOHA output includes a reference to serious injury to indicate that 3.5 psi represents a point where injuries are likely due to the overpressure, but this reference does not consider the potential for flying debris or structural damage. There is a potential for a red zone to appear on the graphic when there is an overpressure greater than 8.0 psi. This pressure is known to cause significant structural damage to structures and buildings as well as direct harm to human. The ALOHA output includes a reference to destruction of buildings to indicate the severity of this overpressure.

Examination of the results obtained from the modeling of the hydrocarbon class (consisting of C1-10 alkanes and alkenes) identified no red zones (8 psi). This agrees with other data reviewed prior to modeling which suggested that it was unlikely that typical hydrocarbon releases within process units would achieve this degree of overpressure.

The zone represented by the 3.5 psi overpressure seemed like a conservative selection for determining the potential area impacted by the overpressure, since this overpressure was judged to be significant enough to cause direct injury and to result in flying debris. It was also felt to be more intuitive for the correlation since the qualitative consequence severity descriptions include the potential for injury and fatality. Selecting a higher overpressure for the correlation would have resulted in a smaller impact area, but a more severe outcome within the area.

To correlate overpressure damage to differing levels of consequence severity, it was necessary to establish a relationship between zone of damaging overpressure (impact area) and area of occupancy (unit area). The premise is that as the impact area becomes larger there is greater likelihood that personnel in the unit or in surrounding units will be impacted. Using 400 foot by 400 foot as a typical unit footprint, an effective unit area was determined. The ratio of impact area derived from the model (3.5 psi zone) and the theoretical unit area was used to scale the overpressure damage to the different consequence severities as shown in Table 5.

Figure 1

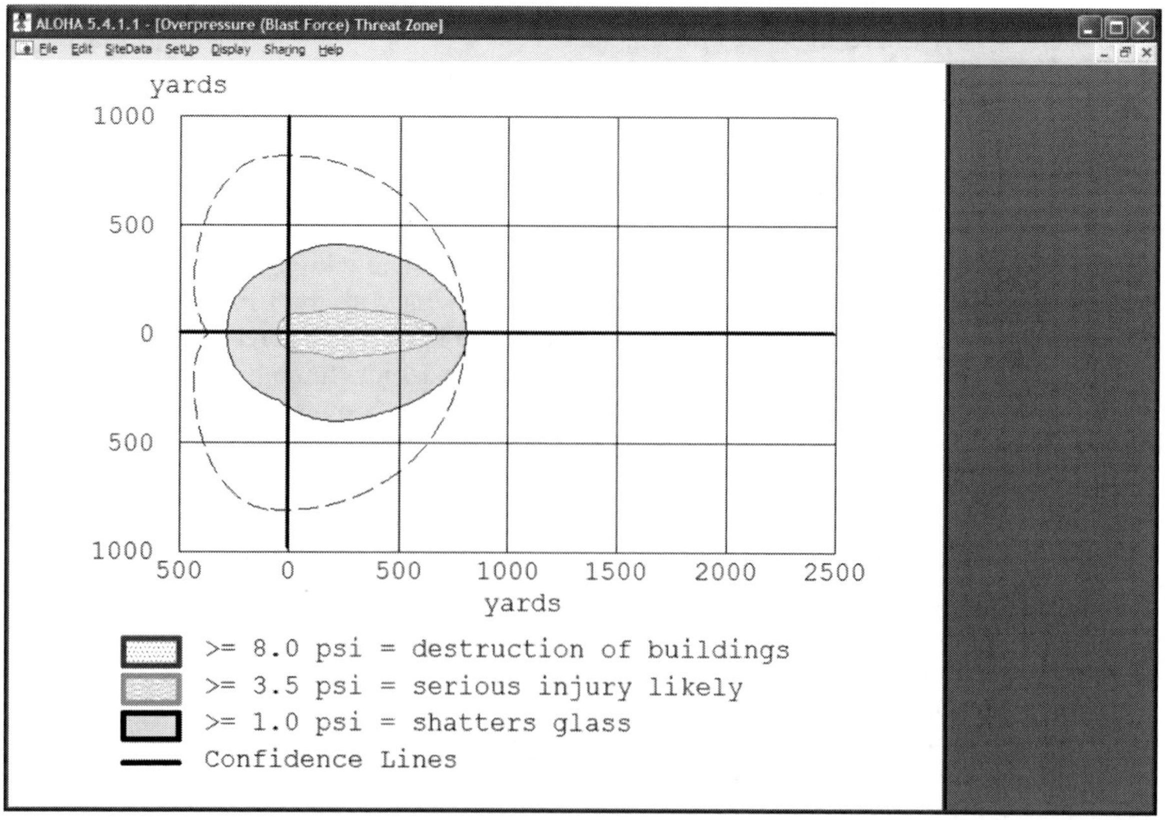

Table 5. Potential Safety Consequence Severity.
Release of flammable gas or flashing liquid within an operating unit.

Leak Rate		Impact area / Unit area	Severity
10,000 lb/min	And Over	>1.00	5
1,000 lb/min	10,000 lb/min	0.66-0.99	4
100 lb/min	1,000 lb/min	0.33-0.66	3
10 lb/min	100 lb/min	0.16-0.33	2
0 lb/min	10 lb/min	<0.16	1

7. Worked Example

A LOPA team was considering a potential pump seal leak (Figure 2).

Tower intercooler pump, 15 HP, pumping a rich oil saturated with C4.
Scenario – Deadheading pump resulting in a seal leak.
Leak pressure – Pump head (30 psi) plus tower pressure (60 psig).
PHA severity – 3 (severe injury).

Figure 2. Example Diagram Showing the Scenario under Review

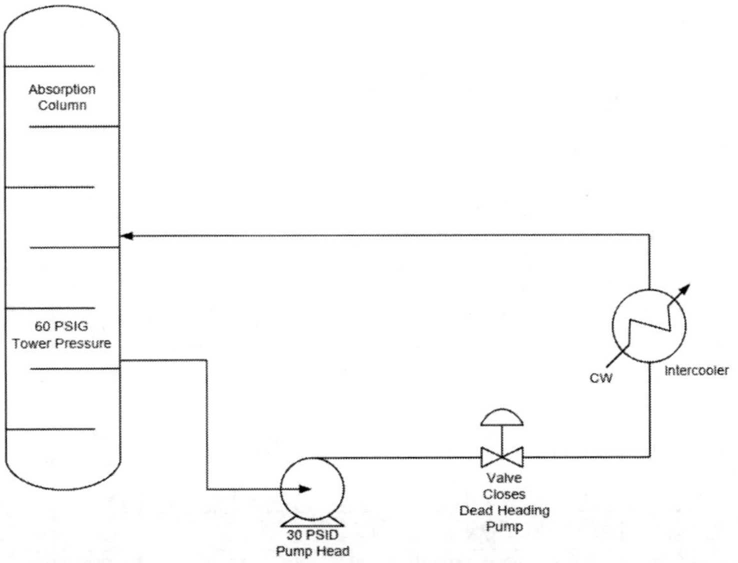

The PHA team had previously ranked this seal leak as a serious injury event or a 3 on a 1-5 severity scale shown in Table 1. As described by LOPA team members who were also in the PHA, the severity had been heavily debated and in the end no one wanted to underestimate the hazard. The consensus was that this scenario involved a hydrocarbon release and other hydrocarbon releases had been ranked as a consequence severity of 3 by the team.

The LOPA team was introduced to the draft consequence severity guidance. It was emphasized that they had full authority to overrule the results based on their experience or on site specific factors, such as location of the pump.

7.1 Step 1

Estimate the hole diameter expected from a mechanical seal failure. The pump shaft size was not available to the team, but the pump horsepower, 15 HP, was shown on the P&ID. This yielded an equivalent hole diameter of ¼ inch.

Table 7. Table 3B with the Team Choice Highlighted

Horsepower vs Equivalent Hole Diameter							
HP	<5	5-10	**10-50**	50-150	150-300	300-600	>600
Hole (inc)	1/8	3/16	**1/4**	3/8	1/2	5/8	3/4+

7.2 Step 2

Find the intersection of leak diameter and highest credible pressure. The team used 100 psig (the next selection higher than the 90 psig expected), the intersection with ¼ inch gives an expected leak rate of 98 pounds per hour.

Table 8. Table 4 with the Team Choice Highlighted

Liquid Release Pressure vs. Hole diameter. Rate in lb/min.										
Psig / inches	1/16	1/8	**1/4**	1/2	1	1.5	2	3	4	6
10	2	8	31	120	500	1100	2000	4500	7900	17900
50	4	17	69	280	1100	2500	4400	9980	17700	X
100	6	25	**98**	390	1600	3500	6300	14100	X	X
150	8	30	120	480	1900	4300	7700	X	X	X
300	11	42	170	680	2700	6100	10900	X	X	X
500	14	55	220	880	3500	7900	X	X	X	X
1000	19	77	310	1200	5000	11200	X	X	X	X
2000	27	110	440	1800	7000	X	X	X	X	X
6000	47	190	760	3000	12100	X	X	X	X	X
10000	61	250	980	3900	X	X	X	X	X	X
30000	110	420	1700	6800	X	X	X	X	X	X
50000	137	550	2200	8800	X	X	X	X	X	X

7.3 Step 3

Use the expected leak rate to determine the consequence severity. The team chose severity 2 based on the calculated leak rate. Though the 98 pounds per hour is close to 100, the team discussed other factors associated with the release to determine whether severity 2 was appropriate versus severity 3. The scenario involved the release of a lean oil mixture containing light and heavy hydrocarbons. The heavier fractions in the lean oil are not expected to flash to a vapor as easily as the general hydrocarbon class. Consequently, the team selected consequence severity 2.

Table 9. Table 5 with the Team Choice Highlighted

Leak Rate		Severity	
10,000 lb/min	And Over	5	
1,000 lb/min	10,000 lb/min	4	
100 lb/min	1,000 lb/min	3	
10 lb/min	100 lb/min	**2**	← 98 lb/minute
0 lb/min	10 lb/min	1	

8. Moving Forward

LOPA continues to be a great tool for the semi-quantitative analysis. A case can be made for alternatives to LOPA but in most organizations these alternatives are applied to specific hazards or applications. LOPA teams, which include supervisory, technical and worker participation, can often address many hazardous events in a single meeting. All involved gain an appreciation of the thoroughness of the analysis and the suitability of the IPL and recommendations.

These teams apply LOPA to risks that range from relatively minor issues to scenarios that could potentially destroy facilities and the lives of employees and neighbors. Different levels of protection are appropriate for different levels of risk, but inconsistent levels of protection for the same level of risk can bring the LOPA process into question. Given the importance of achieving consistent results it is important that tools be provided to the teams to assess risks in an accurate and consistent manner.

Attempts to improve consistency using standard LOPA templates or conditional modifiers introduce new variables into the assessment but do not address the root of the problem, which is often an inconsistent consequence severity assessment. This paper proposes that PHA and LOPA teams be provided with look-up tables that give guidance on consequence severity based on equipment type and release conditions.

LOPA can be substantially improved by implementing consequence estimation tools that assist team members in understanding the flammability, explosivity, and toxicity of process chemicals. When these tools can reliably and simply distinguish different size events they can be used to separate the severity of large loss of containment events from small. The consequence severity tool has proven to be a welcome development to the teams that have tried it. By increasing consistency in severity classification, the consistency of the final result is likewise improved.

9. References

[1] CCPS/AIChE, *Guidelines for Hazard Evaluation Procedures,* Third Edition, Wiley-Interscience, New York (2008).

[2] CCPS/AIChE, *Layer of Protection Analysis: Simplified Process Risk Assessment*, Concept Series, New York (2001).

[3] CCPS/AIChE, *Guidelines for Process Safety Metrics*, Wiley-Interscience, New York (2010).

[4] CCPS/AIChE, *Guidelines for Safe and Reliable Instrumented Protective Systems*, Wiley-Interscience, New York (2007).

[5] OSHA, "Process Safety Management of Highly Hazardous Chemicals; Explosives and Blasting Agents, 29 CFR Part 1910." *Federal Register 57, 36*, Washington, DC (1992).

[6] CCPS/AIChE, *Guidelines for Developing Quantitative Safety Risk Criteria*, New York (2009).

[7] Environmental Protection Agency (EPA), "Accidental Release Prevention Requirements," 40 CFR Part 68, Federal Register 61, 120, Washington, DC (1996).

[8] Environmental Protection Agency (EPA), National Oceanic and Atmospheric Administration, "Aloha Users Manual," The CAMEO Software System, February (2007).

6TH GLOBAL CONGRESS ON PROCESS SAFETY

Instrumented Protective Systems for Distillation Operations

Jennifer F. Mize
Senior Chemical Engineer
Eastman Chemical Company
P.O. Box 511, Building18
Kingsport, TN 37662
jfmize@eastman.com

J. Wayne Chastain
Engineering Associate
Eastman Chemical Company
P.O. Box 511, Building18
Kingsport, TN 37662
chastain@eastman.com

Prepared for Presentation at
American Institute of Chemical Engineers
2010 Spring Meeting
6th Global Congress on Process Safety
San Antonio, Texas
March 22-24, 2010

UNPUBLISHED

Instrumented Protective Systems for Distillation Operations

Jennifer F. Mize
Senior Chemical Engineer
Eastman Chemical Company
P.O. Box 511, Building18
Kingsport, TN 37662
jfmize@eastman.com

J. Wayne Chastain
Engineering Associate
Eastman Chemical Company
P.O. Box 511, Building18
Kingsport, TN 37662
chastain@eastman.com

Keywords: Distillation, overpressure, Layers of Protection Analysis (LOPA), High Integrity Protective Systems (HIPS)

Abstract

Distillation is one of the most common unit operations in the chemical processing industry. Using relative volatility as a means of separating chemicals involves adding large amounts of energy to the system at the base of the column, typically through the use of a reboiler, and then removing that energy at the top of the column, via a condenser. Because of the high energy levels present in the column, failures within the system can result in overpressure of the column leading to a potentially catastrophic incident. The typical approach to dealing with these overpressure scenarios has been through the use of pressure relief systems. Layer of Protection Analysis (LOPA) provides a method for the evaluation of the risk presented by these overpressure scenarios. LOPA can be used to specify the required reliability of Instrumented Protective Systems (IPS) used as High Integrity Protective Systems (HIPS) which offer an alternative means of addressing some of these overpressure scenarios. This paper addresses several potential failure modes in distillation columns and then illustrates situations in which HIPS can be used to reach a safe state without reliance on the pressure relief system.

1. Introduction

Distillation is a widely used unit operation used for the separation of materials with disparate volatility. Like most other unit operations, upset conditions in a distillation system can result in overpressure of the equipment. Traditionally and by application of recognized and generally accepted good engineering practice (RAGAGEP), overpressure scenarios are addressed through the use of pressure relieving systems. While a properly designed and maintained pressure relief system will protect a distillation column and the associated equipment from overpressure, the advent of standards for the design and application of reliable instrumented protective systems have increased the options available to designers in how overpressure scenarios can be addressed.

Before determining the best means of addressing the overpressure concerns associated with distillation equipment, the potential scenarios which can result in overpressure must be understood. Once the overpressure scenarios applicable to a system of interest have been identified, designs can be considered for the scenarios that range from the traditional pressure relief system approach, novel pressure relief as a means of protecting the equipment, or application of instrumented protective systems. Factors which should be taken into account when determining the appropriate design solution for a given scenario include the required reliability, cost, potential for secondary or knock on effects, and compliance with RAGAGEP.

Evaluating options for overpressure protection by means other than the traditional approach, which has long been supported by RAGAGEP, requires estimation of the risk imposed by the scenario. A common method for evaluating risk from single cause – consequence scenarios is the Layer of Protection Analysis (LOPA). Use of the LOPA technique can provide the required reliability for instrumented protective systems for overpressure protection.

2. Overview of Distillation

Distillation is a unit operation that uses energy input to separate the components in a mixture based upon the differences in volatility between the components. Some type of heating media is used to introduce the energy required to boil the more volatile components (low boilers), which are separated from the less volatile components (high boilers) that remain in the liquid. The vapors generated in the distillation system are typically condensed and collected as a liquid product (distillate). Distillation can be conducted as a batch operation by introducing a liquid charge into a pot, heating to vaporize the low boilers, and then condensing the vapors. Distillation can also be conducted as a continuous operation by introducing a liquid feed into a column and removing the separated condensed vapor (distillate) and liquid (bottoms) products at a steady rate.

This paper will focus on continuous steady-state distillation conducted in a typical column system, which consists of a distillation column or tower, a condenser, a reboiler (or baseheater), and a reflux pot, which is used to collect the liquid product (distillate). A portion of the distillate is returned to the column as reflux to provide cooling and condensation in the column as well as improve the efficiency of the separation. The vapors flowing upward through the column and the liquid reflux flowing downward through the column contact each other and additional separation of the components occurs as a result of this contact. In order to facilitate the vapor/liquid contact, a series of trays or packing material are present in the column.

Figure 1: Typical Continuous Distillation System

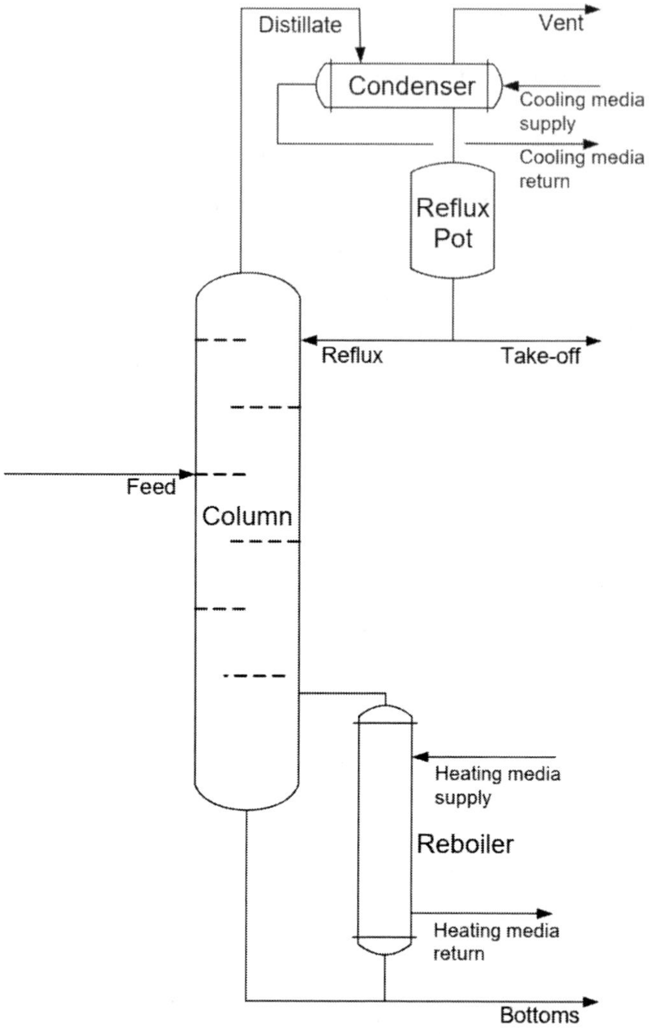

3. Layer of Protection Analysis

Layer of Protection Analysis (LOPA) is a risk analysis method documented in the CCPS text Layer of Protection Analysis: Simplified Process Risk Assessment (CCPS, 2001) [1]. The LOPA method is applied to single cause – consequence events to estimate the frequency of occurrence. The estimated frequency can be then be compared to a risk criterion to determine if the hazard of that event has been adequately controlled. LOPA uses strict rules in its application of safeguards to an event. This approach allows for the simplified approach of the method. Further details on the method can be found in the CCPS text.

In evaluating overpressure scenarios on distillation columns, LOPA allows the adequacy of the controls in place to prevent or mitigate an overpressure event to be determined. In some cases, LOPA can indicate that additional controls other than the normal process control and the relief system are needed to properly manage the risk of the event. In other cases, LOPA can be used to determine if controls other than an adequately sized relief system can be used to address the hazard of an overpressure scenario. When evaluating a potential overpressure scenario using LOPA and considering if a relief system is required, the strictures placed on the operation of a pressure vessel by the ASME code [2] should be taken into account. The LOPAs conducted on the example cases below are included to show how the risk criteria was met without relying on a relief system. In these cases, the risk criteria used in the analysis was a value of $1x10^{-5}$ events per year for scenarios which could cause a single severe injury on the plant site. Different risk criteria could apply to scenarios with more or less severe consequences.

The LOPAs included later in this document are conducted using a certain methodology. The values given for frequencies and probabilities are presented in log format, so that a value of -5 for a risk criteria represents one event in 10^5 years and a value of -0.5 for occupancy indicates that personnel are present no more than $10^{-0.5}$ fraction or 32% of the time which would correspond approximately to 7 ½ hours per day. Another convention used in the LOPA method is allowing fractional values at the log mean average between whole orders of risk reduction. Some companies restrict LOPAs to only using whole order of magnitude values.

4. Overpressure Scenarios

Overpressure is defined as the accumulation of pressure inside a vessel above its rated or maximum allowable working pressure (MAWP). This accumulated pressure can cause vessel failure and release of the contents to the surrounding area. These scenarios are typically referred to as relief cases, because the overpressure that is generated is typically relieved by devices designed to prevent vessel failure (relief devices). The following overpressure scenarios should be evaluated for each vessel or system. A more complete discussion of these relief scenarios can be found in ANSI/API Standard 521, Pressure Relieving and Depressuring Systems[3].

1. Closed Vessel Outlet	Isolating the column system by closing the normal vent path and/or liquid outlets can result in overpressure by feed streams with supply pressure greater than the system rated pressure.
2. Loss of Cooling	Loss of the cooling media can lead to loss of condensing. If this occurs, energy input from the reboiler can no longer be removed from the system.
3. Top Reflux Failure	Loss of reflux can lead to reduction of the energy removal within the column due to vapor-liquid contact. In most cases, loss of reflux is less severe than cooling water failure leading to total loss of condensing.

4. Sidestream Reflux Failure	This case is similar to loss of top reflux in most cases. However, it can occur independently of loss of cooling if the column system receives a sidestream reflux from another vessel that provides a significant condensing duty or if the main feed is very cold compared to the column operating temperature.
5. Non-condensable Accumulation	Accumulation of non-condensables can severely limit the vapor removal capacity of a condenser. This case is similar to Case 2, Loss of Cooling.
6. Introduction of Volatiles	This case covers the introduction of volatiles through a feed stream or process and/or utility leak into a hot vessel resulting in the potential for overpressure due to excessive vapor generation.
7. Vessel Overfilling	An imbalance in the rate of liquid feed and liquid removal can result in liquid accumulation leading to overfilling.
8. Automatic Control Failure	Failure of automatic valves, loss of instrument services, and wind-up or other failure of automatic control loops can lead to upset conditions in the distillation system. Due to the wide range of potential failures, evaluation of this scenario requires analysis on a case-by-case basis. In general, automatic control failure is the initiating failure that leads to other relief scenarios.
9. Abnormal Heat Input	This case covers all sources of internal and external heat input to the system, with the exception of fire impingement. All energy sources must be considered.
10. Split Exchanger Tube	This case covers the failure or rupture of an internal coil or tube that allows a utility stream (or process stream functioning as a utility stream) to enter the distillation system. A vapor stream that enters the system can lead to an overpressure with gas/vapor scenario, while a liquid stream that enters can lead to an overpressure with liquid (overfilling) or introduction of volatiles scenario.
11. Hydraulic Expansion	This case covers overpressure due to expansion of liquid-filled vessels exposed to a heat source.
12. Chemical Reaction	This case covers energy input to the system due to an exothermic chemical reaction, including the potential for a runaway reaction and two-phase relief flow.
13. External Fire	This case addresses heat input to the system due to a fire outside of the equipment heating the vessels through the walls.
14. Power Failure	Power failure can result in other relief cases becoming credible, such as automatic control failure, loss of cooling, and loss of agitation. Due to the wide range of potential upsets resulting from power failure, this case requires evaluation of system-specific failures and analysis on a case-by-case basis. In general, power failure is the initiating failure that leads to other relief scenarios.
15. Deflagration	A flammable mixture can exist inside a vessel and/or system if flammable vapors and an oxidizer are present. This case is typically mitigated by ensuring that air cannot enter the system when flammables are present.
16. Vacuum Relief	Generation of excessive vacuum inside a vessel can lead to damage and the potential for release of the contents. Vacuum relief should be considered for vessels not rated for full vacuum.

The relief cases that are generally applicable to distillation systems result from failures that lead to energy input via the reboiler that exceeds the energy removal capability of the condenser. During normal operation, the energy input and removal are balanced to allow the system to function at its steady-state operating pressure. Potential failures must be evaluated on a case-by-case basis to determine which, if any, could lead to an overpressure event due to the imbalance between energy input and removal. In order for the energy input to overpressure the system, the vapor pressure of one or more of the components must be in excess of the system rated pressure at the temperature of the heating media.

For a typical distillation system, the potential sources of overpressure can be reduced to a subset of the 16 relief cases.

4.1 Case 1--Liquid Overfilling (Closed Vessel Outlet / Vessel Overfill)

Filling the column system with liquid can lead to overpressure from the liquid source, such as a high head centrifugal pump, and/or the static head of the liquid. This scenario can occur during start-up of a distillation system or during an emergency shutdown when feed is entering the column without energy input in the reboiler. It can also occur if valves are positioned incorrectly such that liquid cannot be removed through the normal flow path out of the column base.

4.2 Case 2--Loss of Cooling

Loss of energy removal in the column system due to loss of cooling can lead to overpressure due to continued vapor generation from energy input to the reboiler. In a typical distillation system, the condenser is designed to be the main source of energy removal. The reflux stream(s) typically do not provide a significant source of cooling duty when compared to the condenser. Therefore, reflux failure (top and/or sidestream) is generally less severe than loss of cooling. Loss of cooling can lead to loss of reflux when level can no longer be maintained in the reflux pot.

Improper venting of non-condensables can cause vapor-lock of the cooling water side, leading to a reduction in the condensing capacity. Loss of cooling can also be experienced due to power failure. Case-by-case evaluation is required to determine if local and/or plantwide power outages can lead to loss of cooling. Another potential cause of loss of cooling is overcooling resulting in the freezing of condensed material in the condenser tubes leading to plugging.

4.3 Case 9--Abnormal Heat Input

Abnormal heat input considers the maximum possible energy input into the column system through the reboiler. This can occur due to an automatic control failure that leads to excessive utility feed to the reboiler. If the additional energy input cannot be removed by the condenser, excessive vapor generation can result leading to overpressure.

4.4 Case 10--Split Exchanger Tube

A reboiler tube rupture can lead to the introduction of the heating media into the column system. If the supply pressure of the heating media is greater than the normal operating pressure of the column system, the heating media will enter the column system. Overpressure can result from vapor generation due to contact with the heating media. If the supply pressure of the heating media is greater than the system rated pressure, overpressure can result from the heating media entering the column system without considering any additional vapor generation.

A condenser tube rupture can lead to a liquid overfilling scenario as the cooling media accumulates in the reflux pot and/or column. If the supply pressure of the cooling media exceeds the normal operating pressure of the column system, the cooling media will enter the column system. If the reflux pot and condensing system becomes liquid-filled, loss of condensing could occur leading a scenario similar to Case 2.

If the normal operating temperature in the column system exceeds the boiling point of the cooling media at the system rated pressure, the column system can be overpressured by vaporizing the cooling media. This scenario would be evaluated as an Introduction of Volatiles, Case 6.

4.5 Case 13--External Fire

If a pool of flammable or combustible liquids can accumulate around the vessels in the column system, the potential exists for a pool fire to impinge on the system. The heat input due to this fire impingement can lead to vaporization of the contents and overpressure of the vessels. Due to the high temperatures experienced during a pool fire, high boiling components that may not vaporize due to loss of cooling or abnormal heating can be vaporized by fire impingement. This scenario can be mitigated by fire protection systems, including the use of fire resistant insulation and approved deluge spray (per NFPA 15 [4]). These mitigations reduce the required size of the relief system.

Figure 2: Overpressure Cases for Example Column System

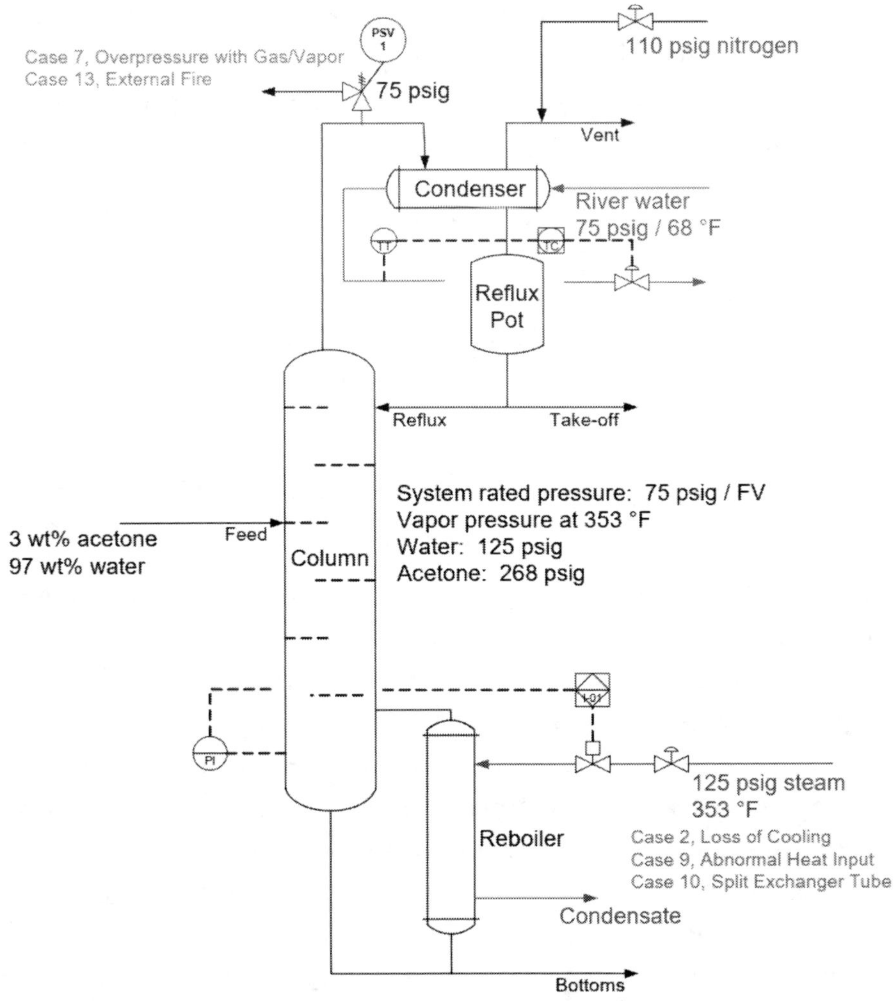

5. Inherently Safer Approach

Typically, the relief system is designed after all the other equipment design has been completed. In some cases, the relief designer is constrained by existing equipment with a set system rated pressure, a limited selection of heating and/or cooling media, and/or the operational requirements of the desired separation. Designing the relief system to handle all of the potential sources of overpressure can result in a large relief device and associated piping, as well as the potential for a significant discharge from the relief system that must be safely vented to an effluent handling system, secondary containment, or possibly the atmosphere.

One inherently safer approach is to design the column system to eliminate one or more potential relief scenarios. In order to eliminate loss of cooling and abnormal heating as relief scenarios, the vapor pressure of all components, including cleaning solvents, must be less than the system rated pressure at the maximum temperature of the heating media. This can be achieved by selecting a heating media with the lowest practical operating temperature and designing the column system to operate at this temperature. The appropriate system rated pressure is then selected based upon the maximum vapor pressure that can be experienced with the selected heating media. The relief designer must consider all operating modes, including maintenance and cleaning operations when selecting the appropriate system rated pressure. The designer must consider the use of all cleaning agents, such as water or solvents, which may generate sufficient vapor pressure to overpressure the system.

In order to eliminate tube rupture as a potential relief scenario, the supply pressure of the heating and cooling media must be less than the system rated pressure. Since cooling media leaking into the column system could lead to an Introduction of Volatiles case, the system rated pressure must also be specified to be greater than the vapor pressure of the cooling media at the column operating temperature.

6. Novel Applications of Pressure Relieving Systems to Address Overpressure Scenarios

In some cases, it may not be possible to achieve the required system rated pressure to eliminate potential relief scenarios. The cost associated with designing vessels to a specific pressure rating may be prohibitive, or the system rated pressure may be limited by an existing vessel. With the exception of external fire, the relief scenarios that govern the required size of the relief system are all related to the maximum energy input to the column via the heating media in the reboiler. For a column system driven by steam heating in the reboiler, the energy input can be minimized by limiting the supply pressure, and subsequently the supply temperature, of the steam. Steam control loops normally control the operating pressure in the reboiler at a fraction of the supply pressure of the steam from the source header. A failure of the steam control valve in the wide-open position can result in steam at the source header pressure entering the shell resulting in an overpressure event due to abnormal heating.

Provided that the required operating pressure in the reboiler is significantly less than the source header pressure for the steam supply, the steam control loop can be designed to limit the maximum steam pressure such that the maximum temperature in the reboiler is less than the boiling point of the components in the column system at the system rated pressure. A relief device can then be installed downstream of the steam control valve to insure that the steam pressure is limited in the event of an automatic control failure. The relief device must be sized to handle the maximum steam flow across the control valve in a failed open condition or through a manual bypass, whichever situation provides the greater flow.

Relieving the steam downstream of the pressure reducing valve instead of relieving the system through the relief device on the column has several advantages. The size of the relief device required can be reduced since the relief system is handling steam instead of relieving organics which are boiled by higher pressure steam. The difference in heats of vaporization of water and typical organics means that this can result in a decrease in the relief rate of 50% to 80%. In addition, since the relieving fluid in this case is steam as opposed to organic vapors, the safety and environmental hazards from a relief event are minimized. In this case, the column will be protected by a relief system identical to the system that acts to limit header pressure in steam systems in most industrial facilities.

An example of this approach is shown below.

Figure 3: Example Column System with Relief Systems

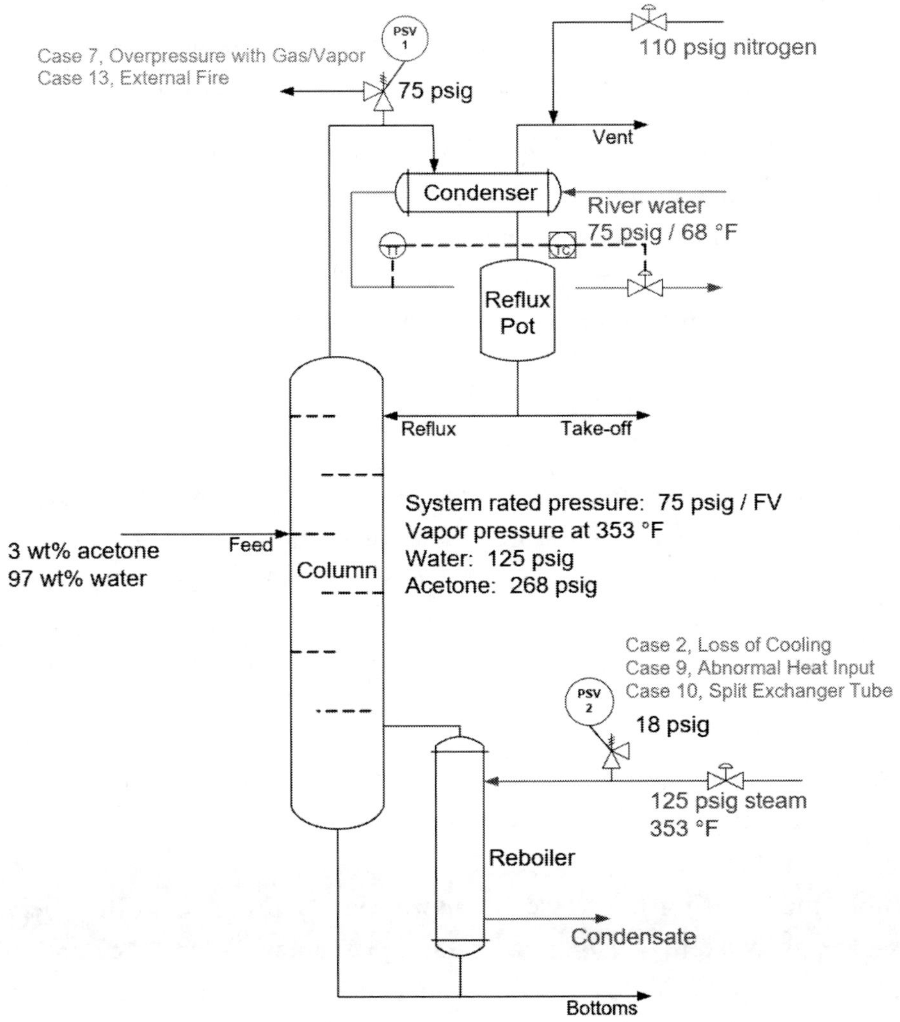

In this case, the set pressure of the relief valve on the steam supply to the reboiler is 18 psig. This set pressure keeps the pressure in the exchanger below 20 psig even when the accumulation of 10% is taken into account. The temperature of saturated steam at this pressure is 259 °F. The vapor pressure of the acetone, the low boiler in this pair, at the temperature of the steam is less than 87 psig, which is the allowable overpressure of 116% of the MAWP specified in UG-140(b) of the ASME pressure vessel code [1].

Reducing the pressure of the steam to the reboiler in a column system will not reduce the temperature of the steam to the saturation value. Even if no superheat is present in the steam prior to the pressure reduction, the pressure reduction will result in some degree of superheat in the steam. In general, this superheat can be ignored in the analysis and the saturation temperature of the steam can be used in the evaluation. The enthalpy of the sensible cooling of the steam to the saturation temperature will typically be much less than the enthalpy from the condensation of the steam. In general, the steam must condense to allow steam flow into the system since steam traps normally will only allow condensate to pass. Superheat generally only has to be considered in situations where reactive species are involved and skin temperatures in a small section of the heat exchanger may be heated to values approaching the temperature of the superheated steam. If this skin temperature of the tubes in the exchanger is higher than the onset temperature of a reactive species in the equipment, an undesired reaction could occur.

One other issue that must be addressed when applying a solution of this type is the presence of low boiling azeotropes. A low boiling azeotrope will result in a higher pressure in the equipment at a given temperature than that predicted by the vapor pressure curves of the components in the mixture. In this case, the azeotrope itself must be evaluated for its impact on the design. In the specific example above, the acetone / water system does not form an azeotrope.

7. Basis for Using IPS to Address Overpressure Scenarios

The use of Instrumented Protective Systems to address overpressure scenarios is included in ASME Code, Section VIII, Division 1, UG-140: Overpressure Protection by System Design [1]. In particular, UG-140(b) is the section of the code applicable to mitigating some scenarios with active controls while using a pressure relieving device to address others. UG-140(b) has several requirements which must be met in order to use system design to address overpressure protection.

- The vessel cannot be in air, steam, or water service unless the overpressure protection of the vessel is critical in preventing the release of a material which could result in safety or environmental concerns.
- The user is responsible for ensuring the vessel is appropriately protected. In order to limit the overpressure by system design, a detailed evaluation of the credible overpressure scenarios as described in ANSI/API Standard 521 must be conducted. This evaluation must be documented to describe all possible credible overpressure scenarios and any safety critical instrumentation used to limit the system pressure.

- Overpressure scenarios protected by system design must be announced so that action can be taken by operators or instrumented systems to limit the overpressure below the MAWP.
- No credible overpressure scenario should exceed 116% of the MAWP times the ratio of the allowable stress at the relief scenario temperature and the rated design temperature, however, the overpressure limit cannot exceed the test pressure.
- The results of the analysis conducted must be documented and signed and include P&IDs, the overpressure scenarios, a detailed description of any instrumentation used to mitigate a relief event and analysis showing the resulting pressure from any event.

ISO 23251, Petroleum, petrochemical, and natural gas industries – Pressure-relieving and Depressuring Systems [3], addresses high integrity protection systems (HIPS) in Annex E. This annex references ASME Section VIII, Code Case 2211-1 (which has been incorporated into the current addition of the Code as UG-140).

7.1 Benefits of using IPS to mitigate overpressure scenarios

The use of Instrumented Protective Systems to mitigate overpressure scenarios has many benefits. During an upset condition, the IPS takes the column system to a safe holding condition prior to reaching the system rated pressure. In most cases, the IPS will mimic actions already taken by the Basic Process Control System (BPCS) in response to high temperature and/or pressure in the column system, such as reducing the heating media flow to the reboiler by closing the control valve. Blocking the flow of heating media to the reboiler will stop vaporization in the system and/or prevent the flow of high pressure heating media into the column base due to a tube rupture in the reboiler. This will eliminate the potential for rapid vaporization of the liquid in the base of the column or the flow of high pressure heating media into the base that could result in significant damage to the trays due to uplift as the vapor travels up through the column. The relief system will not open if the pressure in the column system does not reach the set pressure, preventing the loss of valuable product and the significant environmental and safety concerns associated with the release of hazardous materials.

Using Instrumented Protective Systems to mitigate overpressure scenarios allows the relief designer to size the relief system for the external fire case only. Utilizing fire protection to mitigate the heat input will allow the designer to minimize the required size of the relief system, reducing the amount of material released during a relief event, the required size of the effluent handling or secondary containment system, and the environmental and safety concerns associated with the release of hazardous materials.

7.2 IPS Solutions Addressing Individual Cases

Instrumented Protective Systems can normally be used to address all credible overpressure scenarios for distillation systems with the exception of external fire. These systems are independent of the basic process control system (BPCS) that is used to conduct the normal operation of the column system. The BPCS includes the logic solver, instrumentation, and field elements such as control and block valves. As previously documented, the potential sources of overpressure generally applicable to column systems are liquid overfill, loss of cooling, abnormal heat input, and split exchanger tube.

7.2.1 Sensing Devices

In general, it is preferable to install indicators rather than switches when choosing a sensing instrument for initiation of an Instrumented Protective System. Indicators will provide continuous data on the column operating conditions. Failure of an indicator can be detected by monitoring the data it provides, which makes it an revealed failure. Switches do not provide data until the trip point is reached. Therefore, failure of a switch cannot be detected by monitoring the process, which makes it an unrevealed failure.

7.2.1.1 High base pressure

High base pressure can be used to sense liquid overfill, abnormal heat input, and tube rupture in the reboiler, which can all lead to high pressure in the column base. For liquid overfill, the pressure in the base will increase due to the static head of liquid accumulating in the column. For abnormal heat input, the pressure in the column base will increase due to vapor pressure of the contents at the temperature of the heating media. For tube rupture, the base pressure will increase to the supply pressure of the heating media. Therefore, care must be taken to select the trip point for the high base pressure instrument to ensure that it provides protection for all credible overpressure scenarios.

High base pressure may be an unreliable indication of the process condition in applications where the material at the base of the column is plugging or fouling. In this case, special measures such as pad mounted pressure measurements or flushes on pressure measurement taps must be implemented to use base pressure as an indicator to protect the system from overpressure. Otherwise, alternative sending devices should be selected.

7.2.1.2 High top pressure

High top column pressure can be used to sense loss of cooling in the condenser or a blocked vent line. While abnormal heat input and tube rupture in the reboiler will lead to a high top pressure, this is an indirect indication of the upset condition due to the pressure drop through the packing or trays. Additionally, damage to the trays and/or packing supports could result from rapid vaporization in the base that is not promptly detected.

High top pressure can also protect against liquid overfill if the column system is designed and rated to be liquid filled. Even when this design criteria has been used, however, filling a distillation column with liquid can result in severe damage to the column internals if the column is operating. The high pressure drops through the column in this event could result in overpressurization of the base of the column, even though the system is designed to be filled with liquid.

7.2.1.3 High vent temperature

High vent temperature can be used to sense loss of cooling in the condenser. The location of the high temperature instrument must be carefully chosen to ensure that it can accurately measure the operating temperature of the column system. If the high temperature instrument is located a significant distance from the condenser in the vent line, the instrument may be exposed to ambient conditions rather than the column operating conditions. In this configuration, it may not be an effective measurement of upset conditions occurring in the condenser and overhead vapor line due to loss of cooling. If the selected trip point is within the range of the ambient conditions due to environmental heating and cooling, spurious trips can occur during warm weather.

7.2.1.4 High tray temperature

Depending upon the location of the temperature instruments, one or more high tray temperatures can be used to sense abnormal heating, tube rupture in the reboiler, or loss of cooling. The temperature instrument should be located as close as possible to the location of the upset condition to quickly detect the overpressure event. A high temperature instrument located on the top tray can be effective at detecting loss of cooling, but will be an indirect measurement of abnormal heat input or tube rupture in the reboiler. A high temperature instrument located on the bottom tray can be effective at detecting abnormal heat input or tube rupture in the reboiler, but will be an indirect measurement of loss of cooling. Additionally, using high tray temperature may result in damage to trays and/or packing supports due to vapor generation in the base before the high temperature can detect the overpressure event.

7.2.2 Actions

7.2.2.1 Closure of Heating Media Feed to Reboiler

Mitigation of loss of cooling, abnormal heat input, and tube rupture in the boiler can be achieved by closing the heating media feed to the reboiler. One or more dedicated block valves are required to achieve positive shut-off of the heating media based on the required reliability of the HIPS used to prevent the event and the design of the SIF. Removing the source of energy input to the column system will ensure that the system will go to a safe condition no matter which initiating failure causes the overpressure event.

7.2.2.2 Closure of Liquid Feed

Mitigation of liquid overfilling can be achieved by stopping the liquid feed to the column. One or more dedicated block valves or taking action to stop feed pumps are required to achieve positive shut-off of the feed. The number and type of actions which must be taken by the SIF for liquid overfill are dictated by the reliability of the controls determined by the LOPA and the design of the SIF to address that relief case. If there are multiple feed locations for the column system, each feed source must be closed.

7.3 LOPAs for Example Column

Example LOPAs for column system presented above are shown below, presenting the risk calculation used to determine the required reliability of the IPS to mitigate the overpressure scenario which the relief system on the column is not designed to address.

Loss of cooling

Case	2	Loss of cooling on the column results in high pressure and failure of column
Target	-5	Severe injury to personnel due to shrapnel
Initiator	-1	Loss of cooling due to failure of the BPCS temperature control loop on the river water
IPL	0	BPCS ESD on the column tripped by high vent temperature and high base pressure
	0	Likelihood of ignition
	-0.5	Likelihood of personnel in the area
	-1	Likelihood of injury
Total	-2.5	
Difference	-2.5	SIL 2 interlock to close block valve in the steam supply on high base pressure

The scenario evaluated in this LOPA is injury to personnel due to a physical explosion and the resulting shrapnel. The mixture of vapors even in the top of the column is outside the flammable range, so injury due to a potential flash fire or explosion due to combustion has been ruled out. In this LOPA the initiating failure is a BPCS control loop failure for which the consensus failure rate is $1x10^{-1}$ failures per year or an index value of -1 when the log of this value is taken. A BPCS implemented interlock is in place, but in this case, the initiating failure was in the BPCS. Based on the independence requirement of independent protection layers used in LOPA, this safeguard is not given any credit in the analysis.

Three conditional modifiers are considered in this analysis. The likelihood of ignition in this case is not accounted for since the hazard is the physical explosion of the column. The occupancy in the area has been estimated by operations personnel in conjunction with the LOPA analyst. The likelihood of injury in this case has been estimated at 10%. When likelihood of injury is included in LOPA, often the default value is 50%. In the example methodology injury from shrapnel is considered to be less likely and estimated at 10%. The occupancy and likelihood of injury have to be evaluated together. As the distance from the event decreases, the occupancy normally decreases while the likelihood of injury increases.

Based on this LOPA, an additional risk reduction of 2.5 orders of magnitude is needed. This risk reduction can be provided by an IPS. A SIL 2 safety instrumented function will meet this need if the risk reduction factor used in the design is 316 or greater. In this case, a SIL 2 SIF implemented with a minimum of the log mean risk reduction factor for a SIL 2 system's range to close the steam supply to the reboiler on the column in the event of high base pressure would adequately address the risk. This would allow the relief system on the column to exclude this case as a design criteria as long as all of the requirements of UG-140(b) were met.

Abnormal heat input

Case	9	Abnormal heat input due to steam control loop failure
Target	-5	Severe injury to personnel
Initiator	-1	Abnormal heat input due to failure of the BPCS steam control loop
LOP	0	BPCS interlock on the column tripped by high vent temperature and high base pressure
	0	Likelihood of ignition
	-0.5	Likelihood of personnel in the area
	-1	Likelihood of injury
Total	-2.5	
Difference	-2.5	SIL 2 interlock to close block valve in the steam supply on high base pressure

The abnormal heat input LOPA is identical to the analysis above with the exception of the initiating event. In this case, the failure is the BPCS steam control loop providing too much energy to the column. Although the loop which initiated the event is different, the recommendation is identical.

Tube rupture in the reboiler

Case	10	Overpressure due to tube rupture in the reboiler
Target	-5	Severe injury to personnel
Initiator	-1.5	Tube rupture in the reboiler
LOP	-1	BPCS interlock on the column tripped by high vent temperature and high base pressure
	0	Likelihood of ignition
	-0.5	Likelihood of personnel in the area
	-1	Likelihood of injury
Total	-4	
Difference	-1	SIL 2 interlock to close block valve in the steam supply on high base pressure

Case 10, Split Exchanger Tube, differs slightly from the above examples. In this case, the initiating failure is a mechanical failure of a tube in the reboiler due to erosion, corrosion, vibration, or some other mechanism. Experience, collected data on failures, and values for the frequency of this type of failure from the open literature indicate that it occurs less frequently than a failure of a BPCS loop. In this case, the -1.5 index value used in the analysis is an indication of a failure rate of 1 in 32 years. Since the initiating failure in this case was not related to the BPCS, the interlock present in the BPCS to stop steam flow was taken into account and given a probability of failure on demand of -1 or 10%. Accounting for this in the LOPA as an IPL will place certain requirements on the BPCS interlock including regular testing. The BPCS interlock, in this case, would also be used as a dormant control, only activating if challenged. The efficacy of BPCSs for this type of control may be questionable. For these reasons, the analyst might consider removing this IPL from the LOPA.

Other than these factors, this LOPA is identical to the examples above. Even with the removal of the BPCS interlock from consideration, the results of the LOPA indicate that a SIL 2 SIF with a risk reduction factor of 100 which stops the flow of steam to the reboiler when high base pressure is detected will adequately protect against this event.

Figure 4: IPS to address required risk reduction for example column

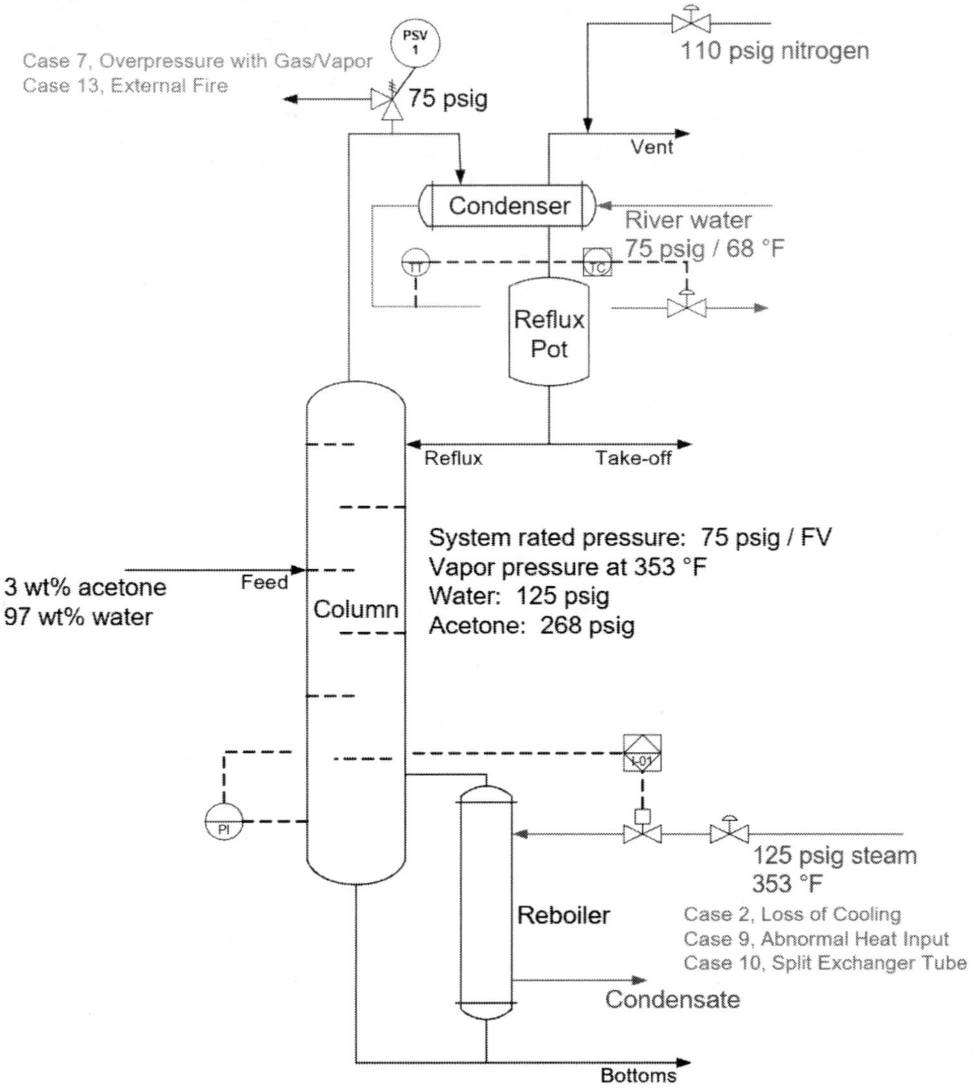

8. Conclusions

Distillation systems are one of the most demanding unit operations from an overpressure protection viewpoint due to the large amounts of energy being exchanged. Traditionally, these systems have been protected by relief systems to address the overpressure potential as indicated by RAGAGEP and the availability of technology. In certain cases, novel applications of relief systems can serve to reduce the overall cost and risk of the overpressure protection for distillation systems. RAGAGEP has been evolving with the advent of modern instrumented protective systems standards and changes in the applicable codes by ASME and standards from API. With these changes to the codes and standards, IPS can be used to address certain relief cases. Use of IPS to address relief scenarios prevents the release of process material through a relief system in the event of an overpressure scenario. Layer of protection analysis can be used to determine the appropriate reliability of the IPS to address overpressure scenarios.

9. References

1. Center for Chemical Process Safety (CCPS), Layer of Protection Analysis: Simplified Process Risk Assessment (CCPS, 2001)
2. American Society of Mechanical Engineers, Boiler and Pressure Vessel Code (BPVC), Section VIII, Division 1
3. ISO 23251, Petroleum, petrochemical, and natural gas industries – Pressure-relieving and Depressuring Systems
4. National Fire Protection Association, NFPA 15 : Standard for Water Spray Fixed Systems for Fire Protection

Using the ISA 84 / HAZOP / LOPA Procedure to Design a Safety Instrumented System (SIS) For a Fumed Silica Burner

Bruce K. Vaughen
Cabot Corporation, 700 E US Highway 36, Tuscola, IL 61953
(217) 253-5775
bruce_vaughen@cabot-corp.com

Jeffry O. Mudd
Cabot Corporation, 700 E US Highway 36, Tuscola, IL 61953
(217) 253-5718
jeffrey_mudd@cabot-corp.com

Bryan E. Pierce
Cabot Corporation, 700 E US Highway 36, Tuscola, IL 61953
(217) 253-5503
bryan_pierce@cabot-corp.com

Prepared for Presentation at
American Institute of Chemical Engineers
2010 Spring Meeting
6th Global Congress on Process Safety
San Antonio, Texas
March 22-24, 2010

UNPUBLISHED

Using the ISA 84 / HAZOP / LOPA Procedure to Design a Safety Instrumented System (SIS) For a Fumed Silica Burner

Bruce K. Vaughen
Cabot Corporation, 700 E US Highway 36, Tuscola, IL 61953
(217) 253-5775
bruce_vaughen@cabot-corp.com

Jeffry O. Mudd
Cabot Corporation, 700 E US Highway 36, Tuscola, IL 61953
(217) 253-5718
jeffrey_mudd@cabot-corp.com

Bryan E. Pierce
Cabot Corporation, 700 E US Highway 36, Tuscola, IL 61953
(217) 253-5503
bryan_pierce@cabot-corp.com

Abstract

A Cabot Team used the ISA 84-2004 (IEC61511-1) standard to evaluate the Safety Instrumented System (SIS) for an existing fumed silica burner and made design changes to reduce the risk. Years prior to this initiative, the layers of protection review used was subjective and resulted in assigning unreasonably high Safety Integrity Levels (SIL) on safety critical interlocks. The Team turned to the ISA 84 approved LOPA procedure and found it to be a more realistic and a less subjective risk evaluation that resulted in a practical, defensible process safety system design. The procedure steps followed were to:

1) Perform a HAZOP,
2) Use Process Hazard Analysis (PHA) software to generate LOPA spreadsheets,
3) Populate LOPA spreadsheets for each risk scenario and compare to the Cabot Risk Matrix,
4) Design an SIS/SIL to bridge the LOPA-identified risk gaps, and
5) Implement and document the changes.

After completing the PHA/LOPA analysis, two scenarios were identified where the process safety risk was marginal and required design improvements. The Team struggled on how to design proper SILs and, with the help of a certified consultant, completed Step 4. Both SIS design improvements were implemented in 2009, reducing the inherent control-related risks with a practical solution.

Keywords: SIS, safety instrumented system, SIL, safety integrity level, LOPA, layer of protection analysis, HAZOP, hazard and operability study, PHA, process hazard analysis

1. Introduction

The OSHA Process Safety Management (PSM) standard, CFR 29 1910.119 requires organizations to provide assurance of the mechanical integrity (the reliability) of all emergency shutdown systems and safety critical controls [1]. From a PSM perspective, the goal of the Safety Instrumented System (SIS) is to prevent a process-related hazardous event – to prevent fires, explosions, and/or toxic releases. This paper describes the procedure used to design an SIS for a Cabot Fumed Metal Oxide site's burner process. The SIS is designed to the U.S. ANSI/ISA-84.00.01-2004 and the international IEC6151 1 standards, which are recognized by OSHA as a means to comply with the PSM regulation [3, 4, 6].

The steps described in this paper to assure the mechanical integrity of the SIS include performing a HAZOP with software that automatically generates LOPA spreadsheets, populating the LOPA spreadsheets for each risk scenario, comparing these results to the Cabot Risk Matrix, and then designing an SIS/SIL to bridge any risk-related gaps. A recent report noted that combining ("integrating") a LOPA with the HAZOP Team effort helps save time and ensures decision-making objectivity [2]. The Team for this project overcame a "roadblock" when they identified the need for and subsequently located a Certified Functional Safety Expert (C.F.S.E.) who facilitated the final stages of the LOPA procedure and then helped with the SIL design. Once the C.F.S.E. resource completed his work, the Team bridged the LOPA-identified risk gaps and implemented appropriate SIS/SIL designs within a couple of months.

2. The ISA 84 / HAZOP / LOPA Procedure

This section provides an overview of the three major parts of the procedure used by the Team to successfully implement an appropriate SIS/SIL design. This methodology provided the Team with a procedure that prevents both under- and over-design of the engineering controls.

2.1 ISA 84 Overview

The Instrument Society of Automation (ISA) issued a methodology for designing a Safety Instrumented System (SIS) with its industry standard ANSI/ISA 84.01 in 1996 to promote compliance with the PSM regulation [3]. It was reissued in 2004 and is recognized by OSHA to meet its PSM mechanical integrity requirements as "recognized and generally accepted good engineering practices for SIS" [4]. Note that ISA 84 is essentially identical to IEC 61511-1, the specific international SIS design standard for the process industry, which is backed by the Health and Safety Executive (HSE) [5, 6].

An SIS is a system of Safety Instrumented Functions (SIFs) composed of sensors, logic solvers, and final elements for the purpose of taking the process to a safe state when predetermined conditions are violated. A schematic of an SIF is shown in Figure 1. Each SIF detects for out-of-limit (abnormal) conditions, bringing the process to a functionally safe state. Defining the Safety Integrity Levels (SILs) for each SIF defines the level of controls required to ensure detecting for and responding to potentially hazardous conditions.

Figure 1. Parts of a Safety Instrumented Function (SIF)

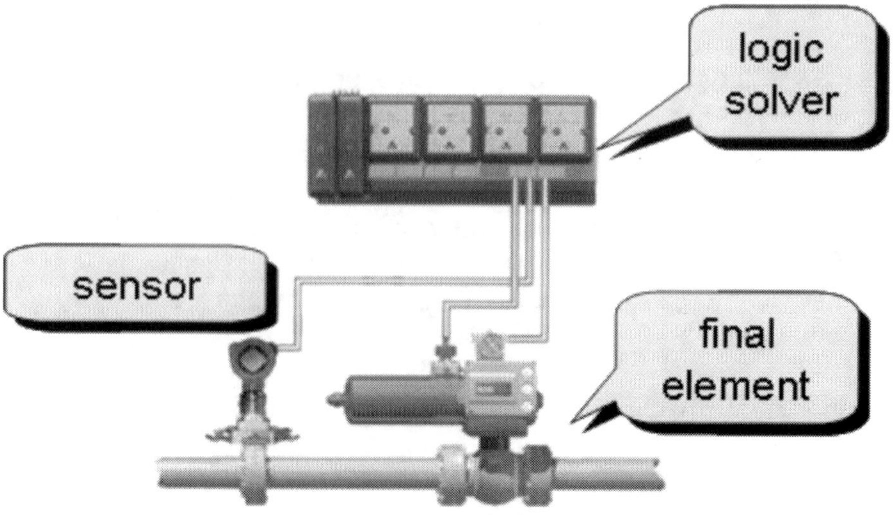

When an ISA 84 analysis is complete, it has defined the Distributed Control System (DCS) architecture, identified the safety critical instrument loops, defined the reliability and design of safety critical loops, and has defined the requirements to maintain these safety critical loops. The ISA 84 methodology addresses the complete safety lifecycle of the SIS, including: performing a hazard and risk assessment; evaluating the layers of protection; design, installation, operation, maintenance, and decommissioning.

The methodology also addresses

2.2 HAZOP and Tolerable Risk

Process Hazards Analyses (PHA) are designed to understand the process hazards, to determine appropriate hazard reduction safeguards, and to propose risk reduction strategies that bridge gaps identified between the existing or proposed process design and a process designed with the lowest tolerable risk. A HAZard and OPerability Study (HAZOP) is one of many tools that can be used in a PHA [7]. One goal of the ISA 84 / HAZOP / LOPA procedure is to answer the following question: "Is the Safety Instrumented System (SIS) for our process adequate?" In other words, to determine the reliability required for the safety instrument loop that reduces the risk to a tolerant level as defined by the company's risk matrix.

The HAZOP procedure used by this Team included the following: start with accurate and up-to-date Process and Instrument Diagrams (P&IDs) which show all Safety Instrumented Functions (SIFs); define the process nodes from the P&IDs; perform the HAZOP analysis to identify and document each process hazard and their associated safeguards (i.e., administrative, engineering, SIFs, etc.). The Team used commercial software to document the HAZOP, and as is noted below, this software automatically generated the LOPA spreadsheets for the tolerable risk gap analysis [8]. A part of the HAZOP table used by the Team is shown in Table 1.

Table 1. HAZOP Spreadsheet Example

Deviation	Cause	Consequence	Cat	S	L	Risk
High concentration of acid	AYT-742-4 fails closed or out of anticipated position	Possible emission violation	S_E	S3	L3 - Occasional	C

Since the success of the final SIS design hinges on defining acceptable or tolerable risk, the management must clearly define "tolerable risk." Without a clear understanding of the tolerable risk, the Team may under specify the SIS design and not reduce the risk sufficiently (and hence, be unsafe), or may over specify the SIS design which may add significant equipment and operational costs that don't add much additional benefit to the overall risk reduction effort. Over-design of a SIS can be characterized by adding multiple, redundant instruments that provide the "assurance" that the risk is sufficiently mitigated. Note that both the under- and over-design scenarios may be caused by not knowing the target's acceptable risk – by not defining the tolerable risk.

As is shown in Figure 2, the approved "tolerable risk" matrix is based on the fundamental definition of risk, as is represented by this equation: Risk = Likelihood x Severity. The goal is to reduce the risk by reducing the likelihood (frequency) with engineering or administrative controls and/or reducing the severity (consequences) with inherently safer process design. The engineered design of an SIS focuses on reducing the likelihood term associated with the process risk. With the definitions of the Likelihood Levels, shown in Table 2, and the Severity Levels, shown in Table 3, the Team determined the "Tolerable Risk" Level through the Cabot Risk Matrix (Figure 2) and the guidance in Table 4. Hence, if the final Risk Level generated through the HAZOP is an "A" or a "B," the Team must recommend a suitable SIS that will reduce the risk to a tolerable level. Conversely, an "E" Risk Level conclusion means that no additional risk reduction efforts are needed (the Team should spend its time on other efforts!).

2.3 LOPA Overview

The next step in this procedure used the HAZOP risk results with the Layers of Protection Analysis (LOPA) approach to quantify the Likelihood of every Severity level S-3 and S-4 event. The LOPA is a simplified risk assessment tool that is used to "evaluate the effectiveness of Independent Protection Layers (IPLs) that are designed to reduce the likelihood or severity of an undesirable event" [9]. The Team identified the IPLs, assigned a frequency to both the "Initiating Event" and the "Failure of Safety Device" (refer to Table 5), and then performed a LOPA on every S-3 and S-4 to quantify the Likelihood (see Figure 2). An example of an automatically-generated LOPA spreadsheet is shown in Table 6 [8]. The LOPA spreadsheet provided the Team with a red "+" field when additional IPLs are needed to reduce the Likelihood. As is seen in Table 6, the recommended IPL provides a green "=" LOPA result field, representing an acceptable Frequency. The Team used this LOPA spreadsheet to identify gaps between current SIF design and the tolerable risk.

Figure 2. The Cabot Risk Matrix

Severity (S)

Likelihood (L)

	S-1	S-2	S-3	S-4
L-5	D	C	B	A
L-4	E	D	C	B
L-3	E	E	D	C
L-2	E	E	E	D
L-1	E	E	E	E

Table 2. Definitions for Likelihood w/ Safeguards in Place

L-5	Once a year or more
L-4	Once in ten years or more
L-3	Once in the life of the process
L-2	Once in the life of 10 processes
L-1	Once in the life of 100 processes

Table 3. Definitions for Severity

S-4	Multiple worker fatalities	Public disability or fatality	Over $1M damage
S-3	Work disability of fatality Public injuries	Public injury (hospitalization); sufficient env. damage	$100K – $1M damage
S-2	Worker injuries (lost time)	Minor public injury (first aid)	Up to $100K damage
S-1	Minor worker Injuries	No public injury Exceed discharge limit	No significant loss

Table 4. Definitions for Tolerable Risk

A/B	Team is obligated to make recommendation (s)
C	Team is strongly encouraged to make recommendation (s)
D	Team is encouraged to make recommendation (s)
E	Team is discouraged from making recommendation (s)

Table 5. Definitions for the LOPA Likelihood

Type	Description	Frequency (event/yr)	PFD
Initiating Event	Small Fire	10^{-2}	
Initiating Event	Human Error	10^{-1}	
Safeguard	Relief Valve Failure		0.001
Safeguard	Typical Instr. Loop		0.01

2.4 SIL Design

The concept of Safety-Integrity Levels (SILs) involves ranking hazardous failures into safety-integrity groups, with specific requirements for each group based on the system's configuration and contributions of other risk reduction approaches [10]. The SIL is a measure of the system's safety-integrity, incorporating random hardware, systematic and human-factor related failures with a "Probability of Failure on Demand (PFD)" term. Based on the LOPA spreadsheet results, the Team determined the reliability required for safety critical loops. The guidance included receiving a maximum of 2 Credits for non-SIL rated SIFs that feed the same processor.

How to use Table 7 (SIL based on PFD). Once the SIL Level is determined, the Team located a SIL Designer, a C.F.S.E (Certified Functional Safety Expert) [11], to design an instrumented loop with the required SIL rating. The C.F.S.E. is an expert certified to design systems to ISA 84 standards and is certified by the TUV (Europe). The SIL Designer reviewed the HAZOP and the LOPA, designed all SILs identified in the LOPA, and documented the SRS (Safety Requirements Specification report).

An SRS document defines all the components of ISA 84, contains details on SIL design (Probability of Failure on Demand and Common Cause Failures) and addresses SIS maintenance and utilities. Once the SRS is developed, any qualified engineering group can design the SIS.

Table 6. Layer of Protection Analysis (LOPA) Spreadsheet Example

Scenario Results

Frequency (/yr)	Case	Results
1.00E-01	Unmitigated (without IPLs)	+
1.00E-03	Mitigated (with existing IPLs)	+
1.00E-04	Projected (with recommended IPLs)	=

Impact Event

Severity Category	Description	Comment
S4	Unreacted hydrogen, combustion air downstream of burner leading to explosion; fire, explosion hazard.	

Initiating Event

Frequency (/yr)		Description	Comment
1.00E-01		Burnback (Flame shift inside burner barrel) DUE TO Low/no combustion air flow DUE TO Flow control malfunction	DCS only

Independent Protection Layers (IPLs)

PFOD		E/H/L	Description	Comment
		E: Existing, H: HAZOP, L: LOPA		
1.00E-01		E	Self-checking flame eye with alarm and interlck to shut down burner fuels	
1.00E-01		E	Burnback thermocouple with alarm and interlock to shutoff burner fuels.	DCS input and PLC output (via serial link) - team believes that there is adequate independence from
1.00E-01		L03	Provide redundant flame-eye	

Table 7. Safety Integrity Levels (SIL)

Safety Integrity Level	Probability of Failure on Demand (PFD)
SIL 4	10-4 > PFD > 10-5
SIL 3	10-3 > PFD > 10-4
SIL 2	10-2 > PFD > 10-3
SIL 1	10-1 > PFD > 10-2

3. Conclusion and Implementation

After completing the LOPA analyses, the Team identified two scenarios with the process safety risk level of "C," requiring additional efforts at mitigation (refer to Table 4, "Team is strongly encouraged to make recommendations"). These two scenarios were: 1) the potential for un-reacted fuel through the silica burner, and 2) the potential for un-reacted fuel through calciner burners. The Team implemented the following SIS design improvements based on the results from the ISA 84 / HAZOP / LOPA procedure:

1) Silica Burner Design Improvement:

> A SIL 2 level design was implemented where the DCS outputs for burner fuels are hardwired through the flame-eye relay. This hardwire system meets the SIL 2 requirements and is much simpler than installing a SIL rated logic solver such as a DeltaV SIS.

2) Calciner Burner Design Improvement:

> The calciner was designed where the mechanical linkage for setting the air:fuel ratio was at the header level. Several scenarios were identified where if this linkage malfunctioned un-reacted fuel could accumulate in the firebox. The calciner design was modified where a linkage for the air:fuel ration was installed for each of the 10 calciner burners. Now, multiple linkage failures would have to fail for un-reacted fuel to accumulate. This modification eliminated the need for a SIL-rated SIF, saving equipment, operational and maintenance costs.

In addition, the Team discovered another advantage for using the ISA 84 procedure. The existing silica burner and the calciner burner flame-eyes did not meet the NFPA standard requirement for UL-rated logic solvers for burner management interlocks [12]. Both burner flame-eyes use a Honeywell flame relay with self checking flame-eyes, units that are not UL Listed for combustion safety service. However, since the NFPA recognizes the ISA 84 standard as an alternate means for complying with its Safety Instrumented Systems (SIS) requirements, completing this LOPA-based procedure met the NFPA SIS requirements without having to change the existing equipment, as well.

4. Summary

The Team improved the design of a fumed silica burner SIS by implementing two practical risk-based recommendations for reducing SIS control-related risks. The ISA 84 / HAZOP / LOPA safety evaluation procedure succeeded due to the Team's quality risk-based hazards evaluation and its resourcing of a Certified Functional Safety Expert for the final SIL design and review.

5. References

1. OSHA 29 CFR 1910.119(j) Process safety management of highly hazardous chemicals – Mechanical Integrity Section, http://www.osha.gov/pls/oshaweb/owadisp.show_document?p_table=STANDARDS&p_id=9760

2. D. Baum, N. Faulk, and P.E. John Perez, Improved Integration of LOPA with HAZOP Analyses, Process Safety Progress 28 (2009), 308-311.

3. The Instrumentation Society of Automation (ISA), American National Standard, ANSI/ISA-84.00.01-2004 Part 1 (IEC 61511-1 Mod), Functional Safety: Safety Instrumented Systems for the Process Industry Sector – Part 1: Framework, Definitions, System, Hardware and Software Requirements, Research Triangle Park, NC 2004.

4. OSHA Letter of Interpretation, Use of ANSI/ISA S84.00.01-2004 Parts 1-3 (IEC 61511 MOD) to comply with OSHA's Process Safety Management standard, 11/29/2005, http://www.osha.gov/pls/oshaweb/owadisp.show_document?p_table=INTERPRETATIONS&p_id=25164

5. International Electrotechnical Commission (IEC), IEC 61508-0, Functional safety of electrical/electronic/programmable electronic safety-related systems – Part 0: Functional safety and IEC 61508, http://www.iec.ch/zone/fsafety/fsafety_entry.htm

6. International Electrotechnical Commission (IEC), IEC 61511-1, Functional safety – Safety instrumented systems for the process industry sector – Part 1: Framework, definitions, system, hardware and software requirements, http://www.iec.ch/zone/fsafety/position.htm

7. Center for Chemical Process Safety (CCPS), Guidelines for Hazard Evaluation Procedures, 3rd Edition, American Institute of Chemical Engineers, New York, NY 2008.

8. ABS Consulting, Risk Consulting Division, *Hazard Review Leader* software , LEADER Software Team, Knoxville, TN 2009, http://www.absconsulting.com/leadersoftware/HRL.html

9. Center for Chemical Process Safety (CCPS), Layer of Protection Analysis: Simplified Process Risk Assessment, American Institute of Chemical Engineers, New York, NY 2001.

10. The Instrumentation Society of Automation (ISA), Safety Integrity Level Selection: Systematic Methods Including Layer of Protection Analysis, E. Marszal and E. Scharpf, editors, Research Triangle Park, NC 2002.

11. Mike Schmidt (C.F.S.E), Bluefield Process Safety LLC, (314) 420-9350, "Mike Schmidt" bluefieldsafety@gmail.com

12. National Fire Protection Association (NFPA), NFPA 86: Standard for Ovens and Furnaces, Quincy, MA, 2007, http://www.nfpa.org/aboutthecodes/AboutTheCodes.asp?DocNum=86

6TH GLOBAL CONGRESS ON PROCESS SAFETY

Implementing an Effective Conduct of Operations and Operational Discipline Program

James A. Klein
DuPont, North America Operations SHE
James.A.Klein@usa.dupont.com

William M. Bradshaw
ABSG Consulting, Inc., Knoxville, TN
WMBradshaw@absconsulting.com

Lee N. Vanden Heuvel
ABSG Consulting, Inc., Knoxville, TN
LNV@absconsulting.com

Donald K. Lorenzo
ABSG Consulting, Inc., Knoxville, TN
DLorenzo@absconsulting.com

Gregory Keeports
Center for Chemical Process Safety
gregkeep@comcast.net

Prepared for Presentation at
American Institute of Chemical Engineers
2010 Spring Meeting
6th Global Congress on Process Safety
San Antonio, Texas
March 22-24, 2010

UNPUBLISHED

Implementing an Effective Conduct of Operations and Operational Discipline Program

James A. Klein
DuPont, North America Operations SHE
James.A.Klein@usa.dupont.com

William M. Bradshaw, Lee N. Vanden Heuvel, and Donald K. Lorenzo
ABSG Consulting, Inc., Knoxville, TN

Gregory Keeports
Center for Chemical Process Safety

Keywords: Conduct of Operations, Operational Discipline, Process Safety, PSM

Abstract

This paper introduces the concepts of conduct of operations (COO) and operational discipline (OD), the attributes of effective COO/OD systems, and the steps an organization might take to improve its COO/OD programs, based on CCPS Project 0808. In general, the COO program encompasses the management systems developed to encourage performance of all tasks in a consistent, appropriate manner. OD is the deliberate and structured execution of the COO management systems by personnel throughout the organization.

1. Introduction

Excellent process safety management systems, practices, and guidelines require detailed understanding and documentation of process design and operating procedures, evaluation of process hazards, training and qualification of operating personnel, proper maintenance of equipment and facilities, and continued analysis of process performance through auditing and incident investigation [1,2,3]. However, these systems are only as effective as the day-to-day commitment of everyone in an organization to rigorously follow system requirements correctly every time. The failure of just one person in completing a job task correctly, just one time, can initiate or contribute to potentially catastrophic incidents. In fact, the design, implementation, and daily execution of process safety management systems are dependent on workers at all levels in the organization doing their job tasks correctly every time.

Despite the implementation of well-designed process safety management systems, some of which are centuries old [4], serious accidents still occur far too frequently. While continuous improvement of systems to account for new methodologies and identified problems is always necessary, increased focus on human factors, such as conduct of operations (COO) and operational discipline (OD), helps ensure that system requirements are consistently followed, laying the foundation for excellent safety, health, and environmental performance. According to some estimates, human error contributes to 90% of all accidents [5]. This number considers all possible sources of error, including front-line operating personnel, engineers, and supervision. Certainly, major accidents at Bhopal, Seveso, and Texas City, among others [6], highlight the importance of developing effective COO and OD programs for accident prevention and improved process safety management performance.

The ability to implement or improve effective COO/OD programs relies both on the commitment of leadership and the strength of the safety culture in supporting and sustaining changes over the long term. What are the common elements of effective COO and OD systems? How can current company performance in these areas be evaluated? How can effective programs be implemented? These topics are discussed in the following sections.

2. Background

2.1 What Are COO and OD?

The terms COO and OD have often been used interchangeably to describe programs intended to influence workers to rigorously follow policies, standards, procedures, and practices, which helps improve performance while preventing serious injuries and accidents. Certainly, military programs have long emphasized conduct of operations for weapons development, manufacturing, handling, and storage. Nuclear energy programs have similarly promoted COO. The Department of Energy [7] first published *Conduct of Operations Requirements for DOE Facilities* in 1990, and the International Atomic Energy Agency [8] published *Conduct of Operations at Nuclear Power Plants* in 2006. A checklist for evaluating COO programs has also been published [9].

In the chemical industry, OD has been emphasized for many years. DuPont, which was founded in 1802 for the manufacture of gunpowder, has always stressed OD as an important piece of the company's safety culture and programs. Reflecting on the hazards of the gunpowder business in 1808, E. I. du Pont wrote [10]:

> The safety of our family, the safety of the farmers who live in our neighborhood, has imposed upon us the absolute duty of making choice of steady, sober men and of establishing the most rigid discipline among our workmen.

DuPont established formal OD programs as part of its process safety management standards in the 1980s. It was based on 11 characteristics [11,12], such as leadership by example, sufficient and capable resources, and no shortcuts. In 1990, Dow Chemical described its "14 Steps of Operational Discipline," which included defining control strategies and systems, documenting detailed operating procedures, and specifying remedial actions for problems that can be anticipated [13]. Many other chemical companies similarly stress the importance of COO/OD in their process safety management programs

CCPS included COO as an element of Risk-Based Process Safety (RBPS) in 2007 [3]. The *RBPS Book* defined COO as:

> Conduct of operations is the execution of operational and management tasks in a deliberate and structured manner. It is also sometimes called "operational discipline" or "formality of operations," and it is closely tied to an organization's culture. Conduct of operations institutionalizes the pursuit of excellence in the performance of every task and minimizes variations in performance. Workers at every level are expected to perform their duties with alertness, due thought, full knowledge, sound judgment, and a proper sense of pride and accountability.

CCPS Project 0808 is currently developing a book on COO/OD. In this project, and also in this paper, we have chosen to differentiate COO and OD as follows:

> **Conduct of Operations (COO):** The company values and supporting management systems that are developed, implemented, and maintained in an effort (1) to perform operational tasks in a deliberate and structured manner consistent with underpinning risk assessments, (2) to ensure every task is correctly performed, and (3) to minimize variations in performance.
>
> **Operational Discipline (OD):** The personal commitment of individuals to carry out tasks correctly and safely every time, but in a thoughtful manner that recognizes unanticipated situations and encourages the involvement of wider expertise to ensure personal and process safety.

COO activities, along with other process safety management systems, specify how tasks should be completed safely and correctly; OD focuses on the performance of individuals consistently conforming to standards, policies, procedures, and rules. It is important to note, though, that OD encourages "thinking compliance." Thinking compliance is not a euphemism for workers "winging it." Rather, it is an expectation, given experience and knowledge, that all workers will be aware of an unexpected situation and recognize when existing rules do not apply or standard procedures will not work. This expectation includes that work will stop or a process will be brought to a safe state pending further evaluation. These requirements do not apply only to front-line workers – effective COO/OD programs are intended for all levels of an organization. Individuals may have different tasks, but these tasks must be completed correctly and safely every time:

> Every accident is due to human error: someone, usually a manager, has to decide what to do; someone, usually a designer, has to decide how to do it; someone, usually an operator, has to do it. All of them can make errors but the operator is at the end of the chain and often gets all the blame. We should consider the people who have opportunities to prevent accidents by changing objectives and methods as well as those who actually carry out operations [5].

Consider, for example, an operator who did not use a checklist and initial it as each step was performed, but later completed it to provide the required documentation. While this appears to be a clear case of an operational discipline violation because the procedures were not followed, more detailed analysis may indicate a number of other problems. Does site management emphasize and enforce that procedures must always be followed, or is this shortcut behavior tolerated or even encouraged as long as deviations don't lead to incidents? Did operator training emphasize proper use of the checklist or that it must be "turned in?" How useful and up-to-date was the checklist? When was the procedure last reviewed and last performed by the operator? Could the process have been designed in a way to minimize the hazards, eliminating the need for a checklist? Further analysis may indicate several issues related to COO and OD as well as other process safety management systems involving the operation, supervision, and engineering design.

2.2 What Are the Benefits of a COO/OD Program?

The emphasis in an effective COO/OD program is on completing all tasks correctly and safely each time, regardless of the role of individuals in the organization. Certainly some tasks have more safety impact than others, but the benefits of an effective program include:

- Equipment is properly designed
- Equipment is properly operated
- Equipment is properly maintained
- Management systems are properly executed
- Errors and deviations are consistently addressed

Therefore, in addition to improving safety, COO/OD programs should improve overall operational and business performance through higher productivity, higher quality, reduced waste, and lower costs. The emphasis of COO/OD programs on doing tasks right the first time is fundamental for any company that hopes to remain competitive or to achieve world-class manufacturing operations.

COO/OD programs help ensure that well-designed safety systems are successfully implemented and that they translate into consistent, desired, day-to-day results. The Safety Triangle, as shown in Figure 1, shows that significant consequences, such as catastrophic incidents or serious injuries, are often the result of, or indicated by, a larger number of smaller, undesirable, or unsafe acts or behaviors. A focus on minimizing or eliminating problems at the bottom of the triangle helps to prevent the more serious events at the top. Thus, an effective COO/OD program works to reduce the number of unsafe acts and undesirable behaviors in the organization, thereby preventing more serious incidents and injuries from occurring.

Another useful way to consider the impact of effective COO/OD programs on safety is to qualitatively modify the risk equation as follows:

$$\text{Risk} = f\,\frac{(\text{Frequency})\ (\text{Consequences})}{(\text{Operational Discipline})}$$

Risk is normally expressed as a function of the frequency of a possible event, often determined by the effectiveness of process safety management systems and safeguards, and the potential consequences of the event, characterized by the inherent substance and process hazards. OD, expressed in fractional form, has been added to the denominator of this equation, suggesting that the actual risk may be higher than the normally calculated risk if OD problems are present. As OD performance increases, meaning that the process safety management systems and safeguards

Figure 1: Safety Triangle Showing Focus of OD Program (adapted from [14])

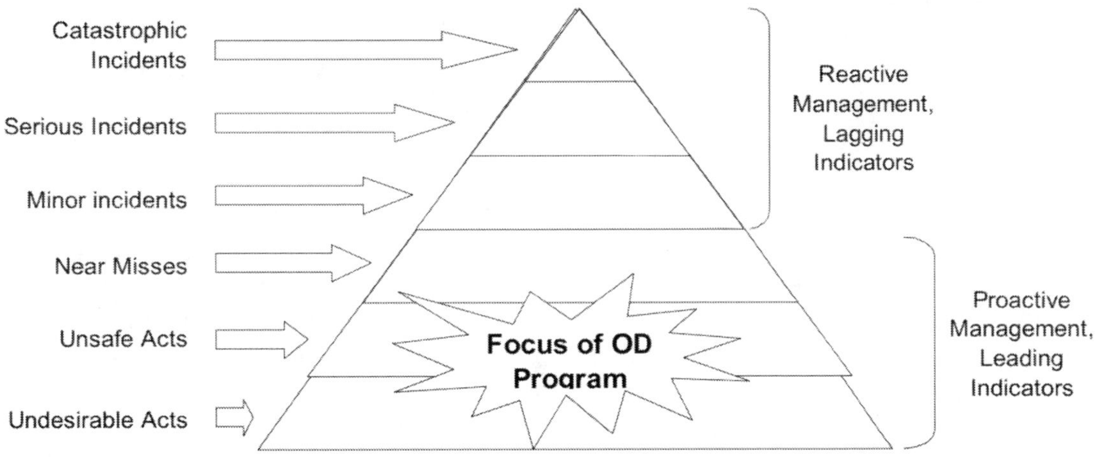

are well maintained, then risk decreases. If OD is not implemented, maintained, or followed rigorously, then overall risk increases.

This is shown in Figure 2, where poor OD can increase the potential frequency of an event, the potential consequences, or both, increasing the risk from level R_1 to a higher level. The resulting higher risk can lead to greater likelihood of serious incidents and injuries than might be considered by process hazard analysis teams evaluating the hazards of a process. An example of this is the Bhopal incident [6] in 1984, where multiple process safeguards designed to prevent or mitigate a large release of methyl isocyanate failed to work or were unavailable. Thousands of people died in the surrounding community due to leadership gaps and OD failures at the facility.

Figure 2: Poor OD Can Increase Risk (Typical Qualitative Risk Matrix)

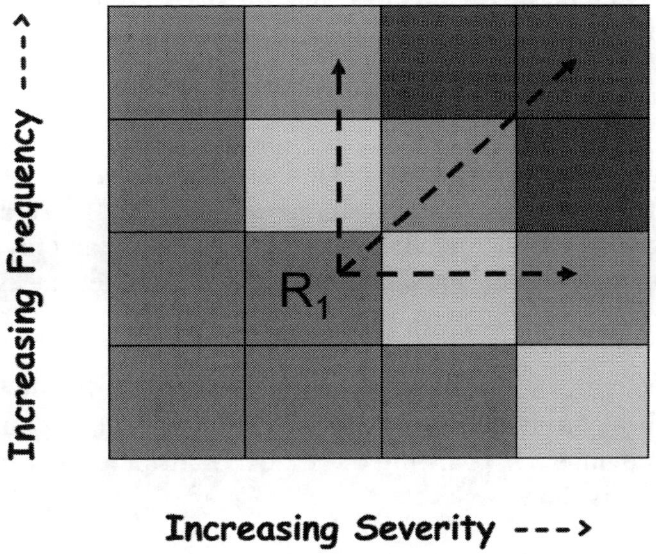

3. Key Attributes

3.1 Conduct of Operations

COO is a collective term for the management systems that encourage consistent, appropriate, and reliable performance. At its core, a COO system seeks to positively influence what people do or avoid. COO helps establish the conditions necessary to achieve highly reliable performance, which is reflected in predictable, consistent, and proper actions by members of the organization, capable and stable processes, and reliable plant equipment. Therefore, it is a set of attributes that helps ensure people make the right choices, at the right times, and that the organization operates in a safe, predictable, and reliable manner. The COO system starts by identifying the standards (including policies, procedures, and practices) that are most critical and helps ensure compliance with those standards. For example, there would likely be several safeguards to help prevent a single human error from purchasing and installing critical process equipment that does not conform to specifications, whereas there would likely be no specified means to ensure that non-critical items are not fit for their desired purpose.

COO is behavior-oriented, and organizations that strive to ensure that activities are conducted in a predetermined manner already have some level of a COO system. The effectiveness of a COO system can be measured in terms of overall organizational reliability. For example, "How well do we follow specified procedures and does following our procedures achieve the intended result?" In fact, an effective COO system cannot coexist with unreliable implementation of a process safety management system. Likewise, it cannot coexist with an unreliable quality management system, or any other similar system. The mere fact that these systems are unreliable is an indication of an ineffective COO system.

There are limitless opportunities to improve COO. However, some COO attributes are likely to provide significantly greater benefits than others. For example, a facility may already have a very sound program that addresses a particular attribute, or an attribute may not apply at all. Thus, organizations considering implementing or improving a COO system should critically evaluate each attribute to determine if it is likely to provide value for their facility, and judge whether the expected value justifies the effort that might be required to implement a new system or improve an existing system.

Key activities of effective COO programs generally include:

- Understanding risk significance and paying attention to what matters
- Establishing standards that support the organization's mission and goals
- Understanding what can be directly controlled, and what can only be influenced
- Providing required resources and time necessary to complete tasks within standards
- Ensuring organizational competency

The result of these activities is the development of a number of COO practices that positively influence individual behavior in the organization to follow system requirements and meet system goals, ultimately improving OD. A listing of key COO attributes that form the foundation of an effective COO program is shown in Table 1.

Table 1: Key Attributes of Effective COO Programs (adapted from [3])

People	1. Clear authority/accountability 2. Communications 3. Logs and records 4. Training and skill maintenance 5. Learning, evaluation, and critique 6. Thinking compliance 7. Safe/productive work environment 8. Aids to operation – the visible plant 9. Intolerance for deviations 10. Task verification 11. Supervision/support 12. Occupancy/access control 13. Routines 14. Worker fatigue/fitness for duty
Process	1. Process capability 2. Safe operating limits 3. Limiting conditions for operation
Plant	1. Asset ownership/control of equipment 2. Equipment monitoring 3. Condition verification 4. Management of subtle changes 5. Control of maintenance work 6. Maintain capability 7. Control intentional impairments
Management Systems	1. Supporting programs • Sound management systems • Leadership • Human factors • Operational discipline 2. Related programs • Standards of conduct • Evaluation/performance assurance 3. Necessary conditions • Hazard evaluation • Safe work practices • Management of change • Planning for/responding to emergencies • Audits, inspections, and critiques

3.2 *Operational Discipline*

OD is the personal commitment of individuals to carry out all tasks correctly and safely every time, but in a thoughtful manner that recognizes unanticipated situations while encouraging the involvement of wider expertise to ensure personal and process safety. An effective OD system leads to very predictable behavior and actions that closely conform to the actions that have been prescribed. OD complements COO. While an effective COO program provides structure and systems, an effective OD program helps deliver reliable performance. The OD program is focused on group and individual attributes that help influence and determine behavior. In short, it is focused on how people work, and the discipline they apply to their work.

An effective OD program demands that in all but the most extraordinary situations, members of the organization work within system requirements. Moreover, OD programs provide workers with the knowledge and skills necessary to determine when the established system practices are not relevant to the immediate situation as well a means for recognizing when additional review and authorization may be needed before proceeding. The goal of an effective OD program is not to create human robots that perform tasks the same way every time, regardless of the outcome. Such an organization would not last long; it would be unable to cope with external and internal changes and be quickly overtaken by more creative competitors. Instead, an effective OD program promotes, some would say demands, that the organization continuously learn and improve, but in a disciplined manner. For example, improvements are based on sound engineering principles, not merely on the fact that "it worked out OK last time" or "it only shortens the prescribed heat-up cycle by 10%, and surely the designers allowed for that much variance." If there is a better way to conduct a task or procedure, even a shortcut, then it must be reviewed and authorized, with training provided for everyone to follow in the same way to minimize the potential for unexpected consequences.

No program will completely eliminate human error, but the likelihood of error can be greatly reduced. The number of errors can be decreased by training workers thoroughly, managing worker stress and fatigue, providing logical human-machine interfaces, and so forth. In addition to these measures, an effective OD system should measurably reduce the likelihood that workers will accidently or intentionally violate policies, procedures, and practices. An effective OD system will also produce an organization that is intolerant of intentional violations, regardless of the intent or outcome.

OD programs focus on both the organization and the individual. The first are OD principles that apply to the organizational responsibilities of workers: what the organization expects of its personnel and the standards it sets for the work environment. The second are OD principles that apply to individuals completing their own work tasks: attributes that shape behavior and help determine what workers do, or fail to do, on a day-in, day-out basis. A summary of key attributes is provided in Table 2.

Table 2: Key Attributes of Effective OD Programs(adapted from [12])

Organizational	1. Leadership 2. Team building and employee involvement 3. Intolerance of deviations 4. Maintenance of a safe, productive, work environment
Individual	1. Knowledge 2. Commitment 3. Awareness

4. Implementing an Effective COO/OD Program

4.1 Leadership Commitment

Leadership commitment is the first step toward implementing an effective COO/OD system. However, unlike some management initiatives, this one cannot simply be assigned a budget and delegated to others for execution. Truly effective COO/OD programs start at the top with upper management personally demonstrating the behaviors they want others to emulate and also making the organizational commitment to initiate and maintain effective COO/OD systems. The purpose of any COO/OD system is to reliably accomplish the mission of the organization within the framework of its core values. Thus upper management's most fundamental responsibilities are to (1) identify values consistent with the organization's mission and (2) to translate them into business principles, establish policies consistent with those principles, communicate them throughout the organization, and consistently uphold business standards and practices based upon them [15,16].

Implementing a COO/OD program is a pragmatic way for management to ensure that the organization's values are embedded in its daily activities and supported by the company's safety culture. Even if good management systems already exist, OD is essential for good performance (following the procedures that comprise the good system). Improving OD impacts all aspects of the business (quality, reliability, profitability, reputation, occupational safety, environmental impact, etc.) as well as safety. Key activities for implementing or improving a COO/OD program that require the attention of both upper and lower levels of management include:

- Defining expectations and acceptable performance
- Consistently enforcing expectations
- Verifying implementation status and progress
- Monitoring performance data
- Sustaining performance

Companies everywhere deliver products and service to their customers using a multi-cultural workforce, even if they are not global companies. When implementing COO/OD programs, it is a fundamental error to believe what works in one location will work everywhere. While the organization should have one guiding set of values and principles, the implementation of COO/OD must be tailored to the reality of specific businesses, locations, and safety cultures. For example, some sites may have workers in collective bargaining units with contract requirements that are different, or irrelevant, at other sites. A "world-class" COO/OD system is not simply an exemplary model superimposed on a global workforce without regard to local differences.

Effective implementation of COO/OD requires strong leadership with cultural sensitivity and a willingness to adapt the policy to achieve the intended results despite regional or cultural differences. Despite these differences, the fundamental steps are the same. Management must select people who understand the local culture and who can work productively with people who may be different from themselves. These leaders then provide the COO/OD "vision" for the workforce, define expectations, and model the expected behaviors. The organization's COO/OD policy statement sets the baseline for global workforce practices, but the vision should reinforce the relevance of the local culture. These leaders can also confront the reality of the needs of people from the local culture and define acceptable deviations from the global standards.

Achievable goals must be established, and progress toward those goals must be monitored. Effective leaders know how to motivate their people to achieve those goals and know how to use the initial successes to entice others to embrace the principles of COO/OD. Thus, the detailed system implementation should be tailored for each region and operated, more or less, on a stand-alone basis. However, relevant performance-monitoring data should be transferred to a central reporting database for the whole organization. The key is that while approaches and tools used for achieving high levels of COO/OD in an organization may vary, the goal of high COO/OD performance is consistent everywhere. It is also essential that consideration be given to the ability of the corporate and local safety culture to support and sustain changes being made for COO/OD programs. Communicating with and involving all levels of the organization in planning, implementing, evaluating, and improving the COO/OD program will help ensure its success.

4.2 Getting Started

In addition to leadership commitment and vision, it is also important, of course, to recognize your starting point. As shown in Figure 3, this will vary depending on the current programs that may have been implemented, if any. It may also vary for different locations in a large company or even different operating units at a large location. Overall, the cycle for continuous improvement of COO/OD is the same, but the specific goals and action steps for getting started may vary.

To help determine the starting point, both leading and lagging indicators can be used to evaluate current performance. In particular, incident and audit data can be analyzed to determine if COO/OD issues have already been identified. OD failures contributing to incidents are usually evident if incident investigations have considered human error as possible root causes. If they are, and they indicate significant variability in performance, OD is likely an issue. If human performance is not routinely and effectively considered, COO issues likely exist. Similarly, audit findings showing overdue or incomplete system implementation or performance can indicate COO/OD issues. A number of other metrics can be used to provide insight on the current COO/OD status in the company or at a specific site, such as:

- Number of overdue hazard evaluations
- Number of overdue procedure revisions
- Number of scheduled training sessions that have been missed
- Number of missed routine audits or inspections
- Number of overdue or extended preventive maintenance tests
- Number of disciplinary actions
- Higher levels of defects or quality problems
- Lower process productivity

Once the current situation has been evaluated, specific implementation planning for new or improved COO/OD programs can begin.

Figure 3: Getting Starting on Implementing an Effective COO/OD Program

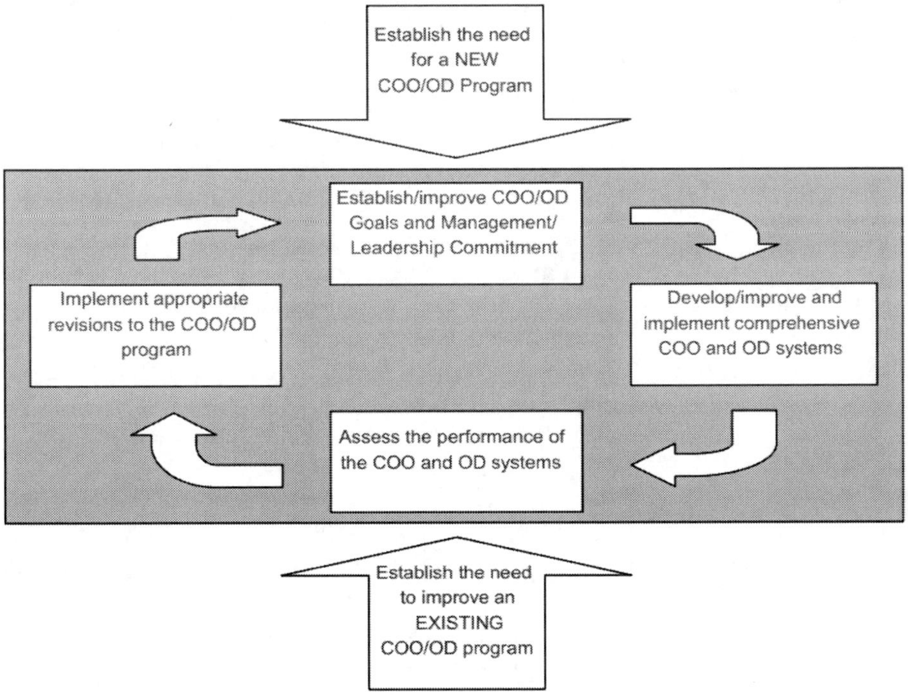

4.3 Implementing and Maintaining a COO/OD Program

A good approach for implementing, improving, and maintaining an effective COO/OD program is based on the Plan-Do-Check-Adjust (PDCA) process, as shown in Table 3. Based on the starting point for the company or location, appropriate steps following the PDCA steps can be developed. What are the goals and initial milestones for the program? What benefits can be expected or are desired? What specific action steps need to be taken to both develop the expertise to initially implement the program and then to sustain the improvements that have been achieved? What barriers exist to successful implementation? Who needs to be involved for the program to be successful? Once the program has begun, data must be collected to evaluate the impact of the program and to allow for adjustments to be made to continuously improve performance.

4.3.1 Develop the Plan

The first task is to establish expectations. What are the ultimate goals? What is the vision of success? What realistically can the culture support? Who should be involved to help develop, implement, and support the plan? Once the goals are understood, managers must then develop action plans that they believe will achieve the organization's goals, considering the reality of the current situation. The organization must develop a strategic plan with realistic, achievable goals, set targets and timelines, develop measurable performance standards, and consider how all levels of the organization can best be involved. Some specific tasks involved at this stage include:

- Considering the benefits
- Setting consistent performance expectations
- Focusing on management leadership and commitment
- Providing and developing required resources throughout the organization
- Focusing on long-term sustainability and consistency
- Setting a few milestones and planning to achieve them

Table 3: PDCA Process Applied to COO/OD Implementation

PLAN: Analyze the situation and develop plans	Set a measurable milestone toward the goal for the COO/OD effort Identify the processes impacted by COO/OD Select where to apply COO/OD List the steps in each process as it currently exists Map each process Identify issues related to COO/OD implementation Collect data on the current process Generate implementation plans Gain approval and support
DO: Implement the plan	Implement the chosen solution on a trial or pilot basis (first pass through the PDCA cycle) Implement the change throughout the organization (subsequent passes through the PDCA cycle)
CHECK: Evaluate the results	Gather data on the modified system results Analyze the results data
Was the desired goal achieved?	If YES, skip the Adjust step, revise the goal to the next milestone, update the plan, and repeat the PDCA cycle If NO, proceed to the Adjust step, modify the implementation plan and repeat the cycle
ADJUST: Standardize the implementation (and continually improve)	Identify systemic changes and training needs for full implementation Plan ongoing monitoring of the COO/OD system Continue to look for incremental improvements to COO/OD

4.3.2 Implement the Plan

The second step in the PDCA cycle is the most difficult because it truly tests whether management has the courage of its convictions. The time for discussion is past – management must press forward to do what it deemed necessary to achieve the organization's goals. Change inherently requires a leap of faith that the long-term results will be worth the near-term costs and risks. Consistent and persistent management leadership will be required to convince others of the necessity and benefits of change. Management must effectively communicate new performance expectations and manage the changes in process flow, job function, tasks, and activities that will be required to meet the new performance standards. Management must also be prepared to adapt the implementation plan to site-specific realities without compromising the core values.

Once new performance standards have been established, management must enforce them. Issues that occur during the transition to new performance standards normally arise because of poor communication between supervisors and employees. When there is confusion regarding performance standards and employees ask questions about them, the supervisor should address these concerns quickly. By promptly and accurately resolving questions, supervisors earn the trust and respect of their employees and eliminate issues before they become major problems. Workers at all levels need to be involved in supporting the COO/OD program and ideally to provide honest feedback to management about changes that are being made. Some specific tasks involved at this stage include:

- Communicating performance standards
- Involving the organization and workers
- Implementing and enforcing performance standards
- Adapting approach to site-specific conditions

4.3.2 Check Progress

The Check element of the PDCA cycle provides the objective measures by which progress can be monitored and judged. Thus, it is essential that the plan identify such measures and that the supporting data be collected. Inspections, audits, management reviews, investigations, and self-assessments are all means by which data can be collected. Some metrics, such as the number of late/missed inspections, may be collected directly; others may result from other management systems, such as incident investigations and audits. Employee feedback also provides useful information on how well COO/OD implementation is proceeding and how additional improvements can be made.

A robust information system infrastructure focused on quality and safety is essential to the Check step. Systems that support that framework include digital process archives, activity tracking boards, checklists, a dashboard of metrics on process operations and safety, computerized maintenance and inspection records, and data mining for regular reports of progress on key safety and quality indicators. To be useful these processes must ensure data accuracy and timeliness. Management can also advance the COO/OD system by being transparent when reporting metrics on system performance. Everyone should be able to see the organization's areas of growth and success as well as those areas that need more attention and improvement. Sharing this information is important in transforming the organization, and it will improve workers' ability to understand how COO/OD impacts every aspect of the business. Some specific tasks involved at this stage include:

- Using appropriate metrics
- Using audit and incident investigation results
- Conducting self-assessments either formally via survey or informally via observation

4.3.3 Periodically Adjust the Plan

The fourth step in the PDCA cycle is to analyze what is or is not working and make corrections or other necessary changes to the plan with respect to current goals. If the plan is meeting current goals, then the goal should be advanced to the next milestone in the quest for continuous improvement. If goals are not being met, then additional changes can be made to the implementation as needed to promote further progress toward the goal. In some cases, progress at some sites may be excellent, and progress at other sites may be slow. Adjustments to the plan can be made as appropriate to support and sustain both overall and location-specific results. Some specific tasks involved at this stage include:

- Scheduling periodic management reviews
- Evaluating current status and potential gaps
- Adjusting implementation and improvement plans as needed

5. Conclusion

Excellent safety systems have been implemented at many companies with the goal of reduced injuries and accidents. The effectiveness of these systems is dependent on the implementation and support of COO/OD programs to ensure that system requirements are rigorously followed every day to achieve desired performance levels. COO activities along with other process safety management systems are intended to specify how tasks should be completed safely and correctly, while OD involves the actual performance of individuals in following these specifications, procedures, or rules each time when conducting job tasks. Effective COO/OD programs contribute to overall operational and business performance and help reduce risk and improve safety.

Attributes of effective COO/OD programs have been identified, but one approach may not work everywhere because of differences in company practices and local customs. Rather, COO/OD attributes should be considered a collection of different methodologies or targets that can be used as needed at different locations or at different times to drive continuous improvement. Implementation of effective COO/OD programs involves upper management commitment, evaluation of current status, and a repeating PDCA cycle to introduce and sustain performance improvements.

References

[1] U.S. Occupational Safety and Health Administration standard 29 CFR 1910.119, 1992

[2] Center for Chemical Process Safety, *Guidelines for Implementing Process Safety Management Systems*, AIChE, 1993

[3] Center for Chemical Process Safety, *Guidelines for Risk Based Process Safety*, John Wiley & Sons, 2007

[4] Klein, James A., "Two Centuries of Process Safety at DuPont," *Process Safety Progress*, Vol. 28, No. 2, 2009

[5] Kletz, Trevor, *An Engineer's View of Human Error,* 3rd Ed., IChemE, Rugby, UK, 2001

[6] Atherton, John and Frederic Gil, *Incidents That Define Process Safety*, Center for Chemical Process Safety, John Wiley & Sons, 2008

[7] US Dept. of Energy, *Conduct of Operations Requirements for DOE Facilities*, DOE 5480.19, 7-9-09 (revised 10-23-01)

[8] IAEA, *Conduct of Operations at Nuclear Power Plants*, Draft Safety Guide DS347, 2006

[9] Haas, Paul and Guy Hager, "Conduct of Operations: A Control System for Your Most Important Safety Component," *Process Safety Progress*, Vol. 19, No. 1, 2000

[10] du Pont, E. I., letter to his father, 1808 (source: Hagley Museum & Library, Wilmington, DE)

[11] Klein, James A., "Operational Discipline in the Workplace," *Process Safety Progress*, Vol. 24, No. 4, p. 228-35, 2005

[12] Klein, James A. and Bruce K. Vaughen, "A Revised Model for Operational Discipline," *Process Safety Progress*, Vol. 27, No. 1, 2008

[13] Trask, Murray N., "Operating Discipline," *Plant/Operations Progress*, Vol. 9, No. 3, 1990

[14] Heinrich, H. W., *Industrial Accident Prevention: A Scientific Approach*, McGraw-Hill, 1931

[15] Collins, Jim, *Good to Great*, HarperCollins, 2001

[16] Kotter, John P., *Leading Change*, Harvard Business School Press, 1996

Process Safety in the Face of Significant Change:
Commitment to Consciousness of Consequences

Leslie May, Ph.D.
The Sinclair Group
1400 Woodloch Forest Drive
Suite 500
The Woodlands, Texas 77380
lmay@sinclairgroup.com

Don Pomraning
The Sinclair Group
1400 Woodloch Forest Drive
Suite 500
The Woodlands, Texas 77380
dpomraning@sinclairgroup.com

Prepared for Presentation at
American Institute of Chemical Engineers
2010 Spring Meeting
6th Global Congress on Process Safety
San Antonio, Texas
March 22-24, 2010

UNPUBLISHED

Process Safety in the Face of Significant Change:
Commitment to Consciousness of Consequences

Leslie May, Ph.D.
The Sinclair Group
1400 Woodloch Forest Drive
Suite 500
The Woodlands, Texas 77380
lmay@sinclairgroup.com

Don Pomraning
The Sinclair Group
1400 Woodloch Forest Drive
Suite 500
The Woodlands, Texas 77380
dpomraning@sinclairgroup.com

Keywords: behavior-based, process safety, nonconscious

Abstract

Development of an effective process safety culture requires employees to look beyond intended performance, to their own nonconscious assumptions about process safety. When firmly in place, effective process safety culture instills an environment where employees differentiate between events which, while detrimental to performance, have low probability of generating an incident, and those with seemingly minimal impact on performance, yet high probability of escalation into full-blown process safety incidents.

How does an organization instill personal accountability for process safety in the face of significant change? One major manufacturing firm needed to rapidly improve their competitive position, yet they worried about process safety performance. Shifting their management style to gain personal commitment, they engaged the organization in changing assumptions about how key process decisions were made, both technically and socially. They were surprised by the results: personal commitment led to personal accountability, which in turn, maintained process safety performance in a time of intense change, yet without constant management oversight.

To develop effective process safety cultures, underlying assumptions and strategies must be examined, and often recreated. The organization's preferred management style will be a potentially hidden source of strength, or weakness, in its success.

1. Introduction

Once there was a very large chemical company, one of the largest on the planet. Known for its attention to detail of both its chemical processes and work practices, its competitors considered it well above average in process safety practice and performance. One year, as Christmas approached, one of its facilities in a developing country was operating at only a third of its capacity due to lack of demand for the pesticides it produced. This facility developed a catastrophic leak, releasing a deadly gas; thousands of the surrounding populace were killed or injured by this gas (Alexander & Stoler, 1984).

The company investigated the incident and paid reparations to the country, the victims and the surrounding community. Beyond that individual company's response, the chemical industry as a whole responded by voluntarily developing policies, procedures and evaluations to ensure that such a tragedy never occurred again. As these words are written, the 25th anniversary of this event is upon us.

In addition to driving many of today's process safety regulations, policies, and procedures, this event catalyzed industry to explore the human factor in process safety events. As a result, behavior-based approaches to process safety were developed and implemented by many companies in their efforts to create and ensure safe working environments. Nonetheless, process safety incidents continue at all levels of tragedy.

Thus begins another tale, of another prominent company which recognized challenges to personal safety and so worked with its employees, particularly process operators, to develop better behavioral approaches to it. This company experienced significant improvements in its overall personal safety as evinced by its OSHA recordable rate. Beyond its effect on personal safety, this company's choices also generated a learning experience related to its process safety practices. An employee was overcome and ultimately killed by a gas release, and the resultant investigation revealed breakdowns not only in the process safety practices which should have prevented such a release, but also in the behaviors and attitudes related to individual employee protection against such a tragic loss of life (Isom & Cummings, 2005).

In both companies in our tale, leadership and employees were well-intended and active in their pursuit of process and personal safety. Yet these good intentions alone were not enough, any more than compliance with published policies and procedures were. To take process safety to the next level, organizations must inspire and challenge individual- and organizationally-generated attitudes about process safety, especially among those employees most often affected by serious incidents--those actively operating specific units.

Our definition of the "next generation" organizational approach to process safety is: to create an environment in which all stakeholders, including operators, technical contributors, staff, and leadership, are actively engaged in mutual learning. They share process-safety-related concerns beyond not only what they see in operations, but what they see changing in the world at large. The belief is that this type of approach will not only prevent known hazards, but help to anticipate and to respond to as-yet-unidentified risks -- to anticipate incidents with low frequency yet potentially huge impact on the health and safety of employees and / or the community.

This paper builds on current behavior-based process safety practices, suggesting the "next generation" is a commitment-based approach to process safety. Consider this definition of *behavior*-based process safety: an approach to process safety in which employees intervene with each other at all levels of the organizational hierarchy and with stakeholders to identify and modify behavior which could significantly affect the health and safety of employees and/or the community. To successfully achieve this vision, unconscious predispositions and nonconscious drivers of process-safety-related-behavior must be raised to consciousness and explored at all organizational levels.

2. The Human Factor

For purposes of this paper, consider the possibility that awareness occurs at three levels:

- Consciousness is the light-on state in which we are aware of our surroundings and aware that we are aware of them (Carter, 2002). With consciousness, we have the opportunity to intentionally articulate, model, and reinforce expected process safety behaviors. However, as human beings, much of what drives our behaviors and habits resides in our nonconscious and unconscious.

- The Nonconscious is unaware learning, memory or information processed outside of conscious awareness (Bargh, Gollwitzer, Lee-Chai, Barndollar, & Trotschel, 2001). The nonconscious is where automatic cognitive and emotional processing occurs after we have extensive experience with a particular task, situation, or type of information. These automatic mental and emotional processes free our limited conscious attentional capacity, but at a cost. Unfortunately, while automatic processing is occurring, our behaviors and habits are influenced by beliefs, values, and assumptions outside of conscious awareness, typically not taken into account. This automatic processing is further complicated by our unconscious predispositions.

- The Predisposed Unconscious includes genetically-defined cognitive, emotional, and behavioral predispositions. Residing as this level is our tendency to pay extra attention to negative information, to see the world through rose-colored glasses, to imagine what might have been, to avoid thinking certain thoughts, and to assume one's thoughts can influence the physical world in a manner not governed by the laws of physics.

Without our awareness and exploration, much of what we think, see, and feel is processed automatically, without consciousness. As a result, our reactions and behaviors are often not appropriate to the situation in which we find ourselves. Ultimately, this distorted filter often sabotages many of our best efforts to create and sustain effective commitment-based process safety environments.

3. The Typical Influence of the Nonconscious on Organizations

Recognizing that human beings are often influenced by internal processes of which we are unaware, behavioral-based safety programs are often developed through the lens of control and/or soft leadership and working together styles. Interestingly, as shared by our clients, these styles typically co-exist in most organizations.

3.1 Control

This model is particularly well known:

- The underlying assumption is that exclusive knowledge about chemical processes, business drivers, and economic constraints resides within leadership and senior staff.
- Interpersonal challenge (the need to define and contract challenge with conflict), especially between leadership and the workforce, but often within leadership or within the workforce, is typically viewed as nonproductive and is therefore avoided in most situations.
- Avoidance of negative consequences is a strategy used throughout the organization. Specific actions are chosen to avoid repercussions, discipline or loss of employment. In the extreme, fear becomes a significant driver of behavior. This may extend to leadership's fear of community outrage and government action if process safety events occur.
- For the community and government entities outside the chemical production area, negative consequences to be avoided include loss of economic viability, tax base and well-paying jobs if the company loses the right to operate due to process safety issues.
- Specific to process safety, the orientation is "control of the known" via compliance: policies, procedures, checklists, and other documentation. This orientation is inherently extrinsic to the individual employee and does not therefore consider nonconscious drivers of choices and behavior. As a result, this approach fails to tap into and transition these drivers into those which could better drive personal commitment to process safety.

3.2 Soft

Because of the limitations of the control model, management sometimes shifts to this model:

- The underlying assumption is that every employee is well intended and will therefore figure out how to achieve what needs to be accomplished.
- There is a tendency to promote individuals into leadership based on "niceness" without additional, requisite experience and skills already in place. In the extreme, this "niceness" may ultimately devolve into simply doing what upper leadership asks without consideration of its ultimate impact on the organization and business results.
- Often the tendency is to create an environment safe from the demanding realities of the "outside world", such as community and other public opinion, economic changes, shifts in cost structure, and other challenges of a highly competitive global marketplace.
- In the soft model, interpersonal challenge, especially between leadership and the workforce, but often within leadership or within the workforce, is typically viewed negatively as questioning good intent, and is therefore avoided in all but the most critical situations.
- In this orientation, individuals in the organization often assume they have a right to choose how they perform their work, even if their choices impact others and ultimately, business outcomes and results.

- Often in this orientation, there is a minimalist, compliance attitude toward structure and supporting systems such as information technology, human resources, training and employee development, and other organizational best practices.
- There tends to be little documentation of specific processes and procedures; these reside in individual employees.
- The tendency is to create a pleasant work environment based on the belief that "a happy workforce is a safe workforce."
- Related to the "happiness" assumption is a belief or assumption that "scaring" employees will result in stress-related mistakes, particularly in demanding workplace conditions. With this basic grounding in these two models, let's examine how they influence process safety, often resulting in unintended consequences which may actually promote incidents.

With this basic grounding in these two models, let's examine how they influence process safety, often resulting in unintended consequences which may actually promote incidents.

3.3 Influence of the Control Model on Process Safety

In many organizations, a control model drives employee process safety underground because of the difficulty in challenging leadership on specific concerns, observations, and issues. Surprising outcomes or near misses are often ignored. Employees may practice "vicious compliance", in contrast to authentic commitment to process safety, in which they follow management's policies, procedures and other mandates "to the letter", even in situations where other choices would make more sense. The workforce may act unilaterally based on their own assumptions, for example:

- "They talk a good game about process safety, but I know it's really about getting the work done. I'll operate to get the desired production rate, even if this pushes the equipment to its limit." Indeed, the oft-ignored employee assumption and resulting belief is that "production trumps process safety".

- Employees' reliance upon personal experience to evaluate risk of low-frequency events: "as an employee, my experience teaches me that I've worn my personal protective equipment 100 times and there has never been an incident. So if I don't wear it this one time, especially since I am so busy, it seems to me that the associated risk of an incident is quite low. So why does management constantly 'get on me' about wearing it? Aren't there more important issues facing us?"

- Routine behavior often includes silence or compliance, but with little challenge of issues or conversation about what could or might happen. As a result, management may perceive that "all is well" when in fact employees have significant concerns and issues.

3.4 Influence of the Soft Model on Process Safety

In the soft model, employee concerns may be expressed. While compliance structures exist, there is little commitment to adhering to these structures to drive action and follow-through. The soft model also discourages challenge in the belief that any conflict is inherently negative and will result in unhappy workers, so the focus may be on more frequent but less conflict-laden events, since there is no disciplined process to consider low-frequency, highly significant ones. Even more significantly, employee accountability and skill may suffer from an attitude of minimal structure, resulting in low expectations and scanty procedural detail. Consequently, the workforce may act unilaterally based on their own assumptions. For example:

- "I have process safety concerns. I may mention them in our next quarterly process safety meeting, but in the meantime I will operate in a more conservative fashion to alleviate my own concerns, without consulting others."
- "My experience guides me in my actions. I've never seen an incident in my 20 years here, so why should I wear my personal protective equipment when I go out into the process area?"
- Routine behavior may include conversations in the control room revolve around setting the shift schedule, not talking about potential process safety issues.

4. The Next Generation: Commitment-Based Process Safety

This model is unique, not an integration of the Control and Soft Models:

- Commitment-based process safety is built on an assumption of mutual learning, an expectation of discipline, and recognition of the value in structure and systems based on fairness and integrity.
- Acknowledgment of the critical need for specific knowledge and skills to deliver results. This leads to a commitment to not only training, but to setting performance expectations, coaching and follow-through to ensure that employees can perform using their knowledge and skills in potentially hazardous situations, both known and unknown. An expectation is set to anticipate possibilities, to be curious about unusual events, in order to advance knowledge of low-frequency events with potentially significant outcomes.
- An underlying assumption of practitioners of this model is that leadership and the workforce both have knowledge and skills to be shared in order to generate results safely, reliably and productively.
- The model assumes that leadership sets an expectation and asks for individual commitment to appropriate interpersonal challenge, especially between leadership and the workforce, but often within leadership or within the workforce. This environment generates in each employee the ability to transcend his or her natural human defenses to talk about potentially significant outcomes in a safe and open environment. Leadership builds trust through openness, clear expectations and the avoidance of blame. In return, employees are open to challenge each other and leadership with no fear, reprisal or other repercussion.

- A characteristic of this model is long-term, or "infinite game", thinking throughout the organization to consider how to manage potential changes, security threats and other low-frequency, high significance events.

- Specific to process safety, the orientation is "anticipation of the unknown" via both compliance with policies, procedures, checklists, and the process discipline to consider the potential for future issues.

- This model seeks to create an environment which provides for focus on the unknown through curiosity, the strength to challenge, mastery of knowledge and skills, achievement of results in a safe and reliable way, and the disciplined discovery and understanding of unknown threats to process safety.

- Beyond the immediate organization, the model encompasses attitudes toward the community and government entities outside the chemical production area. There is an expectation of openness about potential issues and a collaborative approach to their resolution. There is organizational clarity around the importance of process safety in light of economic viability, tax base and well-paying jobs if the company loses the right to operate due to process safety issues.

- Example Assumption: If the Control assumption is that "learning from past PHAs and RCAs will prevent all future incidents", then in times of intense change in external realities including physical, economic and social ones, the organization may miss future threats to process safety which are currently unknown. For example, electronic and physical security is a new source of potential process safety issues. If past RCAs and PHAs do not address this, then the ***committed*** organization will engage employees in mutual learning about how security relates to process safety in order to avoid potentially significant incidents.

4.1 Transformation

Industry has clearly demonstrated that successful change requires some level of dissatisfaction with the current state, articulation of a preferred future, and definition of strategies to move forward. As previously noted, what is often missing is a deeper exploration of nonconscious beliefs, values, and assumptions. A transition to a commitment-based process safety environment requires this exploration at all organizational levels, allowing for choices about, and transition to, preferred drivers.

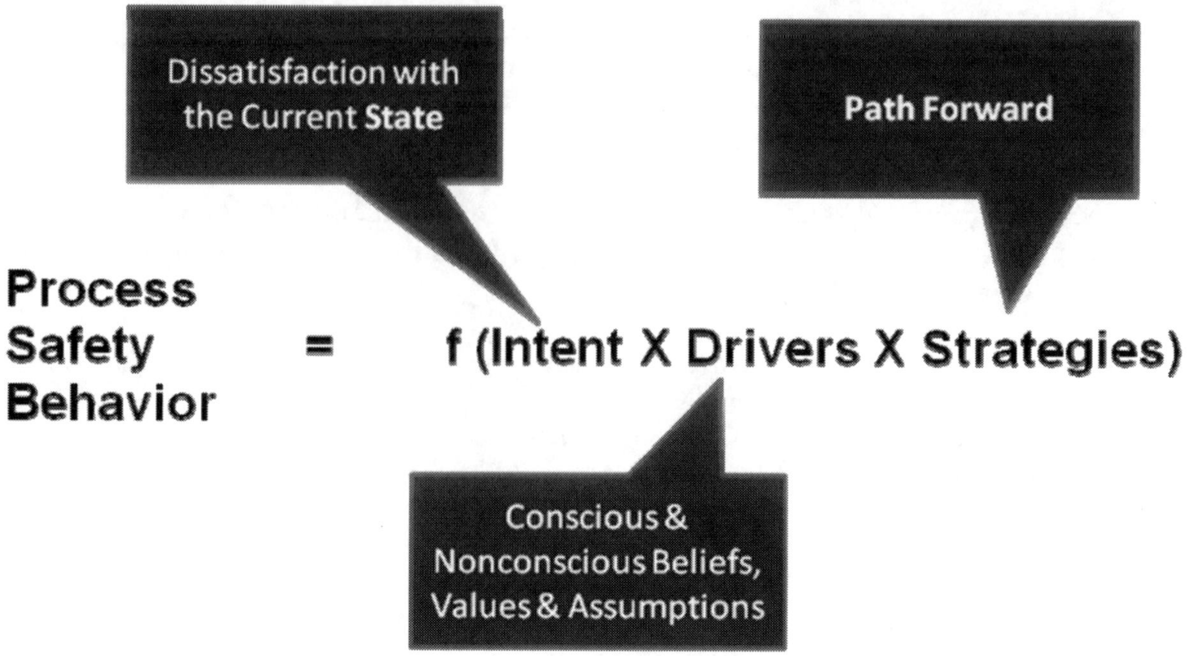

4.2 Formula for Action

The exploration of the conscious and the nonconscious is a key element of our Formula for Action. Using the formula to build authentic organizational commitment to process safety requires a structured series of conversations among all organizational roles. These conversations occur in a series of steps with these desired outcomes:

1. Full understanding and alignment on the intent to transition to a commitment-based process safety environment.
2. Identification and understanding of current conscious and nonconscious drivers, then alignment on preferred drivers which may be different.
3. Agreement on strategies and expected behaviors to support the transformation from current to preferred organizational drivers.
4. Discussion of anticipated intended and unintended consequences.
5. Agreement on how to understand and work through unintended consequences as results become known. This typically requires re-exploration of nonconscious drivers.

This approach can also be applied to support processes for process safety, whether investigations, responses to near misses, or design of new policies and procedures to respond to changes in working, economic or social conditions.

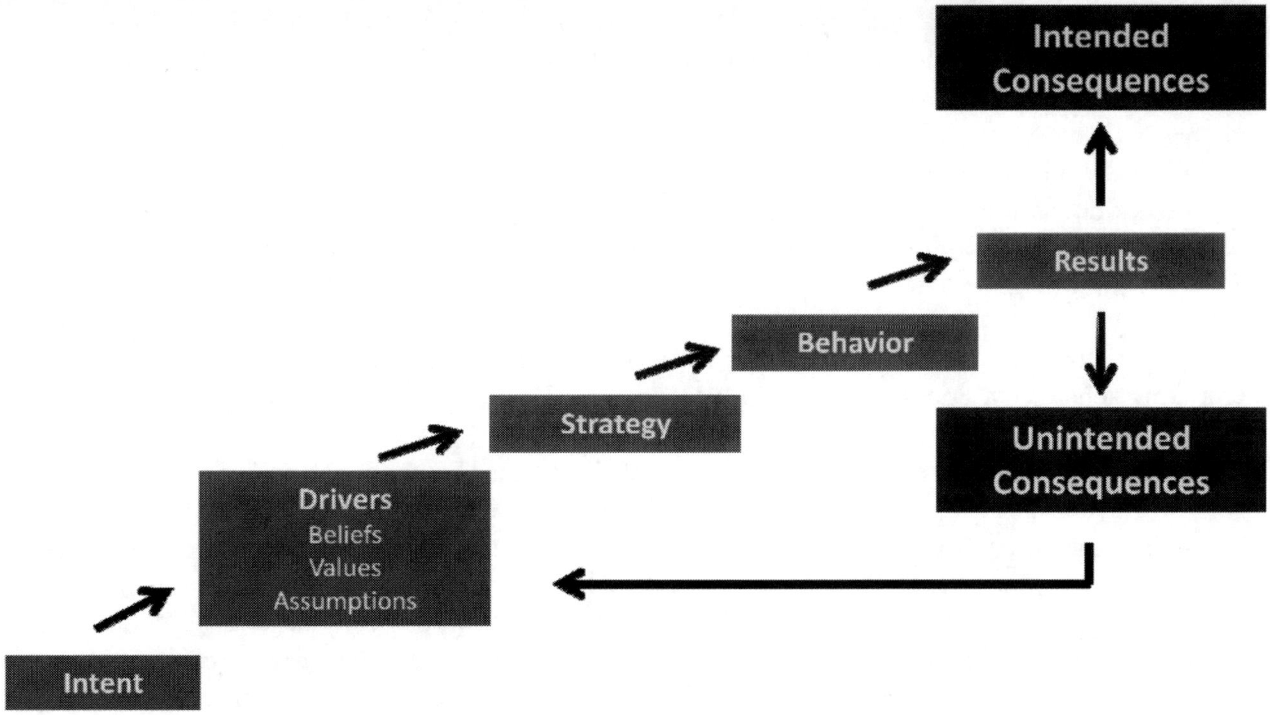

The transition to, and sustainability of, a commitment-based process safety environment only occurs by shifting current drivers that do not support this intent, these agreed-to strategies, and these expected behaviors. This may be the most demanding of change management initiatives as these steps are much easier said than done.

For example, often in chemical processes we introduce reactants in a different molecular matrix than is ultimately desired in the finished product. The reaction might go most to completion in an organic matrix, but we may actually want the finished product in an aqueous matrix. It is the same for human beings. Even though we desire specific behaviors, such as attention to safety, those behaviors are ultimately "produced" in a very different matrix -- the human unconscious/nonconscious. As with chemicals, so with human beings, the only way to get to a different finished product is to change the input matrix. So, in order to change and sustain new behavior, ultimately an individual must change the antecedents upon which it rests – their unaware beliefs, values and assumptions, often built up over years of experience, trial, and error.

In most models of leadership and working together, the underlying matrix of the nonconscious typically never fully engaged or assessed. These models drive to change behavior via extrinsic levers such as fear of negative consequences, rewards (in the worst case, bribes) or superficial training. Yet without deep conversations and challenge by employees throughout the organization, the underlying matrix of the process safety environment never changes. The current approach is as if we decided to synthesize polypropylene instead of polyethylene, but nevertheless insisted on continuing to use ethylene as the raw material.

As most organizations are fundamentally hierarchical by design, this work typically starts with leadership.

4.3 Leadership

The behavior of leadership in the transition to a commitment-based process safety environment is of great importance. Leaders must:

- Articulate Commitment: Talk the talk.
- Model Commitment: Walk the talk.
- Reinforce Commitment: Ensure an emotionally safe environment for others to talk and walk the talk.

To fully realize the benefits of these behaviors requires visionary focus, realistic expectations, empathic sensitivity, courageous truthfulness, and authentic commitment to process safety:

- Empathetic: Be sensitive to context; seek first to understand; show that you care; forgive quickly; show up non-defensively; actively listen; demonstrate compassion.
- Courageous: Speak the unspeakable to the right people, and in a way that it can be heard; go first; be vulnerable and imperfect; tell the truth; say what things mean; confront painful truths; fall down and get back up; proceed with passion and fiery aliveness.
- Realistic: Stretch, but do not break; understand limits, but be very careful about setting limitations; test your sense of reality with others; accept the unchangeable, change the changeable, and do not be a part of the unacceptable.
- Visionary: Stretch your imagination; have a higher purpose; commit to the long term; have passion for the possibilities; face forward.

5. Summary

To avoid future incidents, both known and unknown, industry can take process safety to the next level by:

- Exploring and challenging deeply-held beliefs, values, and assumptions by examining the underlying antecedents of process safety behaviors and choices more deeply in an environment of mutual learning.
- Using thoughtful conversation and active listening to managing the consequences of infrequent, yet potentially severe, process safety incidents (Duffey & Saull, 2008).
- Achieving the desired operational discipline, including documentation, setting expectations and defining critical procedures.
- Practicing disciplined, proactive anticipation of infrequent events.

6. Conclusion: A Commitment-Based Process Safety Environment

The benefits of a committed approach to process safety include not only rapid response to economic, social, employment, and demographic change, but also increased employee satisfaction and reduced dependence upon span of control to manage infrequent events. By releasing the latent talent in the organization, ultimately this focus on commitment-based process safety could generate extraordinary performance in the face of increasingly frequent change.

A key element of this shift comes from theologian James P. (1986), "to be prepared against surprise is to be trained. To be prepared for surprise is to be educated" (p. 23). Commitment becomes the mindset to create a process safety 'learning organization" in which surprise is welcomed as an opportunity to learn more of the unknown possibilities for process safety excursions and to prepare FOR them.

In this framework, process safety becomes both process- and self-awareness. It shifts our view of others, to respect others as full, concrete individuals, not as their roles, to keep every person in play (Carse, 1986). Most importantly, the time horizon extends far, far into the future, and with no time boundaries, both leadership and the workforce learn to play for keeps. Rules change to deal with specific threats to playing for keeps.

Commitment-based process safety relies upon scrupulous self- and organizational awareness of the potential for any unit to become a Bhopal. It relies upon practicing skills to envision those situations which could lead to such an outcome, and then to methodically, with discipline, prepare for them. In the words of Mark Twain, "it is better to be careful 100 times than to get killed once."

7. References

Alexander, C. P., & Stoler, P. (December 17, 1984). Environment: A calamity for Union Carbide [Electronic Version]. *Time*, from http://www.time.com/time/magazine/article/0,9171,923804,00.html

Bargh, J. A., Gollwitzer, P. M., Lee-Chai, A., Barndollar, K., & Trotschel, R. (2001). The automated will: Nonconscious activation and pursuit of behavioral goals. *Journal of Personality and Social Psychology, 81*(6), 1014-1027.

Carse, J. P. (1986). *Finite and infinite games: A vision of life as play and possibility*: Free Press.

Carter, R. (2002). *Exploring consciousness*. Berkeley: University of California Press.

Duffey, R. B., Saull, John W. (2008) Managing risk: The human element: Wiley & Sons.

Isom, K., & Cummings, B. (September 9, 2005). Application for accreditation of safety programs based on the principles of behavior, Marathon Petroleum Company LLC [Electronic Version], from http://www.behavior.org/cba/Marathon_0907.pdf

8. Additional References Not Cited

Haesle, J., Devlin, C., & McCavit, J. (2007). *Improving process safety by addressing the human element.* Paper presented at the AIChE Annual Global Congress on Process Safety, New York.

Merritt, C. W. (January, 06, 2005). Despite progress, chemical spills persist [Electronic Version]. *Chemical Safety Monitor*, from http://www.csmonitor.com/2005/0106/p09s02-coop.html

How to Evaluate Process Safety Culture

Jerry J. Forest
Lyondell Chemical Company
a LyondellBasell company
2502 Sheldon Road, Channelview, TX
jerry.forest@lyondellbasell.com

Prepared for Presentation at
American Institute of Chemical Engineers
2010 Spring Meeting
6th Global Congress on Process Safety
San Antonio, Texas
March 22-24, 2010

UNPUBLISHED

How to Evaluate Process Safety Culture

Jerry J. Forest
Lyondell Chemical Company
LyondellBasell Industries
2502 Sheldon Road, Channelview, TX
jerry.forest@lyondellbasell.com

Keywords: Culture, Survey, Analysis

Abstract

At first glance it may seem that organizational culture is subjective and therefore difficult to measure. Indeed, even the definition of culture falls into the soft side of process safety. This paper will show that an objective approach can be taken to survey and define process safety culture. With this data gap analysis can be performed that compares survey data to CCPS essential elements of process safety. This gap analysis can be used to identify strengths and weaknesses so that process safety performance can be improved.

The survey method used starts with the CCPS process safety pillars and categorizes questions according to the pillar elements, principles and essential features required for excellence. The appropriate data collecting technique is shown that will give statistically meaningful results. Examples are given of grouping results by safety pillar categories and demographics of the survey population. Finally, a gap analysis method is shown that compares the survey results to CCPS essential elements so that improvement plans can be made.

1. Introduction

Process safety is an encompassing term used to describe the activities that help prevent the catastrophic release of hazardous chemicals. These activities include: Collection of process safety information, process hazard analysis, management of change, pre-startup safety reviews, mechanical integrity and much more. There is a finite list of activities that describe process safety. Although these activities can be easily defined, process safety culture seems a bit more esoteric. The U.S. Chemical Safety and Hazard Investigation Board (CSB) has stated that organizational culture and structure can contribute as much to process safety events as immediate causes [5]. Indeed, the purpose of the Baker Panel Study of the Texas City event of 2005 was not to find root cause, but rather identify the values, beliefs and management assumptions that led to the event; their purpose was to evaluate process safety culture [1].

The term culture is often used to describe management systems, company behavior and process safety without really defining its meaning. How do we measure culture? How do we know if we have achieved a desired culture?

This paper will provide a clear definition of culture in terms of a commonly held set of values, norms and beliefs. A method to objectively define process safety values and beliefs is given by benchmarking with published material from the Center for Chemical Process Safety (CCPS) is presented. A methodology to conduct a culture survey is then presented that allows for objective measurement of a company culture. This is followed by a description of how to analyze and interpret the results.

2. Definition of Culture

Countries around the globe exhibit unique cultures. Groups of people within those countries sometimes have their own culture. Organizations are said to exhibit a culture. People hired into those organizations are often screened to ensure that they fit into the culture. Managers strive to develop a culture. We use this soft term frequently, but do we consider a definition when using it? Similarly, if the term isn't defined, how can we develop it or measure our progress toward it?

A useful definition that can be used to measure organizational culture is "a commonly held set of values, norms and beliefs that operate within the organization [4]". Culture means that *something* must be held in common.

This feels intuitively correct. People within a country are often identified with that country. They might look similar, talk similarly, and have common backgrounds. They are identified with that country. They may have physical traits and habits in common. Beyond these outward things, they may also have a common world view that has been shaped by similar personal, institutional, and historical perspectives. This background defines common beliefs and common values; it makes up the culture.

With this posit that culture is a commonly held set of values, norms and beliefs, what then is a process safety culture? How do we define a commonly held set of values and beliefs regarding process safety?

3. The Objective Benchmark

Process safety professionals often deal with quantitative risk assessments that apply science, statistics and probability to determine if processes have acceptable risk based on an acceptance criterion. This is hard science and the results are knowable and repeatable. In order to develop the same repeatability with respect to development and measurement of culture, a methodology is required to convert the subjective definition of culture to something that can be objectively measured.

CCPS categorizes activities for process safety into four broad groups that are useful as an objective measure: Commit to Process Safety, Understand Hazards and Risks, Manage Risks and Learn from Experience [3]. Each pillar is broken into various elements that must be in place to achieve excellence in process safety. Furthermore, the CCPS has defined underlying principles for each element and essential elements for each principle that represent objective actions that can be taken at a manufacturing site level so that process safety excellence is achieved. The process safety pillars, elements, principles and essential elements can be used to develop employee surveys that objectively measure values, norms and beliefs. Survey results are compared to the CCPS pillars to complete a gap analysis of process safety culture.

4. CCPS Pillars

The first CCPS pillar, *Commit to Process Safety*, defines culture elements as: "Develop a process safety culture, compliance to process safety standards, develop a process safety competency, and involve the workforce". Consider that by themselves the elements do not represent the whole of process safety activities and culture. Development of an objective process safety culture measurement considers all of the pillars as they represent the total of process safety management. A summary of the pillars, elements and principles is shown in table 1. Appendix 1 gives a more detailed version of the summary with essential activities shown for each principle. Please note that this summary deviates slightly from the principles and essential elements shown in the CCPS book, *Guidelines for Risk Based Process Safety*. In the development of an appropriate objective basis upon which to measure process safety, certain principles and essential elements have been eliminated or modified, as they were either redundant or were not applicable to my organization. However, the point should be clear. CCPS has gone beyond the traditional 14 elements of process safety suggested by OSHA in the 1910.119 regulation and grouped those elements, as well as others recognized by industry that contribute to success in process safety, into four categories or pillars. Each pillar has several elements and each element has several principles. The activities that we perform to ensure that the principles are in place and effective become the objective standard we can use to measure our process safety culture. These activities are compared to the essential elements to complete a gap analysis. Since culture is a set of commonly held values and beliefs, understanding the workforce's thoughts and attitudes against this objective measure should provide a clear indication of the process safety culture.

Table 1 – CCPS Pillars, Elements and Principles

CCPS Pillars & Elements	Principles
Commit to Process Safety	
Develop a Process Safety Culture	PS as a core value; strong leadership; high-performance standards; zero tolerance for willful violations; reward PS performance
Compliance with Process Safety Standards	Standard owners
Process safety Competency	Improve knowledge; up-to-date PSI; apply learning's
Workforce Involvement	Competent personnel; work activities
Understand Hazards & Risks	
Process Knowledge Management	Catalog process knowledge; protect & update knowledge; use process knowledge
Hazard Identification & Risk Analysis	ID hazards & evaluate risks; assess risks and make risk-based decisions; complete assessment results
Manage Risk	
Operating Procedures	ID procedures required; develop procedures; use procedures to improve human performance; ensure procedures are maintained
Safe Work Practices	Control non-routine work
Asset Integrity and Reliability	ID in-scope equipment; maintain knowledge, skills, procedures and tools; ensure continued fitness for purpose; address equipment failures, analyze data
Contractor Management	Conduct work activities; monitor contractor management system
Training	ID required training; provide effective training, monitor worker performance
Management of Change	ID potential change; evaluate impacts; decide whether to allow the change; complete follow-up activities
Operational Readiness	Conduct appropriate reviews; follow through on decisions, actions, and use of readiness data
Conduct of Operations	Control operations activities; control the status of systems and equipment; develop required skills & behaviors; monitor organizational performance
Emergency Management	Prepare for emergencies; periodically test the adequacy of plans and level of preparedness
Learn from Experience	
Incident Investigation	ID potential incidents for investigation; use appropriate techniques to investigate incidents
Metrics	Conduct metrics acquisition; use metrics to make corrective action decisions
Audits	Conduct element work activities; use audits to enhance RBPS effectiveness
Management Review & Improvement	Conduct review activities; monitor organizational performance

5. Survey Considerations

Before selecting the objective survey questions with the pillars as the basis, a few mechanical items regarding the survey should be addressed. These include the length of the survey, target audience, participation reminders, length of time for open response, method of delivery, how data is analyzed, anonymity of response and several others.

A first consideration in the survey design is convincing the organization that the results will be meaningful, anonymous, shared, and action taken on the outcome if appropriate. How this is accomplished is a function of the size and scope of the survey. A common consideration is simple communication to effective parties that a survey is coming, the contents, and how the data will be used. Depending upon the size of the survey and population of the organization, a steering team might be considered that can then have influence over leaders in spreading the word.

As engineers we sometimes default to gathering as much data as we possibly can so that we can analyze, trend, draw conclusions and make the best decisions. However, in a survey we must be considerate of the data source - people. If it takes too long to complete a survey, people either will not participate, or will not put a lot of thought into their answers. A good rule of thumb is to keep the survey length such that the average person can complete it in 20 – 30 minutes, with 45 minutes being the maximum. Depending upon the type of questions in the survey, consider that it may take the average person 30 seconds to 1 minute per question to complete. Of course, participants could be *forced* to participate, but that approach in itself might be a good indicator of process safety culture!

Another key consideration in survey participation is identification of the target audience. For which group do you want to measure perception of culture: management, office workers, craftsmen, union employees, contractors, other or all? In making this decision, a rule of thumb to consider is that a 20 – 30% participation rate should be expected. This brings up another survey design consideration – participation reminders and length of time that the survey should remain open for response.

The survey should be held open long enough to ensure the target participation audience has a chance to participate, but not so long that it is perceived as not important or can be delayed (and forgotten about) until much later. Another rule of thumb is to keep the survey open for approximately three weeks. Weekly reminders should be sent to survey participants (and their supervisors) to ensure maximum participation. In recent corporate surveys, reminders have helped boost participation rates to >75%.

Assuming that we are interested in understanding the perceptions of the entire demographic of the organization, the next decision is to identify who actually takes the survey. That decision is best left to your particular organization. However, there are a few things to consider. If the delivery method is electronic or web-based, opening the survey up to all employees may be feasible since all data is collected electronically. While this approach leads to the most data being collected, it may lead to different proportions of participation among the various demographic groups. Comparison of data among those groups to determine independence would involve hypothesis testing. If your organization has a master black belt, they can help design the data collection and analysis. Another approach is to select a statistically meaningful sample of workers from each group to complete the survey. This approach may lead to a smaller sample size, but one that you might have a little more control over in terms of participation. A final word about method of delivery; consider that if paper ballots are used, people have a tendency toward illegible handwriting. Gathering paper ballots and consolidating to electronic media for analysis can be a huge task and may open up the process to error in transcription. A single web-based electronic database should be used if at all possible.

The type of data analysis of the survey should be decided upon before the survey is designed. At the most basic level, descriptive statistics and graphical analysis is done in order to draw conclusions on the survey data. However, if your objective is to compare results across demographic groups, then inferential statistics must be used. A six-sigma black belt should be consulted in survey design and for the data interpretation.

As indicated there are a number of mechanics questions that should be addressed in the survey design phase. The list briefly discussed here is not inclusive, but rather is presented to give you a few things to think about that might be unique to your organization. The main point here is that these items should be addressed before the survey is pushed to the participants.

6. Survey Question Design

As with any good survey, the design is critical to data analysis. Therefore as Steven R. Covey states in *7 Habits*, begin with the end in mind [4]. In this context, first define the objective of the survey and then define the questions.

There are many process safety key process indicators that make up great data to describe the health of the process safety system. The CCPS has recently published <u>Guidelines for Process Safety Metrics</u> [2]. Examples include number of open PHA items, number of past due audit items, time involved for various MOC phases, and many, many more. If that data is valuable and part of your objective, then indeed by all means collect it. However, keep in mind that this type of objective data only infers the process safety culture by identifying gaps in the management system. In the first section of this paper process safety culture is defined as a set of values and beliefs about process safety. That is the primary objective of this survey.

7. Question Selection

An objective basis from which to select questions are the CCPS process safety pillars. Appendix 1 summarizes the essential elements appropriate for an organization to use in order to test perceptions and beliefs about how the organization is performing in that area of process safety. It is a simple matter to develop questions based on the particular essential element/ principle being tested. For example, Figure 1 shows the elements underlying *Commit to Process Safety* as: "Develop a process safety culture, compliance with process safety standards, process safety competency, and workforce involvement." Consider "develop culture", for example; the principles CCPS defines are: "Establish process safety as a core value, provide strong leadership, establish and enforce high standards of performance, zero tolerance for willful violations of process safety policies, and process safety is rewarded." Finally, the essential elements for high standards are defined as "establish, communicate and monitor process safety goals and establish responsibility for process safety roles." We might choose to test those essential elements to develop questions with a Likert[1] scale such as "My site has clearly defined roles and responsibilities defined for process safety." There is more detail about Likert scales below.

Thus, by systematically addressing each essential element of the principles associated with each CCPS process safety element, an objective question bank can be developed.

Figure 1 – Relating questions to essential element

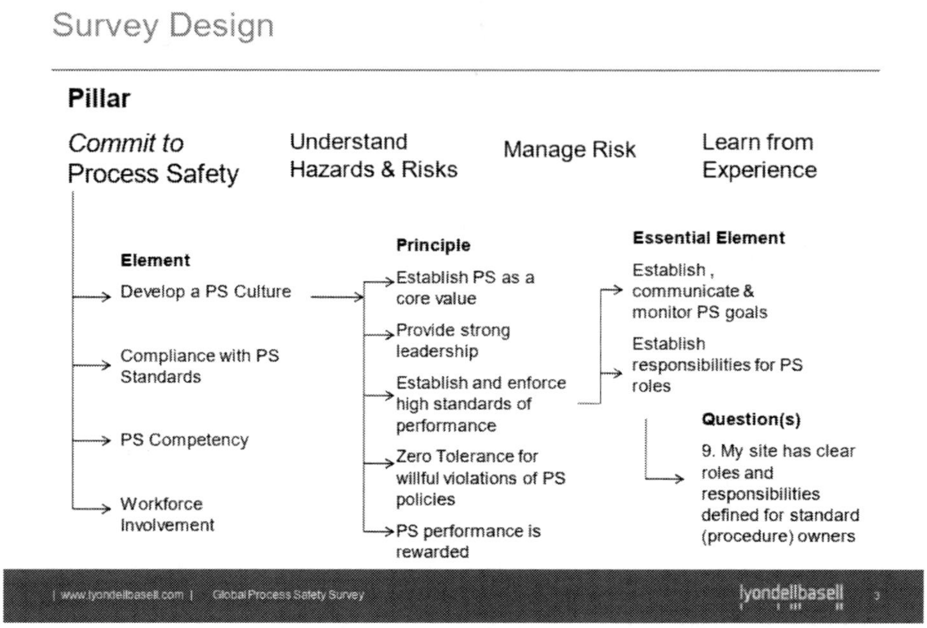

Appendix 2 is an example of one such bank of questions that surveys across each of the four pillars. The questions are clearly tied to essential elements, principles, elements and pillars for data analysis. Also note that the appendix shows the questions broken down into two sets of surveys in order to keep the length of each survey manageable. When measuring the process safety culture of your organization, the essential elements of each principle can be used to develop questions for analysis.

8. Question Design

A simple question design uses a five-point Likert scale: strongly agree, agree, neither agree nor disagree, disagree, or strongly disagree. Participants are asked to rate their level of agreement with the question. While question selection allows us to objectively organize the data into process safety elements, the Likert scale allows us to objectively analyze subjective notions such

[1] A Likert scale is commonly used in questionnaires to measure level of agreement or disagreement with subjective data. Level of agreement is measured by selection of phrases such as: Strongly agree, agree, disagree, and strongly disagree.

as perception of agreement with process safety. In designing Likert scale questions, mixing question types is not recommended. As stated above, you may also wish to collect a second class of data such as process safety metrics. A third type of data to collect is open-ended questions.

Many people prefer to give their opinion, so when faced with rigid questions in Likert format they can become frustrated because they have not been given the opportunity to say what they *really* think. There may be value in adding three open-ended questions: What should we start doing that we presently are not, what should we stop doing and what should we continue to do? These questions originate from Steven R. Covey quality seminars given in 1990. The questions are straightforward, to the point, and tell us much information about trends in thinking. Analysis of this data is discussed below.

A final category of data is demographic data. Again, beginning with the end in mind, we need to know how we'll use demographic data in the analysis so that we ask the right questions. But also note, asking too much demographic data might give the appearance that the survey isn't anonymous (if that is your intent.) This type of data usually isn't a problem and is quite valuable in comparing group perceptions.

9. Data Analysis – Likert Scale Questions

An objective analysis of the data can be made by summing the number of responses in each category and converting to a percentage by category. Subsequent graphical analysis of the percentages will then be a fair representation of degree of agreement for each question. Graphical analysis can be done by demographic segment, for each question, process safety principle, element and pillar. Table 2 is an example table that sums the data and calculates percentage.

Table 2 – Data Analysis of Likert Scale

Commit to Process Safety: Process Safety Culture: Core Value 1. My Site has a documented process safety vision, mission, and strategic plan										
Sum of Responses						Percent of Response				
Strongly Disagree	Disagree	Neither Agree nor Disagree	Agree	Strongly Agree	Total	Strongly Disagree	Disagree	Neither Agree nor Disagree	Agree	Strongly Agree
4	7	15	26	8	60	7%	12%	25%	43%	13%

This is done for each question. Since each question is already categorized by principle, element and pillar, the data can be rolled-up for each level of analysis. Herein demonstrates the value of a database. Responses are categorized by demographic group and questions are categorized by principle, element and pillar. A well-designed database can easily manipulate the responses to calculate the percentages, convert to descriptive statistics, and graph results. Note, even Microsoft Excel can be used with pivot table or force table manipulation to accomplish the same result.

Once pivot or data tables are created, graphical analysis can be done. For a Likert scale, simple 100% stacked bar charts visually represent percent agreement of questions by demographic grouping rolled-up to whatever objective level of the pillars is desired. Appendix 3 shows examples of stacked bar charts for these various data groupings.

10. Data Analysis – Open Ended Questions

Analysis of the open-ended survey questions is a bit more difficult, but can be done. When asking the Steven R. Covey survey questions, the responses can typically be categorized into groups. Read the responses a few times until obvious groupings can be made. Count the number of responses in each group. The number of responses in each group represents a frequency that can be graphed. Table 3 is an example of the group roll-up and Figure 2 shows an example graph of that data.

Table 3 – Open Ended Question categorization

Group/ Category	# Responses in this category
Metrics	1
HSEQ Integration	2
Mechanical Integrity	2
Audits	2
Training	2
Process Safety Systems	3
Safety Critical Variables	3
Incident Reporting	3
Leadership	4
Awareness	5
MOC	5
PHA	6
Other	7
Employee Participation	14

Figure 2 Example of open-ended graphic

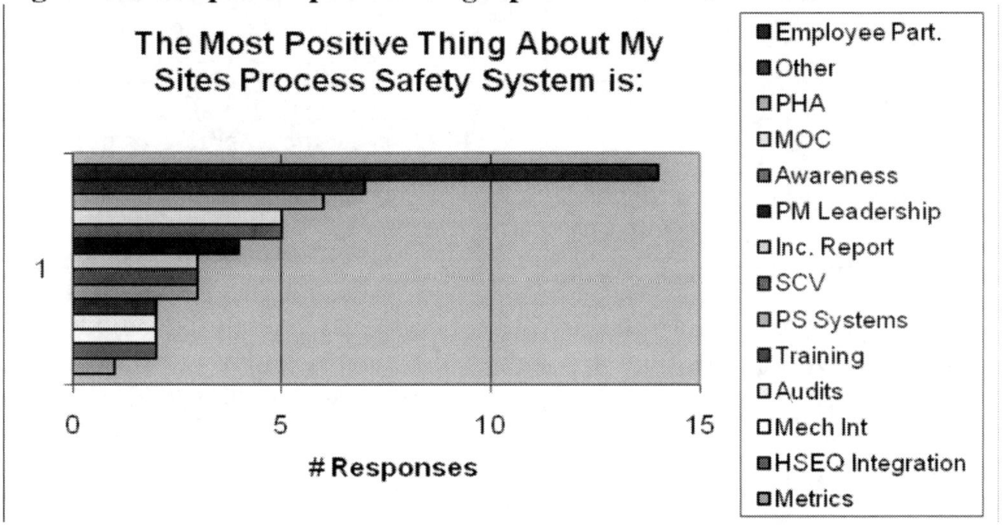

Of course, with the Likert questions in tabular format descriptive statistics such as mean, mode, median, standard deviation and others can easily be calculated by CCPS group or demographic, or any combination thereof. As mentioned above, inferential statistics such as Chi-squared tests could also be performed to determine independence among group perceptions regarding process safety culture. The inferential statistical discussion is beyond the scope of this paper.

11. Gap Analysis & SWOT

Once the descriptive statistics are complete, a second type of analysis can be done with the data that helps management draw conclusions and make action plans for improvement. The second analysis is an objective gap analysis of survey responses to CCPS pillars, followed by a strategic planning tool called SWOT, an acronym that stands for Strengths, Weaknesses, Opportunities, Threats.

Since we have grouped our questions by principles, elements and pillars, the gap analysis is very straight forward. The first step is to add and order the percent positive responses defined as the sum of strongly agree and agree, and sort the answers ascending. In the sorting of the data, be sure to include each questions attribute of CCPS pillar, element and principle. When this sorting exercise is complete, the data can then be grouped by quartile or percentage, or individual principle to determine the perceived gap in process safety culture. When doing this type of analysis, the data will logically group by element and principle clearly showing management direction for improvement. Table 4 shows a hypothetical gap analysis of survey response to CCPS essential feature. In this example, the survey responses have the lowest percent positive response from essential features underlying commit to process safety pillar.

Table 4 – Gap Analysis

Essential Feature	Question	%Positive Response	Gap
Develop, document, and communicate vision and principles	1.1	30	Sites don't have consistent documented vision and

			principles
Training/ communication to front line supervisors	1.15	37	Demographic group 1 response is more negative than demographic group 2
Establish and administer reward system	1.7	45	No consistent reward system
Educate managers in process safety culture	1.2	48	Demographic group 1 response is more negative than demographic group 2. No corporate training

Appendix 4 shows an expanded hypothetical simplified summary of a gap analysis completed for the CCPS pillars, elements, and principles done by sorting the most negative responses of the essential feature survey response questions. It's clear that the majority of responses in this hypothetical example came from the pillar *Commit*, indicating cultural gaps in management commitment to process safety culture.

12. Conclusion

Culture is a soft term used to describe an organization's commonly held values, norms and beliefs. While values and beliefs may seem subjective in nature, objective descriptions can be defined for the various elements and principles of process safety that are widely recognized by industry. An objective benchmark is the CCPS process safety pillars of: Commit to process safety, Understand Hazards & Risks, Manage Risks, and Learn from Experience. Each pillar has defined elements and each element has defined principles. Underlying each principle are various essential features to achieve excellence in process safety.

In order to measure people's views on culture the CCPS essential feature activities associated with each process safety pillar can be used as a basis to develop survey questions. Survey questions designed with a Likert scale that measures degree of agreement to the question produce objective data. These numbers can then be used to obtain descriptive statistics and graphical analysis.

Since the basis of the objective questions is industry-recognized activities from CCPS pillars, gap analysis of the responses to essential features can be performed. Data is organized by sorting survey responses by most positive to least positive. The least positive responses indicate that the organization perceives gaps in those areas of process safety. These responses become a basis for continuous improvement in process safety.

13. References

[1] The B.P. U.S. Refineries Independent Safety Review Panel, Baker et al, The Report of the BP U. S. Refineries Independent Safety Review Panel. January, 2007

[2] Center for Chemical Process Safety, Guidelines for Process Safety Metrics. American Institute of Chemical Engineers, 2010

[3] Center for Chemical Process Safety, Guidelines for Risk Based Process Safety. American Institute of Chemical Engineers, 2007

[4] Covey, Steven R, The 7 Habits of Highly Effective People. Simon & Schuster, New York, 1990

[4] Lawyer, John W., John J. Lawyer, A Systems Approach for Ministers. Loyola's Institute for Ministry, New Orleans: 1986

[5] U. S. Chemical Safety and Hazard Investigation Board, Investigation Report Refinery Explosion and Fire. Report No. 2005-04-I-TX, March 2007

14. Appendix 1 – Summary of CCPS Pillars, Elements, Principles, & Essential Elements

Pillar	Element	Principle	Essential Feature
Commit to Process Safety: Organization fully supports	Develop Process Safety Culture	Establish PS as a core value	Develop, document, and communicate vision, CSF's, and principles
		Provide Strong Leadership	Educate managers in PS culture w/ formal training program
			Share PS responsibilities among managers - provide standard owners
		Establish and enforce high standards of performance	Establish, communicate & monitor PS goals
			Establish responsibilities for PS roles
		Zero tolerance for willful violations of PS policies	Establish formal policy deviation procedures
		PS performance is rewarded	Establish & administer reward system
	Compliance with PS Standards	Develop standard owners	Develop procedures list & document management system
			Assign & educate procedure owners on roles & responsibilities
			Conduct periodic performance-based audits and annual system audits. Report to management
	PS Competency	Continuously improve knowledge	Create technology stewards
			Document tech manuals including process history, incident learning, process reports
		PSI available & up to date	Create PSI matrix, ID gaps, and close. Hyperlink with operating manuals
		Consistently apply learning's	Unit specific training for managers & superintendent/ supervisors on Hazards & Controls
			Evergreen Incident Learning periodic review

Pillar	Element	Principle	Essential Feature
			Tailgates: Present incident review, incident learning review, PHA risk review
	Workforce Involvement	Involve Competent Personnel	Senior people involved in PHA
			Training/ Communication to PSM supervisors
		Work Activities	Performance Based PSM inspections
			PSM/ PHA scenario based drills
			Web -based PS survey
Understand Hazards & Risks	Process Knowledge Management	Maintain a Dependable Practice	Formal written policy for governing knowledge element
		Catalog Process Knowledge	Standardize & catalog PSI
		Protect & Update Process Knowledge	Periodically review PSI
			Manage Change to PSI
		Use Process Knowledge	Train, insure PSI awareness
	Hazard identification & Risk Analysis	Maintain a Dependable Practice	Document the intended risk system
			Involve competent personnel
		Identify Hazards and Evaluate Risks	Select appropriate risk methods
			Ensure reviewers have expertise
			Document & report
		Assess Risks and make risk-based decisions	Apply the risk tolerance criteria
			Select appropriate risk control measures
		Follow through on the assessment results	Communicate results to management
			Document the residual risk
			Resolve recommendations and track completion of actions
			Communicate results internally
			Communicate results externally
			Maintain risk assessment records
Manage Risk	Operating Procedures	Identify What Operating Procedures are Needed	Address all operating modes
		Develop Procedures	Appropriate format

Pillar	Element	Principle	Essential Feature
			Address safe operating limits and consequences of deviation from safe operating limits
			Address limiting conditions for operations
			Supplement procedures with checklists
			Make Effective use of pictures and diagrams
			Address non-routine or temporary operations
			Validate procedures
		Use the Procedures to Improve Human Performance	Use Procedures when training
			Hold the organization accountable for consistently following procedures
			Ensure procedures are available
		Ensure the Procedures are Maintained	Manage Changes
			Correct errors
			Periodically review
	Safe Work Practices	Control non-routine work activities	Develop safe work procedures
			Train Employees and contractors
			Performance Based safe work audits
	Asset Integrity and Reliability	Identify equipment that is in scope	Identify the PM program
		Maintain knowledge, skills, procedures and tools	Develop procedures for inspection, test, repair, and other critical maintenance activities
			Train Employees and contractors
			Ensure inspectors hold appropriate certifications
			Provide the right tools
		Ensure continued fitness for purpose	Conduct initial inspections and tests as part of plant commissioning

Pillar	Element	Principle	Essential Feature
			Conduct test and inspections during operations
			Execute calibration, adjustment, preventive maintenance, and repair activities
			Plan, control, and execute maintenance activities
			Ensure the quality of repair parts and maintenance materials
			Ensure overhauls, repairs and tests do not undermine safety
		Address Equipment Failures and Deficiencies	Promptly address conditions that can lead to failure
			Examine results to identify broader issues
			Investigate chronic failures using a structured methodology
			Plan maintenance and repair activities
		Analyze data	Collect and analyze data
			Adjust inspection frequencies and methods
			Conduct additional inspections or tests as needed
			Plan replacements or other corrective actions
			Archive data
	Contractor Management	Conduct Element work activities	Appropriately select contractors
			Establish expectations, roles, and responsibilities for safety program implementation and performance
			Ensure contractors are trained
		Monitor the Contractor Management System for Effectiveness	Audit the contractor selection process
	Training	Identify What Training is Needed	Conduct a job/task analysis

Pillar	Element	Principle	Essential Feature
			Determine minimum requirements for job candidates
			Determine what training is needed
			Group training into logical programs
			Manage Changes
		Provide Effective Training	Develop or procure training materials
			Interweave related topics
		Monitor Worker Performance	Qualify workers initially
			Test workers periodically
	Management of Change	Identify Potential Change situations	Define scope of MOC system
			Manage all sources of change
		Evaluate Possible Impacts	Provide appropriate input information to manage changes
			Apply appropriate technical rigor for the MOC review process
			Ensure that MOC reviewers have appropriate expertise and tools
		Decide whether to allow the change	Authorize changes
			Ensure that change authorizers address important issues
		Complete follow-up activities	Update PSI
			Communicate changes to affected personnel
	Operational Readiness	Conduct Appropriate Readiness Reviews as Needed	Ensure appropriate elements are in PSSR
			Document PSSRs
		Make Startup Decisions Based on Readiness Results	ID & Communicate risk based analysis of startup decisions
		Follow through on Decisions, Actions, and Use of Readiness Results	Enact risk control measures
	Conduct of Operations	Control Operations Activities	Follow safe work practices
			Use qualified workers

Pillar	Element	Principle	Essential Feature
			Assign adequate resources
			Formalize communications between workers
			Formalize communications between shifts
			Formalize communications between work groups
			Adhere to safe operating limits and limiting conditions for operation
			Control access and occupancy
		Control the Status of Systems and Equipment	Formalize equipment/asset ownership
			Monitor equipment status
			Maintain good housekeeping
			Maintain labeling
			Maintain lighting
			Maintain instruments and tools
		Develop Required Skills/ Behaviors	Emphasize observation and attention to detail
			Promote questioning/ learning attitude
			Train workers to recognize hazards
			Train workers to self-check and peer-check
			Establish standards of conduct
		Monitor Organizational Performance	Maintain accountability
			Strive to improve
			Maintain fitness for duty
			Conduct field inspections
			Correct deviations immediately
	Emergency Management	Prepare for Emergencies	Identify accident scenarios based on hazards
			Provide facilities and equipment
			Determine when unit operator response is appropriate
			Train ERT members
			Plan communications
			Inform and train all personnel

Pillar	Element	Principle	Essential Feature
		Periodically Test the Adequacy of Plans and Level of Preparedness	Conduct emergency evacuation and emergency response drills
			Conduct tabletop exercises
			Practice crisis communication
			Critique exercises, drills and actual responses
Learn from Experience	Incident Investigation	Identify Potential Incidents for Investigation	Monitor all sources of potential incidents
			Ensure all incidents are reported
			Initiate investigations promptly
		Use Appropriate Techniques to Investigate Incidents	Collect appropriate data during the investigation
			Interface with the emergency management element
			Use effective data collection methods
			Use appropriate techniques for data analysis
			Investigate causes to an appropriate depth
			Demand technical rigor in the investigation process
			Provide personnel with appropriate expertise and tools
			Develop effective recommendations
	Metrics	Conduct Metrics Acquisition	Implement appropriate element metrics
			Collect & refresh metrics
			Summarize & communicate metrics in a useful format
		Use Metrics to Make Element Corrective Action Decisions	Use the metrics to improve process safety
	Audits	Maintain a Dependable Practice	Ensure consistent implementation
			Involve competent personnel
			Identify when audits are needed
		Conduct Element work activities	Prepare for audits
			Determine audit scope & schedule

Pillar	Element	Principle	Essential Feature
			Assemble the team
			Assign responsibilities
			Gather advanced information
			Plan onsite activities
			Conduct audits
			Document the audit
			Address audit findings and recommendations
		Use Audits to Enhance RBPS Effectiveness	Share best practices
	Management review & Improvement	Maintain a Dependable Practice	Define roles and responsibilities
			Establish standards for performance
			Validate program effectiveness
		Conduct Review Activities	Determine scope of review
			Conduct & document reviews
		Monitor Organizational Performance	Strive to continuously improve
			Conduct field inspections

15. Appendix 2 – Sample Survey Questions

15.1 Survey No. 1 – Commit, Understand, Learn & Open

Element	Principle	Essential Element Question
Pillar: **Commit to Process Safety**		
PS Culture	Core Value	1. My site has a documented process safety vision, mission, and strategic plan
	Leadership	2. Leaders at my site receive initial and periodic Process Safety Training
		3. Leaders at my site have defined Process Safety Responsibilities
	Performance Standards	4. My site has established Process Safety goals
	Zero Tolerance of violations	5. My site has formal written Process Safety Policies.
		6. My site has a formal policy deviation process
	PS is rewarded	7. A formal reward system is in place to recognize process safety accomplishments, goals, and performance
Compliance with Standards	Standard Owners	8. My site has Process Safety standard (procedure) owners
		9. My site has clear roles and responsibilities defined for standard (procedure) owners
PS Competency	Improve knowledge	10. My site has a defined technology steward(s)
		11. Processes at my site have written technology manuals
	PSI	12. Process Safety information is defined, controlled, and up to date
	Learning's	13. Learning's from incidents are archived and periodically reviewed with operators, technicians, craftsmen, and contractors
Workforce involvement	Competent Personnel	14. What is the average experience level in years of your sites PHA participants

Element	Principle	Essential Element Question
		15. At my site, all personnel are trained in all aspects of process safety
Pillar: **Understand Hazards & Risks**		
Hazard Identification & Risk Management	Identify Hazards and Evaluate Risks	16. A risk based approach is used for my site's PHA's. (Risk based = defining frequency and magnitude of consequence in order to mitigate to acceptable frequency/ magnitude)
		17. My site uses a HAZOP, Fault-Tree, OSS, or more rigorous methodology for PHA
		18. PHA's are redone or revalidated every 5 years
	PHA Results	19. Residual risk from the PHA are documented
		20. PHA results are communicated to all affected personnel.
Pillar: **Learn from Experience**		
Incident Investigation	Identify incidents	21. At my site, all near misses and incidents are reported
	Investigate incidents	22. Regarding incident investigation, at my site:
Metrics	Metrics	23. My site has defined process safety metrics, collects data and periodically reports results to management for corrective action
Audits	Work activities	24. My site keeps a multi-year schedule of process safety audits and follows that schedule
		25. Process Safety audit findings are documented. Each recommendation has an owner and assigned due date.
		26. Management periodically reviews progress toward completion of audit recommendations
Management Review & Improvement	Review Activities	27. Management at my site conducts a formal periodic review of all process safety activities

Element	Principle	Essential Element Question
	Monitor Organizational Performance	28. Periodic field inspections of process safety systems are in place
"Steven R. Covey" Open-ended questions		
		29. The most positive thing about my sites process safety system is:
		30. The single thing I would stop doing or change about my sites process safety systems is:
Demographic & Misc Information		
		34. Site
		31. How is Process Safety Defined at your site? 32. This site has defined knowledge and experience requirements for line management positions reporting to the site (plant) manager in operations, technical, and maintenance functions 33. This site has a process in place to check that work orders not requiring a process safety review are simply repairs or "replacement in kind."
		35. Enter the number of full time process safety resources at your site: 36. Enter the number of full time equivalent process safety resources at your site: 37. Estimate percent of time part-time process safety resources spend on process safety:

15.2 Survey No. 2 – Manage Risks

Element	Principle	Essential Element Questions
Pillar: **Manage Risks**		
Operating Procedures	ID What procedures are needed	1. My sites operating procedures address all operating modes including: startup, shutdown, normal operation, emergency operation, startup from turnaround, etc. (Answer Strongly Disagree if your site does not have operating procedures)
	Develop procedures	2. Safe operating limits and consequences of deviation from safety operating limits are defined in operating procedures.
		3. Non-routine and transient operating conditions are addressed in the operating procedures.
		4. The content of my sites operating procedures is validated annually for correct content.

	Human Performance	5. Operating procedures are used for operator/ technician training
	Procedures are maintained	6. Changes to operating procedures are included as part of the MOC and PHA process.
Asset Integrity & Reliability	Maintain KSA	7. My site has written procedures for inspection, test, repair and other critical maintenance activities
	Continued Fitness	8. My site performs and documents initial inspections of new equipment as part of commissioning
		9. My site conducts a structured test and inspection program during operations
Contractor Management	Work Activities	10. Contractors are treated like our employees with regard to process hazard analysis, management of change, access to process safety information, and incident reporting and investigation
Management of Change	Identify changes	11. All changes at my site are managed through a process safety management of change system
	Evaluate Impacts	12. An appropriate level of risk assessment is always done for management of change (e.g., HAZOP, etc)
	Follow-up	13. Process Safety information is updated in a timely manner as a result of management of change
		14. All affected personnel are notified and trained on changes
Operational Readiness	PSSR	15. A Pre-Startup safety review is required for all management of change
Conduct of Operations	Control of Operations Activities	16. My site has a formalized communication process between operating shifts
		17. There is a process to adhere to safe operating limits
	Control the Status of Systems and Equipment	18. Our site has a formal effective housekeeping program
		19. All equipment at my site is clearly labeled
	Develop skills/ behaviors	20. There is a formal hazard recognition program at my site

	Monitor performance	21. There is a formal fitness for duty program at my site.
Emergency Management	Prepare for Emergencies	22. Emergency drills are based on scenarios identified in the process hazard analysis
		23. My site has a fully staffed emergency response team
	Periodically test preparedness	24. Emergency evacuation and response drills are periodically performed at my site.
		25. Tabletop exercises are routinely performed at my site.
		26. There is a formal critique system of emergency drills and exercises

16. Appendix 3 – Example Bar Charts

16.1 High Level Responses by Pillar

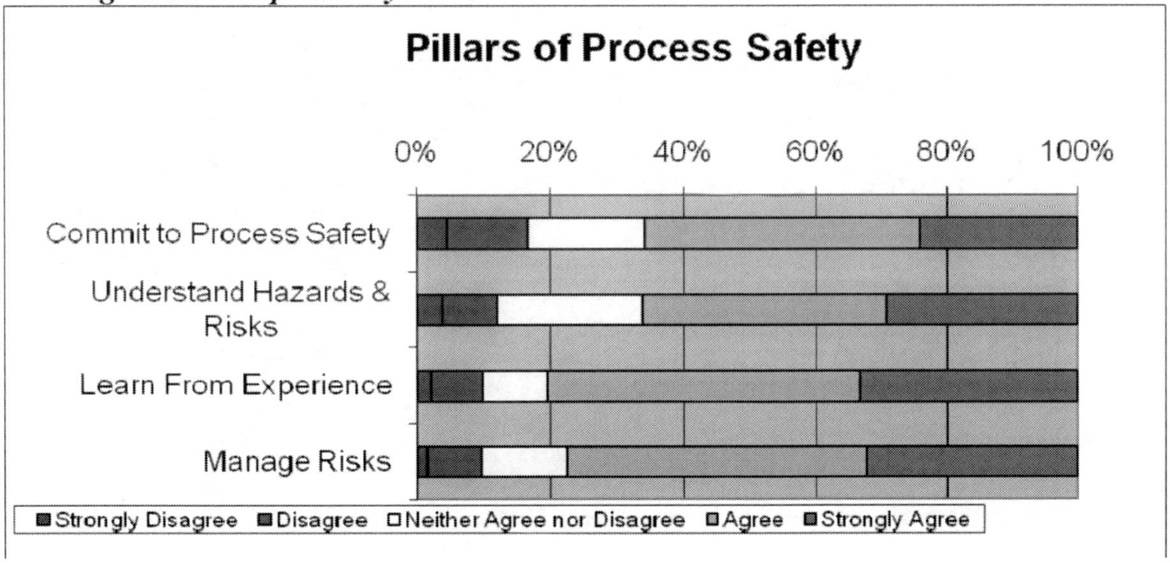

16.2 Second Level Responses by Element

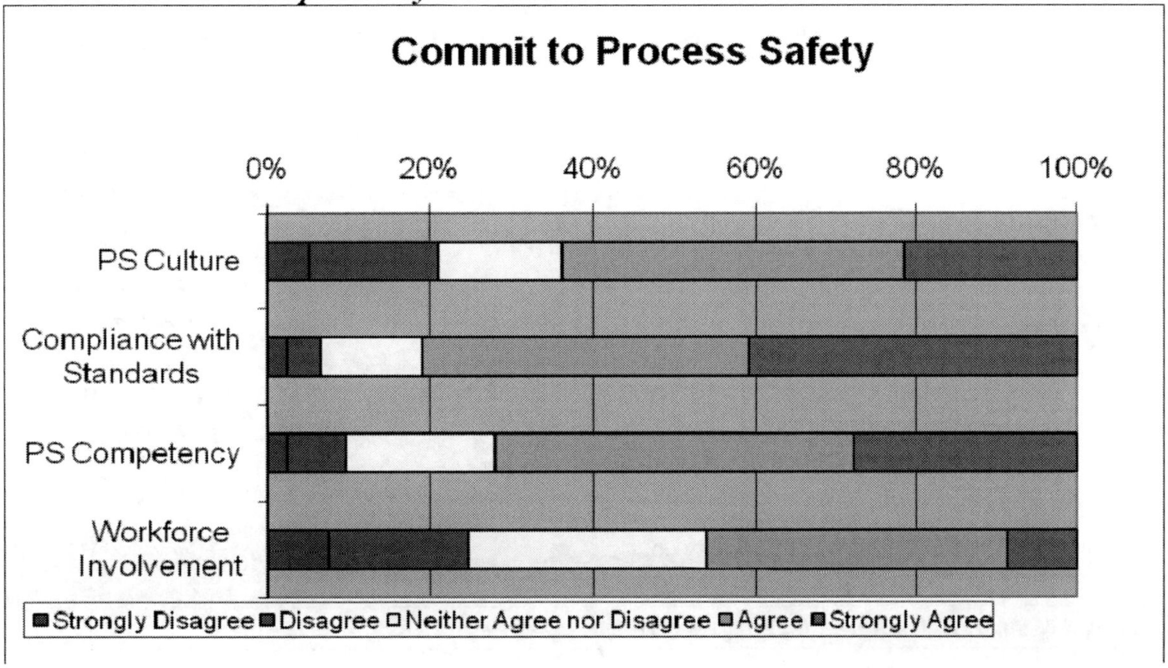

16.3 Third Level Response by Principle

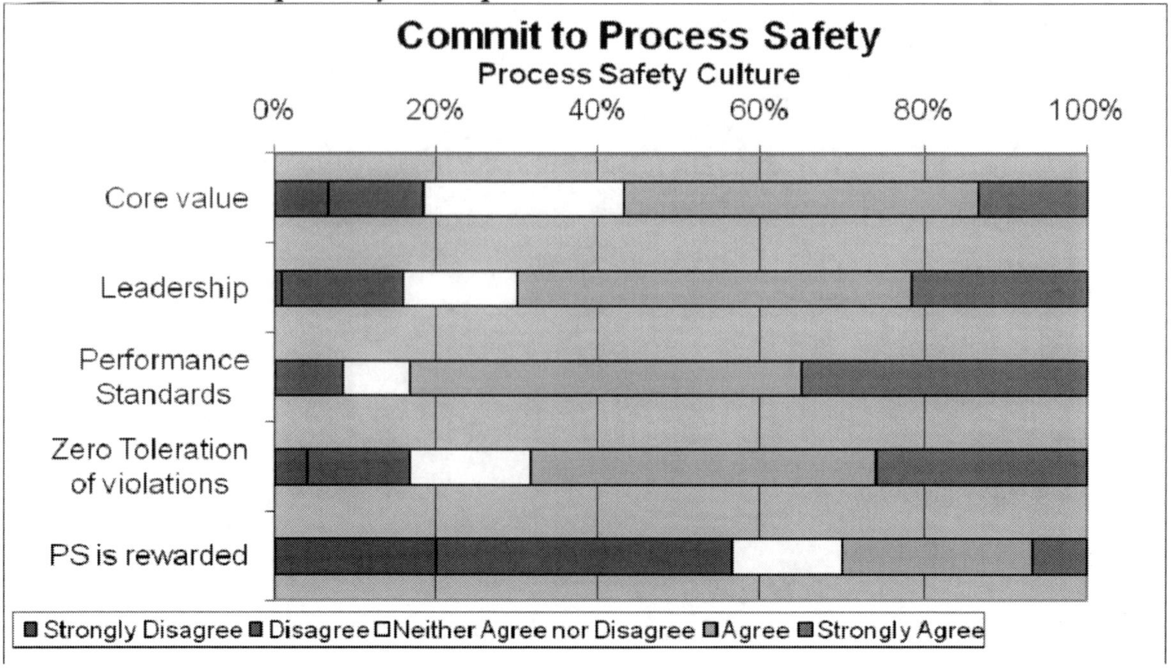

16.4 Fourth Level Response by Essential Element Question

16.5 Responses by Demographic

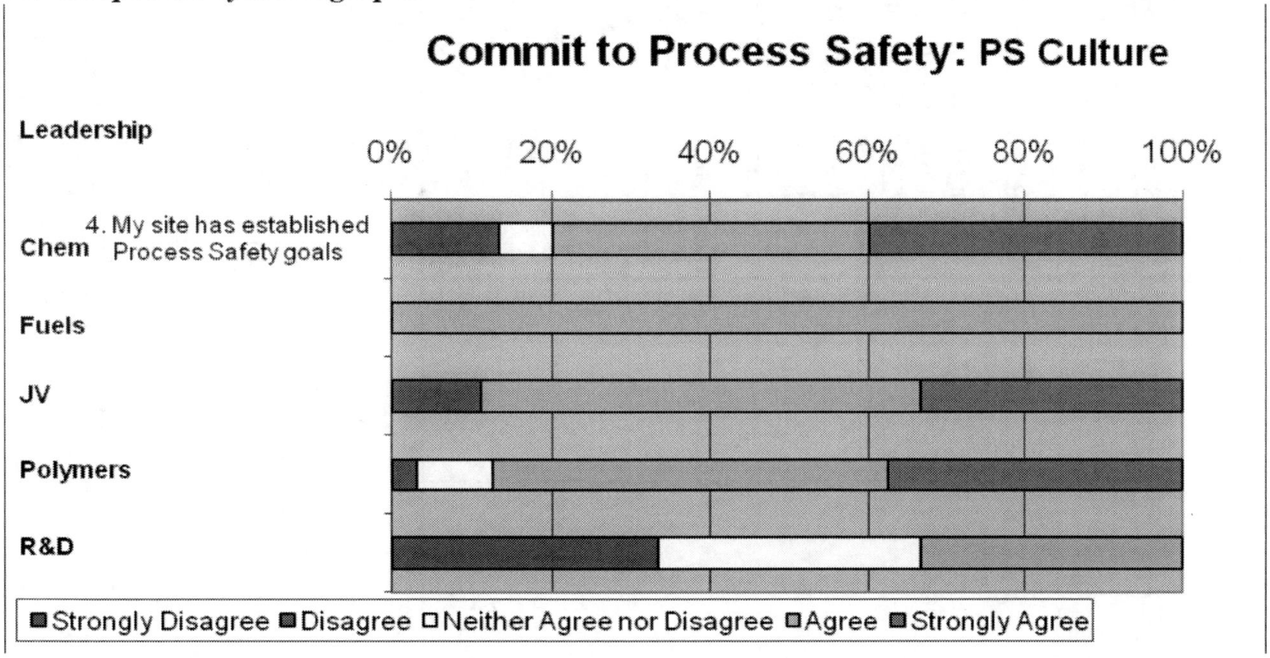

17. Appendix 4 – Survey Gap Analysis

Pillar	Element	Principle	Essential Feature	Question	Gap/ Comment
Commit to Process Safety: Organization fully supports	Develop Process Safety Culture	Establish PS as a core value	Develop, document, and communicate vision, CSF's, and principles	1.1	Sites don't have consistent documented vision, CSF's, and principles
Commit to Process Safety: Organization fully supports	Workforce Involvement		Training/ Communication to PSM supervisors	1.15	Demographic group 1 response is more negative than group 2 response on PS training
Commit to Process Safety: Organization fully supports	Develop Process Safety Culture	PS performance is rewarded	Establish & administer reward system	1.7	This question has the largest negative response indicating that rewards are not attached to process safety
Commit to Process Safety: Organization fully supports	Develop Process Safety Culture	Provide Strong Leadership	Educate managers in PS culture w/ formal training program	1.2	Demographic group 1 sites are more negative than Demographic group 1. No corporate training.
Commit to Process Safety: Organization fully supports	PS Competency	PSI available & up to date	Create PSI matrix, ID gaps, and close. Hyperlink with operating manuals	12	Demographic group 1 response is much more negative than group 2.
Learn from Experience	Management review & Improvement	Conduct Review Activities	Conduct & document reviews	1.27	Site management review of PS activities
Manage Risk	Management of Change	Complete follow-up activities	Update PSI	2.13	~50% of sites reporting PSI not updated in a timely manner
Manage Risk	Operating Procedures	Develop Procedures	Validate procedures	2.4	~50% of sites not validating procedures
Manage Risk	Emergency Management	Prepare for Emergencies	Identify accident scenarios based on hazards	2.22	PHA results used as emergency scenarios
Manage Risk	Emergency Management	Periodically Test the Adequacy of Plans and Level of Preparedness	Conduct tabletop exercises	2.25	Low tabletop drill participation
Understand Hazards & Risks	Hazard identification & Risk Analysis	Identify Hazards and Evaluate Risks	Select appropriate risk methods	1.16,1.17,1.18	30% of sites are not validating PHA's in 5yr period or less.

6TH GLOBAL CONGRESS ON PROCESS SAFETY

Using Root Cause Failure Analysis Results to Estimate Failure Rates and PFDavg for Spring Operated Pressure Relief Valves

Julia V. Bukowski, Ph.D.
Villanova University
Dept of Electrical & Computer Engineering
Villanova, PA 19085
julia.bukowski@villanova.edu

William M. Goble, Ph.D.
exida, LLC
64 N. Main Street
Sellersville, PA 18960
wgoble@exida.com

Prepared for Presentation at
American Institute of Chemical Engineers
2010 Spring Meeting
6th Global Congress on Process Safety
San Antonio, Texas
March 22-24, 2010

UNPUBLISHED

Using Root Cause Failure Analysis Results to Estimate Failure Rates and PFDavg for Spring Operated Pressure Relief Valves

Julia V. Bukowski, Ph.D.
Villanova University
Dept of Electrical & Computer Engineering
Villanova, PA 19085
julia.bukowski@villanova.edu

William M. Goble, Ph.D.
exida, LLC

Keywords: spring operated pressure relief valves, root cause of failure analysis, estimated failure rate, estimated time to failure, probability of initial failure, proof test, PERD

Abstract

In a companion paper [1], root cause of failure analysis (RCA) was used to estimate the times to failure of a set of spring operated pressure relief valves (SORV) that were found in the fail-to-open (FTO) condition upon proof test. We now use these estimated times to failure to estimate the failure rate, $\lambda_D(t)$ of the SORV population with respect to the FTO mode of failure. The average probability of failure on demand (PFDavg) is then computed using the estimate for $\lambda_D(t)$. Additionally, we compute PFDavg directly from the proof test data and estimated times to failure and explain why this new approach provides a more accurate characterization of process-wide safety than does PFDavg computed from $\lambda_D(t)$ and probability of initial failure (PIF). We also demonstrate that the changes instituted as a result of the RCA study at the data-source facility have lead to measurable improvement in plant safety.

1. Introduction

As presented in IEC61511[2], calculating PFDavg and the resulting safety integrity level (SIL) requires first calculating or estimating the failure rate, $\lambda_D(t)$, for the failure mode of dangerous failure. As applied to computing the SIL for pressure relief valves (PRV), PFDavg must also account for the possibly non-zero probability of initial failure (PIF). There have been a number of efforts [3, 4, 5, 6, 7] to analyze proof test data for pressure relief valves (PRV) with the objective being to characterize $\lambda_D(t)$ and PIF for the Process Equipment Reliability Database (PERD) [8] failure mode "fail-to-open" (FTO), also known as stuck shut. However, in [7] it was demonstrated that proof test data consisting solely of operating hours on a PRV and the result of the proof test (pass or fail) with respect to the FTO condition, is insufficient to deduce a single model for $\lambda_D(t)$ and for PIF. Rather, one must necessarily admit multiple PRV failure rates and PIF all of which explain the observed proof test results. This happens because proof tests cannot provide the *time to failure*, only the *time to discovery of failure*. Further, each possible failure rate explanation results in a potentially different value for PFDavg.

In [7] it was proposed that performing root cause of failure analysis (RCA) on failed PRV might provide a basis for estimating, from proof test data, the time to failure of a PRV. This additional information could then be used to characterize more accurately $\lambda_D(t)$ and PIF and narrow the range of possible PFDavg. Nevertheless, based on current techniques, PFDavg remains a quantity computed indirectly from proof test data under assumptions that often are not met in practice.

In this paper, we use the estimated times to failure that resulted from RCA of a set of proof tests performed on used and new SORV originally reported in [1] to estimate λ_D and to compute PFDavg from it for the facility that provided the data. We discuss why current calculation methods for PFDavg do not accurately characterize the process-wide safety being achieved. We propose a new method for estimating PFDavg directly from proof test data which incorporates the realistic conditions found in practice and provides an ongoing profile of safety performance.

2. Notation

FIT	1 failure/10^9 operating hours
FTO	fail to open, also known as "stuck-shut"
PERD	Process Equipment Reliability Database
PFDavg	average probability of failure on demand
PFD(t)	instantaneous probability of failure on demand
PIF	probability of initial failure
PRV	pressure relief valve(s)
R^2	coefficient of determination
RC	root cause of failure
RCA	root cause of failure analysis
SIL	safety integrity level
SORV	spring operated pressure relief valve
$\lambda_D(t)$	failure rate of the FTO mode of failure

3. Background

The reader is encouraged to refer to [1] for a fuller understanding of the RCA study that forms the foundation for this paper. However, to allow this paper to be self-contained, a summary of the RCA study is provided here.

The extensive PERD data set, originating from proof tests conducted over a nearly seven year period between 2003 and 2009, that formed the basis for the RCA study came from Savannah River Nuclear Solutions – US-DOE Savannah River Site (SRS) which has a dedicated valve test and repair facility. SRS has a policy of testing all PRV (new and used) in its facilities not more than six months prior to installation. The proof test database contained a total of 5632 valve tests performed on 1388 used and 4244 new SORV rated at or greater than 35 psig. The used valve data was reduced to 1150 tests; some test data were eliminated because either they were judged to be incorrect or the test record was missing information required for the estimates of $\lambda_D(t)$. No failures of the FTO type were eliminated.

The RCA study investigated the root cause of failure of a total of 44 SORV (including new and used valves) and considered seven of the ten PERD failure modes. Based on the information obtained from the RCA, a technical expert estimated a range of probable times to failure for each of the failed SORV included in the study. In addition, each failure was further categorized as random or systematic and systematic failures were further categorized as originating with the manufacture or the user. Of relevance to this paper are the valves whose conditions were revealed by proof test to be FTO. Of the used valves, 15 tested FTO. The root cause of failure (RC) was identified for each of these 15 valves. Of the new valves, 12 tested FTO and the RC was identified for 7 of these 12 valves.

Table 1 lists the 22 used and new valves that tested FTO along with the results of RCA and the estimated range of times to failure for each of the FTO valves. Please refer to [1] for a detailed explanation of how the times to failure were estimated.

Table 1. Summary of RCA for FTO Valves

RCA #	Valve Origin	Physical Cause	Root Cause (RC) R = random; M = manufacturer; U = user	Est time to Failure(yrs)
1	Used	oxides formed	R: process overheated	0 - 2.7
2	Used	same as RC	R: micro-embedment; surface creep	1.5 – 1.8
3	Used	same as RC	R: micro-embedment; surface creep	2 - 2.5
4	Used	same as RC	R: micro-embedment; surface creep	2 - 2.5
5	Used	same as RC	R: micro-embedment; surface creep	2 - 3
6	Used	Corrosion	U: exposed to weather	2 - 2.4
7	Used	Corrosion	U: poor choice materials; exposed to weather	2 - 2.5
8	Used	Corrosion	U: poor choice of materials	2 - 2.5
9	Used	Corrosion	M: carbon steel screw in stainless steel valve	2 - 2.5
10	Used	Corrosion	U: in service too long	3 - 3.1
11	Used	Corrosion	U: in service too long; exposed to weather	3 - 4
12	Used	Corrosion	U: in service too long	3 - 4
13	Used	Corrosion	U: poor choice of materials; in service too long	4 - 5
14	Used	Corrosion	U: design/engineering specs inadequate	4.5 - 6
15	Used	Corrosion	U: in service too long; exposed to weather	5 - 6
16	New	same as RC	R: micro-embedment; surface creep	0
17	New	same as RC	R: micro-embedment; surface creep	0
18	New	same as RC	R: micro-embedment; surface creep	0
19	New	same as RC	R: soft seat embedment	0
20	New	not set as stamped	M: quality control inadequate; manf did not test	0
21	New	set point high as received	M: quality control inadequate	0
22	New	no discharge port	M: quality control inadequate; manf did not test	0

4. Estimating $\lambda_D(t)$

Because SRS tests all new valves prior to installation, the new valves found to be FTO were not installed in that condition. Therefore, the new valve data do not play a role in estimating $\lambda_D(t)$. Thus this section concerns only the used FTO valves.

The failure rate, $\lambda_D(t)$, in its most general form, usually referred to as a "bathtub curve," is characterized by three distinct regions corresponding to different underlying general causes of equipment failure. These are

- A region of declining failure rate corresponding to infant mortality failures due to weakness in the equipment
- A region of constant failure rate corresponding to failures due to stresses exceeding the norm of the process and causing failure of the equipment
- A region of increasing failure rate corresponding to failures due to aging of the equipment

Experience indicates that it is best to fit a $\lambda_D(t)$ curve separately to each of the three regions and then to combine the three separate characterization into a single $\lambda_D(t)$ curve.

In [7] it was proposed that infant mortality failures could explain the non-zero PIF found in previous studies. However, careful scrutiny of the RCA study results did not identify any used FTO failures corresponding to infant mortality.

One random failure (RCA # 1) could reasonably be assigned to the category of constant failure rate. In this case, a process upset created temperature stresses in excessive of the intended range of normal operation. The excessive temperatures in the presence of moisture allowed for the formation of oxides which caused the valve to test as FTO. Although the formation of oxides may take several weeks to a month, because process upsets are random in nature and can occur at any time, the resultant state FTO due to oxide build up can effectively occur at any time after the first few weeks or month of operation. Generally we could model this type of failure by a constant failure rate and we, in fact, estimated the constant failure rate as 1 failure/(total operating hours on all valves in population). This resulted in a constant failure rate of 1 failure/38,815,680 hours = 26 FITS = $26 * 10^{-9}$ failures/hr = $2*10^{-4}$ failures/yr for this SORV population.

The remaining used FTO failures generally belong to the category of equipment aging. But we need to distinguish two types of aging. In the first type, equipment aging takes place even if the valve is not installed in the process. Thus, in computing a failure rate one must take into account the accumulated storage time since the valve was known to be in "like new" condition. In the second type, equipment aging takes place due to the conditions the equipment encounters because it is installed in the process; little or no aging takes place in storage.

Aging of the first type clearly includes the four remaining random failures which involve micro-embedment or surface creep [1 - Appendix B]. These phenomena most commonly occur when the valve's surface interfaces, under pressure of a spring or other load, deform and adhere to one another. This aging takes place even if the valve is not installed in a process. (This is why several new valves tested FTO with identified root causes of micro-embedment or surface creep. It is likely they had been in storage for some time. In addition, soft seat embedment falls into this aging classification of failures and one such example occurred in the new valve tests.) Some corrosion may also develop while equipment is stored. However, if equipment has been properly stored so that it remains in 'like new" condition until the time of installation, then corrosion represents the second type of aging that takes place primarily when the valve is installed in the process.

Typically, $\lambda_D(t)$ is estimated by computing the operating hours on each valve and then treating all valves as being installed at t = 0 in "like new" condition. The time interval [0, largest operating time of valves tested] is then divided into non-overlapping intervals. The total number of failures in an interval is divided by the total number of operating hours in the interval and that value is used as an estimate for $\lambda_D(t)$ at some t in the interval. Curve fitting is then applied to the estimated points to find a model for $\lambda_D(t)$.

Initially we tried to fit a growth curve to the aggregate of the aging data. This was not successful. However, when we separated the micro-embedment and creep failures from the corrosion failures and fitted those data separately over two different time frames, we achieved satisfactory straight line fits (consistent with linear aging as assumed in [9]) with statistically significant R^2 values. The embedment/creep failures seem to cluster between 1.5 and 3 years. The corrosion failures occur over a period of 2 to 6 years.

Figure 1 shows the estimates for $\lambda_D(t)$ based on the minimum (dash-dot line) and maximum (solid line) estimated failure times. Note that the $\lambda_D(t)$ based on the maximum estimated failure times eventually intersects the $\lambda_D(t)$ based on the minimum estimated failure times because ultimately, all the models must account for the total number of failures by seven years. The solid $\lambda_D(t)$ curve accounts for fewer failures earlier on and then must increase at a faster rate than the dash-dot curve to account for the remaining failures later.

It must be emphasized at this point that the $\lambda_D(t)$ curves as shown apply **only** to this data set. Remember, this data set incorporates a particular mixture of SORV which may be susceptible to different causes of failure. For example, carbon steel valves are more susceptible to corrosion, but not to embedment-creep failures. Similarly, stainless steel valves are more susceptible to embedment-creep failures, but not to corrosion failures. Further, the data also incorporate the particular practices and procedures of the facility, for example, the testing of all new valves prior to installation and the retesting of any stored valve whose date since last test exceeds six months. These practices change the times to failures of valves after installation where "aging on the shelf" can take place compared to a facility that does not support these practices. If the mix of valves or any practices and procedures changed, we can expect $\lambda_D(t)$ to change as well.

Based on these $\lambda_D(t)$ estimates, PFDavg for this data set is shown in Figure 2. Note that PFDavg computed in this manner places the safety performance, estimated from the data set by the techniques described, in the SIL 3 range for the first 2.5 years and in the SIL 2 range for most of the rest of the plot. However, this would not be an appropriate interpretation because, in fact, we are not dealing with a static valve population all installed at the same time but rather a population which is undergoing proof testing and repair on a variety of maintenance intervals over a period of seven years. PFDavg as computed from $\lambda_D(t)$ does not account for these realities.

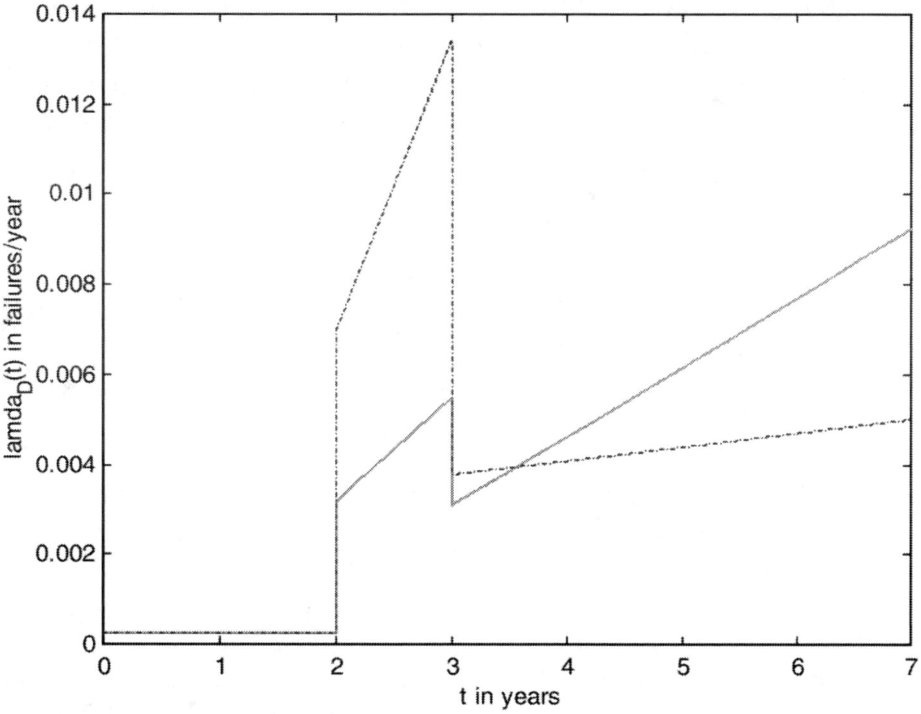

Figure 1. Plots of $\lambda_D(t)$ based on minimum (blue dash-dot line) and maximum (green solid line) estimated times to failure. The two plots of $\lambda_D(t)$ overlap on the time interval 0 to 2 years.

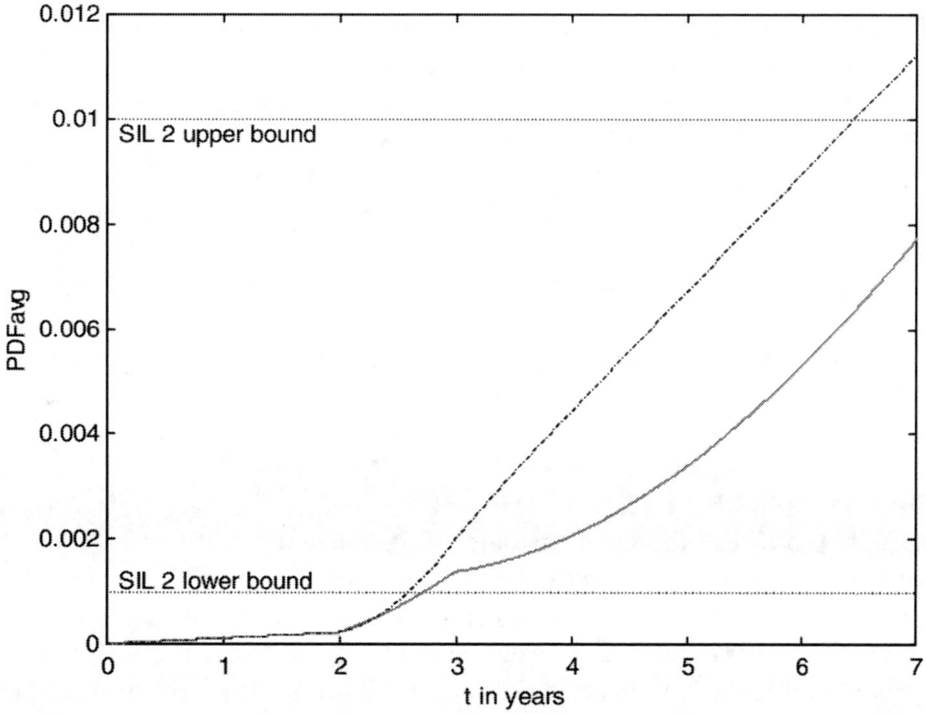

Figure 2. Plot of PFDavg computed from estimated $\lambda_D(t)$ based on minimum (blue dash-dot line) and maximum (green solid line) estimated times to failure. The two plots of PFDavg overlap on the time interval 0 to 2 years.

For this reason, and contrary to our original beliefs, we now believe that the above approach to computing PFDavg by seeking to model $\lambda_D(t)$ from proof test data is not the best way to proceed. Next we discuss a new method for estimating PFDavg directly from proof test data which is easier to implement and more accurately characterizes PFDavg without requiring a number of assumptions that are unlikely to be met in practice.

5. A New Method for Computing PFDavg

PFDavg as originally defined by IEC61511 [2] was intended to apply to electronic equipment whose failure rate is reasonably characterized by a constant. Under the assumptions that the electronic component is working when installed, and that after failure it is restored to a "like new" state, PFDavg serves as a means to set the inspection interval that would maintain the electronic equipment within a certain SIL.

This concept of PFDavg when applied to a *particular* valve would serve the same purpose. However, the $\lambda_D(t)$ used to compute PFDavg must then also apply to that valve specifically. This is not the case for $\lambda_D(t)$ or PFDavg derived in some manner from a proof test database populated by many different types of valves of different sizes, material compositions, and designs, in different fluid services, on different maintenance schedules, and maintained under different policies and procedures. As applied to the chemical process industry, it seems that a more informative approach to the use of PFDavg would be to compute PFDavg on a calendar basis for a given facility to determine that facility's safety performance in a way that incorporates the specifics of the facility including the characteristics listed above.

To accomplish this, we considered data on valves in-service at any time between January 1, 2003 and September 15, 2009. This calendar range encompasses 2450 days. On any day, we identified the number of valves in the proof test database that were in-service on that particular day. We then identified how many of those valves were FTO on that day. We considered three cases. In the worst case, given no information about estimated time to failure from RCA, we assume that a valve was FTO during its entire time in service, so if it were in-service on the day of interest it would count as a failure. For the other two cases, we assume that the valve failed on a day corresponding to either the minimum or maximum estimated time of failure. If on the day of interest the valve were in-service and the day corresponding to the estimated minimum or maximum failure time were less than or equal to the day of interest then the valve was deemed failed on that day and counted as a failure. Now on any given day, (the number of valves failed)/(number of valves in service) is an estimate of PFD(t) for that day.

To estimate PFD(t) day by day under the worst case scenario for each valve requires only the calendar date of last proof test, the calendar date of the current proof test and whether the valve passed or failed the proof test relative to the FTO failure mode. In addition, if one can estimate a range of failure times for the FTO valves, one can refine the PFD(t) estimate. Figure 3 shows the PFD(t) for our data set under the worst case scenario (red dotted line) and the assumption of minimum (blue dash-dot line) and maximum (green solid line) estimated failure times.

Note the general downward trend of the PFD(t) curves for all three times to failure assumptions. We interpret this to mean that the on-going changes instituted by SRS as a result of the RCA study have in fact contributed to improved safety.

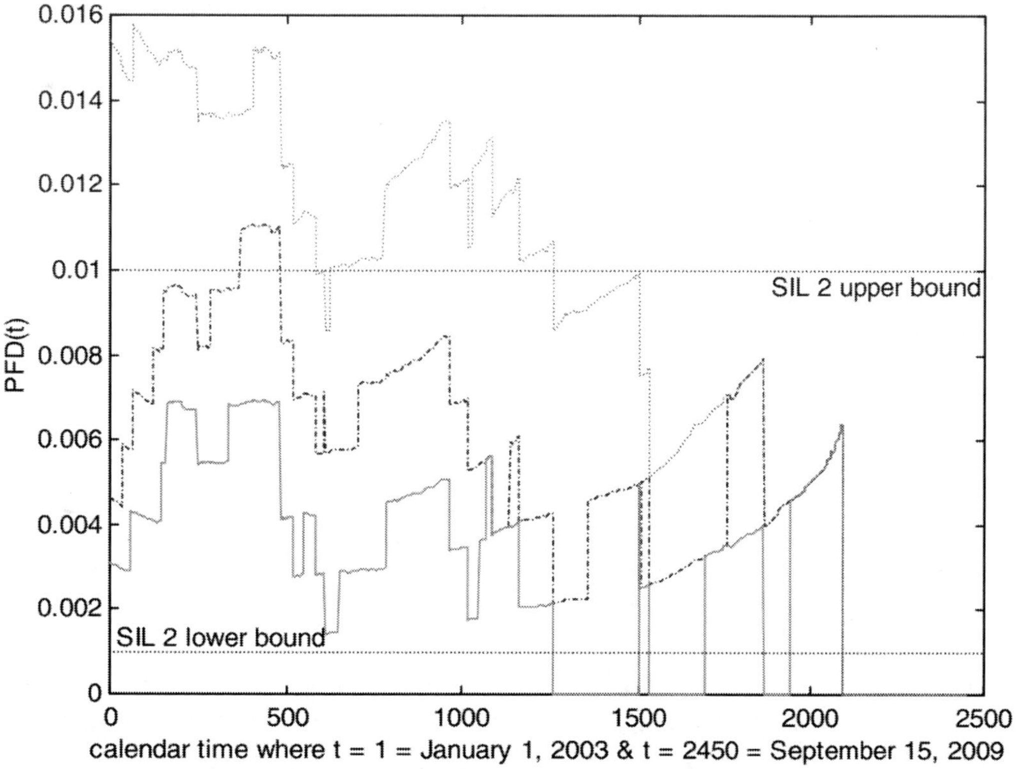

Figure 3. Plot of PFD(t) as estimated for each calendar day between January 1, 2003 and September 15, 2009 based on three assumed times to failure: failed when installed (red dotted line), failed at the minimum estimated failure time (blue dash-dot line) and failed at the maximum estimated failure time (green solid line).

To relate the information in Figure 3 to PFDavg, we considered averaging PFD(t) in two ways. First, we defined

$$PFDavg = \left(\frac{1}{t}\right)\int_{0}^{t} PFD(t)dt. \qquad (1)$$

This means that we are averaging PFD(t) from January 1, 2003 to the date represented by t. The results of this averaging are shown in Figure 4. Note, as t becomes larger, we are averaging over ever larger time intervals and it is not clear that the average is meaningful beyond a period of 5 years (t=5*365+1) as scheduled inspection intervals are 5 years or less. However, it is instructive to note the general downward trend of PFDavg indicating that, as time progresses, the changes instituted in response to RCA findings in prior years are in fact producing measureable changes in the facility's safety performance.

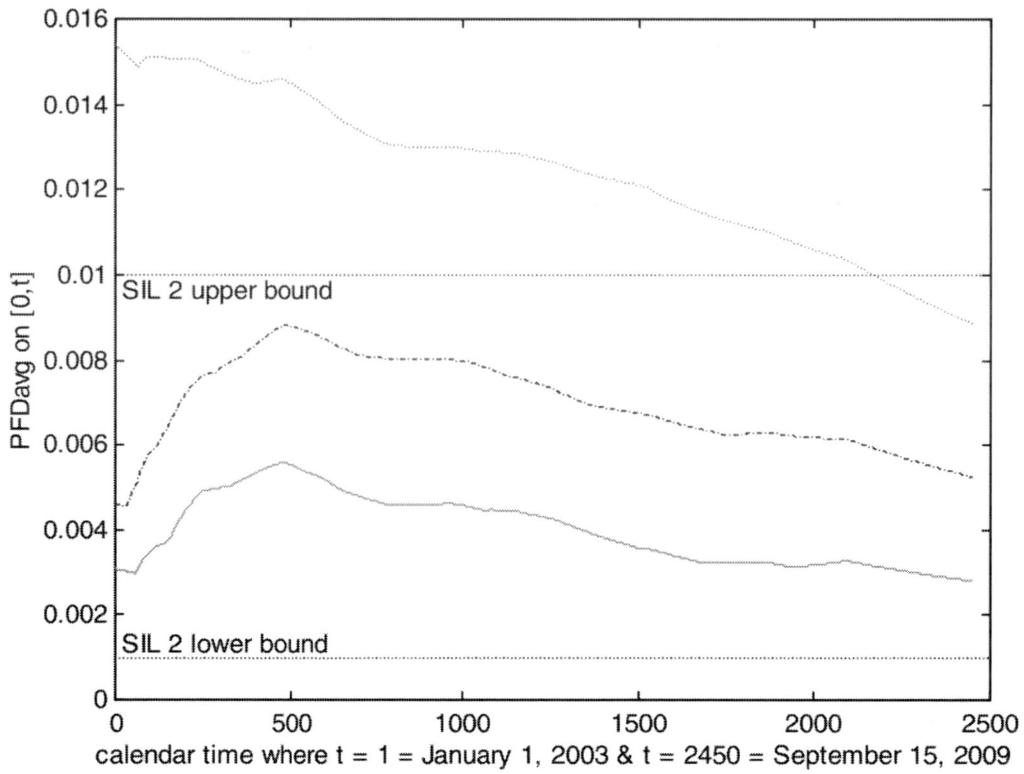

Figure 4. PFDavg based on PFD(t) in Figure 3 and averaged over the time interval [0, t].

In order to compute a more meaningful PFDavg, we note that at SRS, inspection intervals range from 1 to 5 years with an overall average of about 3.5 years. We decided to compute

$$PFDavg = \left(\frac{1}{t}\right) \int_{t-3*365}^{t} PFD(t)dt. \qquad (2)$$

The results of this averaging are shown in Figure 5. In this, case on any given day, we are averaging PFD(t) over the last three years. Thus, the plot does not begin until t = 3*365 days (~December 31, 2005) because we need that long to accumulate three years of PFD(t) data points for the first average. Again note the general downward trend of PFDavg under this scenario indicating the facility's improving safety performance.

6. Estimating PIF

Although we can assert that the data set analyze in this paper has a PIF = 0 because of the SRS practice of testing all new valves on-site prior to installation, in general practice, this is not the case and we would be remiss if we did not address the general case of PIF. The discovery of 12 new valve failures that would otherwise have been installed, if not for SRS test policy, makes it indisputable that, in general, PIF is a non-zero quantity. With an installed population of about

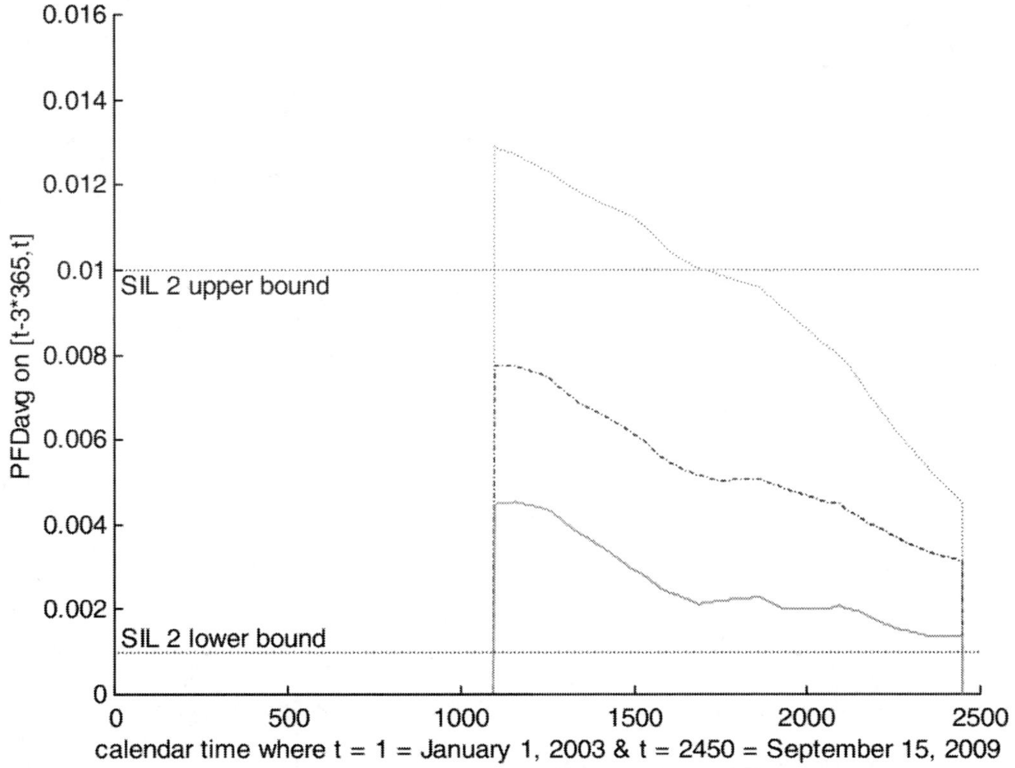

Figure 5. PFDavg based on PFD(t) of Figure 3 averaged over the time interval [t-3*365, t].

5000 valves, 12 new valves installed in the FTO condition would mean a PIF = 0.0024 or 0.24% for this data set.

To understand the impact of not testing new valves on the facility's safety performance, we recomputed PFD(t) and PFDavg based on the facility's average 3 year proof test interval but this time included the 12 new failed valves in the data set. The 12 valves were assumed to be installed on the test date when they were found failed in new valve proof test, and assumed to have remained in service until their next scheduled proof test determined by their assigned preventive maintenance frequencies. Figure 6 shows PFD(t) under the assumption that new valves are not tested. Figure 7 shows PFDavg over 3 year intervals under the assumption that new valves are not tested and compares it to PFDavg under the condition that new valves are tested prior to installation. Note that PFD(t) and PFDavg with the 12 new valve failures installed are only shown up to day 2000 (June 22, 2008) because beyond that day the values rise dramatically and cannot be shown on a scale comparable to PFD(t) and PFDavg without the 12 new valve failures. Clearly not testing new valves would dramatically change the PFD(t) and PFDavg achieved. With its present policy of testing new valves prior to installation, and given the changes instituted due to RCA findings, this facility can claim SIL 2 performance.

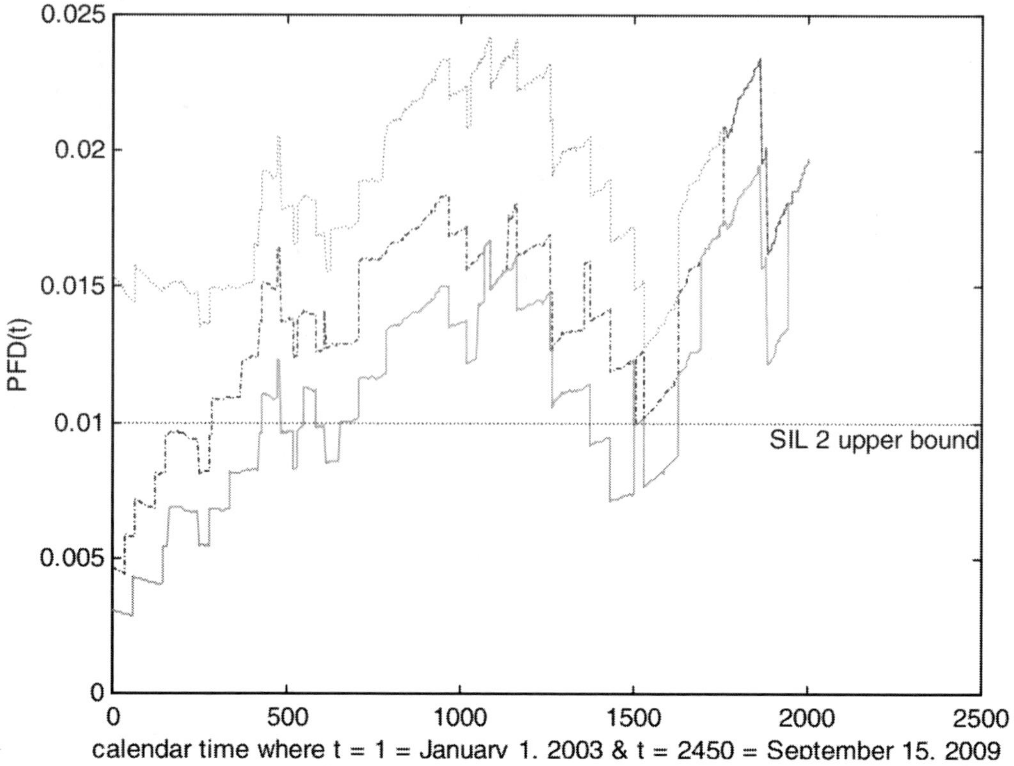

Figure 6. Plot of PFD(t) (with 12 FTO's from new valve tests included) as estimated for each calendar day between January 1, 2003 and September 15, 2009 based on three assumed times to failure: failed when installed (red dotted line), failed at the minimum estimated failure time (blue dash-dot line) and failed at the maximum estimated failure time (green solid line).

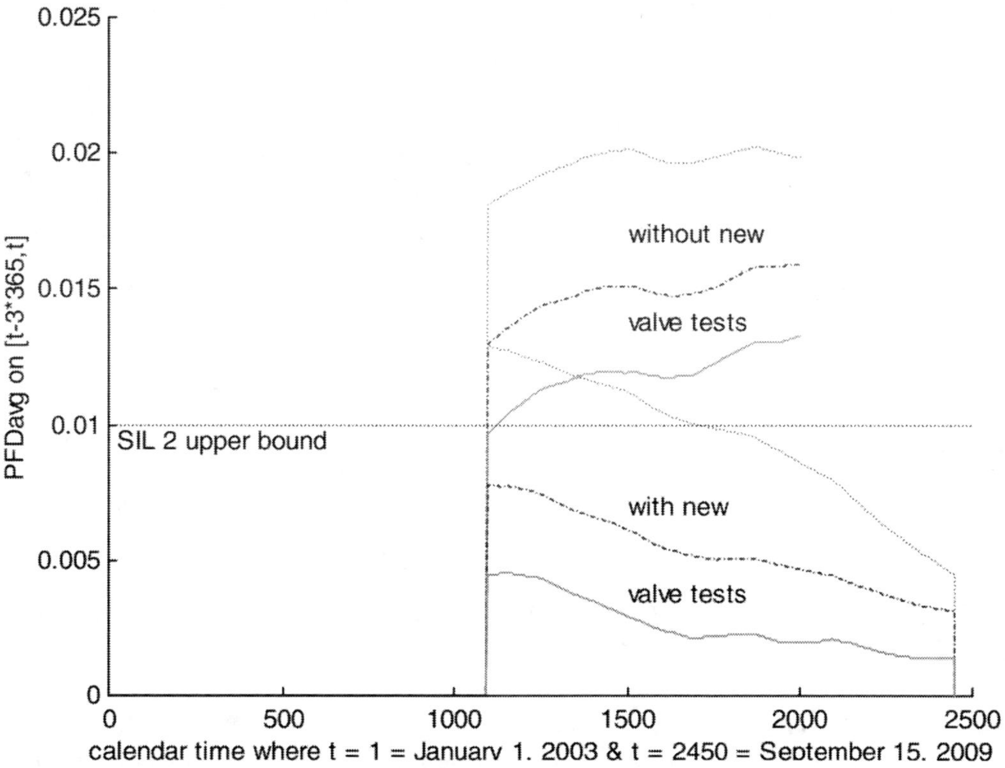

Figure 7. PFDavg based on PFD(t) in Figures 3 & 6 averaged over the time interval [t-3*365, t]. Note that the upper lines represent PFDavg assuming new valves are not tested and the lower lines represent PFDavg when new valves are tested.

7. Conclusions & Recommendations

After working with proof test data for some time now we have come to the following conclusions and recommendations. Even minimal proof test data can provide important information as to the level of safety actually being achieved at a given facility. This can easily be measured in terms of PFDavg and related to corresponding SIL levels. Furthermore, a modest RCA program targeting only FTO failures can, with limited investments of time and other resources, provide significant payback in understanding where to focus efforts to improve safety over many PERD failure modes. This return on investment is especially significant when applied to systematic failures which likely account for significantly more than half of all FTO's. An RCA program along with additional data collection can produce realistic estimates for times to failure that can further refine estimates of PFDavg and SIL.

We also have concluded that it will be difficult to extract useful failure rate information from proof test data encompassing many different types of valves, installed at different times, in different services and maintained under different maintenance plans. Estimating failure rates really requires concentrating on data for a specific valve or at least a specific subset of similar valves. It is unlikely that a single facility will have enough data for each of the specific valves or subsets of valves of interest to easily reap failure rate results from proof test data. However, if companies were to pool their data, the aggregate may well contain sufficient information about particular valves to yield useful failure rate estimates encompassing both systematic and random failures. Of course, this approach will require that companies collect similar data in compatible formats.

Finally, we emphasize the importance and value of being able to track the history of an individual valve. To this end we strongly recommend that all relief valves be identified with a manufacturer serial number that incorporates information about the assembly date. Further, we recommend that, in addition to other information being collected about valves in proof test, the installation and de-installation dates be recorded.

8. References

[1] J. V. Bukowski, Gross, R. E., "Using Root Cause Failure Analysis Results to Estimate Failure Rates and PFDavg for Spring Operated Pressure Relief Valves," *Proceedings of the AIChE 6th Global Congress on Process Safety, 12th Process Plant Safety Symposium*, San Antonio, TX, March 2010.

[2] ANSI/ISA 84.00.01:2004 (IEC 61511 Mod.), Application of Safety Instrumented Systems for the Process Industries, ISA, Raleigh, NC, USA, 2004.

[3] Sheesley, J. H., Thomas, H. W. and Valenzuela, C. A., "Quantal Response Analysis of Relief Valve Test Data," *ASQC 49th Annual Quality Congress Proceedings*, 1995, pp. 741-748.

[4] Bukowski, J. V., and Goble, W. M., "Analysis of Pressure Relief Valve Proof Test Data," *AIChE Journal Process Safety Progress*, March 2009.

[5] Bukowski, J. V. and Goble, W. M., "Unexpected Results from the Analysis of PERD Test Data & the Implication for Pressure Relief Valve Safety," *Proceeding 64th Annual Instrumentation Symposium for the Process Industries*, January 2009, College Station, Tx.

[6] Bukowski, J. V. and Goble, W. M., "Validation of a Mechanical Component Constant Failure Rate Database," *Proceedings Annual Reliability and Maintainability Symposium*, January 2009, Fort Worth, TX.

[7] Bukowski, J.V. and Goble, W.M., "New Insights into Computing PFDavg for Pressure Relief Valves Based on Proof Test Data," proceedings 5th Annual Global Conference on Process Safety, 11th Process Plant Safety Symposium, April 2009, Tampa, FL.

[8] Process Equipment Reliability Database (PERD) Initiative, "Spring Operated Relief Valve Taxonomy," Center for Chemical Process Safety, 2006.

[9] Harris, S. P., and Gross, R. E., "Adjusting Spring Operated Pressure Relief Valve Proof Test Intervals Using Statistical Modeling and Comparing the Resulting Financial Risk," *Proceedings of the AIChE 6th Global Congress on Process Safety, 12th Process Plant Safety Symposium*, San Antonio, TX, March 2010.

Acknowledgement

The authors wish to thank Robert Gross of Savannah River Nuclear Solutions for his helpful suggestions and review of this paper.

Results of Root Cause Analyses of Spring Operated Pressure Relief Valve Failures

Julia V. Bukowski, Ph.D.
Villanova University
Dept of Electrical & Computer Engineering
Villanova, PA 19085
julia.bukowski@villanova.edu

Robert E. Gross, P.E.
Savannah River Nuclear Solutions
US-DOE Savannah River Site
ms 704-2H
Aiken, SC 29808
robert.gross@srs.gov

Prepared for Presentation at
American Institute of Chemical Engineers
2010 Spring Meeting
6th Global Congress on Process Safety
San Antonio, Texas
March 22-24, 2010

UNPUBLISHED

Results of Root Cause Analyses of Spring Operated Pressure Relief Valve Failures

Julia V. Bukowski, Ph.D.
Villanova University
Dept of Electrical & Computer Engineering
Villanova, PA 19085
julia.bukowski@villanova.edu

Robert E. Gross, P.E.
Savannah River Nuclear Solutions
US-DOE Savannah River Site

Keywords: root cause of failure analysis, proof test, spring operated pressure relief valve, PERD, estimated time to failure

Abstract

A subset of 44 failed spring operated pressure relief valves (SORV) from a set of over 5600 Process Equipment Reliability Database (PERD) proof tests was subjected to root cause of failure analysis (RCA). This paper reports on the findings of those 44 analyses and includes specific examples of root causes (RC) identified. We considered SORV failures detected in both new and used valves over a period of almost seven years. We categorized the failures using the PERD taxonomy and, additionally, divided the failures into systematic (where reasonable changes to processes or procedures might prevent future failures) and random failures based on each RC. Further, we identified a number of subcategories of RC because they suggest possible actions to significantly reduce failure rates and improve safety performance. We considered how remedial attention to some root causes has markedly improved proof test results at the facility that provided the proof test data. We provide recommendations for conducting such RCA as well as for initiating a RCA program and discuss how time to valve failure (important information for computing valve failure rates) can be estimated based on the RCA findings. Non-Newtonian fluids are outside the scope of this paper.

1. Introduction

Savannah River Nuclear Solutions (SRNS) – US-DOE Savannah River Site (SRS) has collected and analyzed proof test data for a number of years and began to investigate root causes of failure in some individual valves. Some patterns of failure causes quickly emerged and from the lessons learned, procedures were instituted with the intention of improving safety. Recently, we decided to revisit the RCA reports to look at the "big picture" to see if we could find clear evidence of safety improvements. When we began this study, our intention was only to consider RCA for a set of "fails-to-open" (FTO) or "stuck shut" failures because of their importance with respect to safety standards and performance indices such as average probability of failure on demand.

However, in addition to performing RCA on FTO valves we also looked at a number of other failure categories per the PERD SORV Taxonomy [1]. This paper describes the work undertaken and the results obtained.

2. Notation

ASME	American Society of Mechanical Engineers
FCR	fails to completely reseat
FRRC	fails to relief required capacity
FTO	fail to open
NDE	nondestructive examination
NEW	SORV that has not previously been installed in any process
OAS	opens above set pressure
OBS	opens below set pressure
PERD	Process Equipment Reliability Database
Proof Test	the practice of pressurizing the inlet of a new or used relief valve on a test stand; relieving pressure and seat tightness are tested and the as-found values are compared to initial set pressure.
RC	root cause(s)
RCA	root cause of failure analysis
SL	seat leakage
SO	spuriously opens
SORV	spring operated relief valve(s)
SRNS	Savannah River Nuclear Solutions
SRS	Savannah River Site
TE	Technical Expert
US-DOE	United States Department of Energy
USED	SORV that has previously been installed in a process

3. PERD Failure Taxonomy

We used the ten failure modes for SORV as defined in [1]. Of the four modes of complete failure, two are represented in our data, viz., fails to open (FTO) and spuriously opens (SO). Of the six modes of partial failure, five are represented in our data, viz., opens above set pressure (OAS), fails to relieve required capacity (FRRC), opens below set pressure (OBS), fails to completely reseat (FCR), and seat leakage (SL).

4. The Data

4.1 Data Source

Data for this study came from SRS. The data, collected during an approximate 7 year period from 2003 to 2009, consist of a total of 5632 valve tests performed on 1388 used and 4244 new SORV rated at or greater than 35 psig. While the valves tested represent more than 55 different manufacturers, eight manufacturers account for 4850 of the valve tests with each of the eight accounting for 350 or more tests. SRS conducts all of its valve tests at one dedicated test and repair facility on the SRS which insures consistency of test and repair facilities and personnel, test procedures, management oversight, and data records. It is the policy of SRS to proof test all valves, including new valves, prior to installation. The criterion for "prior to installation" is that the valve be subjected to proof testing by SRS personnel at most six months prior to installation.

4.2 Proof Test Procedures that Generated Data

A full description of the proof test procedures as practiced at SRS is provided in [2]. A brief description is provided here. When a new or used valve is received in the valve repair shop, it is checked for evidence of external physical damage, corrosion, and deposits. The manufacturer, the model, and, if present, the serial number are recorded. Large valves are lifted to the inspection test stand by chain hoist. Following the external visual inspection, valves are first tested in the "as-arrived" or "as-found" condition. Test pressure is increased on the test stand until the valve lifts or "pops" open. This activity is believed to simulate field performance. Typically, a valve will fail its proof test if it lifts above or below the American Society of Mechanical Engineers (ASME) tolerance on the valve's tagged set pressure (set point). Another common mode of failure is seat leakage which occurs if there is a flow of bubbles or bulging of tape at the outlet while the test pressure is held at 90% of set pressure. If a valve fails its proof test, the valve is disassembled and additional parts inspection is performed. All parts are cleaned, either mechanically or chemically. In some cases, parts will be replaced, lapped to ensure a leak-tight seal, or machined if the seat and nozzle have experienced chemical or mechanical deformation.

5. Root Cause Analysis

5.1 Performing a Root Cause Analysis

In order to have first hand information to perform RCA, it is important that relief valve testing procedures and day-to-day work practices be under the supervision of a senior, experienced Technical Expert (TE). At SRS, the role of TE is filled by a licensed Professional Engineer (Mechanical) with more than 20 years of experience in valve design, manufacture, inspection, repair and reliability who provides technical oversight for the relief valve repair shop. SRS experience indicates that the number of improvements that can be made in SORV reliability is directly related to how much time the TE spends with valve shop technicians, management, and technical oversight committees. It is also important that technicians / mechanics feel confident that the TE and the company are interested in their opinions and concerns that arise from testing and rebuilding SORV.

When a valve is found to be stuck shut, i.e., it has failed to open when pressurized to 1.5 times its set pressure, a TE is contacted by the shop to investigate. Once the TE arrives, a discussion takes place reviewing the particular valve's fluid service, set pressure, operating temperature, field location, and time in service. A technical manual on the particular manufacturer and model is reviewed to determine body, bonnet, trim materials, and type of seat. The valve manufacturer or local dealer may be called for advice. The proof test procedure is then executed again on the valve to demonstrate whether it is stuck shut at 1.5 times set pressure; the TE may want to pressurize the valve to 2 times set pressure or more, if the valve body design safely allows such pressurization, to see if the valve will open. Since most fluid piping systems are leak tested per ASME Code to 1.5 times the design pressure prior to operating, and the safety factor for Code piping is in the range of 3-4, it is important to evaluate whether the piping system or vessel pressure boundary was threatened by the SORV's in-service condition.

Once the proof test is complete, technicians and the TE disassemble and inspect the valve, taking photographs as it is being torn down. In many cases, especially carbon steel valves, the failure mode is easily revealed by inspecting the spring, spring washers, valve stem and seating surfaces. However, discovering the failure mode of valves that are not composed of carbon steel, but rather of other materials such as stainless steel or bronze, may require deeper investigation. For example, in cases where the disc and nozzle surfaces are stuck together, nondestructive examination (NDE) by qualified individuals should be requested for appropriate testing. NDE includes visual stereo microscope examination of surfaces and deposits, determination of the metal alloy and surface deposit chemical makeup by x-ray diffraction spectroscopy and measurement of critical parts such as nozzle, disc, spring and spindle. Once this data is collected, the exact physical cause and root cause of failure may be determined. A technical report is issued to all the program (pressure protection) owners with recommendations on how best to eliminate future failures.

5.2 Categorizing Failures: Random vs. Systematic

The generally accepted notion of random failure is one where random stresses exceed the strength of a component causing it to fail. In the case of a SORV, an example of a random stress would be one where the process in which the SORV is installed, strays from its intended range, e.g., overheats, causing the SORV to fail. Failures of this type are difficult to predict or control. IEC61511 [3] describes systematic failures as those that can be corrected by reasonable change in process or procedure. Generally, they can be thought of as having some human error at their root. A few examples of systematic failures are manufacturing defects such as assembling a valve using the wrong part, leaving a valve in service too long and ordering and installing a valve made of materials inappropriate for the intended environment. Because systematic failures are more amenable to correction, after reviewing each RCA we considered whether the RC could most reasonably be categorized as random or systematic. If the RC were systematic, we sought to further categorize it as originating with the manufacturer or the end user. Such categorization provides indications as to what steps might reduce such failures in the future.

5.3 Estimating Time to Failure

Proof tests allow us to determine the *time to discovery of failure* but proof tests, of themselves, give no indication of *time to failure* (except that failure occurred sometime before discovery). When performing an RCA, it is by no means necessary to estimate time to failure. The RCA alone gives very useful information. However, if a significant database from proof testing is available, it may be possible to estimate the time to failure for the valve or at least a range of possible times to failure. Such time to failure information is useful in adjusting the proof test interval and might also be used to estimate the valve failure rate [4].

In order to estimate time to failure, detailed records should be kept on the internal condition of each valve and results of any NDE performed during the failure investigation. This background information and trending is used to predict times to failure for a particular valve. By comparing the results of proof tests in the proof test database, one can quickly collect all like valves at the facility, site, or even at a company level. The proof test, leak test, visual conditions and parts repaired/replaced data are analyzed to estimate a range of likely times to failure.

For example, suppose a valve failure is determined by inspection to be due to corrosion. We have found over the years that the galvanic cell and the oxygen concentration cell seem to cause the most damage. Corrosion rates can range from a uniform 3-5 mils/year to 15 mils/year (pitting) depending on a number of variables. If we know the condition of the valve at installation and can estimate the rate of corrosion formation and the width of the gap that must be filled by corrosion to cause the valve to fail to open, then we can determine a time range for the valve's failure. In Table 1, we see that a given valve which failed due to corrosion generally has an estimated time to failure variability of six months to a year (e.g., 2.5-3 or 4-5 yrs), but different valves have quite different minimum estimated failure times. This is because valves of different sizes require different amounts of corrosion to accumulate before they reach the FTO condition. If the maximum estimated failure time is larger than the actual in-service time, the maximum is reduced to the in-service time, since clearly the valve failed before the proof test which showed it to be failed. Another example of estimating failure times is for the RC of micro-embedment or surface creep. The amount of time to cause these types of failures depends on the material composition and surface area of the interface surfaces under pressure, the set pressure, and the storage and process temperatures. Therefore, we see in Table 1 some embedment/creep failures are estimated at 2-2.5 years while others are estimated at 2-3 years. These types of estimation require the expert knowledge of an experienced engineer and an extensive database to support the estimates made. A few of the estimated failure times are given not by ranges but by a single number. This occurs when the time to failure is known. This can happen, for example, in the case of a demand failure, human damage to the valve during removal, or the observance of a valve in service that is beginning to leak. Used valves found to be failed with physical causes that necessarily prevented the valve from ever working have failure times of 0. New valves found to be failed also have a failure time of 0.

5.4 Summary of Root Cause Analysis Results

Appendix A contains three examples of RCA performed on specific valves that were later categorized as random failure, systematic failure originating with manufacturer and systematic failure originating with user. Table 1 contains a summary of the results of RCA for 44 valve tests. All 15 used-valve tests conducted between 2003 and 2009 that resulted in FTO are accounted for. Seven of twelve new-valve tests resulting in FTO are included. The remaining 22 valve tests provide a sample of five other PERD failure modes encompassing both new and used valves as well as one valve at the end of life. (End of life implies not only that the valve is used but also that it will be scrapped after this proof test. SRS has a policy of testing valves at end of life for the data they provide, and if the valve fails the proof test it is disassembled for additional failure information though it will not be repaired.) Numerous other PERD failures were recorded during testing but not subjected to RCA.

Table 1. Summary of RCA

RCA #	PERD Mode	*	Physical Cause	Root Cause (RC) R = random; M = manufacturer; U = user	Est time to Failure(yrs)
1	FTO	U	oxides formed	R: process overheated	0 - 2.7
2	FTO	U	same as RC	R: micro-embedment; surface creep	1.5 – 1.8
3	FTO	U	same as RC	R: micro-embedment; surface creep	2 - 2.5
4	FTO	U	same as RC	R: micro-embedment; surface creep	2 - 2.5
5	FTO	U	same as RC	R: micro-embedment; surface creep	2 - 3
6	FTO	U	corrosion	U: exposed to weather	2 - 2.4
7	FTO	U	corrosion	U: poor choice materials; exposed to weather	2 - 2.5
8	FTO	U	corrosion	U: poor choice of materials	2 - 2.5
9	FTO	U	corrosion	M: carbon steel screw in stainless steel valve	2 - 2.5
10	FTO	U	corrosion	U: in service too long	3 - 3.1
11	FTO	U	corrosion	U: in service too long; exposed to weather	3 - 4
12	FTO	U	corrosion	U: in service too long	3 - 4
13	FTO	U	corrosion	U: poor choice of materials; in service too long	4 - 5
14	FTO	U	corrosion	U: design/engineering specs inadequate	4.5 - 6
15	FTO	U	corrosion	U: in service too long; exposed to weather	5 - 6
16	FTO	N	same as RC	R: micro-embedment; surface creep	0
17	FTO	N	same as RC	R: micro-embedment; surface creep	0
18	FTO	N	same as RC	R: micro-embedment; surface creep	0
19	FTO	N	same as RC	R: soft seat embedment	0
20	FTO	N	not set as stamped	M: quality control inadequate; manf did not test	0
21	FTO	N	set point high as received	M: quality control inadequate	0
22	FTO	N	no discharge port	M: quality control inadequate; manf did not test	0
23	OAS	U	oxides formed	R: process overheated	0 - 3
24	OAS	U	corrosion	U: poor choice of materials	2.5 - 3
25	OAS	U	corrosion	U: outdoor exposure	3 - 3.1
26	OAS	N	set point not achieved	M: quality control inadequate	0
27	FCR	U	corrosion	U: poor choice of materials	2.5 - 2.75
28	FCR	U	corrosion	U: poor choice of materials	3 - 4
29	SO	U	damaged in service	U: excessive vibration	1 - 2
30	SO	U	damaged during removal	U: rough handling	4
31	SO	N	disc not touching nozzle	M: disc out of specification	0
32	FRRC	U	plastic plug in inlet	U: inadequate training or attention to detail	0
33	FRRC	U	flow passage too small	U: miscalculation of required size	0
34	OBS	N	not set as stamped	M: quality control inadequate; manf did not test	0
35	OBS	N	not set as stamped	M: quality control inadequate; manf did not test	0
36	SL	U	corrosion	U: in service too long; exposed to weather	3 - 3.3
37	SL	U	corrosion	U: in service too long	5 - 7
38	SL	U	insect nest	U: vent plug missing	1- 2
39	SL	U	wrong valve	U: human error; insufficient technical review	1.55
40	SL	U	damaged during removal	U: poor workmanship	1.9
41	SL	N	same as RC	R: soft seat embedment	0
42	SL	N	wrong nozzle	M: factory assembly error	0
43	SL	N	poor seating	M: manufacturing defect	0
44	SL	N	nozzle imperfections	M and/or U: improper handling & storage	0

*N = New; U = Used

6. Discussion

6.1 General Observations

Random failures account for 11 of 44 (25%) of failures. This is broadly consistent with findings of another study on failures of control systems [5] which found 15% of failures to be random and the rest systematic. Clearly, systematic failures originating with the user form the largest category of failures accounting for 52.3% of failures, while systematic failures originating with the manufacturer account for 25% of failures. Table 2 provides a list of subcategories for the three RC failure categorizations along with a failure count for used and new valves. Note that RCA #44 could have originated with the manufacturer, the user, or both and is counted in both categories. Consequently, the total number of failures listed in Table 2 is 45 instead of 44 and the sum of the percentages given above exceeds 100%. The subcategories listed under the two systematic failure categories are self explanatory. Appendix B provides additional descriptions for the subcategories for random failures.

Table 2. Subcategories of Random and Systematic Failures

Root Cause Failure Category	Subcategories	Used #	New #
Random failures	• Micro-embedment; surface creep • Oxide formation • Soft-seat embedment • Process deviation/upset	6	5
Systematic failures originating with Manufacturer	• Inadequate quality control • Inadequate or no testing	1	9-10
Systematic failures originating with User	• In service too long • Exposed to weather (without proper protection and/or without proper choice of materials) • Valve not cycled/exercised • Inadequate design/engineering specifications • Inadequate training or inattention to detail	22	0-1

As one would expect, when testing new valves, systematic failures originating with the manufacturer dominate and when testing used valves, systematic failures originating with the user dominate. But this observation is true only because SRS tests all new valves. A company that fields new valves without testing them would expect to see a mix of manufacturer and user originated systematic failures among its used valve tests. Clearly, the user's best first option to improve safety is to address the user originated systematic failures.

6.2 Using RCA Results to Improve Plant Safety

RCA has been an ongoing effort at SRS and steps have been taken during the course of this data collection to use the available results to institute improvements. Choosing appropriate materials for the intended environment, protecting valves from weather exposure, adjusting maintenance and testing intervals, and choosing manufacturers with better quality records for new valve purchases have been improvements relatively simple to implement. Once changes are introduced, improvements will take some time to become evident in subsequent proof test data.

The only failure mode completely analyzed was used-valve proof test data that indicated FTO. Consequently, this is the place to look for improvements. Table 3 compares the number of FTO's found during the period 2003-2007 with the number found during the period 2008-2009, relative to the total number of valves tested during those two periods. In 2008-2009, the overall percentage of FTO relative to the total number of valve tests shows a 63% decrease over the 2003 – 2007 period and, similarly, the percentage of user originated FTO's shows a 70% decrease. Thus, it seems plausible that the lessons learned from RCA and acted upon by SRS have proven beneficial to overall safety.

Table 3. Comparison of FTO Discovered During Different Periods of Time

Years	2003-2007	2008-2009
Total Used Valves Tested	976	412
Total # FTO	13	2
# RC: Random	5	0
# RC: Manufacturer	0	1
# RC: User	8	1
Total FTO/ Total # tested	1.33%	0.49%
Total FTO (User)/ Total # tested	0.82%	0.24%

6.3 Suggestions for Initiating a RCA Program

SRS has an obvious advantage in that all testing and repair takes place on-site under controlled conditions. SRS also has an extensive database of test results. It is easy to see that if a company has not cultivated a data driven safety culture the thought of beginning a RCA program is daunting. However, we would like to suggest that there are potentially significant benefits in even a modest RCA effort.

Consider what would happen if one limited RCA only to used-valve proof tests that indicate FTO and concentrated primarily on finding systematic RC of failure. First note that while initially untested new valves would clearly introduce FTO into the process, these FTO's would be found subsequently in proof test. Now consider the RC in Table 1. The first 22 entries would be the RC eventually uncovered in RCA of FTO's and 12 of those would be systematic causes. Of the remaining 22 of 44 failures that would not be subject to RCA, at least 11 have the same root causes as the FTO's have and would be addressed by fixing RC discovered through analysis of FTO's. Further, to the extent that a valve tested FTO and is to be repaired (as opposed to scrapped), the necessary disassembly for repair follows the RCA steps except that now the disassembly and findings would need to be overseen as well as documented. With experience, this should not take more than an hour per valve. So in this case, RCA would require minimal additional effort. If a valve tested FTO and was designated to be scrapped, then a little more effort would be needed for the RCA to disassemble and document the valve's condition even though it will not be repaired. However, again, this should not take significant time.

Putting this in context, in the case of SRS, which has a total in-service valve population of about 5000 valves, a total of 15 used and 12 new valves tested FTO over a period of 7 years means an average of 4 valves per year would undergo RCA if all FTO were so evaluated. (Recall that in this study only 7 of 12 new valve FTO's underwent RCA). The benefits of performing RCA for these valves would include not only decreasing the number of FTO's in the future but also the "spill over" benefits of decreasing the numbers of other PERD failure modes covered by the discovered systematic RC. When seen from the viewpoint of this larger picture, instituting a modest RCA program makes good sense from both an economic and safety point of view.

7. Summary & Recommendations

The abstract provides an adequate summary of this work and we will not repeat it here. We do want to add one recommendation. We have been working with valve databases from both SRS and other sources for some time and believe there is much benefit to being able to trace the history of a particular valve beginning with the time of manufacture/assembly. To this end we strongly recommend that all relief valve manufacturers assign a unique serial number to each valve and embed information about the assembly date within the serial number.

8. References

[1] Process Equipment Reliability Database (PERD) Initiative, "Spring Operated Relief Valve Taxonomy," Center for Chemical Process Safety, 2006.

[2] R. Gross, "Reliability Testing of Pressure Relief Valves," Design and Analysis of Pressure Vessels, Heat Exchangers, and Piping Components – 2004, PVP-Vol 477, Paper # PVP2004-2610, San Diego, CA, July 2004.

[3] ANSI/ISA 84.00.01:2004 (IEC 61511 Mod.), Application of Safety Instrumented Systems for the Process Industries, ISA, Raleigh, NC, USA, 2004.

[4] J. V. Bukowski, W. M. Goble, "Using Root Cause Failure Analysis Results to Estimate Failure Rates and PFDavg for Spring Operated Pressure Relief Valves," *Proceedings of the AIChE 6th Global Congress on Process Safety, 12th Process Plant Safety Symposium*, San Antonio, TX, March 2010.

[5] Out of Control: Why Control Systems Go Wrong and How to Prevent Failure, U.K: Sheffield, Health and Safety Executive, 1995.

Acknowledgements

This document in part was prepared in conjunction with work accomplished under Contract No. DE-AC09-08SR22470 with the U.S. Department of Energy. Portions of this work were prepared under an agreement with and funded by the U.S. Government. Neither the U. S. Government or its employees, nor any of its contractors, subcontractors or their employees, makes any express or implied: 1. warranty or assumes any legal liability for the accuracy, completeness, or for the use or results of such use of any information, product, or process disclosed; or 2. representation that such use or results of such use would not infringe privately owned rights; or 3. endorsement or recommendation of any specifically identified commercial product, process, or service. Any views and opinions of contributing authors expressed in this work do not necessarily state or reflect those of the United States Government, or its contractors, or subcontractors.

Appendix A

This appendix contains three examples of RCA performed on specific valves that were later categorized as random failure, systematic failure originating with the manufacturer and systematic failure originating with the user.

A.1 Example of RCA for Random Failure

Figure A1 shows a bronze valve disc with deposits on and to either side of the raised area seating surface. Figures A2 and A3 provide additional views. The seal was evidently compromised and was leaking or sizzling for some time. This allowed the oxides to form on the seating surfaces between disc and nozzle. This failure was determined to be random because the 350-400 degrees F exposure was a process upset, i.e., off normal air/water temperature excursion. The causes of failure and recommendations are summarized below and additional information from the RCA report follows.

Cause of failure

- Moisture, causing corrosion byproducts
- Micro-motion
- Leaking past the seat
- High heat

Recommendations

- Avoid source and formation of corrosion products
- Use lubricant or anti-seize on seating surfaces
- Use stainless steel or nickel based valve internal components

Figure A1. Bronze valve disc with deposits on seating surface.

Figure A2. Lead oxide deposits on the guide and copper oxide deposits on the disc surface.

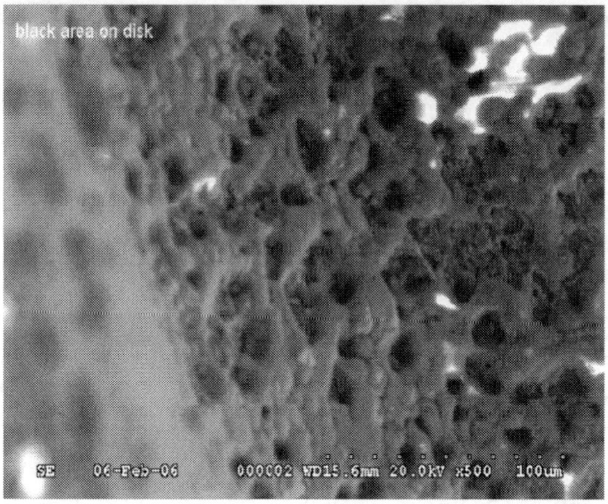

Figure A3. Scanning electron microscope image of surface deposits at 500x.

Analysis revealed the coating to be oxides that plated out as indicated by x-ray diffraction spectroscopy. The results of the spectroscopy are shown in Table A1.

Table A1. Results of x-ray diffraction spectroscopy

Spectrum processing Peaks possibly omitted: 0.268, 3.670 keV		
Processing option: All elements analyzed.		
Standards		
O SiO2	1-Jun-1999	12:00 AM
Cu Cu	1-Jun-1999	12:00 AM
Zn Zn	1-Jun-1999	12:00 AM
Pb PbF2	1-Jun-1999	12:00 AM
Measurement		
Element	Weight %	Atomic %
O	33.50	67.52
Cu	45.87	23.28
Zn	17.72	8.74
Pb	2.91	0.45

X-ray analysis of the seat deposits indicates the presence of lead, copper and zinc oxides on the mating parts. Lead leached from a free machining bronze blow-down ring into the seat area. The Savannah River Materials Laboratory examined the deposits and indicated the surface bond that formed was either adhesive (metal-metal oxide) or clean metal-to-ceramic. Operators recall that this particular compressor ran without cooling water last year due to failure of the cooling tower. Using standard textbook formulas for the adiabatic compression of air with K=1.4, a compression ratio of 4, and a summer inlet air temperature to the second stage of 125-130 degrees F, the resulting air compressed to the after cooler and relief valve would have been in excess of 400 degrees F. This is most likely the reason for the accelerated oxide formation and the "baked on" appearance.

Additional evaluation was provided by SRS Materials and Coatings consultants. Depending on the level of oxygen available, the compounds that form can vary. For example, zinc can oxidize to zinc oxide but can also form zinc hydroxide or carbonate (white rust) in lower oxygen environments (reaction with CO_2). For this valve, there are probably a mixture of metal oxides, hydroxides, carbonates and other compounds. The bond strength for this case could not be determined without measuring release torque or pull-off adhesion or some other indicator of adhesive forces. An example of mixed metal/ceramic bonding is the adhesion of inorganic zinc (zinc dust in ethyl silicate binder, IOZ) coatings on steel. IOZ coatings are somewhere between galvanized zinc metal coatings and organic coatings. IOZ coatings are not monolithic layers but rather a mix of zinc oxides, hydroxides and carbonates within a silicate network. Adhesion values for IOZ coatings on prepared steel can range from 300 to 1000+ PSI. In comparison, some organic coatings (even with zinc dust) can achieve adhesion bond strength values of up to 2000-2500 PSI.

In any case, the source and formation of corrosion products should be avoided. One would expect that if the valves are being used within their rated or "normal" service environments, they would not fail in this manner. Any cleaning operations or exposure conditions where moisture can get between valve surfaces could have been a factor. One option to consider is the use of a lubricant or anti-seize product to inhibit moisture or prevent corrosion product from forming in the first place. Teflon, graphite, MoS2, Krytox grease (Teflon filler), copper, or nickel-based products could all be possible options. Of course, since corrosion of brass or bronze alloys has already occurred, use of copper anti-seize products is probably not the way to go. Testing of different anti-seize products and contact/surface release properties may be needed. Otherwise, the use of stainless steel or nickel-based components should be substituted to avoid corrosion product formation of this type.

A.2 Example of Systematic Failure Originating with the Manufacturer

The new SORV was sent to the valve shop for pre-installation testing. The set pressure was checked and found to be within acceptable limits however, the valve was not leak tight after reseating. A work order was generated to repair the valve, and the valve was disassembled to repair the cause of leakage. The disassembly identified that an incorrect nozzle was installed. The nozzle installed was for a metal to metal seat not a soft seat as was identified in the model number. Mixing of a metal to metal nozzle with soft seat disk materials caused the valve to leak upon reseating. The proper nozzle was ordered and installed. This incident reinforces the importance of pre-installation testing of new valves as well as repaired valves.

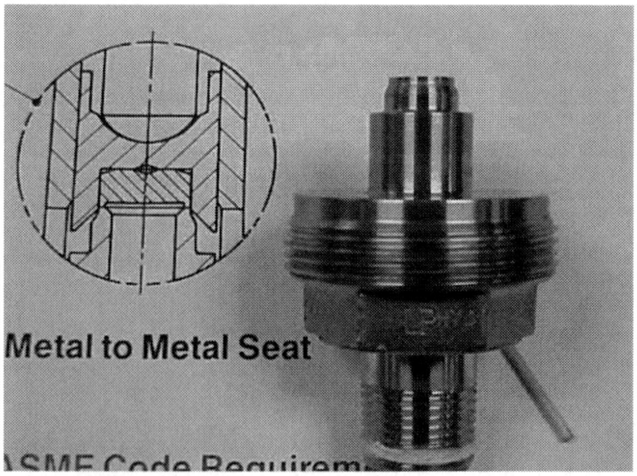

Figure A4. Installed nozzle was designed for metal to metal seal but disc was provided with O-ring.

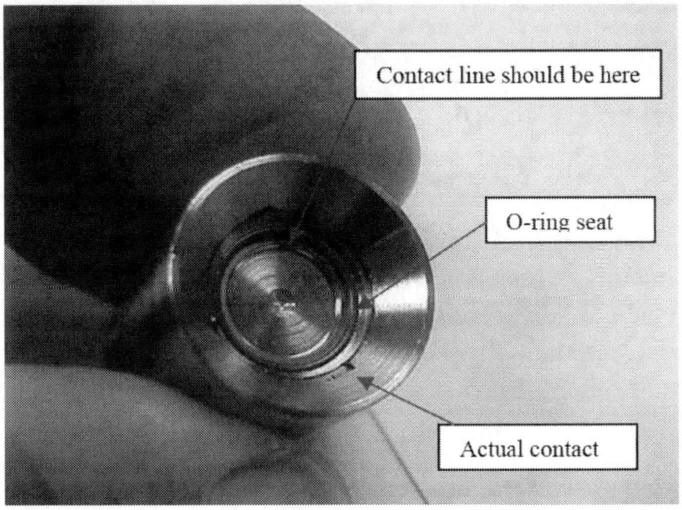

Figure A5. The blue circle illustrates the hard contact point between the nozzle and the disk. The contact point was well above and not making contact with the O-ring seal.

A.3 Example of Systematic Failure Originating with the User

This SORV was installed outdoors on a nitrogen service system. The relief valve was removed from service and sent to the valve shop for periodic testing. The "first pop" proof test showed the valve to open at 237 psig while the set pressure was 125 psig. Repeated exercise and the use of penetrating oil brought the set pressure back within tolerance. The valve was disassembled to determine the cause of the failure to open at set pressure. Examination determined the spring retaining washer had corroded / bonded tightly to the spindle thus preventing the spindle from "sliding" to allow the disc to rise at the set pressure. The valve was tested every three years but was not disassembled on the last cycle. Future outdoor installation of this type of valve should include installation of a weather hood to cover the opening allowed for the lifting lever; consider more frequent testing and disassembly; minimize corrosion by using stainless steel trim materials.

Figure A6 Corroded internal components.

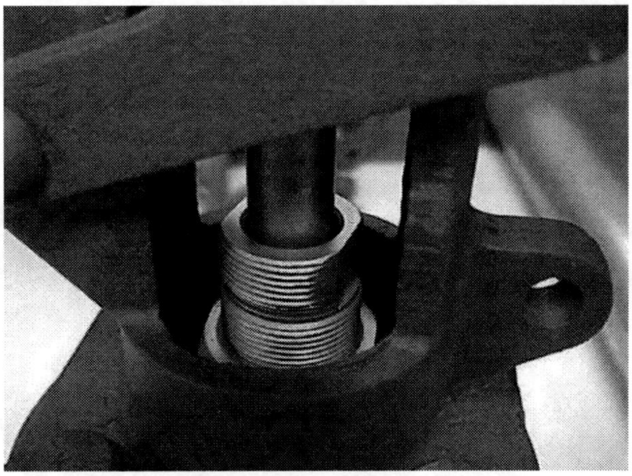

Figure A7 Bonnet and stem where rain water entered the valve.

Appendix B

This appendix contains more detailed explanations of the subcategories of random failures.

- **Micro-embedment; surface creep**

  These are two separate phenomena that are very hard to differentiate without significant investment in time and NDE. Embedment is the mechanism when load on the metal to metal joint varies (e.g., due to vibration or thermal expansion). The protruding peaks of the imperfect surfaces become stress concentrations and yield until the stress is relieved. **Creep** occurs when the surface to surface interface deforms slowly under constant stress. Asperities on each surface under constant stress (load) tend to be reduced until both surfaces are exact mirror images of each other. Both of these random failure mechanisms express themselves as surfaces adhering to one another.

- **Oxide formation**

  This is the formation of metal oxides on the metal surface itself due to combinations of moisture, chemical compounds in the surround air, and high temperatures which increase the rate of deposit.

- **Soft-seat embedment**

  This failure phenomenon applies to SORV with non metallic seats, e.g., nitrile rubber, Teflon, and neoprene. As the spring pressure presses down on the joint between the soft seated disc and the hard seated metal nozzle, over time, the softer material takes a "set" and reduces the flexibility needed to seal or reseat the SORV once it is pop tested.

- **Process deviation or process upset**

  This occurs when normal variation in temperature, pressure, flow, etc. of the process is exceeded for an amount of time. For example, the steady state cooling water temperature of a hydraulic pump or air compressor after-cooler is exceeded sufficiently to cause damage or accelerated wear and deposit formation.

6TH GLOBAL CONGRESS ON PROCESS SAFETY

Use of Multiple Non-destructive Examination (NDE) Methods in Life-cycle Evaluation of Process Equipment and Systems

F. Russ Davis CSP
Solutia, Inc.
575 Maryville Centre Dr.
St. Louis, MO 63141
314-674-1146
frdavi@solutia.com

Shane Ardoin
Mistras Services Group
3214 Pasadena Freeway
Pasadena, TX 77503
shane.ardoin@mistrasgroup.com

Prepared for Presentation at
American Institute of Chemical Engineers
2010 Spring Meeting
6th Global Congress on Process Safety
San Antonio, Texas
March 22-24, 2010

UNPUBLISHED

Use of Multiple Non-destructive Examination (NDE) Methods in Life-cycle Evaluation of Process Equipment and Systems

F. Russ Davis CSP
Solutia, Inc.
575 Maryville Centre Dr.
St. Louis, MO 63141
314-674-1146
frdavi@solutia.com

Shane Ardoin
Mistras Services Group
3214 Pasadena Freeway
Pasadena, TX 77503
shane.ardoin@mistrasgroup.com

Keywords: Mechanical integrity, damage mechanism, non-destructive testing/examination method, x-ray testing, ultrasonic testing, acoustic emission testing, eddy current testing, magnetic testing

Abstract

Equipment reliability is essential to the safe operation of process equipment. The health of process equipment must be analyzed by competent personnel using proper methods to identify current damage and predict anticipated life. The use of proper testing and inspection methods is basic to identifying damage mechanisms. The correct inspections and tests must be performed to identify equipment damage. There is no one non-destructive examination method that can be used to identify all damage mechanisms that may impact process equipment and systems.

This paper will examine various non-destructive examination methods for detection of specific damage mechanisms. The use of multiple NDE methods will be proposed for evaluating various equipment and process systems to determine suitability of service and predict the life cycle of the process systems and equipment.

1. Introduction

In October 2009 a leaking pipeline in a storage facility in Jaipur, India led to a fire and explosion which spread throughout the facility. Over 500,000 people were evacuated. There was an estimated loss of $65 million in property and product. The facility will be out of service for 12 to 18 months to be rebuilt. The environmental impact is devastating.

In November of 2009, Silver Eagle Refinery in Woods Cross Utah suffered a devastating explosion and fire when a 10" pipe attached to a reactor ruptured. The explosion and fire caused onsite as well as offsite impact. The Chemical Safety Board recommended a complete shutdown of the refinery while failure of the mechanical integrity program was evaluated. Inspection was being performed on this line but failed to recognize significant wall loss.

Process systems are continuing to fail causing unacceptable hazards that can and have affected safety, the environment, production and asset loss. These failures often are occurring not because we do not inspect the equipment but because we do not *correctly* inspect the process systems.

Mechanical integrity programs have been in place in refineries and chemical plants for several years but equipment failures continue to severely impact safety and the environment and result in millions of dollars lost each year. Non-destructive examination (NDE) is being used as the primary method for detecting damage to equipment.

There is no single non-destructive examination method that will properly detect all damage mechanisms. Many process systems require the use of multiple NDE methodologies to detect the various damage mechanisms. This paper will discuss many of the current NDE methods available for use in the process plants, discuss several frequently encountered damage mechanisms and suggest various NDE methods for the detection of these damage mechanisms. This paper will conclude with a discussion of the benefits provided by the use of multiple NDE methods in evaluating the life-cycle of process systems.

2. Non-destructive examination methods

2.1. Radiographic testing (RT) has been used for many years in the evaluation of welds. Profile radiography is excellent in evaluation of wall thickness for small bore piping up to 14" diameter. In some circumstances radiographic exam can also be used for detecting cracks. The benefits of radiographic examination include the visual picture documentation of the examination. The advent of computed radiography to determine wall thicknesses has proven advantageous in obtaining accurate thickness readings as well as higher image quality. The ability to vary lighting in digital radiography is a benefit in crack and internal pitting identification. RT is a very good method for weld inspection. Corrosion of welds and heat affected zones (HAZ) can be visualized in RT pictures. RT does have drawbacks that have to be considered, however. The piping size limitation of profile radiography in wall thickness determination is restricted to smaller piping. Cracking and delamination must be cross oriented to the direction of the x-rays in order to be seen.

2.2. Ultrasonic testing (UT) has also been used for years in refineries and chemical plants. UT thickness measurements taken on piping and vessels have been the primary method for determining wall thickness. There are many ultrasonic methods currently available for damage detection. This paper will review only four of these methods with wide spread applicability:

2.2.1 UT Thickness testing

The main benefit of UT thickness examinations is the ease with which the data can be collected. Technicians using small portable battery powered equipment can measure and document thickness readings at several locations on many different pieces of equipment in a day. This ease of application makes UT thickness examination economical. UT thickness is a very accepted testing method and is widely recognized. The drawbacks of UT thickness examination are in the application and the understanding of the results. UT thickness is very operator dependent and the UT technician must be dedicated to documenting the best results possible. A UT location point or thickness measurement location (TML) is typically a very small point, diameter less than .5". When considering that one point is taken on a 6" pipe that is 10' long having a surface area of 2300", one can quickly see that a UT point does not measure a significant portion of the pipe. Ultrasonic signals have a tendency to reflect the good wall thickness in any corroded area of pipe wall. The strongest signal is returned from the non-corroded wall. This fact requires that UT technicians be very well trained and knowledable in interpreting signal reflections.

2.2.2 Automated Ultrasonic Testing (AUT) is a very good tool for analyzing corroded areas for a fitness for service evaluation. AUT can also be used for cracking evaluation using angle beam signals. AUT with computer assistance can give a topographic map for analysis of equipment life-cycle. AUT is non-intrusive and can save the downtime and preparation costs associated with opening equipment for internal inspections. The on-line inspection ability of AUT can serve as a shutdown/turnaround preparation tool. AUT may be considered expensive.

2.2.3. Long Range UT or guided wave UT may be used for rapid scanning of large piping systems. Many feet of piping can be inspected at one time depending on piping configuration and surface conditions. Long range UT is intended as a screening tool to pinpoint areas requiring further inspection. Long range UT is dependent on piping configuration to be efficient. If the piping has several bends, attachments or injection points long range UT may be less effective. Long range UT is most effective for long straight runs of pipe.

2.2.4. UT Shear wave is a very good method for detecting cracks and for the evaluation of weld corrosion and wash out. The method can be performed from the inside or outside of a vessel or piping. The surface must be relatively clean and flat in order to properly use shear wave inspection. This methodology is highly technician dependent and requires that technicians are adequately trained and experienced. Shear wave inspection can be performed on thin wall piping or small bore piping. Corroded piping can be extremely difficult to inspect using shear wave because of sound scattering at the corroded surface.

2.3. Acoustic emission (AE) testing is a powerful non-intrusive screening tool for crack and active corrosion detection. Vessels are typically pressured to 110% of the highest operating pressure of the last 6 months. The one-time over stress causes slight crack propagation and movement within a corroded area. This movement causes noise which can be detected through the use of acoustic sensors. Computer analysis locates damaged area(s) for follow up inspection. AE on-stream testing methodology has been developed to identify cracking damage without having to take vessels out of service. AE testing is used in conjunction with other non-intrusive NDE methods to determine vessel fitness for continued service without the need for costly cleaning, clearing and entry for internal inspection. AE testing is costly and requires additional NDE testing to evaluate findings.

2.4. Eddy current testing (ET) is used extensively to detect damage, corrosion and cracking in heat exchanger tubes. This testing methodology uses the principle of magnetic field disruption. A probe is inserted into the exchanger tube and then withdrawn. Corrosion and cracking are readily identifiable by trained technicians but is highly dependant on the skill and experience of the technician interpreting the output. Eddy current testing must be performed by well trained and experienced technicians capable of interpreting the data from the ET equipment. Eddy current testing can be equipment specific and using ET results as applicable to the system in which the exchanger operates is difficult. Use of remote field testing on carbon steel tubes can have high false indication returns.

2.5 Robotic application is finding a place in NDE testing especially for above ground storage tanks. Robots with UT probes attached are being inserted into tanks to scan the bottom plates to obtain UT thickness readings. The method does not require emptying, cleaning and decontamination of the tank. The testing is recognized as sufficient to forgo opening tanks for internal inspection depending on test results. Robotic testing is service sensitive and cannot be used for every service. The testing is expensive but is cheaper than having to open and prepare tanks for internal inspection.

2.6 Magnetic particle (MP) and dye penetrant testing (PT) are proven methods for detecting surface cracking in vessels, tanks and piping. It is necessary to properly prepare the surface in order to properly perform these methods. You must have access to the surface being tested, which means insulation removal for external exams and confined space and cleaning for internal exams. These methods do have many benefits and have a proven history of detecting cracks open to the surface. It does not require a huge amount of training and experience for a technician to be proficient in these methods. One of the primary drawbacks is the need to properly prepare the surface for testing. The methods are intrusive and require vessels and piping to be clean, prepared for entry and in some cases sandblasted to prepare the surface.

2.7 Visual inspection by a competent inspector has been part of NDE inspections for as long as there has been equipment to inspect. Visual inspection with photo documentation is a very good tool for recording damage mechanisms for future evaluations. There are however many draw backs to visual inspections. Inspectors cannot see through insulation and must make entry into possible hazardous confined spaces in order to perform the inspection. Internal visual inspection can be very expensive. Cleaning, clearing and preparing a vessel or storage tank for a visual inspection can cost as much as $10,000 per vessel. Visual inspection augmented with additional NDE is still the tried and true inspection method of choice.

3. Process systems/equipment and damage mechanisms

Process systems are composed of various pieces of equipment. While inspection is typically specific equipment oriented the analysis of the results should be viewed as applicable to the system. An example of a process system may include a column, a reflux drum, condenser, and pumps. The storage tanks, feed pumps and day tanks are another example of a process system.

Figure 1

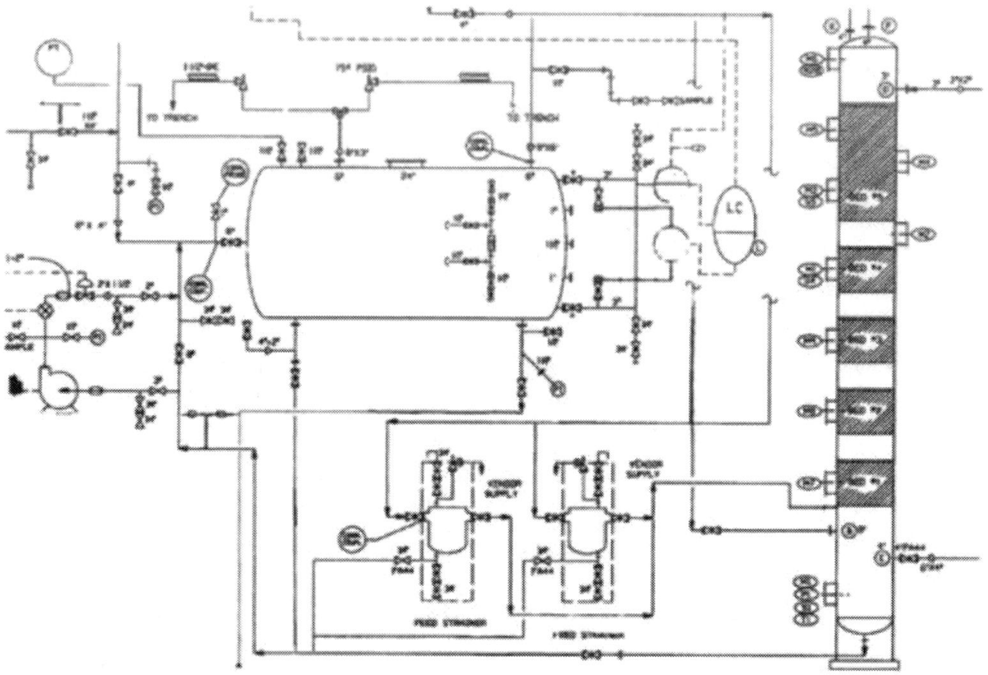

Equipment damage mechanisms apply to the system. In figure 1, damage discovered in the column can be expected to affect the drum, filters, pump, and piping.

In this section we will discuss some of the widely experienced damage mechanisms and the equipment and systems that are affected by these damage mechanisms. It is not the intent of this paper to examine the causes of the damage but to examine NDE detection methodologies.

Damage mechanisms that affect equipment are caused by the operating environment in which the equipment resides. Internal corrosion may be caused by the chemicals being handled and external corrosion may be caused by the environment. Internal erosion can be chemically induced as in the case of sulfuric acids in carbon steel systems operating above 3' per second or caused by physical erosion when the service is a stream of abrasive slurry. Cracking can be caused by a multitude of reasons from chloride presence in stainless steel systems to mechanical fatigue caused by pressure cycling or vibration.

3.1 A pressure vessel is typically defined as any vessel that operates with an internal or external pressure and is not configured to be a flat bottomed storage tank. As can be seen in figure 1 the vessel associated with this column may operate at 2 psig internal pressure but by its physical configuration it will be best to inspect the vessel in the same way that a vessel operating at 50 psig internal pressure would be inspected.

3.2 Above ground storage tanks are typically defined as the flat bottom, cone or floating roof type of tanks operating at low atmospheric or a slight internal pressure. A process system with above ground storage tanks is illustrated in figure 2

Figure 2

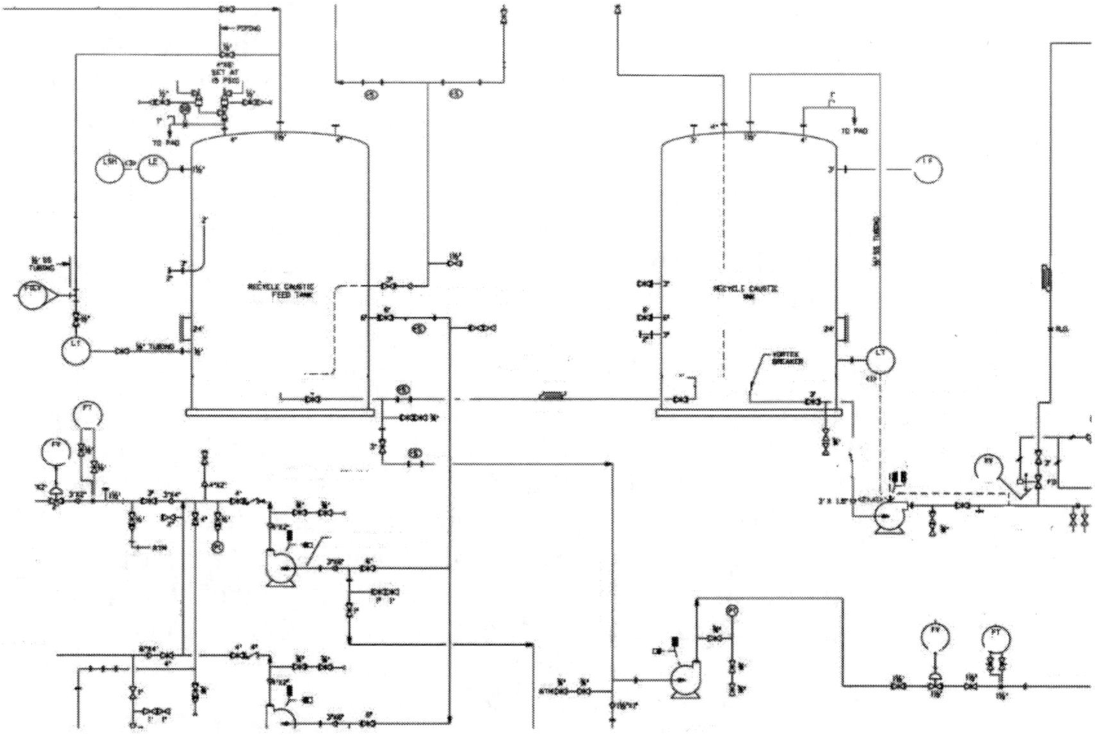

3.3 Piping systems include the pipe, valves, flanges, bolting and other piping attachments. Piping systems as shown in figures 1 and 2 should be analyzed as part of the process system whole.

3.4 Heat exchangers should be evaluated as a part of two process systems. The process flow that passes through the exchanger has its unique potential damage mechanism and the heating or cooling medium has its own potential damage mechanisms. Each of these systems may require different detection methods. Data collected from NDE testing of heat exchangers should be utilized to evaluate the system in which it operates.

4. Process system, damage mechanism, NDE methodology and life cycle evaluation

4.1 Corrosive metal loss caused by service and/or environmental conditions and can be either localized or general in its affect. Pitting can be difficult to find; whereas general corrosion covering a large area of a vessel, tank or piping is typically easier to identify.

Inspection of pressure vessels has typically involved on-stream UT thickness inspection at established thickness measurement locations (TML) and an internal visual inspection. The current industry practice is to establish a life cycle for the individual vessel based on the findings of the inspections. The process can be better served by life cycle evaluation of the process system and the use of multiple NDE tools to identify damage. Identifying pitting corrosion is a prime application for this practice. Good results have been obtained by the use of acoustic emission testing on the vessel to supplement data collection. The AE inspection identifies areas of active corrosion. UT thickness measurements are still taken at any areas identified by the AE data. By using multiple NDE techniques the inspection group is not wasting time and money performing UT on random TML points.

Piping attached to the vessel is part of the process system. Inspection results from NDE performed on the piping should be used in the evaluation of the process system whole. With this in mind the NDE techniques used for evaluation of the system can be expanded. Corrosion found in any part of the system should be considered as affecting the system whole. If corrosion is identified in a vessel or piping then it is important that the reliability group anticipate that corrosion is affecting all the equipment in the system. The use of long range UT is proving successful in identifying corroded areas in piping systems. Corrosion identified through long range UT can be readily followed up with additional NDE methods such as RT profile in order to facilitate a fitness for service analysis.

Automated UT should be considered for evaluating vessels with identified internal corrosion. The AUT will provide data to be used in fitness for service calculations. The topographic presentation of data collected by AUT is valuable in evaluation of pressure vessels especially where pitting corrosion is suspected. Angle beam AUT is also an effective method of identifying weld corrosion and heat affected zone corrosion.

Corrosion in heat exchanger tubes can be detected using eddy current testing techniques. ET is well suited for detecting internal and external tube corrosion. Corrosion noted by ET testing of the system should be applied to the system whole when the inspection group performs life cycle analysis. For exchanger tubes it is important to remember that there are two systems that require analysis. The tube side of the exchanger is part of one system where the shell side is a part of another separate system. Corrosion noted on a tube external wall can serve as a good indictor additional NDE should be performed on the shell of the exchanger as well as the rest of the process system. Corrosion noted on the internal wall of a tube is an indicator that additional NDE may need to be performed on the remainder of that process system. It must be noted that alternate techniques such as Remote Field Eddy Current and Internal Rotating Inspection (IRIS) must be used to obtain inspection data on carbon steel tubes.

Both manual and automated UT thickness testing can be used for the detection of corrosion in above ground storage tanks (AST). Flat bottom storage tanks are most susceptible to corrosion of the top side and bottom side of the floor plate. UT thickness on the shell can be performed from the outside. Corrosion noted on a shell wall should be used in evaluating the bottom plate. In addition, bottom plates are susceptible to underside corrosion. Applicable NDE inspection techniques for inspecting AST's are UT thickness and magnetic flux testing. Both of these methods will involve cleaning, clearing and opening the tank for inspection personnel. There are additional NDE techniques that can provide the inspection group with data to evaluate the fitness for service and to establish the need for an internal inspection. Robotics has been used to gather thickness data on the bottom plate. Robotic motorized crawlers that are intrinsically safe are being used on in service tanks to gather bottom plate thickness readings. The robots are dropped through a top manway and then crawl across the bottom plate in predefined grids taking thickness readings. External robotic crawlers have been successful in gathering shell thickness readings on un-insulated tanks. Acoustic emissions techniques are also used to evaluate active corrosion and to identify any bottom leaks which may be occur. Identifying the presence of active corrosion presence will indicate the need to further evaluate the AST.
(Table 1)

4.2 Cracking is a damage mechanism that can be caused by many service conditions as well as mechanical and thermal stresses. The causes of cracking are the subject of other papers. Crack detection must be accomplished early enough to prevent complete system failure. A good reference for identifying damage mechanisms and their causes is the RP API 571 Damage Mechanisms Affecting Fixed Equipment in the Refining Industry.

Cracking in vessels can be identified from the outside with the use of UT shear wave either manually or through the use of automated UT. External cracking can be identified through the use of magnetic particle and dye penetrant inspections. Acoustic emission testing is a very good screening tool for identifying cracking in pressure vessels. AE testing is well suited for identifying cracking areas and can maximize the use of UT shear wave to evaluate the extent of the cracking.

Cracking in piping can be identified by the use of RT inspection as well as UT shear wave techniques. Long range UT is also proving valuable in identifying circumferential cracking in piping.

Cracking discovered in any of the equipment associated with the process system should be analyzed as applicable to the entire system.

Cracking in heat exchanger tubes can be identified by using eddy current testing. Analysis of cracking identified on the internal or external tube wall should be followed up through the analysis of the rest of the process system. The use of multiple NDE methods throughout the rest of the process can supply the data necessary for fitness for service evaluations.
(Table 1)

4.3 Inspection of piping systems can serve as a good indicator of the health of the process system on the whole. There are several NDE methods for inspecting piping that are suitable for detecting corrosion, erosion and cracking damage. UT thickness inspection is one of the most widely used methods for the detection corrosion and erosion. As noted earlier in this paper thickness measurement locations do not represent a statistically significant amount of the piece of equipment. RT profile testing does show a larger amount of pipe wall (figure 3) than UT TML's and can be used in conjunction with UT. Long range UT has proven successful in inspection of large amounts of piping from one location and is very good at identifying areas where more UT or x-ray may be required to evaluate piping fitness for service and life cycle planning.

Figure 3

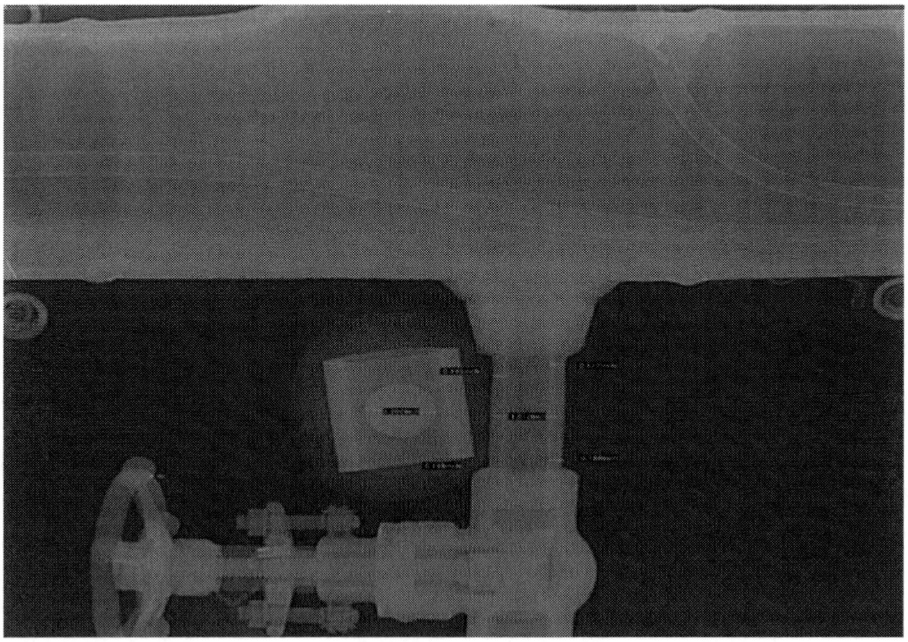

4.4 Corrosion under insulation (CUI) is a wide spread industry problem and is most readily identified in piping. Process systems where CUI is discovered during piping inspections should be further evaluated for CUI throughout the process system. Effective NDE methods should be used for inspecting all insulated columns, towers, drums, heat exchangers and other equipment within the process system. Use of multiple NDE methods has proven effective in identification of CUI and in documenting and evaluating these systems. RT, UT and acoustic emission testing are proven methods for CUI inspection. Insulation removal and visual inspection to identify corrosion has always been an effective method to identify CUI. Figure 4 is an example of a digital profile radiograph showing CUI in a piping system. Real time fluoroscopic inspection is also a useful tool in identifying CUI areas in large sections of piping.

Figure 4

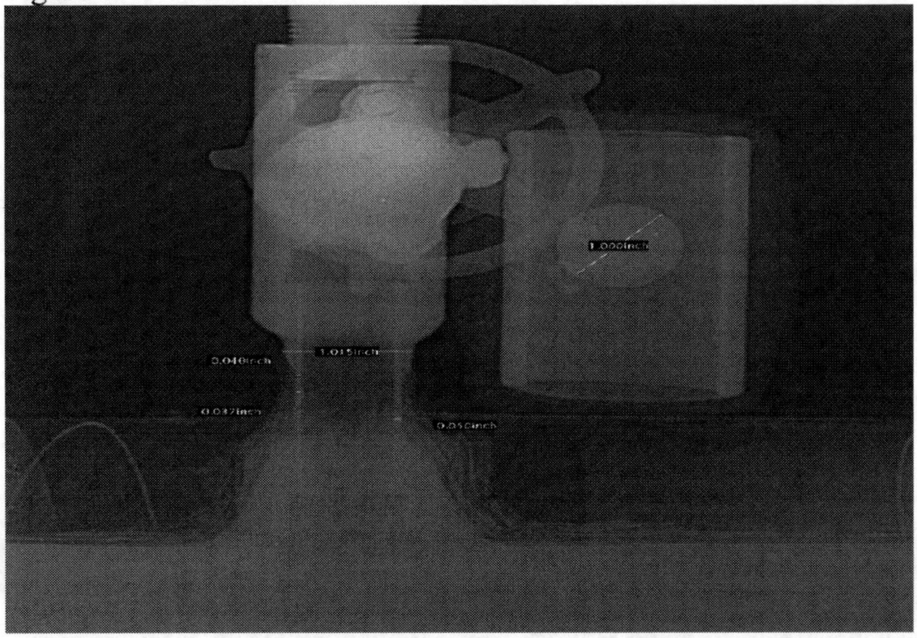

4.5 The plant reliability group should develop plans for process system inspection. These plans should identify the suspected damage mechanism and the suitable NDE methods for the detection of the damage. Table 1 is an example of a matrix identifying damage mechanism. Reliability groups should evaluate the use of Risk Based Inspection (RBI) methodology when developing inspection plans. RBI methodology identifies potential damage mechanisms and inspection methods suitable for detecting damage.

Table 1

	RECOMMENDED NDE METHODS FOR DETECTING DAMAGE									
DAMAGE MECHANISM										
	RT	UT-THICKNESS	AUT straight beam	LONG RANGE UT	UT SHEAR WAVE	AE	ET/ Remote Field	ROBOTIC UT Crawlers	Visual	MT-PT
Corrosion -vessels		X	X			X		X	X	
Erosion-vessels		X	X					X	X	
Corrosion -Storage tanks		X				X		X	X	
Corrosion-piping	X	X	X	X						
Erosion-piping	X	X	X	X						
Cracking-vessels			X		X	X				X
Cracking-piping	X		X		X	X				X
Corrosion under Insulation	X	X		X						
Exchanger tube corrosion							X	X	X	
Exchanger tube erosion							X		X	
Exchanger tube cracking							X			

5. Conclusion

Each piece of equipment is a component of the process system and inspection and analysis of plant process equipment is best done by evaluation of the process system as a whole. Damage discovered in one piece of equipment should flag the entire system as suspect. There are many and varied damage mechanisms that affect process systems. Early detection of the damage must be accomplished in order to evaluate the system fitness for service and for life cycle planning. The use of one inspection method is not adequate to detect all possible damage mechanisms. The use of multiple NDE methods is effective in identifying damage and producing data suitable for performing fitness for service evaluations.

Inspection and reliability groups must evaluate the NDE methods currently being performed on their process systems. Understanding the benefits and limitations of each method is critical to obtaining data suitable for a full life cycle evaluation of process systems.

6. References

[1] "Guidelines for Mechanical Integrity Systems", Center for Chemical Process Safety of the American Institute of Chemical Engineers, 2006

[2] American Petroleum Institute Recommend Practice 571 "Damage Mechanisms Affecting Fixed Equipment in the Refining Industry", First Edition, December 2003

Risk-Based Mechanical Integrity – Beyond Fixed Equipment

Thomas J. Folk P.E.[1]
Lloyd's Register Capstone, Inc.
1505 Hwy 6 S. Suite 100
Houston, Texas 77077
281-649-2260
Thomas.Folk@lr.org

Prepared for Presentation at
American Institute of Chemical Engineers
2010 Spring Meeting
6th Global Congress on Process Safety
San Antonio, Texas
March 22-24, 2010

UNPUBLISHED

[1] The opinions and recommendations expressed in this paper are those of the author and do not necessarily represent the opinions or recommendations of Lloyd's Register Capstone, Inc.

Risk-Based Mechanical Integrity – Beyond Fixed Equipment

Thomas J. Folk P.E.
Lloyd's Register Capstone, Inc.
1505 Hwy 6 S. Suite 100
Houston, Texas 77077
281-649-2260
Thomas.Folk@lr.org

Keywords: Risk-based Inspection, RBI, Mechanical Integrity, Preventive Maintenance, Predictive Maintenance, Condition Monitoring, API 691, PSM, Process Safety Management

Abstract

This paper presents approaches to provide a safer operating plant by extending mechanical integrity to all process equipment using a risk-based approach to better focus resources.

Risk-based approaches for inspection and testing of fixed equipment for mechanical integrity (API 580/581) and for the integrity of Safety Instrumented Systems (ANSI/ISA-84 -IEC 61511 Mod) are well established.

For other equipment types, there are Preventive Maintenance (PM) programs of various types. An effective PM program should:

- improve plant reliability and plant safety – a shift away from reactive maintenance and the resulting operational issues
- provide a rational basis for the development of PM program activities and their intervals based on RAGAGEP (Recognized and Generally Accepted Good Engineering Practice) or sound engineering and reliability principles using a structured logical approach that will provide consistent results
- create and document appropriate PM strategies for plant equipment with their basis (including targeted damage mechanisms and affected components/parts)
- provide risk-based prioritization with resulting improved effectiveness
- integrate EHS and economic driven PM activities
- document results and respond to non-conformances found by PM activities via appropriate review, repairs, or as PM program revisions
- track compliance to the PM activities schedule
- evaluate program results and effectiveness

- collect failure data needed to support and update risk-based PM strategies and equipment selection for replacements and capital projects.

In the absence of codes, standards and recommended practices, manufacturer recommendations or "traditional" approaches for inspection, testing and preventive maintenance may not be sufficient to achieve these objectives. This paper presents a structured risk-based approach to better address the objectives listed above.

In addition, an update will be included on the API 691 Task Group's efforts to address some of these issues.

1. Introduction

What is Mechanical Integrity? To some it covers the inspection of vessels and piping. If you are not overly focused on the word "mechanical," it clearly includes assuring the integrity of Safety Instrumented Systems. But if you expect to have a safe and productive facility, it includes all process and utility equipment.

Further, we have risk-based approaches for the pressure equipment with Risk-Based Inspection (RBI) and for safety instrumented systems in ANSI/ISA-84.00.01-2004 (IEC-61511 Mod). CCPS has provided guidelines for risk-based process safety in *Guidelines for Risk Based Process Safety.* [1]

What about all your other equipment? There is a CCPS Guideline book on Mechanical Integrity that provides background and guidance for inspection, testing, and preventive maintenance of other equipment types. These activities are normally referred to as Preventive Maintenance or PM. Generally, there is no RAGAGEP (Recognized and Generally Accepted Good Engineering Practice) document defining a methodology for the development of PM activities. PM has traditionally been applied based on manufacturer recommendations or traditional practices. Applying a risk-based approach to the prioritization of PM resources is rare. PM activities are not always valued in some organizations; they can be too busy with reactive maintenance. The API has started to develop a risk-based approach to the PM of rotating equipment through the work of Task Group 691.

What is the opportunity to improve safety and plant reliability or uptime? There is a clear correlation of plant reliability to safety. Unexpected breakdowns and plant shutdowns present significant safety, as well as economic, issues. A breakdown culture leads to crisis operations and work-arounds that in time seem normal but can lead to releases in process streams, emergency shutdowns, loss of critical systems, etc. Prudent use of a PM program to minimize such breakdowns can greatly reduce safety risks while significantly improving reliability.

2. Mechanical Integrity Implementations

The methodologies implemented for Fixed Equipment and Safety Instrumented Systems are different as is appropriate for the needs of their application. They are both described in standards documents and implemented with uniformity.

The methodologies used for integrity management of other process plant equipment will likewise have some differences in their application. While, as mentioned above, there is currently no recognized and generally accepted good engineering practice document describing a methodology for the development of PM activities, there are a number of approaches in practice and an effort within API (Task Group 691) to document a methodology.

A brief review of the approach used for fixed equipment will provide some background on what is needed for the other equipment, including their similarities and differences, along with some of the opportunities to leverage information and systems developed for them to other plant equipment.

2.1 RBI for Pressured Fixed Equipment

RBI for pressured fixed equipment is defined by a minimum set of requirements in API RP 580 - Risk-Based Inspection. RP 580 presents the principals and minimum general guidelines for implementing and sustaining an RBI program. Much of the technology in its application is described in API RP 581 - Risk-Based Inspection Technology. The main focus of these efforts is to avoid catastrophic release of hazardous process fluids. The focus is on the integrity of the pressure boundary in providing containment. These inspection programs are performed for the safety of the employees and the community as a part of process safety management, for regulatory compliance, and for improved reliability of the fixed equipment for the business.

Inspection and testing is only one element of Mechanical Integrity (MI) for this equipment, but has proven to be the area requiring the greatest cost and effort. The other areas for MI are:

- design, fabrication and installation
- operation within design limits
- quality assurance of repairs (incl. materials, parts, and work)
- conformance to codes and standards
- deficiencies identified and addressed

- written procedures
- training.

The work-process steps for inspection-plan development, execution, and updating is shown in Figure 1 below. There are generally two major phases to an RBI program. The first is the initial program implementation or set up. These steps are typically done once unless changes (equipment, process service, operating parameters, etc.) require adjustment to the basis information. Any changes of this nature need connection to RBI from the Management of Change process.

Figure 1

Fixed Equipment RBI

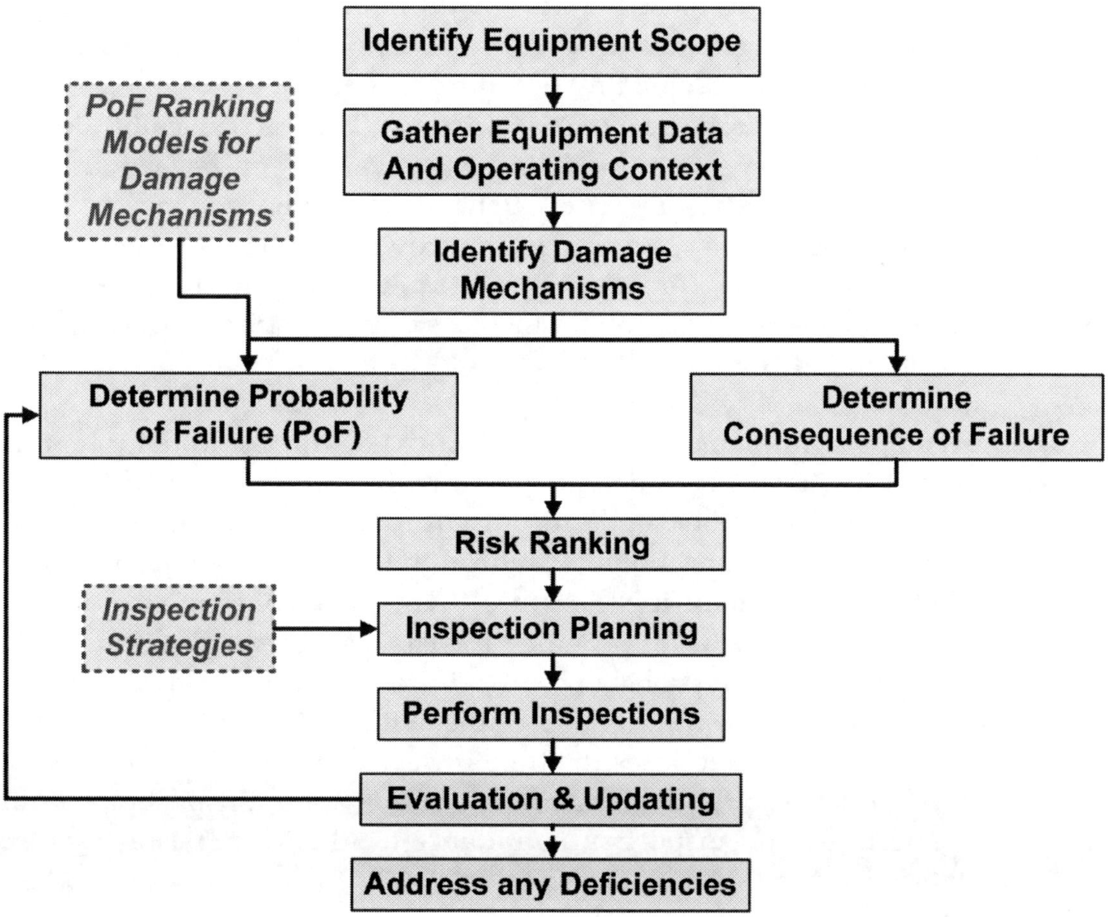

Implementation requires setup of budgets, knowledgeable resources, procedures, software, and a management structure. Implementation includes the following steps (refer to Figure 1):

- **Identify equipment scope** – selection and listing of the equipment assets to be included
- **Gather equipment data** – identification and collection of the specific data required on the equipment design, operation, process service, inspection history and repairs. Sufficient information must be collected to permit the following: establishing acceptance criteria for inspections, modeling loss of containment and business consequences, identifying potential damage mechanisms, and modeling the probability of failure (PoF).
- **Identify damage mechanisms** – identification of known and potential damage mechanisms, and estimation of damage rates. With the focused approach in RBI it is important that all potential damage mechanisms be considered and included for inspection. Clear understanding of the damage mechanisms and rates is important in developing accurate modeling of the probability of failure (PoF).
- **Determine consequence of failure** – This is typically done utilizing simplified dispersion modeling of a variety of hazardous-fluid-release types (toxic, explosion, fire, etc.) for safety consequences. The intent of these "dispersion" models is to provide a rough (order-of-magnitude) result as a basis for relative risk ranking of the equipment. Business-impact consequence (down time, loss of production, clean-up, etc.) can also be included.

The next steps in implementation are not only used in the initial pass through but are repeated for "ever-greening" purposes within an on-going RBI process. Ever-greening is the continuous updating of equipment data, condition assessments, risk evaluations, and inspection plans through the life of the equipment.

- **Determine probability of failure** – This, like consequence of failure, is an estimate intended to provide a basis for relative risk ranking of the equipment. The probability of failure is based on a model for specific damage mechanisms. For example with a damage mechanism like wall thinning, the model uses the time in service (age), the corrosion rate, initial wall thickness and information from prior inspections (number of inspections and inspection confidence). API RP 581 is a basic source for these models. Models are also available for cracking and other damage mechanisms.
- **Risk ranking** – Consequence ranking and probability ranking are combined to give a risk ranking. This is typically shown on a risk matrix, frequently 5 x 5, with four levels of risk providing good discrimination.
- **Inspection planning** – Planning selects the inspection techniques suited for the potential damage mechanisms and the risk levels to determine the

extent (i.e., full or partial) and frequency for the activities. Guidance on the techniques comes from various API or other practices. Guidance on the extent and frequency comes from **inspection strategies** or guidelines developed by subject matter experts or from running various cases and evaluating the impact on future risk versus user-defined risk tolerances. See Table 1 for a sample of an inspection strategy.

- **Perform inspections** – Use qualified inspectors to properly perform and document the inspections to the planned extent and frequency.
- **Evaluate and update** – Evaluate inspection results and update information on equipment condition (e.g., current wall thickness). Damage rates are reviewed and updated. Non-conformances are evaluated and classified for appropriate follow-up (e.g., re-inspection); deficiencies are identified and set up for correction.
- **Address deficiencies** – If any deficiencies are identified, appropriate follow-up for correction is made (e.g., Fitness for Service analysis, repair, replacement, etc.).

Table 1

Sample – RBI Fixed Equipment Inspection Strategy

Pressure Vessel – Visible External Deterioration by External Visual Inspection

	Risk Rank			
	Low	**Medium**	**Medium-High**	**High**
Minimum External Visual coverage [1]	100% External Visual	100% External Visual	100% External Visual	100% External Visual
Maximum Interval (Yrs.), or ½ remaining life	15	10	10	5
Maximum Interval allowed by jurisdiction	Depends on Jurisdictional Requirements	Depends on Jurisdictional Requirements	Depends on Jurisdictional Requirements	Depends on Jurisdictional Requirements
Inspection Confidence: Non-insulated Insulated	 Very High Medium	 Very High Medium	 Very High Medium	 Very High Medium

NOTES:

1. Formal external visual inspections are performed on pressure vessels to prevent failures by identifying those components that are deteriorating externally. External deterioration may include external corrosion, corrosion under insulation, fatigue, factors that may contribute to flange leaks or indications of other failure modes.

The new inspection data and updated age are used to recalculate the probability of failure and risk ranking for use in the next round of inspection planning. Addressing issues and other opportunities for corrections and improvements should be part of the ongoing work process.

Other key Process Safety Management processes, like Management of Change, maintaining Process Safety Information, Audits, Training, Procedure updates, and other supporting processes (e.g., qualification of inspectors and inspection risk analysts) are needed for continuous improvement and continued success.

2.2 *Approaches for Other Equipment Types*

The fixed-equipment program described above focuses on only a part of a process plant's equipment. Fixed equipment, along with safety instrumented systems, are the items that generally have the highest risk and that are most directly addressed by regulatory requirements. However, the reliability of rotating equipment, electrical power distribution, other instrumentation and controls, or functional aspects other than containment of some fixed equipment also have substantial impact on safety and business needs.

Preventive maintenance provides the basic care of equipment including predictive or condition-monitoring tasks. The term "Preventive Maintenance" as commonly used has different meanings to different people. For some it includes predictive and condition-monitoring activities; to others, those are a separate set of activities. For this discussion the broader inclusive definition will be used. Preventive Maintenance is defined as "maintenance carried out at predetermined intervals or according to prescribed criteria and intended to reduce the probability of failure or the degradation of the functioning of an item" per API 689 (ISO 14224).[2]

Preventive Maintenance consists of two major types of activities:

- Scheduled Restorative or Discard Tasks
  - This is renewing a condition, consumable, or a wearing part.
  - Examples include oil changes, greasing, filter changes, cleanings, annual overhauls, quarterly calibration, weekly inspections, other wearing part changes, or adjustments on a scheduled basis.
  - This subpart alone is sometime called Preventive Maintenance.
- Predictive Maintenance (or Condition Monitoring)
  - "An equipment maintenance strategy based on measuring the condition of equipment in order to address whether it will fail during some future period, and taking appropriate action to avoid consequences of that failure,"[3] as defined in the CCPS *Guidelines for Mechanical Integrity Systems*

- Examples include vibration monitoring, oil analysis and thermography.
- A more specialized additional element is Life Prediction - predicting when the next major overhaul or replacement is needed. This may be based on operating history (e.g., loads, temperatures, starts, other running conditions, etc.), as well as traditional condition monitoring activities.

2.2.1 PM System Objectives

The basic objective of a PM program is to improve plant reliability and plant safety. This includes a shift away from higher levels of reactive maintenance and the resulting operational issues. The major focus of this includes:

- reduce failures – prolonged equipment life, lower life cycle costs
- reduce unexpected failures in-service – no surprises
- project remaining life to major maintenance or replacement – for capital-budget planning and turnaround planning.

The objectives are not unlike fixed equipment RBI but perhaps a bit broader. Attributes of a strong PM program that is effective, efficient and continuously improving also have much in common with RBI. These attributes include:

- using effective PM strategies for plant equipment with a documented basis, including targeted damage mechanisms and affected components
- tracking schedule compliance of the PM activities
- documenting results and responding to non-conformances found by PM activities through appropriate review, repairs or as PM program revisions
- providing a documented rational basis for the development of PM program activities and their intervals based on sound engineering and reliability practices and utilizing a structured logical approach with consistent results
- recognizing and addressing the operating context (service conditions that promote damage mechanisms or may otherwise stress the equipment)
- providing risk-based prioritization with resulting improved program effectiveness
- integrating EHS and economic-driven PM activities
- evaluating PM-program results and effectiveness
- collecting failure data needed to support and update risk-based PM strategies, as well as equipment selection for replacements and capital projects.

2.2.2 Differences from Fixed Equipment

There are basic similarities with the fixed equipment RBI process outlined previously but also some significant differences. The objectives are somewhat broader. There are also fundamental differences in the equipment and the types of activities. These need to be recognized and the process modified to address them.

Key equipment differences include:

- a greater variety of types or classes of equipment (e.g., compressor trains, electrical switchgear, a variety of pump types, electrical transformers, etc.)
- a greater diversity in equipment components (e.g., bearings, cases, insulators, mechanical seals, couplings, circuit breakers, filters, etc.)
- a plethora of different primary functions (e.g., compress process gas, switch high voltage power, move process liquid, transform voltage levels, etc.)
- an extensive and varied range of damage mechanisms (e.g., imbalance, fouling, wear, rubbing, arcing, bearing wear, loss of lubrication, overheating, etc., along with the same corrosion, cracking, and other fixed-equipment damage mechanisms)
- a wide variation in damage rates and mean time to failure (MTTF) - Some damage can progress from barely detectable to failure in days or weeks; some equipment has MTTF in months versus years.
- a broad range of additional technologies for PM activities (e.g., vibration analysis, performance monitoring, oil analysis, corona detection, thermography, oil changes, greasing, vibration measures, etc.)
- a wider range in intervals used for PM activities – Intervals are measured in days, weeks, months, or years.
- more in-place spares for some equipment, providing easier access for work and lower economic risk from loss of availability - Spares may also lower the incentive to commit resources to PM and foster run-to-failure decisions.
- generally a more dynamic system (e.g., greater number of activities, greater variety of activities and personnel used, more frequently, more data to evaluate, etc.)
- fewer guidance documents - There are few equivalents to API fixed equipment-inspection-related standards or recommended practices, for example:
    - API 580, 581, 571, & 579 for RBI, damage mechanisms, inspection techniques and fitness for service standards & practices
    - API 510, 570, & 653 for in-service inspection codes
    - API 572, 573, 574, & 575 for inspection practices
    - documents from other organizations (e.g., NACE, MTI and others).
- more complex approaches to assessing remaining life.

These differences warrant consideration of a change in the approach or work processes developed for these equipment types versus those defined for fixed equipment RBI.

2.2.3 Defining an Approach

Development of approaches similar to API RP 581 would involve extensive documents defining PoF ranking models for this much broader scope of equipment types and the greater variety of components, damage mechanisms, etc. as previously described. Clearly a common approach needs to be defined first as a basis for further work whether at a corporate or industry level. Providing a document with a more generic work process for the development of these models and activities that can be used generally will provide broader and more immediate usability. With a work process definition, approaches for different equipment types can follow. These can be developed by individual companies or for broader application by industry-standards organizations. With a common development process these have consistency with other work done.

Three of the initially described objectives that relate heavily to the work process used are:

- provide a rational basis for the development of PM program activities and their intervals based on RAGAGEP or sound engineering and reliability principles using a structured logical approach that will provide consistent results
- create and document appropriate PM strategies for plant equipment with their basis (including targeted damage mechanisms and affected components)
- provide risk-based prioritization with resulting improved effectiveness
- integrate EHS and economic-driven PM activities.

Issues that currently exist with some conventional PM programs need to be considered and addressed. Some of these are:

- Changes in personnel give different results for activities and frequency recommended. These tend to cause a lack of consistency and continuity in activity plans.
- Manufacturers do not necessarily have an understanding or knowledge of the operating context differences of the equipment's use that can affect the appropriateness of their recommendations. In addition, some manufacturers may have economic incentives to recommend more varied and extensive PMs than may really be required.
- Activity types and frequency selections are not connected to risk due to failure.

- Activity types and frequency selections are not connected to operating context.
- No documented basis exists for the activities and frequency selections.

It must always be remembered that PM of equipment, can only address a subset of the failure causes. Failures driven by the following may not be addressed:

- maintenance repair, materials or parts quality issues
- design or installation issues
- operation outside the design window
- operational errors

2.2.4 Consequence Ranking

A methodology is needed to consistently establish Consequence of Failure criteria based on impact to EHS and business needs. The consequence of failure is quite dependent on the specific equipment item's relationship to its actual service and location rather than any generic characteristic of the equipment type. However, there are only a few consequence types of general interest, and these have fairly good correlation to the equipment type. For instance, only certain equipment types have the service characteristics and damage mechanisms to have a potential loss of containment consequence for the process steam. The typical consequence types and the data needed to utilize them are:

- Loss of equipment-service availability – The primary function of this item is not available regardless of the specific failure mode. It could be fails to run, fails to start; it may be due to loss of primary function or loss of any other function that takes or keeps it out of service. This consequence typically translates to a business cost determined by the impact of the loss of service as described by:
    - impact, usually on the unit, train, or process system – may be described as a percentage of full production and is affected by the availability of installed full or partial spares
    - the mean time to repair (MTTR) including clearing for work, acquiring parts, the time to repair, and re-commission
    - typical cost of production loss
    - one time costs for unit or system shutdown, cleaning, purging, heating, re-starting, etc.
- Loss of containment – Models including software already exist for modeling flammable and toxic releases in fixed equipment RBI. These can be leveraged to provide consequence rankings with little more than the selection of an appropriate hole size to characterize the failure. This is typically given in terms of affected area.

It should always be remembered that these are for ranking purposes, and order-of-magnitude results are generally sufficient to provide the needed discrimination for PM-activity planning. A quick review of equipment using a systematic approach with a knowledgeable group of operations, process, and maintenance personnel can normally provide this key information for all equipment in a unit.

2.2.5 Work-Process Methodology

A work-process methodology is proposed on the flow chart in Figure 2: Risk-based Preventive Maintenance, consisting of two separate work processes. The basic program set-up and operation steps are essentially the same as in fixed equipment RBI:

- identify scope
- gather information
- identify/quantify damage mechanisms/rates
- determine probability of failure
- determine consequence of failure
- risk ranking
- PM/inspection planning
- perform activities
- evaluate results and update risk and plan
- address any deficiencies found

There are two related supporting processes to these activities; they are PoF-Ranking models and PM/inspection strategies. The PoF-ranking models provide a PoF ranking of equipment typically based on equipment data for design, construction, time-in-service, operating context and condition assessments. PM strategies or guidelines provide PM activity recommendations at frequencies appropriate to the damage mechanism and risk of the associated failure.

Figure 2
Risk-Based Preventive Maintenance

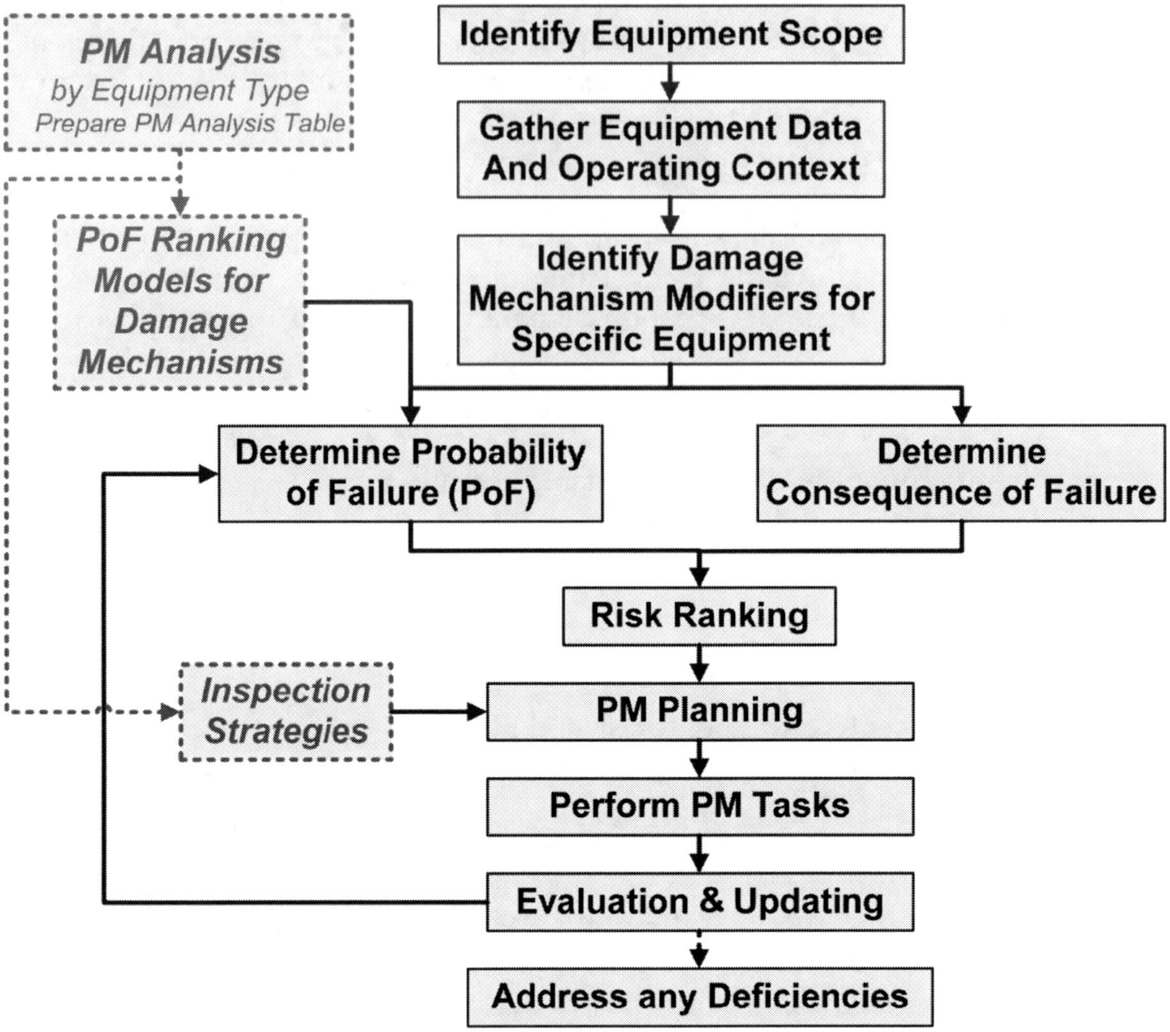

2.2.6 Development of Supporting Processes

Performing this level of analysis one asset at a time will assure that it never gets completed. The nuclear power industry has tried full RCM and found that to be excessive. There are some practitioners in the industry breaking out portions of this development process to classes of similar equipment and doing partial generic development at that level. This approach still applies key elements of RCM as applicable to the PM development and when properly designed and executed, can get the same value with far less effort.

These processes can be defined in part at a generic level for similar equipment types with provision for adjustment to differences in specific equipment of the same type. Appropriate grouping by equipment type allows leveraging of the development and resulting models and strategies to many items. Care must be exercised to provide adequate adjustment for variations in design, construction, installation and operation. Equipment types are generally like those listed in API 689.

For each equipment type it is important to determine an appropriate taxonomy to organize and apply to the PM-planning process. Taxonomy identifies both the component or parts breakdown for the equipment asset and also the boundaries of the equipment for analysis. Use of standard taxonomies will provide uniformity to the PM activity plans allowing for broader common development and usage. One source for taxonomies is ANSI/API Std. 689 (ISO 14224:2006) *Collection and Exchange of Reliability and Maintenance Data for Equipment*. Others are available from the CCPS Process Equipment Reliability Data (PERD) project and OREDA.

A **PM Analysis process** is needed for the development of the probability-of-failure (PoF) ranking model and the PM strategies. This requires an analysis of degradation mechanism, the affected component/part, and the cause of the degradation. The damage mechanism and corresponding affected component/part together define a potential **degrading condition** that will be a candidate for PM activities.

This PM analysis is similar to a Failure, Modes, and Effect Analysis (FMEA) but modified in several respects and will produce a similar table. It is simpler since the specific failure mode (e.g., fails to run, fails to start, etc.) is not needed nor is much information on the failure effects (see section 2.2.4 Consequence Ranking above). The consequence type should be included. Additional information is needed to aid in defining the focus of PM activities and to develop PM strategies to address degrading conditions. The degrading condition is described by the damage mechanisms and the affected component/part, and is matched with causes or drivers of the degradation. To facilitate PM-activity recommendations a listing of applicable PM technologies should be included. Alternately a ranking of the effectiveness of each of the standard PM technologies could be included in the table.[4] Where operating context, installation issue or equipment design will have a significant effect on damage mechanisms, causes, damage rates, MTTF, etc., these conditions should be noted and an appropriate modifying factor recommended.

This PM Analysis process provides the information needed for the PoF ranking modeling and for PM strategy development. This will consist of an FMEA-like table for each equipment type including the following:

- equipment components and damage mechanisms to address – describing a degraded condition

- consequence type(s) as defined above – Secondary damage effects may need to be considered when additional consequence types could be added (e.g., pump bearing failure may cause a seal failure adding a loss of containment consequence).
- cause of damage
- estimation or characterization of damage rate or MTTF of affected components/parts – These are used in assessing the PM-activity-frequency recommendation. Relative times from potential detection of degradation to failure (P-F interval) are also useful.
- service specific conditions that warrant modification of the default (generic) MTTF or damage rates, including differences in:
    - design or construction (e.g., API pumps versus ANSI pumps, special purpose versus general purpose, etc.)
    - installation differences
    - operating condition (e.g., deviation from best efficiency point for a pump)
    - other significant conditions that stress or provide robustness to the equipment component affecting its reliable life
- listing or ranking of effective PM activities/technologies
- equipment data requirements to set up the program and to feed back to the PoF model for updating
- failure-reporting-code definitions to provide for updating probability of failure models – These will need to be checked against the PM Analysis to see if there is sufficient correlation to the damage mechanisms to provide useful information. Potential sources for failure coding are ANSI/API Std. 689 (ISO 14224:2006) *Collection and Exchange of Reliability and Maintenance Data for Equipment*, the CCPS Process Equipment Reliability Data (PERD) project and OREDA.

Table 2

Simplified PM Analysis Table with Sample Entries

PM Analysis Table – Centrifugal Pump						
	Component or Part	**Damage Mechanism**	**Damage Cause**	**Consequence Type**	**Time to Repair (days)**	**PM Types**
	Bearing – Antifriction	Wear	Normal Wear	OoS	1	Vibration Monitoring
	Bearing – Antifriction	Wear	Contamination	OoS	1	Oil Analysis Oil Change
	Bearing – Antifriction	Loss of Lubrication	Oil Leak	OoS	1	Operator Rounds
	Bearing – Antifriction	Excessive Vibration	Imbalance	OoS	1	Vibration Monitoring
	Bearing – Antifriction	Excessive Load	Operational Error	OoS	1	Vibration Monitoring
	Mechanical Seal	Wear	Normal Wear	**LoC** OoS	1	Operator Rounds
	Pump Casing	Thinning	Corrosion	**LoC** OoS	360	Internal Inspection Rebuild UT thickness measuring
	Belt (V belt drive)	Wear	Normal Wear	OoS	0.5	Inspect & Re-tension Replace

OoS – Out of Service – not available for use
LoC – Loss of Containment – Process fluid leakage

See separate tables for Avg. MTTF and Design, Installation and Operating Context Life Modifiers

The **PoF-rank-modeling** process is carried out at the damage mechanism/component level. It utilizes the information developed above and provides a ranking of the probability of failure based on equipment type, the considerations above, and other information (e.g., time in service, condition-monitoring results, etc.). This is typically a probabilistic approach and a variety of techniques may be applied (e.g., Weibull-based mean time to failure model, etc.).

The **PM Activity Strategies** or guidelines provide:

- inspection, test, or PM activity recommendations that correspond to and address the listed damage mechanism for the equipment and component type including consideration of its effectiveness or any limitations or other consideration in it use
- recommended application levels for these activities defining the frequency and extent based on the risk ranking (e.g., Should lube oil analysis be limited to oil condition or should it include wear particle analysis? Should it be done monthly to monitor wear or annually to recommend oil changes or conditioning?)
- definition or guidance of appropriate acceptance criteria to be used based on "as-found" conditions (e.g., What are appropriate vibration levels? - a

general guideline or an advisory that the criteria should be developed at more specific level.)

Table 3

PM Strategy - Generalized Listing for Example Only

Component Types	Damage Mechanism & Component / Part	Activity	Extent - See Activity Procedure for details	High	Med-High	Medium	Low
Centrifugal Pump	Bearing Failure Coupling Failure Seal Failure etc	Vibration Monitoring	Vibration Monitoring of all Components (1)	Daily (1)	1 Month	3 Months	N/R
Centrifugal Pump	Loss of capacity due to internal wear or corrosion	Performance Monitoring	Check actual flow and head developed versus pump curve (2)	1 Year	2 Year	N/R*	N/R
Centrifugal Pump	Oil contamination, Bearing Failure, Seal Failure	Change Oil	Change Oil in Sump	3 Months	6 Months	1 Year	1 Year
Mechanical Coupling Grease Lube	Coupling Failure	Re-grease	Inspect and add new grease.	3 Months	6 Months	1 Year	1 Year
Cog Belt Drive	Belt and/or Sheave failure, High bearing loads	Re-tension or replace belts	Inspect and re-tension or replace belts. (3)	2 Year	5 Year	N/R	N/R

Notes:
1) For High Criticality continuous on-line monitoring is recommended
2) If pumps are run in parallel add Medium criticality pumps at 2 Years
3) For High or Med-High Criticality with only turnaround availability for PM - Replace belts at these intervals

The PM-Analysis process provides the means to create and update the PoF Ranking model and PM-activity strategy. These in turn are key elements in the implementation and evergreening of the on-going process for risk-based PM.

3. Conclusion

The process industry has an opportunity to develop and implement a PM methodology that can provide considerable benefit in safety and reliability. This approach is similar to those in discussion in the API 691 Task Group for rotating equipment, but it is too early to suggest how that will develop. A common practice that establishes the principals and minimum general guidelines or methodologies for the development and objectives of these processes will provide a common foundation for on-going risk-based PM implementations. Utilizing a standard methodology for the PM analysis, PoF-ranking models and PM strategies to support a risk-based PM program will provide an efficient and effective program. This type of program will be capable of accomplishing the specific supporting objectives listed above.

Reliability and maintenance cost improvements are tied to a shift from reactive to a high degree of preventive maintenance including a significant component of condition-based maintenance. Best-in-class distributions are <10% reactive, and ~50% predictive (including follow-up of condition monitoring) with the balance split between the restorative type of PM and proactive maintenance (that addresses root causes including those from design, maintenance or operating practices). Organizations without focus can exceed 50% in reactive work. The transition requires not just a strong PM program, but strong planning and scheduling discipline among other aspects of an asset-management program. The effort is clearly worth it. The best in class have 30 to 50% lower maintenance costs and huge differences in on-stream performance. [5]

4. References

[1] *Guidelines for Risk Based Process Safety*, American Institute of Chemical Engineers, Center for Chemical Process Safety, New York, NY, 2007

[2] ANSI/API Std. 689 (ISO 14224:2006) *Collection and Exchange of Reliability and Maintenance Data for Equipment,* American Petroleum Institute 1st Ed. July 2007

[3] *Guidelines for Mechanical Integrity Systems*, American Institute of Chemical Engineers, Center for Chemical Process Safety, New York, NY, 2006.

[4] D. H. Worledge, discussions related to API 691 development and unpublished work on PM analysis

[5] Moore, R., *Making Common Sense Common Practice – Models for Manufacturing Excellence*, Elsevier, Burlington, MA, 3rd Ed. 2004.

Additional references not cited:

[6] Valbuena, R. R., Aller, J. E., Renner, M., and Feigel, G., *Implementing and Evergreening RBI in Process Plants*, Materials Technology Institute, 2005.

[7] API RP 580, *Risk-Based Inspection*, American Petroleum Institute, 2nd Ed, Nov. 2009.

[8] API RP 581, *Risk-Based Inspection Technology*, American Petroleum Institute 2nd Ed. Sept. 2008.

[9] ANSI/ISA-84.00.01-2004 *Functional Safety: Safety Instrumented Systems for the Process Industry Sector – Parts 1, 2, and 3 (IEC 61511 Mod)*, ISA - The Instrumentation, Systems, and Automation Society, Research Triangle Park, North Carolina, 2004.

[10] 29 CFR 1910.119 "Process Safety Management of Highly Hazardous Chemicals," Occupational Safety and Health Administration,

[11] *OREDA – Offshore Reliability Data Handbook 2002*, 4th Edition, DNV – Det Norske Veritas 2002

Acknowledgements:
To D. Mainwaring who provided valuable inputs, discussion and lots of editing.

Adjusting Spring Operated Pressure Relief Valve Proof Test Intervals Using Statistical Modeling and Comparing the Resulting Financial Risk

Stephen P. Harris, Ph.D.

Savannah River National Laboratory

Aiken, South Carolina
803.725.5304
stephen.harris@srnl.doe.gov

Robert E. Gross, P.E.

Savannah River Nuclear Solutions

Aiken, South Carolina
803.208.8822
robert.gross@srs.gov

Prepared for Presentation at
American Institute of Chemical Engineers
2010 Spring Meeting
6th Global Congress on Process Safety
San Antonio, Texas

UNPUBLISHED

Adjusting Spring Operated Pressure Relief Valve Proof Test Intervals Using Statistical Modeling and Comparing the Resulting Financial Risk

Stephen P. Harris, Ph.D.

Savannah River National Laboratory

Aiken, South Carolina
803.725.5304
stephen.harris@srnl.doe.gov

Robert E. Gross, P.E.

Savannah River Nuclear Solutions

Aiken, South Carolina
803.208.8822
robert.gross@srs.gov

Abstract

A method for adjusting spring operated pressure relief valve inspection intervals is presented based on statistical modeling, Monte Carlo Simulation, financial risk assessment, service severity, and life-cycle testing costs. Intuitively, longer intervals between maintenance and proof testing should tend to increase the likelihood of the fail dangerous mode (valve stuck shut on demand; $r>1.5$). A predicator of this would be conditions contributing to a failure on demand; $r>1.3$. A case study for one production facility is presented based on an actual demand on the pressure relief device, the probability of failing on demand (PFD), and the consequences of the event determined by service severity. Estimates for the PFD were obtained through extensive statistical studies and modeling of Department of Energy's Savannah River Site (SRS) data and use of the Center for Chemical Process Safety Process Equipment Reliability Database. The probability of failure on demand for mild, moderate, and severe Newtonian fluid services is derived from existing risk-based inspection technology and experience at SRS. The paper gives a graphical illustration of the inspection costs versus the associated risks by using a statistical model and Monte Carlo Simulation for test pressure growth. The intersection of both curves indicates a cost effective maintenance frequency balancing both financial risk and inspection cost.

1. Introduction

Since consensus codes allow a company or facility to set and adjust maintenance and inspection intervals for pressure relief valves, methods must be developed to ensure that risk reduction and cost effectiveness are balanced in a safe manner. General industry utilizes strict time-in-service based, performance-based, and risk-based approaches to conduct inspections. The time based method is used primarily at SRS but modified by performance and risk qualitatively. This paper presents Monte Carlo Simulation (MCS) of a combination of the three approaches. Inspection intervals, total life cycle costs, and statistical estimates of test pressure growth over time are used to estimate impacts on risk. The risks we refer to are the probability of failure on demand (PFD) and the assumed demand rate.

2. Notation

ASME	American Society of Mechanical Engineers
B&PV	Boiler and Pressure Vessel Code
MCS	Monte Carlo Simulation
NBIC	National Board Inspection Code
PERD	Process Equipment Reliability Database
PM	Preventative Maintenance
Proof Test	The practice of pressurizing the inlet of a new or used relief valve on a test stand. Pressure relief and seat tightness are tested and the as-found values are compared to initial set pressure.
PFD	Probability of failure on demand
PRV	Pressure relief valve(s)
r	Ratio of proof test pressure to set pressure
SORV	Spring operated relief valve(s)
SRNS	Savannah River Nuclear Solutions
SRS	Savannah River Site

3. SRS Pressure Protection Program (Mechanical Integrity)

3.1 Program Requirements – Major Activities

The Savannah River Site (SRS) program includes six major activities: relief valve and vessel Procurement, Verification, Registration, Fitness for Service, Repair, and Deficiency identification / resolution. The activities are briefly summarized below.

3.1.1 Procurement

All spring operated relief valves (SORV) and pressure vessels must be procured to comply with ASME B31.3 piping and ASME B&PV Codes. [1, 2]

3.1.2 Verification

All pressure relief devices with a set pressure at or above 15 psig must comply with the ASME B&PV Code. All SORV and pressure vessels must have a stand-alone, Engineering-Approved verification record that establishes their adequacy for the service intended. Once an item is actually installed for the first time on a piping system or vessel, it is physically verified to be in the proper place and in compliance with the approved record.

3.1.3 Registration

The vessels and relief valves must be registered, i.e., entered into the Computerized Maintenance Management System database. As part of the registration, the maintenance or inspection interval must be specified and the record set to active. This activity initiates the proof test cycle on the specified frequency (months-years).

3.1.4 Fitness for Service

All pressure vessels except boilers are required to be visually and volumetrically examined on a three to five year basis. New pressure relief valves are bench tested and inspected visually prior to being placed in service and then periodically as discussed below.

3.1.5 Repair

Pressure relief valves are repaired under the ASME-NBIC approved Valve Repair Quality Control Manual in compliance with manual NBIC-NB-23.[3]

3.1.6 Deficiencies and Deviations

For all deficient or unsafe conditions, a non-conformance report is generated and dispositioned by Quality and Engineering under the "Control of Non-Conforming Items" manual.

3.2 Fitness for Service

Periodic examination and testing involves removal of the SORV from the process, initial external visual inspection, and proof testing (as received) at the on-site relief valve repair shop (SRS Valve Shop). Guidelines are provided for disposition of each valve based on the proof test results, but the final decision to disassemble and repair is made by National Board certified technicians. Examination and testing intervals range from one to five years for clean gases and refrigerants, up to three years for steam, and two years for non code soft seated valves < ¾ inch inlet. Actual intervals are determined by the authority having jurisdiction based on service severity, location, failure modes and consequences. [4]

3.2.1 Fluid Service Severities

Using API RP 581 as a basis for classification, the valves included in the scope of this paper involve 56% mild service, 40% moderate service, and 4% severe service for the fail to open case.[5] Examples of severe service are high pressure air and gas, acids, and medium pressure steam >165 psig.

3.2.2 Reasons for Failed Proof Tests

Various earlier studies of proof test stuck shut failures found several common failure modes. The primary cause was found to be galvanic cells and oxygen concentration cells. Corrosion of the spring and/or spring retainers and their adhesion/seizure to the stem prevents the stem from sliding in the retainers. Root causes for the failed stuck case were determined to be exposure to corrosive conditions and the use of sacrificially anodic materials on the discharge side. Soft seated discs and nozzles (elastomer inserts) will take set both on the shelf and in service. Within two to three years the disks will tend to grip the nozzle causing high proof test results.[6]

3.2.3 Most likely Demand Failure

The average examination and test time at SRS for valves used in this paper's MCS is 3.7 years. In some cases that three to four year time frame is sufficient to cause high proof test pressures in air, clean gas, and water service especially if the SORV is fully exposed to ambient conditions year round. Later in this paper we discuss the number of inspections per month and show what impact variable test pressure growth over time has on increasing or decreasing the number of tests. At an average 3.7 years (44 months) between tests for the 864 selected valves, the current number of tests per month is 21.2.

4. Statistical Methodology

4.1 Statistical Model

A statistical model developed by Harris and Gross[7] was used for the Monte Carlo Simulation (MCS) to generate test pressures that are realistic and that include uncertainty and also the impact of specified trends. The model represents proof test pressure (psig) as a function of set pressure and manufacturer. The model is expressed in the natural logarithmic space as:

$$\log_e(y_{Test\ Pressure}) = \beta_0 + \beta_1 \log_e(y_{Set\ Pressure}) + Manufacturer\ Terms + \varepsilon \qquad (1)$$

where $\log_e(y)$ is the natural logarithm of y, $y_{Test\ Pressure}$ is the test pressure (psig), $y_{Set\ Pressure}$ is the set pressure (psig) and ε is the random error assumed to follow a normal distribution with zero mean and constant standard deviation σ. Using Model (1), $y_{Test\ Pressure}$ is represented in the original metric (psig) as:

$$y_{Test\ Pressure} = e^{[\beta_0 + \beta_1 \log_e(y_{Set\ Pressure}) + Manufacturer\ Terms + \varepsilon]} = e^{(\beta_0 + Manufacturer\ Terms)} e^{\varepsilon} y_{Set\ Pressure}^{\beta_1} \qquad (2)$$

The parameters in Model (1) were estimated using the JMP statistical software (SAS Institute, Cary, NC) applied to the SRS new valve database. The estimates and standard errors for Model (1) are: $\hat{\beta}_0 = b_0 = 0.108 \pm 0.011$, $\hat{\beta}_1 = b_1 = 0.988 \pm 0.002$, and $\hat{\sigma}^2 = 0.070$ while the correlation between b_0 and b_1 is $Corr(b_0, b_1) = -0.84$. The *Manufacturer Terms* essentially offset the intercept term β_0 in Model (1).

The aging model was taken to be linear where p is the expected percent increase in proof test pressure per year and t is time (years). The aging model is represented as:

$$y_{Test\ Pressure} = e^{(\beta_0 + Manufacturer\ Terms)} e^{\varepsilon} y_{Set\ Pressure}^{\beta_1} + p \times t \times y_{Set\ Pressure} \qquad (3)$$

To illustrate the use of Model (3), test pressures were simulated for SRS valves supplied by a major manufacturer with set pressure of 350 psig (Plot 1a).

Plot 1a
Simulated Test Pressure (psig) vs. Time in service (Years)
Trend is p=3%/year

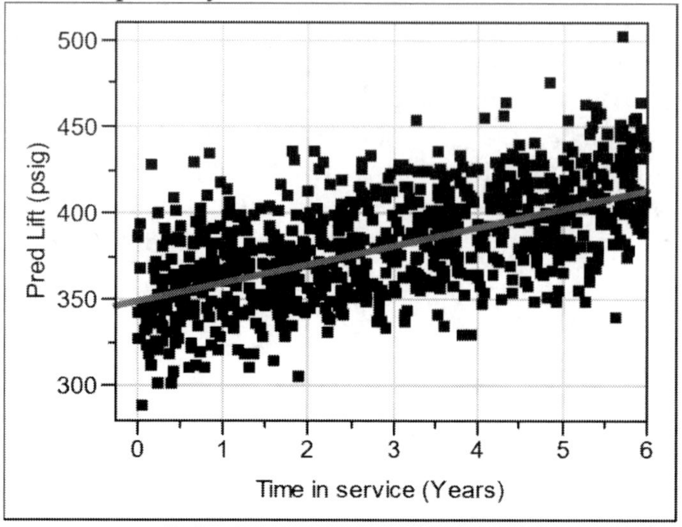

A p=3% per year growth rate was assumed for the proof test pressure trend while time in service ran from zero up to six years. A regression curve fit to the simulated ratio (test pressure/set pressure) versus time resulted in Ratio=1.0 +0.03 x (Years in Service). Model (3) was placed in the used valve database so that the distribution of manufacturers in service at the SRS is represented. Model (2) served as the basis for the MCS after known trends were built into the model. [7] Use of Model (3) in the site-wide simulation enabled realistic variability to be built around pre-specified trends in test pressure (e.g., p=3%/year).

4.2 *Monte Carlo Simulation* (MCS)

The exact impact of aging on valve performance within the PM interval (t_0, t_{PM}) where $t_0 = 0$ is not known. However, it is specified in the MCS. The aging model was taken to be linear where p is the expected percent increase in proof test pressure per year. Seven cases were simulated: p=0%, 1%, 2%, 3%, 4%, 5% and 8%/year. The aging model is represented as in Model (3) above. Time (t) within the MCS time was taken from a uniform distribution over the 0 to 20 years. All other terms in Model (3) were generated from a normal distribution with means, standard errors, and $\hat{\sigma}^2$ taken from the regression output (Appendix 1).

The preventative maintenance (PM) plan in effect at SRS is designated in this paper as "PM Freq0." Approximately 52% (451/864) of the valves (Table 1) within the used valve database are on a 3 year maintenance plan while approximately 34% are on a 5 year maintenance plan. The average inspection time is 3.7 years, and approximately 21.2 inspections (proof tests) are made per month on average. Three other maintenance plans are considered in the simulations: PM Freq0 tightened by 1 year (PM Freq0-1), PM Freq0 extended by 1 year (PM Freq0+1), and PM Freq0 extended by 2 years (PM Freq0+2). To illustrate, there are 52 valves under a 2 year maintenance plan. For PM Freq0+2, the maintenance interval becomes 4 years for those 52 valves.

Each preventative maintenance plan is implemented within the MCS. A random time was generated between 0 and 20 years, the maintenance plan was applied, valves were removed from service at their PM time, tested and replaced with a new valve. The proof test pressure value was generated from Model (3), the ratio (r) of the proof test pressure to the set pressure was calculated and a comparison was made with the valve failure limit of r=1.3. The number of failures was summed within each one year interval ranging from 1 to 20 years, thus enabling the estimation of the probability of failure on demand (PFD).

Table 1
Sitewide Preventative Maintenance Programs Evaluated in the MCS

	PM Freq0-1 (Years)	PM Freq0 (Years)	PM Freq0+1 (Years)	PM Freq0+2 (Years)	Number of Valves (N)
	0.5	1	2	3	3
	1	2	3	4	52
	2	3	4	5	451
	3	4	5	6	68
	4	5	6	7	290
Avg Ins Time (Years)	2.7	3.7	4.7	5.7	864
Avg Tests/Month	31.6	21.2	16.1	13.1	

4.3 Simulation Results

The risk for each percentage increase in proof test pressure per year (*p*) in terms of US dollars was estimated for each year as $Risk_i=PFD_i*Consequence$ where the Consequence was taken to be \$1.25 million and i=1, 2, 3,...,20. The inspection cost was estimated at \$3,000 per proof test. **Plot 1b** displays expected risk and straight-line inspection cost versus the average number of inspections per month (Table 1) resulting from curves statistically fit to the MCS output. The intersection of the risk curve with the maintenance cost curve provides a metric for the sensitivity of various inspection plans for various degradation rates. For example, the PM Freq0 plan starting at 21.2 tests per month provides cost effective protection for up to an approximately 5%/ increase in proof test pressure. Extending the maintenance intervals by 1 year (average test per month=16.1) results in providing cost effective protection for up to an approximately 3-4%/ year degradation rate. A 95% upper bound on risk (with fixed consequence \$) is provided in Plot 1c. The intersection of the 95% upper risk limit with inspection cost provided an upper 95% limit of approximately 31 proof tests per month.

Plot 1b – Risk Curves with *p*=0% to 8%
Expected Sitewide Risk (US $) vs. Number of Proof Tests per Month

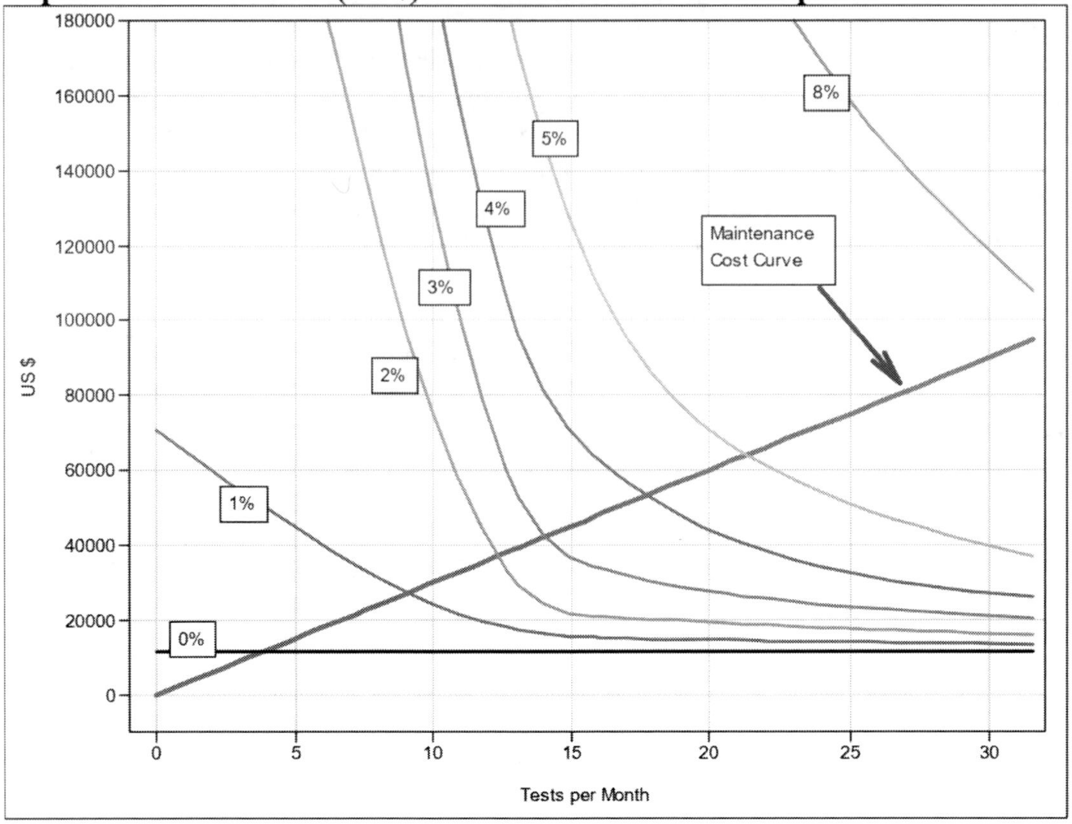

Plot 1c Risk Curves with *p*=0% to 5%
95% Upper Limit for Sitewide Risk (US $) vs. Number of Proof Tests per Month

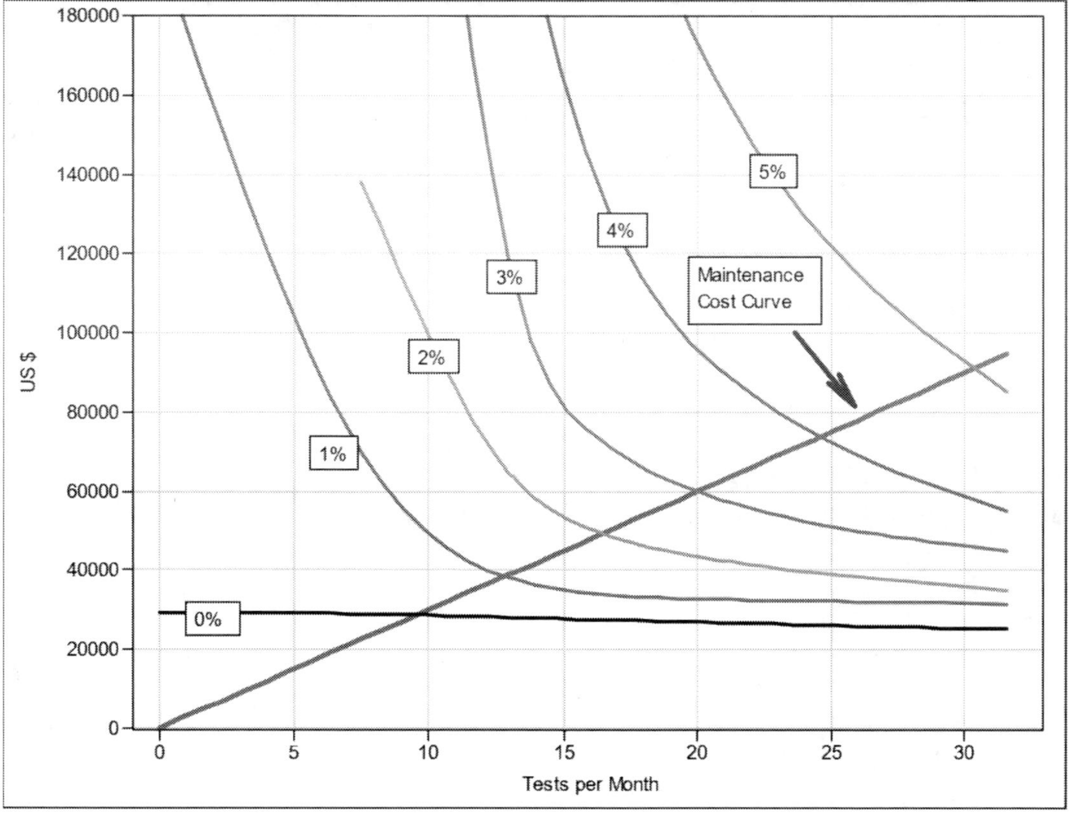

4.4 Special Cases

Two special cases are considered: the first based on 28 steam service valves with set pressure in the 140-175 psig range under a mild service consequence and the second based on 78 nitrogen service valves with set pressure in the 300-1,000 psig range under a moderate consequence.

4.4.1 Special Case 1 – An example of moderate fluid service with mild consequence ($500,000)

There is a pressure control valve on each of these systems at SRS that reduces 300 psig saturated steam to 80-120 psig. The downstream piping is rated to handle 220 psig at 330 degrees F. The system was originally hydrostatically tested to 1.5 times design (100 x 1.5 = 150 psig). In a building or other enclosed space, failure of this system on demand would cause mild material damages, loss of production and injuries but not necessarily fatalities. Approximately 90% of the valves are on a 3 year inspection frequency (Table 2), and the average maintenance time is 2.9 years. There are approximately 0.86 tests per month (10 per year) on average. A ratio r=1.3 was used in the MCS to assess the sensitivity of the maintenance plan for early indications of valve degradation (increased proof test pressure). The MCS results in Plot 2 show that cost effective protection is provided up to a degradation rate of 4%/year and that 2 more proof tests should be conducted to provide cost effective protection up to a 5%/year degradation rate.

Table 2
Preventative Maintenance Programs Evaluated in the MCS for Special Case 1

	PM Freq0-1 (Years)	PM Freq0 (Years)	PM Freq0+1 (Years)	PM Freq0+2 (Years)	Number of Valves (N)
	0.5	1	2	3	1
	1	2	3	4	2
	2	3	4	5	25
	3	4	5	6	0
	4	5	6	7	0
Avg Ins Time (Years)	1.9	2.9	3.9	4.9	28
Avg Tests/Month	1.38	0.86	0.62	0.49	

Plot 2 Risk Curves with *p*=0% to 8%
Expected Risk (US $) for Special Case 1 vs. Number of Proof Tests per Month

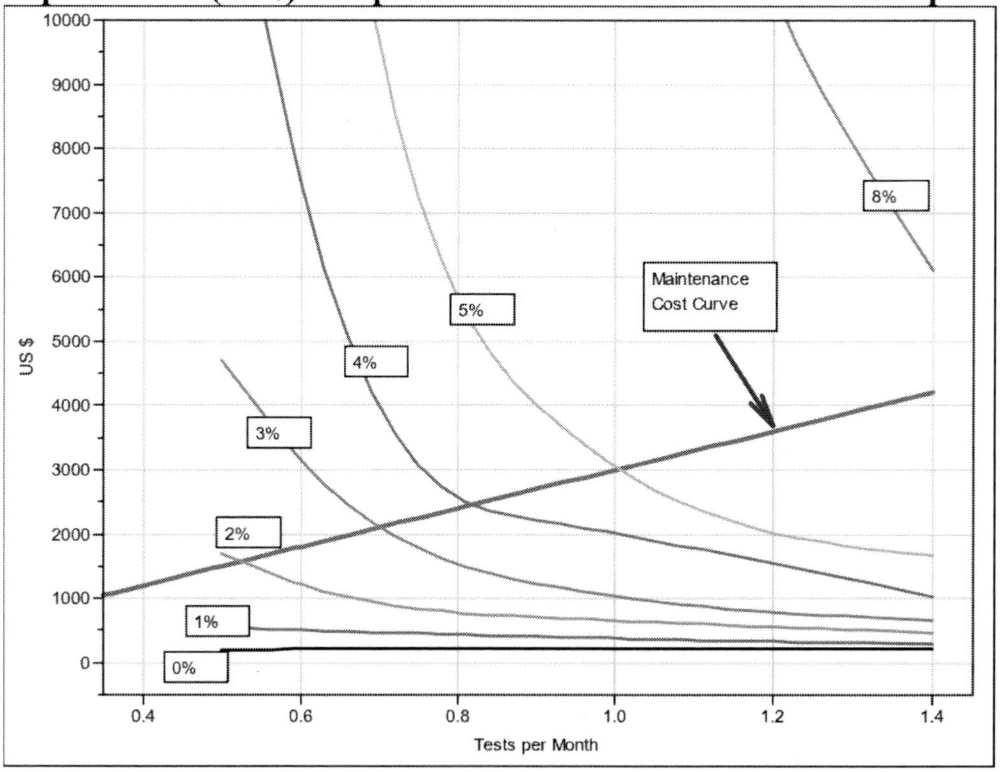

4.4.2 Special Case 2 - An example of mild to moderate service with significant consequence ($2 million)

A 2400 psig compressed nitrogen gas source is regulated to 350 psig at SRS. Should the regulator fail to open and the SORV not lift, downstream piping and components would be over pressurized to 2400 / 350 or a factor of about 7. The weakest point in the system would fail catastrophically due to over pressurization. Approximately 82% of the valves at SRS are on a 3 year inspection frequency (Table 3), and the average maintenance time is 3.3 years. There are approximately 2 tests per month (24 per year) on average. A ratio r=1.3 was used in the MCS to assess the sensitivity of the maintenance plan for early indications of valve degradation. The MCS results in Plot 3 show currently that cost effective protection is provided up to a degradation rate of 3-4%/year. Nine (9) more proof tests per year should be conducted to provide cost effective protection up to a 5%/year degradation rate. If the target of 5%/year is chosen, then the resulting time between inspections would be 78/33 or 2.4 years.

Table 3
Preventative Maintenance Programs Evaluated in the MCS for Special Case 2

	PM Freq0-1 (Years)	PM Freq0 (Years)	PM Freq0+1 (Years)	PM Freq0+2 (Years)	Number of Valves (N)
	0.5	1	2	3	0
	1	2	3	4	1
	2	3	4	5	64
	3	4	5	6	2
	4	5	6	7	11
Avg Ins Time (Years)	2.3	3.3	4.3	5.3	78
Avg Tests/Month	3.03	2.04	1.55	1.25	

Plot 3 Risk Curves with *p*=0% to 5%
Expected Risk (US $) for Special Case 2 vs. Number of Proof Tests per Month

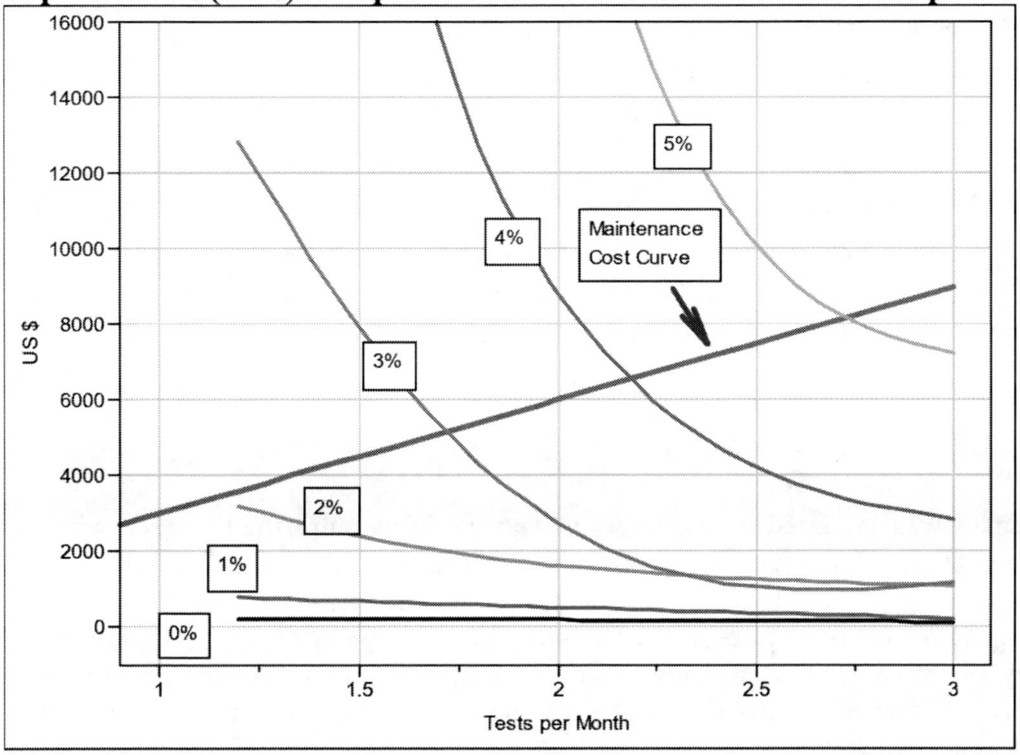

5. Conclusion

Statistical modeling of new valve proof test data with Monte Carlo Simulation was used as a baseline to determine and verify cost effective maintenance frequencies for a sitewide maintenance plan and two special cases within the site plan. A fixed percentage increase in proof test data (p) was incorporated in the simulation. For the method to be useful, the maximum p tolerable in practice must be known. Within the simulation, the test pressure and time (years) at which each valve will open is known. This is not the case in practice. All that is known after a proof test is that 1) if a valve fails successful proof test, it actually failed between the time of installation and the proof test time, or that 2) if the valve passes the test, it will fail at some unknown time after the proof test. The simulation enables the determination of the effectiveness of the maintenance plan and enables the tuning of a given plan. Different functions for proof test increases (degradation) such as an immediate change in valve performance capability due to system contamination could be considered in future work.

6. References

[1] ASME B31.3 Pressure Piping Codes, American Society of Mechanical Engineers, New York, NY.

[2] ASME Boiler and Pressure Vessel Code (B&PV), American Society of Mechanical Engineers, New York, NY.

[3] ANSI-NB-23, National Board Inspection Code, The National Board of Boiler and Pressure Vessel Inspectors, Columbus, OH.

[4] Pressure Equipment Protection Requirements, SRS Engineering Standard 15061, Rev 6, 08/01/2009, Pressure Protection Technical Committee.

[5] API RP 581 Risk-Based Inspection Technology, Section 7 Pressured Relief Devices, American Petroleum Institute (API) Recommended Practice 581, second edition, September 2008.

[6] Gross, R.E. and S.P. Harris, "Extending Proof Test Intervals by using Statistical Analysis of Test Data", PVP2006-ICPVT11-93109, ASME PVP Conference, Vancouver BC, July 2006.

[7] Harris, S.P. and R.E. Gross, "Validation of Spring Operated Pressure Relief Valve Time-to-Failure and the Importance of Statistically Supported Maintenance Intervals," Process Plant Safety Symposium Proceedings, AIChE Spring National Meeting, Tampa, FL, April 2009, pp. 127-141.

Acknowledgements

This document in part was prepared in conjunction with work accomplished under Contract No. DE-AC09-08SR22470 with the U.S. Department of Energy. Portions of this work were prepared under an agreement with and funded by the U.S. Government. Neither the U. S. Government or its employees, nor any of its contractors, subcontractors or their employees, makes any express or implied: 1. warranty or assumes any legal liability for the accuracy, completeness, or for the use or results of such use of any information, product, or process disclosed; or 2. representation that such use or results of such use would not infringe privately owned rights; or 3. endorsement or recommendation of any specifically identified commercial product, process, or service. Any views and opinions of contributing authors expressed in this work do not necessarily state or reflect those of the United States Government, or its contractors, or subcontractors.

Appendix 1
Reference [7] Harris and Gross AIChE April, 2009
Regression Model based on New Valve Data Set

Predicted Test Pressure By Actual Test Pressure

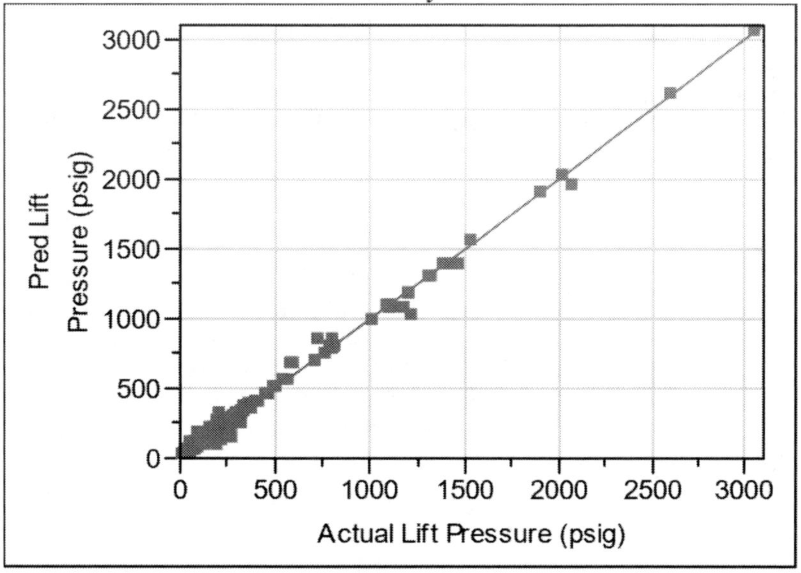

Response $\log_e$(Test Pressure)
Summary of Fit

RSquare	0.99
Root Mean Square Error	0.07
Mean of Response	4.64
Observations	2338

Analysis of Variance

Source	DF	Sum of Squares	Mean Square	F Ratio
Model	45	1483.8	32.97	6722.4
Error	2292	11.2	0.005	Prob > F
C. Total	2337	1495.1		0.0000

Parameter Estimates

Term	Estimate	Std Error	t Ratio	Prob>|t|
Intercept	0.108	0.0108	10.0	<.0001
$\log_e$(Set Pressure)	0.988	0.0019	508.8	0.0000
+ Manufacturer Terms				

Effect Tests

Source	Nparm	DF	Sum of Squares	F Ratio	Prob > F
ln Set	1	1	1270.0	258921.3	0.0000
Manufacturer	44	44	1.3	5.7924	<.0001

6TH GLOBAL CONGRESS ON PROCESS SAFETY

Strategies & Tactics for Mechanical Integrity Budget Battles

Mechanical Integrity Response to Economic Downturn

Daniel A. Long, P.E.
Open Range Engineering PLLC
P. O. Box 2179
217 North Cuyler, Suite 300
Pampa, Texas 79066-2179
Office 806-665-1115
Cell 806-662-6748
oredal@openrangeeng.com

Jack McCavit
JL McCavit Consulting, LLC
1000 Quail Place
Pampa, Tx, 79065
Cell: 806-282-8689
Office phone: 806-665-6948
jlmconsulting@sbcglobal.net

Prepared for Presentation at
American Institute of Chemical Engineers
2010 Spring Meeting
6th Global Congress on Process Safety
San Antonio, Texas
March 22-24, 2010

UNPUBLISHED

Strategies & Tactics for Mechanical Integrity Budget Battles

Mechanical Integrity Response to Economic Downturn

Daniel A. Long, P.E.
Open Range Engineering PLLC
P. O. Box 2179
217 North Cuyler, Suite 300
Pampa, Texas 79066-2179
Office 806-665-1115
Cell 806-662-6748
oredal@openrangeeng.com

Jack McCavit
JL McCavit Consulting, LLC
1000 Quail Place
Pampa, Tx, 79065
Cell: 806-282-8689
Office phone: 806-665-6948
jlmconsulting@sbcglobal.net

Keywords: mechanical integrity; costs; communication; budget; fitness for service

Abstract

Effective mechanical integrity programs are essential to assure safe operation of plant facilities. The current economic downturn is difficult and most manufacturing companies have experienced headcount reductions, early retirement programs and budget reductions. Companies are concerned with short term viability rather long term optimization. How should professionals responsible for the mechanical integrity (MI) programs respond?

Effective communication with management is essential. MI professionals must clearly and concisely present the causes and effects of reducing or eliminating inspections, preventative maintenance, repairs and replacements. MI experts must put recommendations in terms management can easily understand while focusing on safety risks, legal compliance, production impact and budget/cost implications.

This paper will encourage MI professionals to:

1. Make reasonable requests with sound knowledge of business positions and budget restraints.
2. Forecast repairs and replacements two to three years in advance to avoid economic surprises to management,
3. Support increased business risk and reliability risk, not safety risks.
4. Challenge recommendations within the MI group before taking the recommendation to management. Are there any lower cost options to extend the life of the equipment while assuring safe operation?
5. Dialogue with management about the consequences of not doing MI recommendations. Make communication clear, without drama. Avoid language that is watered down or converted to politically correct gibberish that is not clear.

1. Introduction

Recently I made a general process safety presentation to a large group of people at a chemical plant. A few days after the presentation, I received the following email from a mechanical integrity supervisor at that plant.

"I have worked as a supervisor for inspectors and am currently providing general technical support globally. I was wondering if you could elaborate more on the mechanical integrity aspects and what you have seen as "effective" practices vs. just getting by. It also brings to light the point that during economic downturns, in order to control costs sometimes it is too easy to cut back on inspection programs and preventative maintenance to "make the numbers and save money" to the longer term detriment of continually maintaining the plants at some base level of acceptable mechanical integrity. What have you seen as effective strategies to resist this knee jerk response and hold the line on mechanical integrity? Thanks for any insight that you can share."

This paper is a direct result of that email. It is always important to implement a mechanical integrity program that is effective and efficient but during economic downturns the pressure from management to reduce costs is almost always a challenge to mechanical integrity professionals. This paper details ideas that are always important in mechanical integrity programs but adds emphasis to those areas that are particularly important during economic downturns. Mechanical integrity professionals have an obligation to work with management to help get through difficult economic times and if the ideas presented below are practiced, the fights over mechanical integrity budget issues can be minimized.

The employer shall correct deficiencies in equipment that are outside acceptable limits before further use or in a safe and timely manner when necessary means are taken to assure safe operation. (OSHA 1910.119 (j) (5))

This OSHA rule sounds simple but everyone working in MI understands that it takes lot of effort to find all of the deficiencies. If the deficiencies are unplanned repairs and replacements, correcting them before further use can be catastrophic to budgets and profits. A sound MI program that finds and corrects equipment deficiencies is essential not only for the safety of the site but also for economic health. Effective MI programs must have the following results;

- Identify current deficiencies
    - Determine fitness for service; run with deficiency or fix deficiency?
    - Define long term and short term repair options
- Predict future deficiencies
    - Predict future repairs and replacements
    - Predict the end of equipment life.

Of course these activities must be conducted in compliance with applicable codes and jurisdictional regulations.

Finding that one defective elbow in 10,000 miles of piping in a 50 year old facility may seem like an unreasonable expectation, but today that is the requirement. Releases of hazardous chemicals are unacceptable. In order for companies to survive in today's world, MI programs have to be effective and efficient. They must produce great results for a reasonable cost.

There are no silver bullets to help a business survive in the present economic downturn. A cost effective and efficient MI program is not a new idea, but is ever more important in today's economy. If your facility is not using all the methods and tools presented in this paper, please be encouraged to consider these ideas and make improvements.

Effective management systems are essential for great MI performance. To survive in today's economy; they must include the following elements:

- Identify, Forecast and Communicate MI Issues
    - Integrate the MI processes with the other plant management systems
    - Include recommendation logging and tracking processes
    - Support a MI metric reporting system
    - Develop a budget input system
- Support business risks – NOT safety risks
- Challenge recommendations and seek alternate lower cost solutions
    - Minimize waste and inefficiencies
    - Use efficient inspection practices
- Dialogue with management.

Not only are good management systems critical for success but talented, dedicated people are essential. The greatest system in the world is totally ineffective without knowledgeable and well trained people. Likewise talented workers can be slow and ineffective if they have to deal with poor systems.

Overall Effect = (Quality of the Systems) X (Quality of the People)

2. Identify, Forecast, and Communicate MI Issues

Integrated Process

The inspection codes require calculations of corrosion rates, remaining corrosion allowance, and remaining equipment life. Inspection intervals are typically set at ½ remaining equipment life. This should provide most organizations plenty of time to budget and plan for equipment replacements and major repairs. However, many organizations continue to have budget battles with the MI groups. Good communication, a cooperative spirit and a few simple tools can make those funding requests for repairs and replacements or additional inspection resources a more joyful experience.

Because of the economic downturn, there is no more important time than now for the inspection scheduling, execution, and reporting systems to be highly integrated into the budgeting, turnaround planning, and engineering projects systems. Inspectors and engineers in the MI groups must understand budget jargon and project engineering systems well enough to communicate their findings. Typically, projects must be identified to forecast costs three to four years in advance. A sound Predicted Repairs and Replacements (PR &R) System is essential.

A simplified high level overview of the predicted repairs and replacements process might look like this.

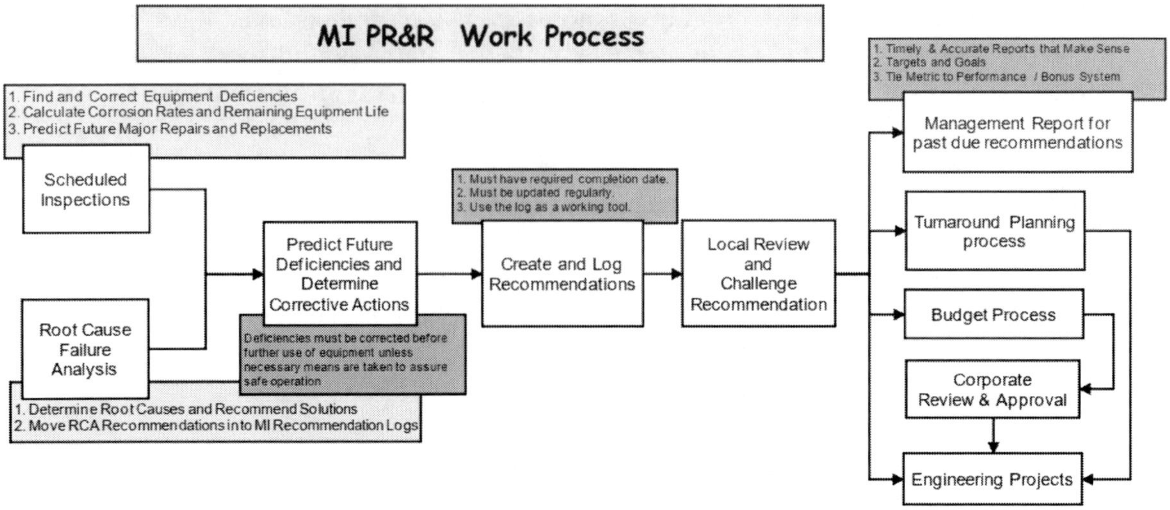

There are two primary sources of data that provide input for the predicted repairs and replacements program, inspections and equipment failures/incidents. Scheduled inspections (which include maintenance and preventative maintenance [PM] activities) should be defined by plant experience, manufacturer's recommendations, and recognized and generally accepted engineering practices (RAGAGEP's). Inspections are designed to find equipment deficiencies so that they may be corrected. Analyzing the data from the inspection program is the first step in predicting the future repairs and replacements. Another source of data is equipment failures and incidents. Incidents due to MI failures should result in recommendations to correct deficiencies as well as recommendations to prevent future failures. In economic downturns it is particularly important to learn from any available source, especially incidents since incidents are actual failures, not potential failures.

Recommendations Logs

It is easy to get behind on paperwork and MI equipment inspectors are no exception. Avoid losing recommendations on notepads or in the detailed inspection reports. Try to note those important thoughts in the recommendation logs as soon as they are discovered. There is no need to wait for an engineering design, just write down a simple recommendation and change it later as details are developed. Always include a due date for completing the recommendation. This will allow the system to sort recommendations that are past due or coming due. Change completion dates as required when new information is available. If the timing is "soft / conservative" change it to accommodate budgets, turnaround timing, and equipment deliveries.

Below is an example of a recommendation log

MI*	NO.	Unit	RECOMMENDATION	DUE DATE for Completion	REC REPORT AUTHOR & Date	STATUS	
						DATE COMPLETED	REMARKS
*	1	Unit #1	Replace HE-454 Tube Bundle. Upgrade tube material to alloy C-276 original material 316ss can't meet the process requirement . MOC should be done before this replace.	8/1/2011	10/2009 Dan Long		Added to 2011 TA List
*	2	Unit #1	Perform 100% Wet MT on V-454 during next TA.	~~8/1/2009~~ 8/1/2011	10/2009 Dan Long		Changed date to match TA Added to 2011 TA List
*	3	Unit #1	Replace Reformer Tubes during 2014 TA	8/1/2014	8/2009 Dan Long		Assigned to JLM Project Engineering
	4	Unit #1	Purchase spare bundle for HE-777. 20% of tubes are plugged	11/1/2010	08/2008 Jack McCavit		Business Risk Capacity Limitation
*	5	Unit #1	Repair the flange face of HE-456. Pit on channel flange and nozzle flange.	8/1/2011	08/2008 Jack McCavit		Added to 2011 TA List

The mark in the MI column indicates if the recommendations are mechanical integrity / process safety critical items. If not, they are a business risk. If a leaking tube in a heat exchanger will allow water into a process stream that will cause a product to go off test and not create a potential safety consequence, it is a business risk to run the tubes to failure. See Section 3 below for more comments about accepting business risk but not accepting process safety risk.

Recommendations logs should be distributed to the operations leaders and managers as well as the maintenance, engineering, turnaround planning, and budget organizations.

MI Metrics

Hopefully your site management team periodically reviews process safety metrics.[1] Two important MI metrics that should be included are past due inspections and the status of equipment deficiency recommendations. The overall status of site recommendation logs is a great way to track the status of past, present and future equipment deficiencies. Past due MI recommendations should be a key plant and corporate process safety metric. This management review can also be a forum to locally challenge recommendations to determine alternate solutions and to understand the associated risks.

Below is an example of a recommendation log scorecard that presents a simple measurement of past due MI recommendations. This can be sorted and presented by plant operational area or by equipment type.

MI Fixed Equipment Recommendation Log - Scorecard											
Today	1/25/2010	Mar-10		Apr-10		May-10		Jun-10		Jul-10	
		Number	Percent	Number	Percent	Number	Percent	Number	Percent	Number	Percent
Unit #1 Recommendations											
	Total Number of Recommendaitons	43		43		43		45		45	
	Number Complete	25	58.1%	25	58.1%	28	65.1%	28	62.2%	32	71.1%
	Number In Progress	16	37.2%	16	37.2%	16	37.2%	16	35.6%	18	40.0%
	Number Past Due	2	4.7%	2	4.7%	2	4.7%	2	4.4%	1	2.2%
Unit #2 Recommendations											
	Total Number of Recommendaitons	56		56		56		56		56	
	Number Complete	16	28.6%	16	28.6%	18	32.1%	18	32.1%	18	32.1%
	Number In Progress	37	66.1%	37	66.1%	37	66.1%	37	66.1%	37	66.1%
	Number Past Due	0	0.0%	0	0.0%	0	0.0%	0	0.0%	0	0.0%
Unit #3 Recommendaitons											
	Total Number of Recommendaitons	123		134		135		135		135	
	Number Complete	67	54.5%	67	50.0%	88	65.2%	88	65.2%	88	65.2%
	Number In Progress	55	44.7%	55	41.0%	55	40.7%	55	40.7%	55	40.7%
	Number Past Due	1	0.8%	1	0.7%	1	0.7%	0	0.0%	0	0.0%
Plant Total											
	Total Number of Recommendaitons	222		233		234		236		236	
	Number Complete	108	48.6%	108	46.4%	134	57.3%	134	56.8%	138	58.5%
	Number In Progress	108	48.6%	108	46.4%	108	46.2%	108	45.8%	110	46.6%
	Number Past Due	3	1.4%	3	1.3%	3	1.3%	2	0.8%	1	0.4%

Another key MI metric is percent past due equipment inspections. If the inspections are not getting done on time there can be undetected deficiencies which can lead to serious problems and surprises for the organization. A high level report should be provided to site management monthly or quarterly. Detailed reports should break this data down by equipment type and by production units so the organization can easily see where the problems are located and identify why inspections are not getting done.

Does your report look like this one!?

Type Equipment	# of Items	# of Items Past Due	% Past due
Fabricated Equipment	2046	0	0.00%
Relief Devices	1059	0	0.00%
Safety Instrumented Systems	32	0	0.00%
Process Interlocks	41	0	0.00%
Electrical	1088	0	0.00%
Analyzer	319	0	0.00%
Instruments	507	0	0.00%
Rotating Equipment	834	0	0.00%
Hoses	136	0	0.00%
Other	1996	0	0.00%
3/1/2010	**8058**	**0**	**0.00%**

Like the recommendations, it is often difficult to sort out those inspections that are process safety critical. A lot of sites take a very strict approach and treat all PM's and inspections as critical. (Remember, when everything is a mountain and nothing is a molehill, you probably have OVERKILL.) Treating a past due oil change on an engine or compressor the same as past due inspection for a hydrofluoric acid tank diminishes the risk associated with the acid tank.

Vessel inspections are regulated by API codes and possibly jurisdictional regulations. A past due inspection of a pressure vessel should be treated the same as an equipment deficiency. It should be corrected before further use unless necessary means are taken to assure safe operation. Only run with a past due inspection if you have evaluated it and are sure that the process can be run safely with the past due condition. The Management of Change (MOC) process is a great tool to perform and document this evaluation. However, be careful not to misuse this tool. A MOC should not change the past due status. It is simply an evaluation tool to see if the process can be safely run in the past due condition and to help identify necessary means to assure safe operation. For example an external on-stream inspection might provide adequate assurance that a vessel has adequate shell thickness to run until the next shutdown when an internal inspection can be performed.

Hopefully your process safety mechanical integrity performance is tied to the plant variable pay system. It should affect several departments and have an appropriate weight with all of the other bonus inputs. This helps the whole organization assume ownership and understand that mechanical integrity really is their job too, not solely the MI group. The most effective organizations always acknowledge that the operations department has ultimate ownership responsibility for the equipment since they are the ones who operate it daily. Maintenance and engineering also need to have a strong sense of ownership for the equipment as well. Accountability for MI performance needs to be spread among all of the appropriate departments. It's a team effort.

What can companies do to lower their inspection costs, stay in compliance with codes, run safely and have great MI metrics during these tough economic times? Look for opportunities in these two areas:

- Decrease the scope or intensity of the inspection.
- Increase inspection intervals

Performing an on-stream external inspection instead of an internal inspection on a pressure vessel significantly reduces outage time, maintenance costs (opening, blinding, clearing, and cleaning) and inspection time. API-510 is very specific regarding the conditions under which an on-stream inspection may be substituted for an internal inspection. Many of your vessels may qualify. Additionally the codes may allow longer period between inspections than what has been historically comfortable. Even if the codes allow an extension, be sure historical inspection results warrant extending the inspection interval.

Budget Input

Many organizations have trouble mining out the budget data from the MI recommendation logs. A separate Predicted Repairs and Replacement (PR&R) budget document will often help the process. The MI group is the likely owner / keeper of this document. They have the best knowledge to summarize the data and with help from the engineering and procurement groups they should be able to put budgetary dollar figures on each item. The development and ownership of an accurate PR&R document should be a team effort. Accurate forecasts of spending are important to most companies and the forecasts are more important in today's economic environment. The PP&R document should focus on when the money will be spent as well as when the projects need to be completed. Consideration should be given to equipment lead times, progress payments, pre-engineering funds, timing to write requests for funding, and timing for review and approval for the funding requests. The further in the future the project is, the less accurate the timing and the cost figures need to be. However, when it is time to get a funding request in an approved budget or capital plan, the timing and dollar forecasts need to be much more accurate.

Every time inspectors make recommendations, they should be asking if the item should be added to the PR&R document PR&R documents need to be informally updated whenever new information is acquired. Update these databases or spreadsheets as soon as new information is obtained from:

- Inspections, inspection data analysis, engineering fitness for service evaluations.
- Design for repairs and replacements
- Cost and delivery information
- Management or business team input

The PP&R document should be formally updated, reviewed with stake holders, and distributed on a routine basis (quarterly). Below is an example of a simple format for documenting and communicating predicted repairs and replacements.

MI Fabricated Equipment Predicted Repairs and Replacements ($M)						
Rec #	**Budget**	**Project Description**	**2010**	**2011**	**2012**	**2013**
		Unit #1				
1	MOB CAP	Replace HE-454 bundle		500		
14	Expense	Repair refractory in reformer	75			75
22	Small Cap	Upgrade furnace burners			800	
19	CAP	Replace reformer tubes in Unit		200	1000	1200
		Sub total Unit #1	**75**	**700**	**1,800**	**1,275**
		Unit #2				
28	Expense	Eddy Current Test & Repair Leaking tubes on HE 456	180			
11	Small Cap	Replace stainless valves in HF Unit		200		
		Sub total Unit #2	**180**	**200**	**0**	**0**
		Unit #3				
32	Expense	Repair trays in tower # 23	30			
41	MOB Cap	Purchase new spare tube bundle for HE 789		150		
42	MOB Cap	Purchase new steam desuperheaters			150	150
		Sub total Unit #3	**30**	**150**	**150**	**150**
		Total Pant	**285**	**1,050**	**1,950**	**1,425**

Important or controversial decisions should be documented. If an important item is reduced, delayed, or removed from the budget; make a note of the date, the decision, and the reasons why. It's easier to tee up again or explain later if you have good documentation. Be clear and concise but do not single out individuals.

3. Support Business Risks NOT Safety Risks

In 1998 an engineer completed calculations that revealed that a 150 foot tall spec product column that had been operated safely for fifteen years was not designed for hurricane wind load forces that are required at its coastal location. What would it hurt to run for 8 or 10 more months (while a solution is engineered) with a deficiency that was just discovered? The product from the tower was a highly profitable strategic product. It was not even hurricane season!

In this actual case, additional engineering analysis revealed that the tower would withstand the high wind forces at ambient temperature but not design temperature. A temporary MOC was done and procedures were put in place to shutdown the tower if a hurricane, thunderstorm, or any other significant winds were forecast for the area. This is a great example of temporary means to assure safe operation. This provided time to engineer and implement a permanent solution.

Consider a heat exchanger with leaking tubes. Does the tube leak pose a process safety risk?

- Is there a large differential pressure that could cause a shell failure?
- Are there any hazards if the 2 streams mix?
- Will it cause any long term damage?

If cooling water leaks into a gas system it may not cause any problems at all. However, there might be corrosion damage 3 or 4 years later. If a tube leak will only cause a process to go off test or cause an efficiency loss, it probably is an acceptable risk.

Consider a shell leak on a heat exchanger. Small leaks of cooling water on the ground typically are not process safety problems. However, even a small leak of hydrofluoric acid is a big problem. Remember that according to OSHA, equipment can be operated with a deficiency if necessary means are taken to assure safe operation.

There are no simple answers or cook book recipes when it comes to sorting out process safety risk from business risk. There needs to be a good understanding of the equipment, the process and the possible failure modes. Additional resources for risk assessments, fitness for service evaluations, additional inspections or nondestructive examination (NDE) should be used if needed. An organization may be willing to take a 20% or even a 30% chance of running a product off test or experiencing a production outage. However taking even a 0.1% chance of experiencing a release of hazardous material which could result in fire, explosion, or toxic release is unacceptable. It is intolerable to have a failure that results in a catastrophic release. Fixed equipment MI programs must assure that the process is contained. However there are often fine lines and gray areas. You must satisfy these two primary requirements:

- Comply with codes, RAGAGEP's and jurisdictional regulations
- Have no process safety incidents

MI professionals take pride in minimizing problems due to mechanical failures. In economic downturns, MI professionals often have heated arguments about reductions of inspections and delays of repairs. If business managers are thoroughly informed and choose to take business risks, it is acceptable. MI personnel should understand that the decision is a business risk and a business decision. If the risk is a process safety risk, the MI personnel must be certain that appropriate measures are taken to ensure safe operation.

4. Challenge Recommendations, Alternate Solutions

The MI Group should challenge their own recommendations before taking the recommendation to management. Typical areas to challenge MI recommendations include:

- Has an engineering analysis been considered using the API-579 fitness for service, (FFS) rules?
- Can on-stream inspections or repairs be made?
- Can temporary repairs or partial replacements be made during a mini shutdown?
- Can the corrosion rates be reduced through minor process changes or by addition of corrosion inhibitors?
- Can the process be operated without the equipment or at a lower rate while the equipment is out of service for repairs?

- Is it cost effective to replace equipment rather than repair it when considering down time?
- Can necessary measures be implemented to assure safe operations with an existing deficiency (like the tower in the above example with the wind loading problem)?

Minimize Waste And Inefficiencies.
Especially in economic downturns, MI personnel should consider ways to streamline inspection programs and lower costs without sacrificing effectiveness and compliance. Consider the 80/20 rule. A relatively low percentage of your equipment population is consuming most of your resources and budget. This 20% needs the most attention from the MI group for those creative and low cost engineering solutions. Practical routine programs, providing code compliance with minimal cost and minimal time need to be applied to the 80%.

Some organizations unnecessarily study a vessel that is in "like new" condition taking hundreds of UT thickness measurements when twenty or thirty will fully meet the codes and reveal any problems with the equipment.

Many organizations struggle with how to handle the following common issues relating to pressure vessel UT thickness readings. UT data that has normal variability due to calibrations and different operators will often show artificial corrosion rates with an artificially short remaining life. Ultrasonic thickness meters typically have a stated accuracy of +- .004 inches. Differences in operator techniques and metal temperatures between surveys can add additional variability in thickness readings. One operator may record a thickness significantly less than the previous survey when no actual metal loss has occurred resulting in an artificial corrosion rate. Likewise there may be unexplained vessel thickness gain. Other times; inspectors struggle with determining the required thickness of a vessel (T_{min}). Implement efficient inspection processes and data analysis techniques that achieve code compliance concerning these issues.

5. Dialogue with Management

It is always important for management and MI groups to dialogue but that dialogue is critically important during economic downturns. Management must have a thorough understanding of the potential consequences of delaying repairs. It's appropriate to discuss failure modes, consequences of failures and even the probability of the failure. Do not inappropriately apply risk analysis in an effort to justify operating a vessel that is not in compliance with the RAGAGEP. The ASME pressure vessel code has a safety factor but one cannot arbitrarily encroach into that safety factor without a serious engineering evaluation. The safety factor is there for a reason. Conversations should focus on maintaining code compliance. If the finishing column is not going to be replaced, focus on the business issues rather than getting bogged down discussing potential big leaks or structural failure. Minimize the drama. Work with technical and operations departments to try to find creative solutions that do not compromise safety. A great response to management probing would be "If we do not replace the tower it will not be in compliance with the codes and we cannot continue to operate it. We will not be able to run our No. 2 LPG unit and our capacity to produce propane will be cut 50%. This will cost us $500,000 per month in net sales and $300,000 per month in economic profit."

Management often needs more information than just the cost and timing of recommendations to effectively make good financial decisions. Details such as project payback or equipment end of life may provide them with the additional data they need to communicate to their superiors and get funds approved. Give the information to them in an easily understandable format.

Sometimes organizations just cannot seem to make that important decision. It's nice to have as much information as possible and have as much time as possible to study budget buster projects. Waiting as long as possible before committing is smart if there are changing business dynamics. Do not let time run out. Push management to make a decision being clear when and why the decision must be made. Understand the consequences of not making a decision.

- Project timing cannot be met.
- Higher costs will be incurred for expediting fees and overtime
- Equipment will have to be shutdown.

Let management know how long they have to make that difficult decision.

Adjust your PR&R document to the ever changing needs of the organization.

6. Conclusions

A cost effective and efficient MI program is not a new idea but it is essential in today's economy and global competition. Struggling with profits might increase temptation to downsize programs. The requirements for an effective MI program are exaggerated during and economic downturn and some aspects of the program need special emphasis. When pushed by budget cuts:

- ✓ Be willing to compromise quality or reliability but do not compromise process safety.
- ✓ Try to find creative solutions that don't compromise process safety.
- ✓ Challenge recommendations within the MI group before taking the recommendations to management.
- ✓ Communicate with management early and often so that they thoroughly understand the potential consequences of their decisions.
- ✓ Do not condone violation of codes and standards.

7. References

1. *Guidelines for Risk Based Process Safety*, New York: American Institute of Chemical Engineers, Center for Chemical Process Safety, 2007.

Understanding Data Requirements

Harold W Thomas
Air Products & Chemicals, Inc
Allentown, PA 18195
thomashw@airproducts.com

ABSTRACT

Too often, there is a lack of sufficient thought with respect to developing data collection systems and work processes within process operating plants. Data is collected due to regulatory requirements, both internal and external, or someone thinks they might need it someday. This paper takes the viewpoint that data collection systems should be intelligently designed so as to support specific activities that are defined in advance of developing the data collection system. It stands to reason that if you do not know for what purpose you are going to use the data, it is hit or miss as to whether the data collected is useful. A brief overview of the CCPS PERD work process is provided to establish a sound foundation and then the paper illustrates the basics of data organization that will improve its usefulness for analysis once it has been collected, allowing for more cost effective analysis designed to provide value. The paper will also briefly discuss how the PERD ladder of success can allow a company to gain some benefit starting with the data they currently have and over time continuously improve the quality and value of their data.

Introduction

It has been said that you cannot manage what you do not measure. It stands to reason then that data collection is important, as it is essential to the measurement process. Understanding what is to be measured is necessary to determine what specific data needs to be collected. There are many reasons why data is collected. In some cases, it is done due to regulatory requirements. For instance, OSHA PSM requires as found, as left documentation as part of mechanical integrity programs. The intent of this requirement was to ultimately enable determination of how reliable various protection systems are with respect to preventing process safety incidents. Companies that choose to do more than the minimum necessary to satisfy the regulatory requirement can turn the intent into a solid foundation for a cost effective mechanical integrity program and to support reliability growth. To do this requires an understanding of equipment failure modes and what data is needed to determine the failure rate for the specific failure modes deemed important.

The Foundation

Once it has been determined what equipment failure modes are necessary for the user to accomplish their "DEFINED" goal(s), a quality system needs to be designed so that the appropriate data is collected. This involves a detailed consideration of the following:

- Denominator (A known population)
  - Whatever population a user decides to track is acceptable
  - Address information defines tracked population
- Population Attributes
  - Whatever data user deems necessary to filter a population for analysis purposes
  - Example: User wishes to compare carbon steel valves to stainless steel valves. Material of construction data would be required
- Numerator (Event data)
  - Data fields necessary to infer failure modes desired by user required
  - Need confidence that those data fields will be captured

Importance of Location Data

A tracked population is necessary to establish the denominator in the failure rate calculations. Process equipment is installed in specific locations and maintenance management systems using enterprise software like SAP or Maximo typically have an address whereby the same equipment Manufacturer and model number can be differentiated, i.e. facility or site code, plant code, unit code, tag number. In some cases, a specific equipment serial number (or system ID) is also available. The address can be thought of much like anyone's home address, while the serial number is much like a person's social security number. Having this information allows the maintenance software to link population data with event data. Figure 1 shows the address relationships that exist within the Process Equipment Reliability Database (PERD).

Figure 1

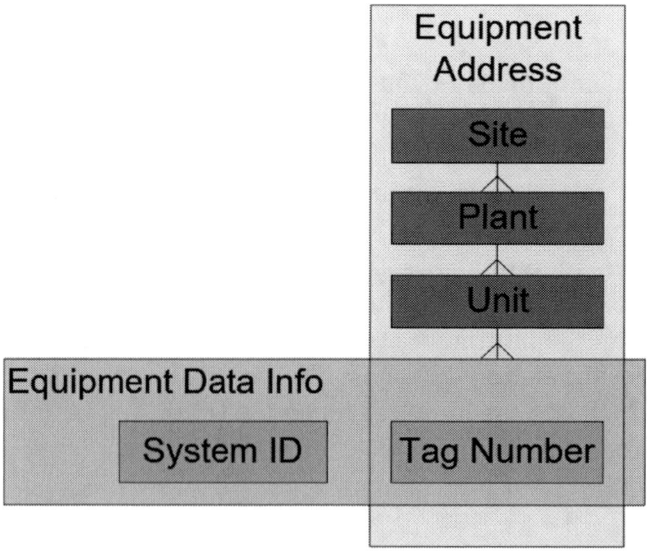

When considering what address location data is sufficient to track a population, bear in mind that there are two extremes. The least data is needed for the case where a single maintenance engineer works solely with equipment located within & dedicated to a specific plant. In this case, only the tag number is required as long as it is a unique tag identifier. However, doing the minimum makes it more difficult to aggregate data from multiple sites having multiple plants. When it is desired to combine data from these disparate location, it makes it almost a given that there will be duplicate tags. As such, this increases the required address information to ensure that the location data is unique for the tracked inventory population.

An example using key fields from SAP shows how company data can be matched within a database application like PERD.

- Example SAP identification number
  - US-15-9999-CATCrackA-SubUnit -Tag
- This code is broken down as follows:
  - 1st section = Country code
  - 2nd section = Business area / Product code
  - 3rd section = Site code
  - 4th section = Plant code
  - 5th section = Unit code (if applicable)
  - 6th section = Tag Number

Figure 2 shows how the data may be displayed to a user with a database such as PERD.

Figure 2

Country	Business Area /Product	Site	Plant	Unit
US	Refinery	New Orleans	Cat Cracker A	Unknown

The SystemID is not part of the location address as it can move from location to location. In many applications however, the SystemID is included as data field that is related to a location.

There are ramifications as to whether both tag number and systemID are tracked or if only the location associated with its tag number is tracked. Tracking the systemID allows the analyst to analyze for the onset of wear-out, i.e. when the failure rate begins to increase, no longer exhibiting the random failure and flat portion of the bathtub curve. See Figure 3.

Figure 3
Bath Tub Curve

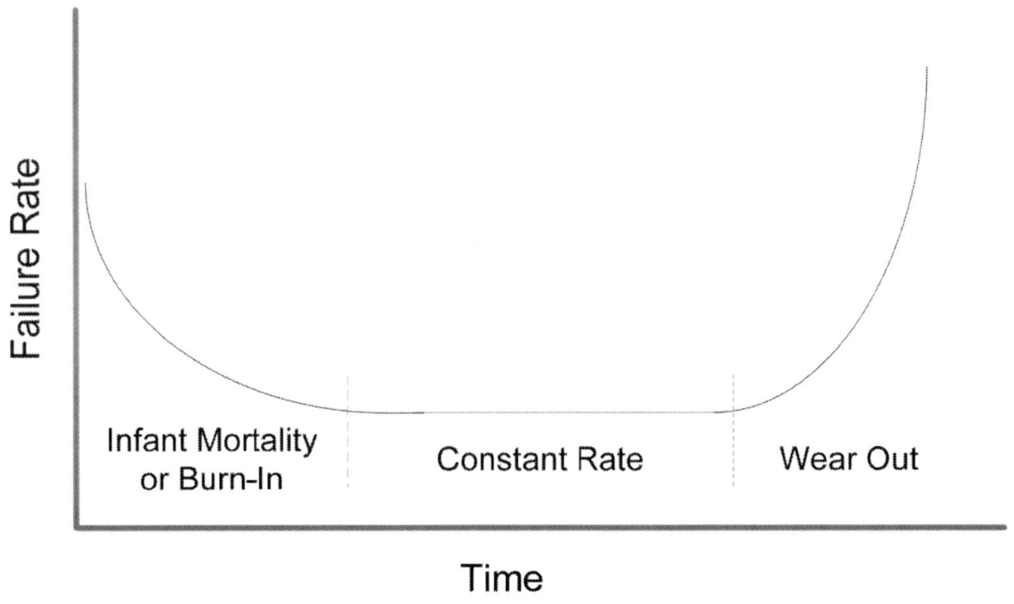

If the systemID is not tracked, then it is important for the analyst to know whether the equipment ever traveled from one tag location to other tag location(s), possibly in different process applications. Any data that potentially traveled where the systemID is not tracked should not be used to analyze wear-out. The other downside of not tracking systemID is that when analyzing equipment at the tag location only, it will potentially include both random (constant failure rate) and wear-out failures with no real means to distinguish. The net result is a likely higher estimate of the random failure rate. Oftentimes, however, tag location data is all that is available. Even so, the data is very valuable when the appropriate assumptions relative to the data quality are known and can be documented as part of the analysis.

Inventory Data – The Population Attributes

In addition to the location information needed to track a population, it is important to consider the equipment attributes. At the tag location, important attributes associated with the process application are known allowing the data to be filtered for additional insight if there is both sufficient inventory and event data. This information includes such data as operating pressures and temperatures, set points, inlet and outlet sizes, fluid properties such as viscosity, relative corrosiveness, fluid phase (vapor, liquid, solid, two-phase, three-phase), etc.

Other potentially useful or important information such as manufacturer, model number, material of construction is really associated with the equipment's systemID rather than its tag number. Although the tag may specify what is supposed to be installed at a particular tag location, the systemID defines what is actually installed. A mismatch of what is supposed to be installed versus what is actually installed can help an analyst gain insight as to the failure's cause if a mismatch is recognized. Not having that information can lead to inappropriate assumptions by the analyst. It can also allow comparison analysis. One example would be to compare different materials of construction installed in the same process application to see if there is any significant effect on the failure rate for failure modes considered important by the user.

Engineering Event Data Forms (Turning Data Into Information)

The final piece to having quality data is to capture data that occurs during events such as:

- Inspections
- Proof tests
- Process Demands
- Failures requiring repair or replacement
- Calibrations
- Etc.

In all cases, what data is captured must be determined by a user who understands the importance of the data and how it can be turned into useful information for their company. A knowledgeable user understands what types of failures are important relative to their company. If that user further understands what purely factual data is required to infer the failure mode or modes of interest, then it becomes relatively straight forward to engineer the various event input forms or the electronic data entry interface. With the use of things like bar codes, the location and population attributes can be related directly to the event with no further effort required by the person recording the event. In some cases, it is possible to automate the entry of some events. In today's programmable electronic systems, automated recording and transfer to an offline database such as PERD is possible for things such as:

- Date/time critical alarm initiated
- Date/time critical alarm cleared
- Date/time interlock bypassed
- Date/time interlock bypass removed
- Date/time demand on interlock occurred
- Etc.

As Found and As Left

Per process safety management, mechanical integrity is important to maintain a safe operating environment in a sustainable manner. One of the fundamental requirements of a good mechanical integrity program is to record as found, as left data. It is important to note that as found data means just that, data that can be used to determine the performance capabilities before doing anything that would destroy the evidence. When a valve in low demand service is asked to close as part of an interlock, it is the first attempt that determines success or failure.

To address this issue, PERD recommends the use of a quality plan, to ensure that the assumptions are valid when audited against actual practice. In one company, they used the quality plan to evaluate their relief valve proof test program. During the evaluation, their first pass look at their data indicated very good performance. Unfortunately, they also discovered that the work practice in place was to skip the "as found" pop test whenever the mechanic's inspection concluded that the valve had to be overhauled anyway. Closer examination revealed that most of their valves would probably not have opened had they been bench tested and included in the data set, impacting their results by at least an order of magnitude.

At another conference, I heard another story where burner fuel valves as part of their proof test for leakage, were stroked multiple times (as many as 30 in some instances) until they passed the leak test. If they were able to get the valve to pass the leakage test without any further repairs, the mechanics considered it a pass, rather than a failure.

In both of these cases, the mechanics thought they were doing the right thing and saving the company money as well, when in reality, it created the situation where bad assumptions could lead to costly events. When engineering support personnel discovered the flaws in their respective mechanical integrity programs, they made the changes and provided the training necessary to correct the issues.

Ready for Analysis

Data that has been input to the PERD software via an auditable quality plan that is in accordance with the concepts previously discussed is immediately available to support analysis as the inventory data is related to the event data via the location data and possibly the SystemID if tracked. Any entity (company, individual plant or functional department, etc.) that can input data to the PERD software is in a position to aggregate their data with any other entity who has done likewise. This relational data can be filtered to create a specific sub-population of interest (i.e. all relief valves with a set pressure greater than 500 psig or a specific manufacturer, or a body material type, etc.) and its associated event data. In the case of the PERD software, the failure modes were inferred during data input, either manually or via a batch electronic upload.

The downloaded data is into a single flat file spreadsheet table that can be used directly by an analyst. For consistency, the downloaded information is recommended to be organized per the tabular format shown in Figure 4.

Figure 4

Address Data					Population Inventory Data					Event Data					Inferred Failure Modes		
Site ID	Plant ID	Unit ID	System ID	Tag	Mfr	Model	Fluid	Set Point	...	Event Type	Date	Time	Open at	...	FM 1	...	FM n

The address location data to the left provides the relational link between inventory and event data. Exporting the inventory data allows the analyst to see what relevant sub-populations might support a statistically sound analysis. The event data contains the dates and times associated with particular events, both the successes and failures as applicable. In the case of the PERD software where the failure modes are inferred, a column for each failure mode to the far right can be populated with a one (1) representing a failure for that specific failure mode associated with that specific event with its own date/time. Where the PERD software is not employed, the inferred failure modes would not exist and the analyst would have to work with the event data to determine what if any meaningful information could be mined from the data, adding cost to the analysis.

PERD Ladder of Success

The CCPS Process Equipment Reliability Database (PERD) Project has focused on developing the necessary taxonomies and infrastructure to support improved information systems that enable cost effective quality data collection and analysis. The PERD project technology allows collection of quality data by using a definable and reproducible taxonomy, providing a fundamentally sound structure to capture data in a format that allows immediate analysis, facilitating the understanding of what the raw data means or represents. Several taxonomies have been developed and implemented within the PERD software.

Figure 5 below shows an overview of the PERD ladder of success. Companies starting out can take advantage of the technical work that has been developed to begin leveraging their information management systems to yield value added information. The PERD ladder of success provides a roadmap to continually improve in a cost effective manner.

Figure 5

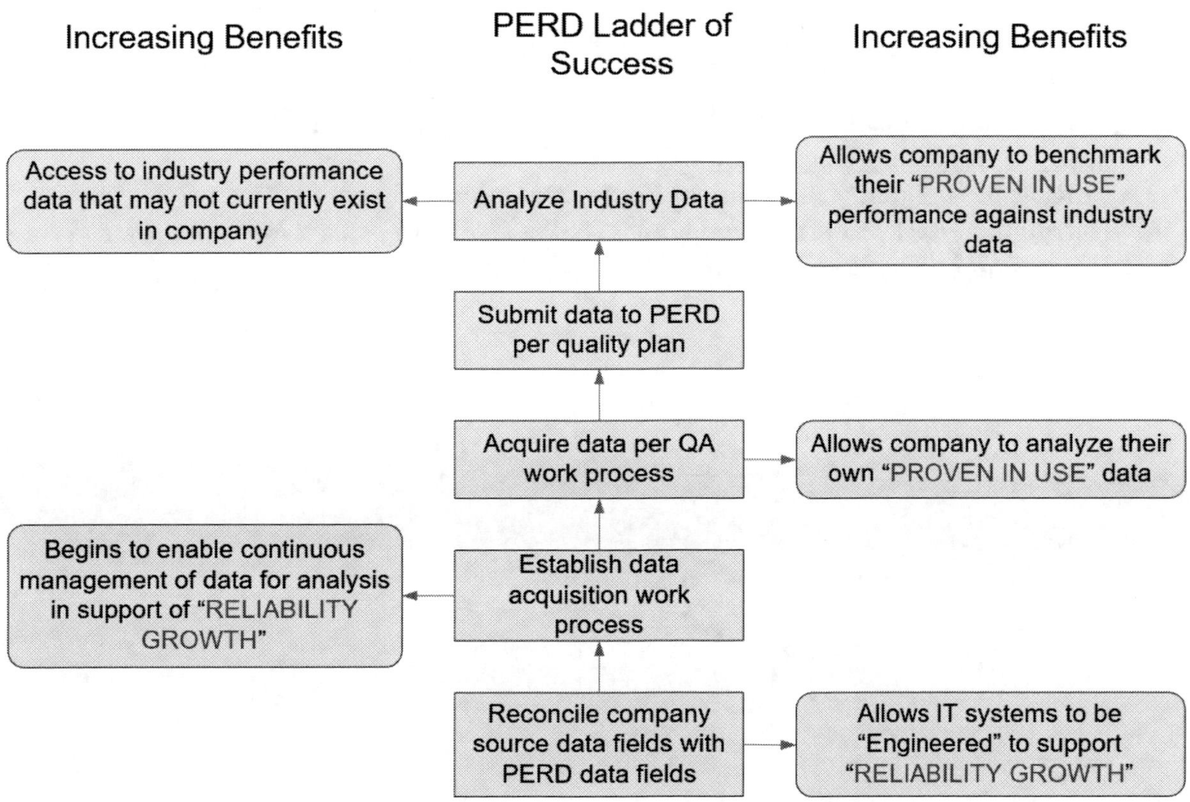

© Air Products and Chemicals, Inc. 2010

Conclusion

Taking the time to understand data requirements provides an opportunity to leverage company data that is routinely captured. This requires some up front thought and implementation of a sound work process that is engineered and implemented via an auditable quality plan to produce desired results. The concepts are not difficult, but it does require cooperation among a number of functional disciplines (i.e. engineering, operations, maintenance, information technology, etc. to maximize the benefits. CCPS PERD provides a foundational infrastructure with associated tools and work processes to help companies pursue a ladder of success.

Those companies that climb the PERD ladder of success position themselves to enable the following benefits:

- Ability to quantify risks and measure the reliability and availability of plants, systems, and equipment
- Ability to benchmark plant performance internally and externally
- Enable improved maintenance optimization
- Support of risk based maintenance planning insight for improvement in the design of new systems/equipment
- Support optimization of life cycle costs
- Provide a foundation for the improvement of existing data management systems

References

1. CCPS *Guidelines for Improving Plant Reliability through Data Collection and Analysis*, 1998.
2. *AIChE CCPS PERD Documentation and Software. PERD website,* http://www.aiche.org/ccps/perd
3. PERD-S5.04, Quality Plan Guidelines for Data Subscribers
4. Harrington , K.H., Kadri, S. & Thomas, H.W; *Using Measured Performance as a Process Safety Leading Indicator*

Semi-automated Work Process For Capturing Relief Valve Proof Test Data To Support Failure Rate Analysis

Jared Gladney
Chevron
Houston, TX 77002
Jared.Gladney@chevron.com

Todd Horner
Air Products & Chemicals, Inc
Allentown, PA 18195
hornermt@airproducts.com

Bob Matthews
Swan Associates, Inc.
Benicia, CA 94510
bmathews@swan-associates.com

Keywords: Relief valve proof test, PERD, bench test analysis

ABSTRACT

Chevron and Air Products have both leveraged an initiative by the Swan Associates test shop that ensures compatibility of their internal software used to manage their operations with the CCPS PERD software, designed to support value added analysis. The work process involves "as found" proof test data being recorded as part of their QA work process by trained technicians, exporting applicable data, and subsequently importing the data to the CCPS PERD industry database where data from multiple companies is aggregated. This paper provides a generic overview of the work process, a specific look at the implementation within the user companies and a summary of the expected benefits as this effort continues.

Introduction

In 2008, Swan Associates became aware of the CCPS PERD initiative whose mission is to operate an equipment reliability database. The purpose of this initiative is to make available high quality, valid, and useful data to the hydrocarbon and chemical process industries. In turn, this useful data will enable analyses to support availability, reliability, equipment design improvements, maintenance strategies, and determination of life cycle costs. Swan recognized the potential synergy between their software and the CCPS PERD vision embodied in the PERD software. Using the Swan software to capture inspection, proof test, and maintenance data, the data can then be exported to the PERD software that is designed to aggregate data in a format that automates inference of failure modes based upon factual information. This in turn provides a fundamental platform for high quality statistical analysis of the data to support a myriad of activities as stated in the PERD mission statement.

Swan Software

As a supplier of services, Swan must meet the requirements of equipment manufacturers and customers, as well as comply with local and national authorities. The rules of ASME Code Section I, ASME Code Section VIII, ASME Code Section IX, and NBIC NB-23 are followed. Documentation of work performed and subsequent records management is an important aspect of compliance. To facilitate improved efficiency, software was developed to assist this work.

Since its inception in 1989, Swan Associates' asset management software has grown from a basic application, to a robust package that is currently programmed using the Microsoft .NET framework. It uses a Microsoft SQL Server relational database for housing the data on the back-end. During conversations with Air Products and Chevron, it was recognized that the software design could potentially add value by leveraging the CCPS PERD industry initiative. By using a secured connection to Swan Associates' servers, each company has direct access to the database used by Swan to house valve records. This allows a customer to check the status, or access information about their valves, from any computer, at anytime. Moreover, in the case where a user company employs multiple test shops, if those test shops were to use the software, the user company would then have access to all of their data from one database location. The Swan database currently stores historical information on over 45,000 assets, and more than 120,000 repair records for multiple accounts.

One of the key benefits of the Swan software is the ability to perform routine calibrations of the testing equipment to ensure the accuracy of captured data. This recorded test information is then housed in Swan's database for later use.

When approached by Chevron and Air Products regarding the CCPS/PERD initiative, Swan Associates began working on a viable export routine, recognizing the potential value that can result from a consistent industry approach as fostered by the industry initiative. At that time, Swan was provided with the various guidelines and technical documentation for the format of data to be used in the PERD industry database.

Building on these data formatting guidelines, Swan analyzed their current database content. While working to achieve results in the proper format, a few tools needed to be developed to assist with the process. This began by ensuring that all the information in the database was compatible with the PERD project. This meant converting all valve information that might cause an issue (fractions to decimals, etc) into valid data. Once this update routine was completed, a method using Microsoft SQL Server Stored Procedures to retrieve only the relevant information required by the PERD industry database was developed. Once this information is located by the stored procedure, the data is temporarily stored, and ready to be converted into the proper format. This formatting is achieved by using XML mapping to sort the data that is presented by the stored procedure. The process for XML mapping happens transparently to the user, preventing corruption of the data and other troubles commonly associated with the manual coding of information. At this point, the mapped data is written to a Microsoft

Access database file. This newly created database, file contains all the properly formatted "as found" data that is required, and is now ready for import into the PERD industry database. Swan Associates provides this export functionality at no additional cost to its users, saving thousands in IT dollars, and many hours of personnel time.

While conforming to the requirements of the PERD industry database, it became possible to provide valuable and accurate data for the analysis of failure rates and causes. The PERD project is one that will continue to develop into something that will have a positive impact industry-wide. Swan is continuing to work with the PERD initiative so as to increase the number of data fields that are fully compatible for export, contributing their knowledge and experience when needed.

PERD/Swan Partnership – Demonstrating the Feasibility

Swan began to work in earnest with PERD in early 2009. The goal of this effort was to map the Swan data fields to the PERD data fields, allowing export of data for input to the PERD software. Working with participating companies in the PERD initiative, feasibility was demonstrated in October 2009 at a PERD general meeting of its participants. From that point, the companies began moving towards implementing an on-going work process.

Development & Implementation of the Air Products Work Process

Capturing high quality data regarding relief device performance is getting somewhat easier as technology advances and we are able to leverage initiatives like CCPS PERD. Air Products has been striving to capture relief device data in a cost effective manner designed to facilitate performance analysis for some time. A number of years ago, a home grown maintenance management system was utilized that was geared towards scheduling, with essentially no thought for useful data capture that would support analysis of performance.

Over the past few years, SAP has been implemented. Although the initial implementation focused more on the scheduling activity, the foundation was laid to better enable performance measurements. Part of this came from having an enterprise information management system whose relational structure is essentially an industry standard. This structure makes it easier to interface with third party software, as they are more willing to invest in programs that can be used to interface with multiple companies.

During the past year and a half, Air Products has worked with Swan Associates to ensure a compatible interface of the software systems involved so that their relief valve bench test as found data can be accessed and leveraged for analysis. Air Products has had a long standing relationship with Swan Associates and a decisive reason Swan was selected is their ability to capture real time, accurate data in their Asset Management System. This is a web-based system that allows customers to track their repairs as a relief valve goes through the repair process. It is part of the QC process that ultimately limits inaccuracies and mistakes by requiring data to be input into the Swan Asset Management System periodically throughout the repair by the technician completing the work.

In addition, Air Products has worked with Swan so that the data collected for Air Product's relief valves can be exported and subsequently imported into the Air Products copy of the PERD software. This approach has been taken since the PERD software acts as an offline database specifically designed to immediately support analysis of the bench test data.

The new work process that has been developed allows Air Products to remotely access and download their data in a format that is compatible with CCPS PERD software. During this development, Air Products worked with Swan to data map their data fields with SAP and the PERD data fields, reconciling any issues as they were identified. This data can then be uploaded into the CCPS PERD industry database, providing value added data that can be used for analysis. The overall work process was first successfully demonstrated in October 2009 and is shown in Figure 1.

From a Mechanical Integrity standpoint, the data being uploaded from Swan's Asset Management System into the CCPS PERD industry data base is important in developing and revising maintenance tasks such as testing frequencies, i.e. whether a less frequent or more frequent maintenance cycle is needed for a particular type or style of valve in a particular process application.

The SAP address and associated valve specification data allows Air Products to filter the tracked valves for the analysis of data using subsets such as carbon steel versus stainless steel valves, or as a function of set pressure or valve size, etc. The initial as found bench test data includes the test date, whether the valve lifted and its lift pressure. Comparing the lift pressure to a valve's set pressure allows the PERD software to infer failure modes such as lift below set pressure, lift above set pressure and valve fails to function.

Now that the entire work process has been successfully demonstrated, Air Products is in the process of formal implementation so that high quality data is captured on an on-going basis. This also involves the process of reconciling additional plants so that the number of plants is expanded, increasing the size of the formally tracked population of valves providing as found data. The anticipation is that the start of a statistically meaningful data set will occur during calendar year 2010, enabling sharing of data with the CCPS PERD operating database, as well as a baseline for internal analysis of the data. As this initiative moves forward and expands, the expectation is that by working with CCPS PERD, increasingly meaningful analyses will be performed providing increasing value.

Figure 1

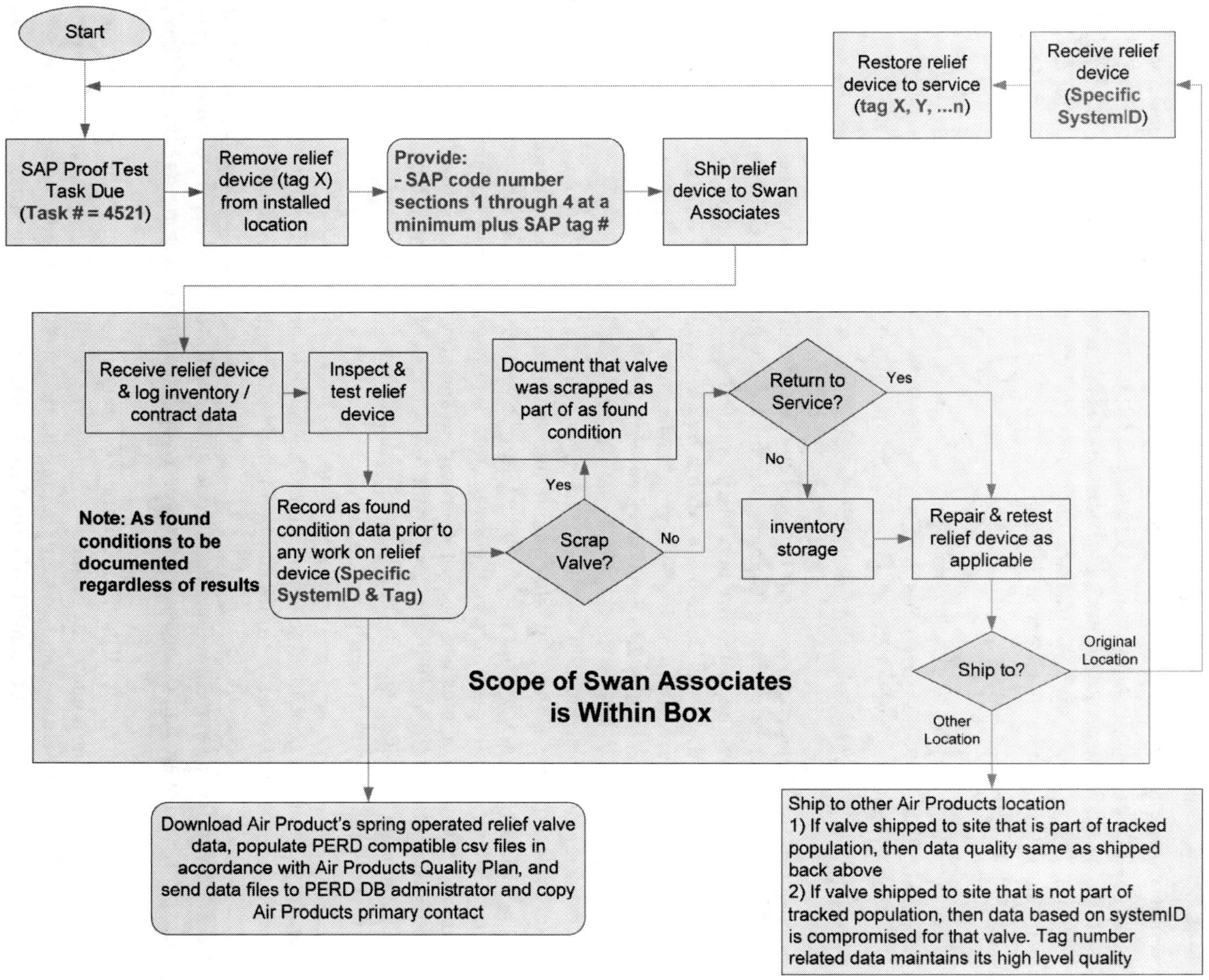

Start
SAP Proof Test Task Due (Task # = 4521)
Remove relief device (tag X) from installed location
Provide:
- SAP code number sections 1 through 4 at a minimum plus SAP tag #
Ship relief device to Swan Associates
Restore relief device to service (tag X, Y, ...n)
Receive relief device (Specific SystemID)
Receive relief device & log inventory / contract data
Inspect & test relief device
Document that valve was scrapped as part of as found condition
Return to Service?
Yes
No
Note: As found conditions to be documented regardless of results
Record as found condition data prior to any work on relief device (Specific SystemID & Tag)
Yes
Scrap Valve?
No
inventory storage
Repair & retest relief device as applicable
Original Location
Ship to?
Other Location
Scope of Swan Associates is Within Box
Download Air Product's spring operated relief valve data, populate PERD compatible csv files in accordance with Air Products Quality Plan, and send data files to PERD DB administrator and copy Air Products primary contact
Ship to other Air Products location
1) If valve shipped to site that is part of tracked population, then data quality same as shipped back above
2) If valve shipped to site that is not part of tracked population, then data based on systemID is compromised for that valve. Tag number related data maintains its high level quality

Development of Chevron Work Process

Chevron Richmond has long strived for reliability in their use of pressure relieving devices. In an effort to facilitate timely data entry for pressure relieving devices managed by routine maintenance, Chevron's Richmond Refinery has adopted a Quality Plan that meets both the requirements of local, as well as national jurisdiction for PRD's serviced on a scheduled frequency and classified as available on the run (OTR). Although Chevron Richmond held a proven system in place to capture quality data, one of the constraints of their existing system was the ability to track the progression and associated costs of pressure relief devices during the inspection process. To improve tracking capability and compliance of pressure relief devices during scheduled servicing, Chevron Richmond embarked on a partnership with Swan Associates in 2007 to enable:

1.) The identification of obsolete PRD's,
2.) Establishment of a spare parts pool,
3.) SWAP out vs. service PRD's, thereby reducing maintenance costs,
4.) Validation and revision of design data.

The Swan Asset Management System has proven effective in improving the ability to gather quality pressure relief valve data at Chevron Richmond. Currently, Chevron Richmond is on schedule to implement the capability necessary to interface between their legacy pressure relief device database and the Swan database, enabling Swan's database to be utilized as the master for uploading safety relief valve data into Chevron's Meridium software in early 2010. Such efforts are with the exception of pressure relief devices serviced during planned outages.

Chevron Richmond introduced the value of the Swan data as it relates to ensuring compliance for pressure relief devices serviced while equipment remains in operation, and identifying obsolete PRD's. In May 2008, Chevron Richmond engaged Swan Associates with developing an interface to extract information from the Swan Asset Management System and upload into Meridium. Swan Associates was eager to participate and saw the need to design a method that would be easy to modify on the run. The Swan Associates concept included SQL queries, and a (.xml) mapping process that would take existing data formats and convert where necessary to comply with the Meridium configuration.

Swan Associates has continued to support the Chevron Richmond effort by developing training materials. This includes study guides, supplemental tests, and keys to the performance of repair, set and testing of pressure relieving devices, as well as testing systems. Chevron Richmond has used the Swan testing system with a standalone database for the past eight years. As a training platform, the software and database were recently updated to the current version.

Plans are in effect to connect Chevron Richmond's standalone system to the Swan servers where qualified personnel throughout the refinery will interface with the data. Chevron Richmond will use the Swan Asset Management System, in conjunction with their legacy pressure relief device database, to manage their population of over $2,000,000 surplus (spare) assets, rotating them through maintenance requirements and emergency needs.

Chevron Richmond recognizes that one of the most effective aspects of the interface is the ability to search and drill down on a given asset. Chevron Richmond's reliability group utilizes this feature every day to access, and print service and test reports which are instrumental in confirming compliance requirements and readying the unit startup process.

With described efforts underway, coupled with similar efforts completed by Swan Associates with Air Products, each in alignment with the Center for Chemical Process Safety (CCPS) PERD initiative, Chevron Richmond has given the green light to Swan to enable integration of the Chevron Refinery's PRD asset management system and Swan's Asset Management System. The interface is expected to be up and running (pending required system configurations) in early 2010. Chevron Richmond recognizes Swan Associates as a partner who will provide a practical, cost effective solution to the requirement of gathering or validating quality pressure relief device data for analysis.

Conclusion

From the success achieved so far, it is clear that as the initiative moves forward, improved, higher quality data will inevitably allow PERD participants such as Air Products and Chevron to make wiser more educated decisions surrounding the installation and maintenance of relief devices, as well as to support proven in use independent protection layer (IPL) values for risk analysis that are more specific to the actual application and specific relief valve attributes.

References

1. CCPS *Guidelines for Improving Plant Reliability Through Data Collection and Analysis*, 1998.
2. CCPS PERD Spring Operated Relief Valve Taxonomy May 2006
3. *AIChE CCPS PERD Documentation and Software. PERD website,* http://www.aiche.org/ccps/perd
4. Air Products Quality Plan

Preventing Loss of Containment through a Systematic Assessment of Hazards, Consequences, and Risks

S. Dharmavaram
DuPont, Engineering Research and Technology
1007 Market St, D12022-2
Wilmington, DE 19898 U.S.A.
seshu.dharmavaram@usa.dupont.com

James. A. Klein
DuPont, North America Operations SHE
james.a.klein@usa.dupont.com

Prepared for Presentation at
American Institute of Chemical Engineers
2010 Spring Meeting
6th Global Congress on Process Safety
San Antonio, Texas
March 22-24, 2010

Preventing Loss of Containment through a Systematic Assessment of Hazards, Consequences, and Risks

S. Dharmavaram
DuPont, Engineering Research and Technology
1007 Market St, D12022-2
Wilmington, DE 19898 U.S.A.
seshu.dharmavaram@usa.dupont.com

James. A. Klein
DuPont, North America Operations SHE
james.a.klein@usa.dupont.com

Keywords: hazards assessment, risk assessment, loss of containment

Abstract

Loss of containment (LOC) of toxic chemicals, flammable chemicals, and/or energy from vessels, pipes, or other equipment can potentially lead to consequences that can range from minor to catastrophic. A fundamental understanding and detailed assessment of the hazards is an important first step, which may be much more involved than just reviewing or communicating material safety data sheets. Next, a complete range of failure scenarios should be analyzed in a systematic manner, including qualitative and quantitative methods, to develop information on potential consequences. With an understanding of hazards and the consequences of LOC events, and ideally some good process data, a process hazards analysis team, using a semi-quantitative risk analysis, can evaluate the adequacy of systems and the layers of protection necessary for preventing and mitigating loss of containment events. This paper will share some of the typical approaches that have worked well within the industry in general and DuPont in particular.

1. Introduction

Preventing loss of containment of materials and/or energy should be a primary goal of industrial Process Safety Management (PSM) programs [1] and government regulations to minimize process risks. This is usually accomplished by simultaneously addressing, implementing, and monitoring performance of a full spectrum of PSM elements [e.g. Process Technology or Process Safety Information (PT/PSI), Process Hazards Analysis (PHA), Mechanical Integrity/Quality Assessment (MI/QA), Training, Emergency Response, Incident Investigation, etc.]. However, in many situations there is a lack of understanding of the importance of all the elements functioning together in a disciplined manner. There is often confusion about the terminology related to hazards, consequences, and risks.

A thorough understanding of the hazards, identification of a complete range of failure events based on detailed analysis of the consequences of these events, and the analysis of process risks accounting for all the safeguards can help significantly in preventing and mitigating loss of containment incidents. In this paper additional details are provided on hazard assessment approaches, consequence analysis of potential events, and risk analysis for minimizing risks.

2. Loss of Containment Incidents

Loss of containment incidents are generally the largest category of process incidents, ranging from small releases of relatively non-hazardous materials to catastrophic events involving large releases of materials and/or energy. The consequences of loss of containment obviously relate to the intrinsic hazardous properties of the materials released as well as the processing conditions, such as high temperature or pressure. A small release of a highly toxic material can have more serious consequences than a much larger release of a non-hazardous material. A small release of a flammable material above its flash point may be more serious than a larger release below the flash point. Therefore careful evaluation of material hazards and potential release scenarios in order to determine the likely consequences and risks associated with loss of containment is always necessary.

Many significant chemical industry events involve loss of containment, but are not generally thought of directly as loss of containment events. The Bhopal incident in 1984, for example, is considered a toxic event [2] that killed or injured thousands of people, but the inadvertent mixing of water and methyl isocyanate (MIC) in a storage tank led to a runaway reaction that increased pressure resulting in loss of containment of the MIC through a stack. Similarly, the Flixborough incident in 1974 involved loss of containment of hot cyclohexane vapors, the Seveso incident in 1976 resulted from loss of containment of toxic dioxin from a burst reactor rupture disk, and more recently, the BP Texas City explosion in 2005 occurred after loss of containment of hydrocarbons from a stack [2]. Loss of containment, therefore, is present in many significant chemical incidents, either directly from equipment failure or operating problems resulting in severe consequences such as toxic exposures or fires/explosions, or indirectly resulting from process events such as pressure build-up or runaway reactions or dust deflagration that lead to emergency venting or loss of vessel integrity and release of contents.

Many loss of containment incidents relate to equipment failures caused by Mechanical Integrity system problems. This includes inadequate equipment installation, poor or lack of preventive maintenance, absence of reliability engineering, and ultimately inadequate analysis of the life cycle of equipment components. Many loss of containment incidents also result from operating and design problems, such as poor procedures, not following procedures (operational discipline), poor training, poor equipment design, etc. Any approach to preventing and mitigating the consequences of loss of containment must therefore consider a wide range of possible system failures encompassing all aspects of process safety management. A systematic understanding and assessment of the hazards, analysis of consequences, and analysis of risk in detailed process hazards analysis reviews can anticipate potential loss of containment failures and scenarios and provide appropriate process controls and safeguards to help prevent them.

3. Hazards Assessment

A hazard can be defined as a physical or chemical condition that has the potential for causing harm to people property or the environment [3]. A hazard assessment is different from consequence analysis, where the impacts of failure events are determined. It is also different from hazards identification and hazards evaluation in PHAs, where the focus is on identifying hazards, conducting a risk analysis and ensuring that sufficient layers of protection exist to minimize process risk.

A systematic assessment of hazards is one of the first steps necessary before conducting any consequence and/or risk analyses. Many of the governmental regulations (e.g. SEVESO II in Europe, COMAH in the U.K., OSHA1910.119 in the U.S.) require a thorough assessment of hazards. A thorough understanding of process hazards is necessary for identification of potential loss of containment scenarios, which provide a pathway for exposure to the hazardous effects of the materials and of the processing conditions involved.

DuPont PSM standards that are applicable worldwide and the OSHA1910.119 rule in the U.S. require that a complete assessment of hazards be conducted as part of the Process Technology or Process Safety Information (PT/PSI) element of the overall PSM program. Such an assessment must be completed and documented prior to conducting a Process Hazards Analysis (PHA). According to OSHA1910.119(d):

> ".....The compilation of written process safety information is to enable the employer and the employees involved in operating the process to identify and understand the hazards posed by those processes involving highly hazardous chemicals."

In the PT/PSI compilation, complete and accurate written information regarding hazards of materials, process design basis, and equipment design basis is considered essential for effective functioning of the overall PSM program and for conducting of comprehensive PHAs. The compilation of PT/PSI provides necessary information to a variety of users including the PHA team as well as those developing the training programs and the operating procedures, contract administrators and contractors who will be working in a process area, those conducting pre-startup reviews; emergency response/preparedness planners, and where appropriate enforcement and insurance officials.

In many cases, a Material Safety Data Sheet (MSDS) prepared by a manufacturer of a specific chemical has information on some of the pertinent data. The quality of information in MSDSs is quite variable. Use of MSDSs only to compile and document information on hazards of chemicals is generally inadequate and must usually be supplemented with additional sources of data.

In a chemical plant, there are a variety of possible process hazards, such as toxicity, flammability, chemical reactivity, thermal radiation, and explosivity, from handling, processing, and storage of chemicals and other materials. Process hazards are different and distinct from general safety and occupational health hazards (e.g. slips, trips, and falls; noise protection). The focus of PSM is on process hazards and preventing events that can potentially have a major impact on-site or off-site.

To systematically assess the hazards, it is necessary first to define process boundaries within a chemical plant. These process boundaries may correspond to those used in PHAs or to other appropriate parts of the facility. Within a process boundary, a complete list of all the raw materials, intermediates, products, utilities, and other materials used should be compiled. List should also track amounts, rates, state (vapor/liquid/solid), compositions, etc. as appropriate. The level and nature of hazard will vary in different process boundaries depending on the quantities of chemicals handled, the intrinsic material properties, and how they are being used or processed. A tank farm that contains a variety of chemicals stored as liquids under pressure will have a significantly different hazard than a manufacturing building that has a few of the same chemicals in pipes and vessels as vapors in mixtures at high temperature and pressure. Similarly, a tank containing anhydrous ammonia as a liquid under pressure has a much higher level of hazard than piping and columns that contain 20 wt% aqueous ammonia. Consideration should also be given to the potential for chemicals and materials from one process boundary inadvertently entering another process boundary, possibly introducing new hazards that must be evaluated.

3.1 Toxicity

Toxicity is the degree to which a chemical upon exposure can cause harm to any living organism [4]. Toxicity effects are based on concentration and exposure time. The main concern from a process risk perspective is the acute toxicity effect resulting from a short term exposure to an accidental release of a chemical. Chronic/subchronic toxicity effects are of greater concern for long term environmental impact or worker exposure, resulting from continuous releases from a chemical plant where the exposure durations are long. In assessing the toxicity hazards, it is most important to focus on developing an understanding of the acute toxicity effects on people, while noting the potential impact on plants and animals, where appropriate.

While the MSDS or an International Chemical Safety Card for a chemical might provide some information about toxicity effects and concentrations of concern such as the Threshold Limit Value (TLV) and/or Immediately Dangerous to Life and Health (IDLH) concentrations, they are not usually sufficient for detailed hazards assessments. The TLVs are chronic toxicity values appropriate for worker exposure in a plant environment. Even though IDLH values were designed originally for providing acute exposure effects they are not currently supported by industrial toxicologists. Sometimes lethal concentration (LC-50) and lethal dose (LD-50) values where 50% of an animal population (e.g. mouse) might have been affected may be provided. However, this information though useful is not complete and helpful for a hazard assessment. No information is usually available in MSDSs related to concentrations that can cause injuries from acute exposure.

A better source of information on acute toxicity concentration and effects is data developed by the American Industrial Hygienists Association called the Emergency Response Planning and Guideline (ERPG) concentrations. Similar to ERPGs, the Acute Exposure Guideline Level (AEGL) concentrations developed by the U.S. EPA can also be used. The ERPG/AEGL concentrations are developed by toxicologists using a wide range of available toxicological data (e.g. LC-50, LD-50 for multiple animals) to conservatively determine the concentration levels that might affect general and susceptible populations. An AEGL-x concentration (like ERPG concentration), is defined as the airborne concentration (in ppm or mg/m3) of a substance above which it is predicted that the general population, including susceptible individuals, could experience:

- 1 mild, short-term reversible adverse health effects
- 2 irreversible or other serious, long-lasting adverse health effects or an impaired ability to escape
- 3 life threatening effects or death

The concentration values that are most appropriate in terms of assessing toxicity hazards are the AEGL-2/ERPG-2 for injury effects and AEGL-3/ERPG-3 for life threatening effects. Where the ERPG and AEGL values are not available then equivalent values should be determined by industrial toxicologists for interim usage. In DuPont, concentrations equivalent to ERPG/AEGLs are developed and made available for use by an internal committee of toxicologists. Composite values are determined for mixtures using Le Chatelier's mixing rule as follows:

$$1/(\text{ERPG or AEGL-x})_{mix} = \sum_{i=1}^{n} [\, y_i / (\text{ERPG or AEGL-x})_i \,]$$

where y_i = vapor mole fraction of chemical i

The National Fire Protection Agency (NFPA) in the U.S.A. publishes the Fire Protection Guide [5] which includes a Health Score that is on a 0 to 4 scale for a variety of chemicals. This score has been very useful from an emergency response perspective and is categorized based on health effects as follows: 0 (None); 1 (Slight); 2 (Moderate); 3 (Severe); and 4 (Extreme). This NFPA Health Score, where available, should be noted for chemicals and mixtures within a process boundary.

In DuPont, a focused process safety management approach has been developed for managing the hazards of highly toxic materials. A highly toxic material is one that has a high substance hazard index (SHI) which is determined as follows:

$$\text{SHI} = \{\text{Vapor Pressure (in atm) x 1E+06}\} / \{\text{ERPG-3 or AEGL-3 in ppm}\}$$

The acute toxicity impacts related to life-threatening effects are dependent on a combination of concentration and exposure duration. A multiple of C x t where C is concentration in ppm and t is exposure time in min is called toxic dose expressed as ppm-min. A multiple of C^n x t, where n is an exponent is called toxic load. Dangerous Toxic Load (DTL) and Significant Likelihood of Death (SLOD) levels have been published for around 200 chemicals by the U.K. Health and Safety Executive [6]. The concept of toxic load is used in determining the probability of fatality. To that end, Probit (Pr) relationships are available for a few toxic chemicals that are expressed as follows:

$Pr = a + b \ln (C^n * t)$ where C is in ppm and t is in min.

$\text{Probability of Fatality} = 0.5\,[1 + \text{erf}\,\{(Pr - 5)/2^{0.5}\}]$

For example, the following probit relationship has been suggested for ammonia [7]:

$$Pr = -35.9 + 1.85 \ln (C^2 * t)$$

For a 10 minute exposure time, the probability of fatality is 50% and 95% if the concentration is 19985 ppm and 31217 ppm, respectively.

When assessing toxicity hazards, information about concentration levels of concern (ERPG, AEGL, or equivalent), NFPA Health Score, SHI, DTL, SLOD, Probits, etc. can all be used to determine the level of toxicity hazard within a process boundary. Once the list of toxic chemicals have been identified, a combination of the levels of concern, the NFPA Health Score, and any other available information should be used to determine the overall hazard level. The hazard level score is on a scale identical to the scale used for the NFPA Health Scoring and can be categorized as follows: 0 (None); 1(Low); 2 (Medium); 3 (High); and 4 (Very High).

3.2 Flammability and Thermal Radiation

Flammability is a property of chemicals that provides an indication of the potential for ignition and burning [8]. The degree of flammability is determined by testing per standard protocols. For chemicals that are flammable, there are a variety of parameters that are useful in assessment of hazards. The focus is on harm to people from 2nd to 3rd degree burn injuries upon exposure.

An MSDS or International Chemical Safety Card may contain much of the pertinent data such as flash points, flammable limits, auto-ignition temperature, and products of combustion. However, it is important to verify and confirm the data from multiple sources.

The NFPA makes a distinction between liquids that are flammable and combustible based on the boiling point and flash point. Flash point is the lowest temperature above which a chemical can vaporize to form an ignitable mixture with air. Liquids below a flash point of 100 °F are considered Class I flammable liquids. Liquids with flash points between 100 to 140 °F are considered Class II combustible liquids. Liquids with flash points above 140 °F are considered Class III combustible liquids. For chemicals in a process boundary, based on the operating temperature and pressure, it is important to determine whether the chemicals or mixtures are in the flammable range.

In addition to the flash point, the flammable limits of chemicals should be reviewed. Upper Flammable Limit (UFL) and Lower Flammable Limit (LFL) are concentrations where the mixture of the chemical and air can ignite and burn. Upon ignition a flame propagates from the ignition source to the release point resulting in a flash fire, pool fire, or a jet fire. Just like toxic mixtures, for flammable vapor mixtures the Le Chatelier's mixing rule can be used to determine the flammability limit.

$$1/(LFL)_{mix} = \sum_{i=1}^{n} [\, y_i / (LFL)_i \,]$$

where y_i = vapor mole fraction of chemical i

The NFPA has also published a Flammability Score that is on a 0 to 4 scale for a variety of chemicals [5]. This score is useful from an emergency response perspective and is categorized based on impacts as follows: 0 (None); 1 (Slight); 2 (Moderate); 3 (Severe); and 4 (Extreme). This NFPA Flammability Score, where available, should be noted for chemicals and mixtures within a process boundary.

If a fire occurs, then in addition to burn injuries from direct contact, injuries and equipment damage can also occur from thermal radiation. Similar to toxicity impacts for lethal effects, the thermal radiation impacts are dependent on exposure time. Thermal radiation is expressed in units of $KW/m^2/h$ or $Btu/ft^2/h$.

When assessing flammability hazards, information about concentration levels of concern (LFL, UFL), NFPA Flammability Score, Flash Point, Auto-Ignition Temperature, etc. can all be used to determine the level of flammability hazard within a process boundary.

For a process boundary, once the list of flammable chemicals have been identified, a combination of the levels of concern, the NFPA Fire Score, and any other available information should be used to determine the overall hazard level. The hazard level score would be on a scale identical to the scale used for the NFPA Fire Scoring and can be categorized as follows: 0 (None); 1(Low); 2 (Medium); 3 (High); and 4 (Very High).

3.3 Chemical Reactivity

Chemical reactivity is a hazard that results from uncontrolled chemical reactions in any equipment within a process boundary [9]. An exothermic chemical reaction leads to release of heat and an increase in temperature and pressure. An endothermic chemical reaction could lead to generation of products that are unstable and or gases that causes an increase in pressure. An example of an exothermic reaction is an uncontrolled polymerization reaction that might occur in a storage container or process vessel leading to a rapid increase in temperature and pressure.

Exothermic reactions are usually of greater concern than endothermic reactions. Therefore, the first step in chemical reactivity hazard assessment is to identify the possible uncontrolled chemical reactions that can occur. These are usually side reactions that may not be normal within the process boundary and that could occur because of even small changes in operating conditions (i.e. pressures, temperature, composition, etc.). Each reaction of concern within a process vessel should be evaluated to determine the potential level of hazard.

Like polymerization, there are many other reactions (e.g. oxidation, nitration, hydrogenation, etc.) that are highly exothermic where as others (e.g. isomerization, electrolysis) are low on the exothermic scale. Based on the reaction type, the exothermic nature can be used to determine the level of hazard.

The NFPA has also published a Reactivity/Instability Score that is on a 0 to 4 scale for a variety of chemicals [5]. This score is categorized based on impacts as follows: 0 (None); 1 (Slight); 2 (Moderate); 3 (Severe); and 4 (Extreme). For each specific reaction, the NFPA Instability Score, where available, should be reviewed for all the reactants and products. Where an NFPA score is not available, an equivalent score can be determined based on the structure of the chemical molecule. The worst (i.e highest) NFPA score for all the chemicals in the reaction under consideration can be regarded as the overall NFPA Instability score.

For any reaction, the heat of reaction is a potential measure of the hazard. In many cases, it can be estimated using the heats of formation of each of the reactants and products. If the exothermic heat of reaction, usually expressed as calories per gram of reaction mass, is greater than 300 cal/g, then the hazard level is regarded as high; if it is less than 20 cal/g the hazard level is very low.

Once the heat of reaction is available through estimation or measurement using calorimetric methods, it can be used to estimate the adiabatic temperature rise. Adiabatic temperature rise can be calculated using the overall heat generated in a reaction, the reaction mass, and the heat capacity of the mixture. In addition to the adiabatic temperature rise, the self heat rate is another important parameter that can be used to determine the level of hazard. The self heat rate has to be measured or can be estimated based on the type of reaction. Based on the potential temperature increase and the rate of increase, the level of hazard can also be determined. The rate of reaction which is a kinetic measure is also important in addition to the heat of reaction which is a thermodynamic measure.

For exothermic reactions, the temperature at which the onset of exotherm occurs is another important parameter. This is a value that has to be measured using calorimetric methods. The ratio of the onset temperature to typical process temperature, which is another possible kinetic measure, can also be used to estimate the level of hazard.

When assessing chemical reactivity hazards, therefore, information about the type of reaction, the NFPA Instability Score, heat of reaction, exotherm onset temperature, adiabatic temperature rise, self heat rate, etc., can all be used to determine the level of chemical reactivity hazard within a process boundary.

For a process boundary, once the list of potentially uncontrolled reactions has been identified, a combination of the NFPA Instability Score and other available information discussed above (e.g. heat of reaction, adiabatic temperature rise, self heat rate, etc.) should also be used to determine the overall hazard level. The hazard level score could be on a scale identical to the scale used for the NFPA Instability Score and can be categorized as follows: 0 (None); 1(Low); 2 (Medium); 3 (High); and 4 (Very High).

3.4 Explosivity or Explosibility

Explosivity or explosibility can be viewed as the measure to which a material is explosive or explosible. Explosion hazards can result from releases of flammable chemicals and/or fine combustible dusts into a confined and/or congested environment [8]. The confinement and congestion causes a flame to accelerate resulting in a deflagration. In some cases, detonation can occur that causes generation of higher overpressure and impulse from the resulting blast wave.

In addition to vapor cloud explosion (VCE) and dust explosions, explosions can also result from bursting of a vessel caused by a rapid increase in pressure. Such a build-up of pressure could be the result of an uncontrolled chemical reaction, from failure of pressure regulators upstream of equipment, or from deflagration of combustible dusts within pipes/equipment. **This is one of the primary ways a loss of containment might occur.** If the pressure inside the vessel exceeds the maximum working pressure of a vessel, then a vessel burst occurs which can result in propagation of a blast wave. A rapid increase in temperature and pressure from a fire under a container can also lead to a boiling liquid expanding vapor cloud explosion (BLEVE) if the material in the container is a liquid under pressure at a temperature greater than its boiling point. If a BLEVE occurs, in addition to a blast wave, a fireball can also form leading to thermal radiation impacts.

For an explosion, the fuel concentration has to be within flammable limits, the oxidant concentration has to be sufficient to support combustion, and an ignition source has to be present. The maximum deflagration pressure and the rate of pressure rise are important parameters that can be measured for a variety of fuels. Once a fuel ignites, the rate of flame acceleration determines the maximum pressure generated in a confined and congested environment. The fundamental burning velocity of a material is a measure of the rate of flame propagation that can occur. Fundamental burning velocities are available for several flammable chemicals [10]. It is a measure of how reactive a material might be in terms of generating of high deflagration pressures and increased rate of pressure rise.

The rate of pressure rise and expansion volume are combined to provide a measure of the extent of explosion. This measure is called the deflagration index that is expressed as follows:

$$K = V^{1/3} * (dP/dt)max$$

Where P is the pressure in bar, t is time in sec, V is volume in m^3, and K is the deflagration index in bar-m/s. K is further expressed as K_G and K_{st} for gases and dusts, respectively, and values are available for several gases and dusts [10]. K_{st} values for a variety of dusts are typically used to classify them into different hazard classes, sometimes referred to as St-1, St-2, and St-3 corresponding to low, medium and high. For dusts, the particle size can have a significant impact on K_{st}.

For assessing explosivity/explosibility hazards, it is important know the flammable mass/volume, particle size in the case of combustible dust releases, laminar burning velocity of chemical/mixtures, levels of confinement and congestion, and deflagration indices. Such information can then be utilized to determine the level of explosivity/explosibility hazard within a process boundary.

For a process boundary, once the list of flammable chemicals and combustible dusts have been identified, a combination of the levels of concern and any other available information should be used to determine the overall hazard level. The hazard level score would be on a scale identical to the scale used for the NFPA Flammability Scoring and can be categorized as follows: 0 (None); 1(Low); 2 (Medium); 3 (High); and 4 (Very High).

4. Consequence Assessment

4.1 *Consequence Analysis*

Consequence analysis consists of evaluating the direct, undesirable impact of potentially hazardous events, resulting from loss of engineering and administrative controls for the process [3]. The purpose of consequence analysis is to help the PHA team understand the type, severity, and number of potential injuries, property damage, and environmental harm that result from the incidents. The systematic hazards assessment discussed in Section 3 provides the basic input information necessary for consequence analysis. For events that result in loss of containment, the data might include the nature of discharge, rate of discharge, composition, temperature, pressure, and toxicity or flammability of chemicals or mixture. For events inside vessels that could lead to loss of containment, such as an unexpected chemical reaction, deflagration of gases, or deflagration of combustible dusts, the useful data would be the change in temperature and pressure with time, the material of construction, emergency vent design, and pressure rating of a vessel.

In conducting the consequence analysis, the team evaluates the impact of possible hazardous events in terms of potential release quantities, affected areas, and injuries or property damage. In determining the events for consequence analysis, especially for loss of containment, a complete range of small to catastrophic failure events should be considered. A small event could be caused by a small-diameter hole in a vessel or pipe or possibly a procedural error such as leaving a valve open or in the wrong position. Catastrophic failure events are those where there is a complete and sudden failure of any equipment, structure, or system resulting in major loss of containment of chemicals or energy. Even though catastrophic failure events are rare, the consequences of such an event could be significant and should be understood.

In the selection of events for consequence analysis, it is important to assume failure of engineering and administrative controls. No credit should be taken for such controls to screen out events selected for analysis. Ultimately, a wide variety of loss of containment scenarios can be developed and evaluated, so that the effectiveness of current process safeguards can be assessed, and additional engineering and procedural controls can be recommended as appropriate to minimize the potential for loss of containment. Analysis of incident data involving loss of containment failures can be used to ensure that the range of scenarios is comprehensive. Failure event scenarios can be classified as minor, moderate, major, or catastrophic. An approach as shown in Figure 1 below is used within DuPont to determine the potential consequences for a full range of failure events.

If the process involves the storage of ammonia, for example, the PHA team would consider small events such as small holes in pipes, pump leaks, or other types of events that will result in relatively small releases of ammonia. The PT/PSI compilation would include information on the nature (e.g. toxicity, flammability, etc.) and level of hazard (i.e. low, medium, high, very high).

Figure 1: Example Consequence Impact Categories

Type of Event/ Impact	Consequence Category C-1 Minor	Consequence Category C-2 Moderate	Consequence Category C-3 Major	Consequence Category C-4 Catastrophic
On-Site Safety and Health	No significant injury or health impact	Medical Treatment Case (MTC) injury or reversible health effects.	Multiple MTC injuries;1-2 Restricted Work Case (RWC) or Lost Work Case (LWC) injuries.	One or more fatalities; multiple LWC's; irreversible health effects.`
Off-Site Safety and Health	No significant injury or health impact	Minor injury or reversible health effects, perhaps requiring minor medical treatment.	Injury or moderate health effects requiring emergency medical intervention and/or hospitalization.	Death or permanent, irreversible health effects.

The PHA team uses the available hazard information and then evaluates what locations could affected by the release, the possible health effects related to exposure to ammonia, and the resulting type, severity, and number of injuries. Similarly, the team would consider failure events that could lead to larger releases of ammonia, such as large holes in pipes, vessel failure, or open valves, and evaluate the possible consequences. As an initial review, bracketing the range of possible releases and their consequences qualitatively in this way will provide the PHA team a good understanding of potential ammonia release events and will provide a list of possible loss of containment scenarios. Appropriate recommendations can then be made as needed to lower the risk of ammonia loss of containment by addressing the specific causes of these events.

The consequence analysis results will be used to help the PHA team evaluate the consequences associated with potential loss of containment events in conducting other parts of the PHA. A hazards evaluation methodology, such as Whatif/Checklist, HAZOP, or Failure Mode and Effect, is also used by the PHA team to identify specific loss of containment failures resulting from equipment failures, procedural errors, or other operational or design problems. The PHA team then considers the possible impact of these failures, event frequency, and existing safeguards for preventing or mitigating the consequences or frequency of the event. The hazards information from the PT/PSI compilation and the results from the consequence analysis should provide valuable information to the PHA team during this review, by having already bracketed the range of possible releases, injuries, and property damage for many of these events. The PHA team can then consider and evaluate additional safeguards to improve the safety of the process, as appropriate.

The development and evaluation of possible loss of containment events in consequence analysis combined with the specific failures leading to these events from the hazards evaluation methodology allow the PHA team to evaluate potential risks associated with various releases and to minimize these risks as appropriate. This is discussed further in Section 5 and includes consideration of design, maintenance, and operational improvements when risk is considered unacceptably high by the PHA team. In some cases, additional consequence modeling should be completed to provide more detailed information on potential causes and consequences of release events and the effectiveness of improved safeguards that are being considered. Consequence modeling is discussed in the next section.

4.2 *Consequence Modeling*

In most PHAs, some level of consequence modeling will be needed to help the PHA team understand the consequences of hazardous events and make good estimates of possible injuries or property damage resulting from loss of containment. This would be particularly true for process boundaries where the level of hazard is high to very high, and the consequences of failure events are expected to be significant in terms of on-site or off-site impact.

Many models are available to calculate the areas impacted by loss of containment releases for events such as spills, holes in pipes, pump leaks, stack releases, etc.[7]. Typically, these models require inputs on physical properties, release conditions, meteorology, and levels of concern for various consequence thresholds. The primary result from modeling is the area impacted by the release for the defined input conditions and levels of concern. This model result must then be interpreted by the PHA team to evaluate the type, severity, and number of injuries or property damage in order to understand the consequences of the event. This will require the PHA team to also consider how many people may be exposed to the release, how long they may be exposed, the warning properties of the material, ventilation, exits, and the effects the material may have on people. For toxic materials, for example, it will be necessary to consider the acute toxicity of the material, how it affects the body, and whether or not it will affect a person's ability to evacuate. Consequence modeling can also be used to consider possible secondary effects resulting from the event, such as broken lines or damaged vessels in other parts of the facility, which can lead to additional injuries or property damage. In addition, the effectiveness of certain preventive or mitigating safeguards can be evaluated for reducing possible injuries or property damage.

Due to the complexity of modeling many different types of releases and the ease with which poor modeling results can be obtained with improper input, it is generally recommended that specialized resources be provided for completing consequence modeling activities. These people may be plant or central resources, typically with an engineering background, who have received training in consequence modeling, meteorology, levels of concern (e.g. toxicity concentrations and effects, flammability limits), and other related topics. Their role in the PHA is to prepare and help interpret model results for the PHA team so that injuries and property damage can be estimated. For toxics, for example, the resource will typically model a range of releases, review the model results and the area impacted by the release, and interpret the possible toxic effects on people, based on event duration, to help the PHA team understand what injuries may occur. Additional modeling may also be completed, as explained above, to consider secondary effects and the effectiveness of existing or proposed safeguards.

In addition to a specialized resource for providing and interpreting consequence model results, it is also recommended that specific, prescribed guidance be developed to provide conservative approaches for obtaining model results. This typically can be accomplished by assigning the many possible release scenarios into a limited number of modeling categories that cover most cases of interest in PHAs, with a straight forward basis for correctly selecting the appropriate category. For each category, specific step-by-step guidance can be developed for selected models, using appropriate conservative simplifying assumptions for release conditions, meteorology, and other model inputs. For initial evaluation purposes, for example, it is generally simpler and sufficient to consider only steady-state or instantaneous releases in most PHAs to obtain an understanding of possible consequences. Events that require specialized modeling, that do not easily fit into generalized categories, or that are not handled well by the model chosen for initial approximate calculations can be assessed separately from this approach using additional specialized models and technical resources. Examples of this may include liquid spills which freeze, vapor releases that rapidly condense, and multi-component mixtures and solutions. In many cases, government agencies throughout the world have specific requirements for analyzing and documenting the potential impact of events that result in toxicity, flammability, thermal radiation, and explosivity impacts. Typically, distances to toxicity levels of concern (i.e. ERPG, AEGL, etc.), flammability levels of concern (i.e. UFL, LFL, ½ LFL), thermal radiation levels [9.5 KW/m2 for human injury to 37.5 KW/m2 for equipment damage], and explosion overpressure levels [0.3 psi for window breakage to 5.0 psi for building collapse] have been established.

As an example, consider a PHA team reviewing the possible consequences of ammonia storage and handling. The level of toxicity hazard is determined to be very high. The health effects of ammonia exposure are shown in Table 1. It is recommended that Emergency Response Planning Guidelines (ERPG), if available, be used to help the PHA team understand the effects of possible toxic exposures resulting from loss of containment events. Consequence model results that provide the distance to these effects can then be used to help the PHA team understand what areas are impacted by a release, the possible injuries that result due to toxic exposures at these levels, and if new or revised safeguards are needed to prevent or minimize the consequences of loss of containment. Ammonia, in particular, is a strong eye and respiratory irritant, but has a strong odor, which helps to alert people that a release has occurred. It is also important to remember that some individuals may be more susceptible to ammonia injury and could develop very serious health effects at lower concentrations than shown in Table 1.

Table 1: Health Effect of Ammonia Exposure

Exposure	Health Effect
150 ppm	serious health effects that could impair ability to escape (ERPG-2)
750 ppm	life-threatening health effects (ERPG-3)
5000 ppm	rapidly fatal

One of the decisions that the PHA team may consider during the PHA is the storage conditions of the ammonia, which could be stored at ambient temperature or could be refrigerated. Ambient ammonia storage is at a pressure of 125 psig and 75 °F, as compared with refrigerated ammonia at 1 psig and -25 °F. Consequence models can be used to help the PHA team assess the number of injuries that could result from leaks for both options. Figure 2, for example, shows the model

Figure 2 – Model Results for Ammonia Release

(a) Ambient ammonia storage

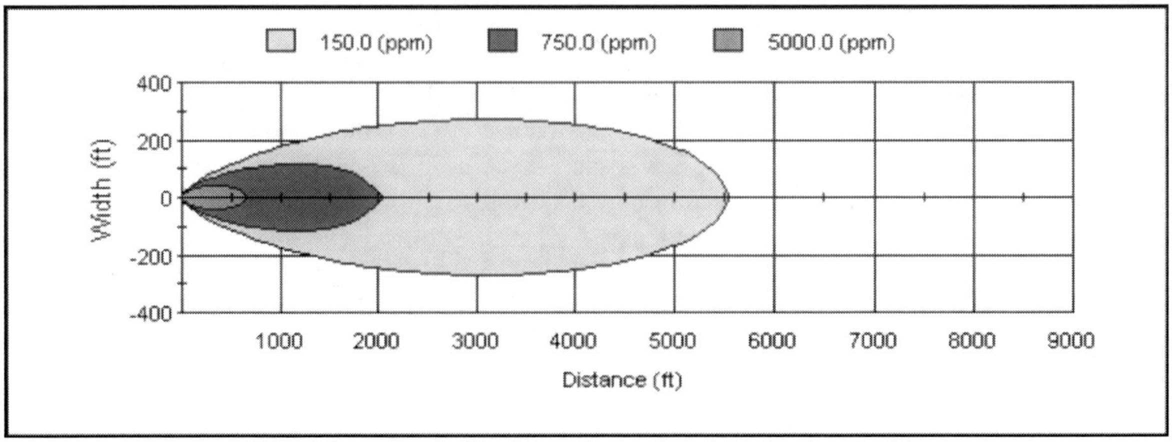

(Liquid ammonia flashing through a 1 inch pipe leak).

(b) Refrigerated ammonia storage

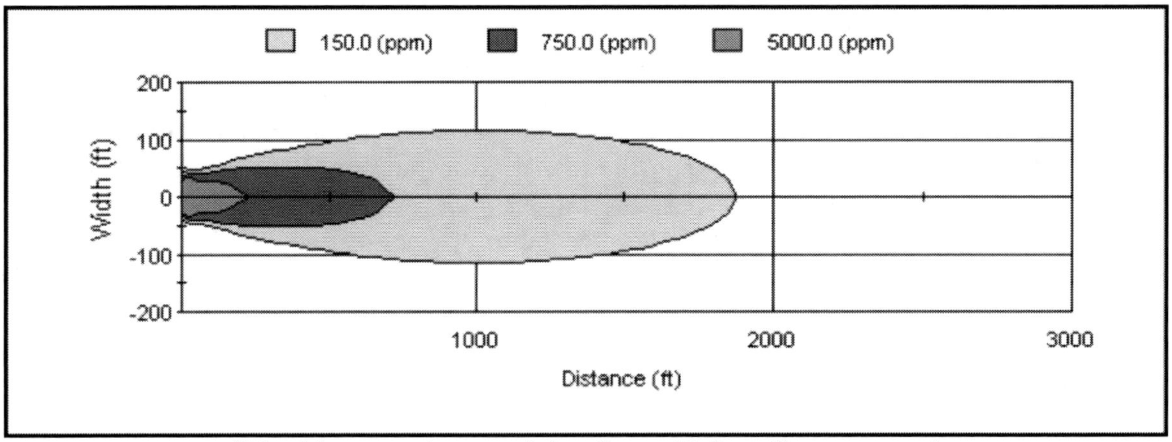

(Liquid ammonia vaporizing from 40 foot diameter pool, resulting from 1 inch pipe leak into diked area).

results for a small leak from a 1 inch diameter pipe for both ambient and refrigerated ammonia storage under one set of meteorological conditions (D atmospheric stability, and 5 m/s windspeed). Clearly, the area impacted by a release of refrigerated ammonia is significantly smaller than the area impacted by the release of the ambient temperature ammonia. For the ERPG-2 level, the ambient temperature ammonia release with a rate of 1405 lb/min impacts an area approximately 5500 feet from the source vs. approximately 1800 feet for the refrigerated ammonia with a vaporization rate of 221 lb/min. For the ERPG-3 level, the ambient temperature ammonia release impacts an area approximately 2000 feet vs. approximately 750 feet for the refrigerated ammonia. By understanding the health effects associated with the ERPG values, the number and location of people that could be exposed to these releases (under multiple meteorological conditions), and the likely duration of the exposure, the PHA team should be able to estimate the number of resulting injuries for both cases. The PHA team then must consider the risk of these alternatives by also estimating the probability of leaks. If it is determined that the ambient temperature storage option could potentially lead to a significantly larger number of injuries, the PHA team may recommend that the cost of installing a refrigerated system is justified in order to help prevent a significant loss of containment of ammonia. Improvements related to failures that lead to LOC for either storage option that can lead to releases can also be made. Risk analysis is discussed in the next section.

5. Risk Assessment

The primary purpose of a PHA is to assess the overall process risk and determine if a process is safe to operate. To evaluate risk, PHA teams use the information available on the nature and level of hazards from the PT/PSI compilation discussed in Section 3, and then conduct a detailed hazards evaluation using tools such as What-If/Checklist, FMEA, HAZOP, etc. An integral part of this risk assessment is the use of consequence analysis results discussed in Section 4 for complete range of loss of containment and other failure events. In defining scenarios for consequence analysis, failure of engineering and administrative controls must also be assessed. The PHA team defines all these events for consequence analysis and will also want to understand the impact of active and passive mitigation systems in risk analysis.

The frequencies or likelihood of all the failure events occurring is determined qualitatively or quantitatively based on site data or industrial data sources. Risk is then determined as:

$$\text{Risk} = f(\text{frequency} \times \text{consequences})$$

If within a process boundary, the hazard level is high and the consequences of a failure event are significant the PHA team must assess in detail the adequacy of the safeguards (i.e. layers of protection) present. The frequency of the event resulting in the consequence is used to determine the risk Often, a qualitative to semi-quantitative risk assessment is adequate, using a risk matrix as shown in Figure 3. In this matrix, different zones are defined based on company risk guidelines that help determine when new or improved safeguards are required to achieve an acceptable level of risk for the process. In some cases, layers of protection [11] or more comprehensive quantitative risk analysis [12] may be required to properly evaluate risk within certain process boundaries.

Use of risk analysis helps to better balance the PHA team understanding of low frequency, high consequence events against higher frequency, lower consequence events. Small releases may occur relatively frequently with relatively lower consequences depending on the materials and processing conditions, but can represent overall higher risk because of the high frequency of occurrence vs. larger, more catastrophic release events.

To minimize the process risk within a process boundary, there are many steps that must be followed. A PHA team must have a thorough and complete assessment of hazards (discussed in Section 3) available for review. The team then identifies the hazards within the boundary of concern and conducts a hazards evaluation. The team is then responsible for determining how the hazards can be avoided, reducing the severity or consequences of failure events, and reducing the likelihood of the consequences occurring. Much of this can be addressed using inherently safer process principles. Segregation of hazards and application of passive or active safeguards can also minimize the frequencies of such events. Application of procedural safeguards is another step that a PHA team must consider. A knowledge of the frequency, consequences, and failure mechanisms helps the PHA team make informed recommendations to ensure process design is appropriate to minimize the potential for loss of containment events, to provide appropriate containment or mitigation systems, to establish appropriate preventive maintenance schedules and procedures, and to determine required operating procedures.

A detailed risk assessment as conducted by a PHA team, before and after the implementation of the recommendations of a team, therefore provides sufficient information to minimize the process risk resulting from potential loss of containment and other failure events. In most cases, a qualitative to semi-quantitative assessment is adequate. However, sometimes a more comprehensive risk assessment [11, 12] may be need to ensure that the risk within a process boundary is as low as reasonably practicable or achievable.

Figure 3: Qualitative Risk Matrix

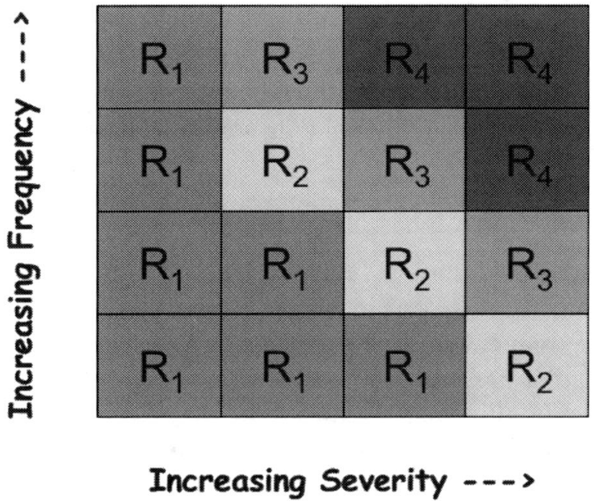

6. Summary

A thorough understanding of the process hazards, identification of a complete range of failure events with detailed analysis of possible consequences, and an analysis of process risks accounting for all the safeguards can help significantly in preventing and mitigating loss of containment incidents. By evaluating the potential causes of loss of containment through use of these methodologies, PHA teams can ensure process design is appropriate to minimize the potential for loss of containment events, can provide appropriate containment or mitigation systems, can establish appropriate preventive maintenance schedules and procedures, and can provide safe operating practices.

References

[1] Center for Chemical Process Safety, *Guidelines for Implementing Process Safety Management Systems*, AICHE, 1993

[2] Atherton, John and Frederic Gil, *Incidents That Define Process Safety*, Center for Chemical Process Safety, Center for Chemical Process Safety, John Wiley & Sons, 2008

[3] Center for Chemical Process Safety, *Guidelines for Hazard Evaluation Procedures*, 3rd Edition, John Wiley & Sons, 2008

[4] Medical Dictionary, 3rd Edition, Wiley Publishing, Inc., 2008

[5] NFPA, Fire Protection Guide to Hazardous Materials, A.B. Spencer and G.R. Colona Editors, 13th Edition, 2002

[6] U.K. Health and Safety Executive, "Assessment of the Dangerous Toxic Load (DTL) for Specified Level of Toxicity (SLOT) and Significant Likelihood of Death (SLOD)," [http://www.hse.gov.uk/hid/haztox.thm]

[7] Center for Chemical Process Safety, *Guidelines for Consequence Analysis of Chemical Releases*, AICHE, 1999

[8] Center for Chemical Process Safety, *Guidelines for Evaluating the Characteristics of Vapor Cloud Explosions, Flash Fires, and BLEVEs* , AICHE, 1994

[9] Johnson, Robert W., Steven W. Rudy, and Stephen D. Unwin, *Essential Practices for Managing Chemical Reactivity Hazards*, AICHE, 2003

[10] NFPA, Fire Protection Handbook, Volumes 1 and 2, 20th Edition, 2008

[11] Center for Chemical Process Safety, *Layer of Protection Analysis: Simplified Process Risk Assessment*, John Wiley & Sons, 2001

[12] Center for Chemical Process Safety, *Guidelines for Chemical Process Quantitative Risk Analysis*, AICHE, 1999

Improving Operational Discipline to Prevent Loss of Containment Incidents

Bruce K. Vaughen
Cabot Corporation, 700 E US Highway 36, Tuscola, IL 61953
bruce_vaughen@cabot-corp.com

James A. Klein
DuPont, North America Operations SHE
James.A.Klein@USA.dupont.com

Prepared for Presentation at
American Institute of Chemical Engineers
2010 Spring Meeting
6th Global Congress on Process Safety
San Antonio, Texas
March 22-24, 2010

UNPUBLISHED

Improving Operational Discipline Programs to Prevent Loss of Containment Incidents

Bruce K. Vaughen
Cabot Corporation, 700 E US Highway 36, Tuscola, IL 61953
bruce_vaughen@cabot-corp.com

James A. Klein
DuPont, North America Operations SHE
James.A.Klein@USA.dupont.com

Abstract

Process safety and risk management systems are implemented to reduce injuries and process incidents. The effectiveness of these systems is dependent on the company's implementation and support for Operational Discipline (OD)-related programs to ensure that the Process Safety Management (PSM) system requirements are rigorously followed day-to-day at all levels of the organization. However, OD-program-related problems continue to contribute to process incidents, including Loss of Containment (LOC) incidents, especially since they may not be formally addressed during incident investigations. This paper provides a brief historical overview of major LOC incidents, describes how poor OD increases process safety risk and can contribute to LOC incidents, and presents an example of how to track and identify weaker OD characteristics – the OD-related "root causes" – using OD characteristic data obtained during incident investigations.

Keywords: operational discipline, incident investigation, conduct of operations, loss of containment, process safety management, risk reduction, root cause failure analysis, near miss, human factors, leading indicator, lagging indicator, process hazards analysis

1. Introduction

Prior work has described how the commitment to Operational Discipline (OD) and reducing at-risk behavior across all levels of the organization can help reduce process safety risk [1, 2, 3]. This paper reviews several major LOC events to emphasize the role of OD in causing LOC incidents, considers how OD affects process safety risk, leading to LOC incidents, and provides an incident investigation tool to help identify OD characteristics as root causes. The framework for both evaluating and improving OD programs is discussed based on the following OD characteristics [2]:

Organizational	1. Leadership Focus
	2. Employee Involvement
	3. Practice Consistent With Procedures
	4. Excellent Housekeeping
Personal	1. Awareness
	2. Knowledge
	3. Commitment

By tracking these OD characteristics, both systemic Organizational and individual Personal OD weaknesses can be identified, OD program improvements recommended and the risk of LOC incidents reduced.

2. Background

2.1 Major Loss of Containment Incidents

Although significant chemical industry events involve Loss of Containment, they are not generally thought of directly as LOC events. Major LOC incidents include toxic events, such as the Seveso incident resulting from loss of containment of a toxic dioxin from a burst reactor rupture disk and the Bhopal incident, resulting from the unmitigated loss of containment of toxic methyl isocyanate (MIC) gas from a storage tank's relief system after its contents were contaminated with water. Similarly, and more recently, the BP Texas City explosion in 2005 occurred after loss of containment of hydrocarbons from a stack during a system start-up after maintenance. Seveso, Bhopal and BP Texas City resulted in severe, irreversible consequences: toxic environmental and personal exposures, fires, explosions and fatalities [4, 5, 6, 7, 8, and 9].

This section provides a brief description of these three major LOC incidents that can be attributed, in part, to poor OD programs. The Seveso incident led to development of Europe's EC Directive on Major Accident Hazards of 1982, also known as the "Seveso Directive" [4]. Bhopal was a "watershed" event for hazards assessments in the chemical industry. The US responded with three legislative efforts around emergency response (EPCRA, 1986), managing hazards and protecting workers within the fenceline (OSHA PSM, 1992), and managing hazards and protecting the communities and environment beyond the fenceline (EPA RMP, 1996) [22, 23, 24]. In response to Bhopal, as well, AIChE formed the Center for Chemical Process Safety

(CCPS) in 1985. The BP Texas City incident led to one of the most comprehensive PSM-focused investigations, both internal to BP and externally through the US Chemical Safety Board and through the "Baker Panel Report. [9]" The BP Texas City incident prompted emphasis on improving a company's "safety culture," including recent AIChE conferences related to the "softer side of engineering" and Human Factors.

The impact of OD on each of these incidents is shown by a matrix in a table for each LOC incident: the incident's PSM-related learnings are listed in the rows of a table; the OD-characteristics are listed in the columns. If there is an interaction (a "relationship) between the PSM learning and a corresponding OD characteristic, the matrix is marked with a "1." The OD characteristics identified for each LOC is discussed in this section, with an overall LOC Organizational and Personal OD-characteristic "trend over time" using a different perspective is discussed in section 2.2 below.

Seveso, Italy

In 1976, a small amount of TCDD, one of the most toxic chemical dioxins known, was released when a rupture disc burst on a chemical reactor. The release dusted the environment with a white solid powder as it drifted across the area downwind. Over the next few days animals died and people in the contaminated area fell ill [4]. After the contaminated area downwind had been characterized and zoned, the most contaminated zone was fenced off, more than 70,000 potentially contaminated animals were killed, and more than 700 people were relocated [10]. As is shown in Table 1, the "Leadership Focus," "Knowledge," and "Awareness" OD characteristics accounted for most of the PSM-related learnings. Studies have shown health and environmental issues in the Seveso area many years later [10, 11, 12, 13, and 14].

Bhopal, India

In 1984, over 40 tons of highly toxic methyl isocyanate (MIC) gas was released when a relief valve lifted on a storage tank contaminated with water. The toxic plume drifted downwind, collected in the valley, and poisoned the residents in the neighborhoods located at the fenceline. More than 2,000 people died within a short period, with tens of thousands injured and countless livestock killed [4]. As is shown in Table 2, the "Leadership Focus" and "Knowledge" OD characteristics accounted for more than half of the PSM-related learnings. Like Seveso, many health and environmental issues, including water contamination, still exist over 25 years later [15, 16, 25].

BP Texas City, USA

In 2005, hydrocarbons overflowed from an oil refinery isomerization unit stack during a system start-up. The hydrocarbons formed a large vapor cloud in the immediate area which subsequently ignited and caused a series of explosions. An occupied contractor's trailer sited in the explosion zone was destroyed, 15 people died and 180 people were injured [5]. OSHA levied severe fines and the CSB concluded that "Organizational and Safety Deficiencies at All Levels of the BP Corporation" caused the disaster [6]. As is shown in Table 3, almost all of the OD characteristics applied to more than half of the PSM-related recommendations. Unfortunately, the site's culture could not change quickly enough, and other PSM-related issues at the BP Texas City refinery prompted more fines after a six month follow-up audit in 2009 [17].

2.2 *A Different LOC Perspective Using OD*

Although these major LOC incidents prompted specific regulatory and industry-wide changes, the specific OD characteristics have not traditionally been formally addressed during the incident investigation process. Root cause failure analysis (RCFA) training techniques do not formally note "OD Root Causes" [18, 19]. In general, RCFA's identify the root causes of events within one or a combination of three broad categories: 1) equipment failure; 2) human failure; and 3) systems failure. Equipment failure may be due to weaknesses in mechanical integrity programs or poor equipment reliability and potential for catastrophic failure. Human failure, attributed to weaknesses in Personal OD characteristics, may be due to poor training or poor decision-making during unfamiliar situations or emergency responses. This human failure category is difficult to identify and address because it deals with humans and their behavior, and is the major focus of minimizing the potential for human error within Human Factors-related risk-reduction strategies. However, it is the third category, the systems-related failure, which is significantly impacted by improving Organizational OD-related weaknesses. Fixing system failures has the greatest overall impact on LOC incidents due to the leveraging effect of system-level fixes. Improving the Organizational OD characteristics – the business' safety culture – impacts all employees in the organization. This includes improving the leadership focus on, support for, and communication of expectations that everyone performs to meet the organization's "code of conduct" (see a definition for OD below).

A different image of the changing nature of our PSM efforts to reduce risk evolve over time can be captured with the familiar DNA double-helix model shown in Figure 1. Think of climbing up the spiral ladder as time (evolution over time; longer molecules are more complicated and have more "understanding"). The rungs represent the building blocks of the ladder (the amino acids); consider that OD could be the weak hydrogen bonds that hold the blocks together. The major LOC incidents are shown over time on our climb where significant gaps occurred in the rung (i.e., previously unknown or un-addressed "weak bonds" caused an incident). Imagine yourself climbing up the ladder, only to look back down at a later date and see our goal (reduced process safety risk) from this different perspective, the view that has more experience. Every major and often tragic LOC has provided us with new opportunities for learning. Hence, over time, our initial qualitative incident investigation tools have evolved to include better and more quantitative techniques [20].

With this different perspective in mind, the OD-characteristics noted in the three major LOC incidents above were compiled and summarized in Table 4, and are shown with the change in OD characteristics "recognition" over time in Figure 2. More Organizational OD weaknesses are being identified today than two decades ago. Hence, as we acquire a better view on improving process safety risk reduction efforts, the trend has shifted from identifying and correcting mostly Personal OD weaknesses (the "individual" human failures) to finding better ways identifying and improving Organizational OD weaknesses (the systemic and cultural failures). It is interesting to note that over time more and better investigation tools have been developed and shared (our building blocks), consistent with the development of the numerous publications by the CCPS and many other publishers during the last two decades. It is hoped that this different perspective supports the case for our need to reducing LOC incidents by measuring and focusing on

Organizational OD-program weaknesses (the gaps, the weak bonds between our building blocks) so that we get closer to our goal: reducing process safety risk.

2.3 The Impact of OD on Risk

The seven Organizational and Personal OD characteristics were listed in the Introduction. These seven characteristics provide a framework for both identifying OD-related problems and for developing appropriate improvements to OD programs at a facility. The concept showing that OD underlies and impacts all aspects of an effective Process Safety Management (PSM) system is shown schematically in Figure 3 [1, 2, 21, 22]. This section of the paper defines OD and describes how OD impacts risk.

What is OD? A definition proposed for "Operational Discipline" is as follows [2]:

> The deeply rooted dedication and commitment by every member of an organization to carry out each task the right way every time.

How does OD affect risk? First, risk must be defined. The overall goal of all PSM efforts is to reduce the risks of the process and its associated hazards (fires, explosions, toxic releases). One useful way of describing the methods to reduce these risks is to modify the normal risk equation and consider the impact of OD with the following equation [2]:

$$\text{Risk} = f\{ (\text{Frequency x Consequence}) / (\text{Operational Discipline}) \} \qquad \textbf{(1)}$$

In this equation, risk is directly related to both a Frequency term and a Consequence term and is indirectly related to an OD term. The concept is simple: decreasing the frequency term reduces the risk; decreasing the consequence term reduces the risk; and decreasing the OD term raises the risk.

The frequency is often determined by the effectiveness of engineering and administrative controls to prevent a possible hazardous event., such as safety systems and operating procedures. Improving engineering and administrative controls will reduce the potential frequency of the event and hence, reduce the risk. The "worst case scenario" consequence term is characterized by the potential LOC event that could result in a fire, explosion, or toxic release. The consequences are determined with the hazardous properties of the materials (e.g., flammable, toxic, etc.), the amount of hazardous material in the system and the processing conditions required store, handle, and transform the raw materials to the final product. Reducing the consequence, such as those defined within inherently safer design principles, will reduce the risk, as well. Note that improving engineering systems and administrative procedures may reduce the consequences, as well. These improvements include more sensitive detectors, earlier and faster event-detection alarms, and faster system shutdowns to a safer state, faster emergency team responses, and improved mitigation procedures.

How does OD affect risk? As is defined in this risk model, risk is inversely related to an OD term. This OD term is expressed in fractional form, where 1.0 represents perfect OD program implementation, and reflects some of the behavior-related aspects within the risk equation, such as the organization's Conduct of Operations (COO) and Human Factors principles. For example, safety systems and operating procedures are designed to reduce the frequency and consequences of the risk. If there is poor OD, the safety systems will degrade and short cuts may prevail, either to personal choice or supervisor decrees. If one-half the time the safety procedure is not followed and the safety systems are not maintained – say 50% compliance to OD - the risk is doubled. Hence, a reduced OD term – poor OD - increases the risk. OD programs must be sustained, maintained and followed to ensure 100% effectiveness of the safety systems and procedures developed to reduce risk.

As is shown in the risk matrix in Figure 4, a system with an acceptable risk ("green") will move toward the unacceptable zone ("red") when poor OD exists. An example for illustration: a process risk is identified as acceptable if it is operated within its safe operating limits. However, the actual process is routinely operated outside of these safe operating limits to meet production demands (and for some reason, this behavior doesn't adversely affect the product's quality!). If this OD characteristic, Practice Consistent with Procedures, is not addressed, the actual risk may be higher due to higher potential (frequency) for the hazardous event and its consequences. Improving this "weak" OD characteristic back toward 1.0, such as enforcing the procedures, will ensure a stronger (higher) OD-related term and prevent the risk from moving toward the unacceptable red zone.

3. Preventing LOC Incidents

Before improving OD programs, not only must a set of OD "metrics" be determined and measured, but any changes proposed must be "knobs" that we can turn (we have some sort of influence over the metric). Product quality improvement tools are designed to find knobs on machines to improve their reliability and measure their product's performance. However, human-related reliability and performance improvement tools are not as so easy to find – which "knob" do we turn? As was discussed in Section 2.2 above, more Organizational OD characteristic weaknesses are being identified today than two decades ago. The trend from the Personal OD focus to a more systemic Organizational OD focus reflects a better understanding that more effective risk reduction efforts will occur when system weaknesses are addressed and resolved. Hence, identifying and monitoring an Organizational OD-characteristic metric will more effectively help reduce risk. To proactively prevent LOC incidents, we need to answer the question: "How do we determine the leading indicators, measure and monitor them, and then improve our OD program?" This section explores the extent of LOC incidents, ranging from near misses to catastrophic events, the importance of identifying leading indictors versus simply measuring lagging indicators, and proposes a "principles" metric that measures "adherence" based on the human actions (Personal OD - at-risk behavior) which may have contributed to the incident. Future work will focus on better Organization OD metric identification and analysis.

The LOC incident pyramid, shown in Figure 5, helps visualize how improving OD to reduce LOC incidents can help reduce major flammable, explosive and toxic releases. This pyramid contains different levels with decreasing number of events represented by smaller areas at higher levels within the pyramid. The largest area at the bottom represents the number of "foundation" level events, such as minor releases or spills with no irreversible personal, process or environmental harm. The smallest area at the top represents a relative smaller number of catastrophic events, such as Seveso, where the LOC caused irreversible harm to personnel and the environment. ("All fires start out small, even the big ones.") The concept to reduce the number of major incidents is relatively simple: if there are fewer minor spills and releases at the lower levels (hence, a smaller "foundational" area), then there should be relatively fewer catastrophic LOC incidents at the top level.

Responding to LOC event that have occurred, a lagging indicator, is known as reactive LOC incident management. Recently, CPPS published guidance on identifying leading indicators to provide metrics for proactive PSM system improvements [26]. Note that this guidance document provides potential leading indicators which are designed to better develop, nurture and sustain all of the PSM Elements (Figure 3). As is described below, focusing on and improving the weaker OD characteristics proactively "attacks the base" of the incident pyramid (Figure 5) and will help reduce the number of LOC incidents.

3.1 The Impact of OD on LOC Incidents

As is described and shown in the risk equation above (Equation 1), the impact of improving OD is to reduce the risk. Poor performance to any OD characteristic, whether it is organizational or personal, will increase the risk. Note that proactive LOC incident investigations should include "near miss" incidents. These types of incidents could have been much worse had conditions or human actions been slightly different: just a matter of a few centimeters to the left or to the right, or just a few seconds earlier or later and a catastrophic LOC could have occurred ("we were lucky"). Effective OD programs will recognize, monitor and track this type of leading indicator as a learning and improvement measurement tool, as well.

It is important to recognize that LOC incident investigations need to identify the contribution of the different OD characteristics to the incident. From this data – the OD program feedback – site-specific OD characteristics that are causing incidents can be identified and targeted for improvement. For example, one site may see "Practice Consistent with Procedures" as the weaker characteristic, whereas another site, reflecting a different safety culture, may identify "Leadership Focus" as its weakest characteristic. Improvements will be and must be based on the site-specific issues. If the different contributions and impact of OD characteristic weaknesses are not being addressed in the incident investigation process, important information needed to reduce LOC incidents will be overlooked.

3.2 Preventing LOC Incidents by Improving OD

This section describes part of an OD improvement incident investigation strategy that is being discussed through Cabot's FMO division to help prevent LOC events and reduce FMO's safety and environmental risks. Although the current incident investigation approach notes the following OD-related leading indicators (the "base" of the pyramid in Figure 5),

1) Is the incident a "Near Miss?"
2) Did we adhere to the "Principle?"
3) Was the incident as a result of an "Unsafe Act?"

this paper describes an approach being proposed to monitor and address the second indicator, the FMO's "Principles of Operation," only. Future work will address data pertaining to near misses and unsafe acts. Connecting the OD characteristics to reducing individual "At Risk" behavior, as is identified with a weakness in an OD-related Principle, will help with improving OD and reducing risk.

The eight FMO "Principles of Operation" were developed in 2007, issued in 2008 and subsequently implemented into the incident investigation procedure in early 2009. The Principles capture and correspond to the basic characteristics of PSM and are linked with the incident investigation procedure, as is shown with the matrix in Figure 6.

The Principles are FMO's code of conduct (a "Conduct of Operations") is as follows:

1. Always operate equipment and processes within designed safety and environmental limits.
2. Always move to a safe controlled condition OR location when a situation is not understood.
3. Always seek assistance and involve people with expertise in changes or improvements that affect equipment or procedures.
4. Always understand and address deviations and ensure proper corrective actions are implemented AND communicated.
5. Always report environmental or safety deviations promptly and accurately.
6. Always follow safe work practices and written procedures and act to stop unsafe conditions and behaviors.
7. Always address abnormal conditions and clarify / understand all procedures before proceeding.
8. Always use tools/equipment/processes for their intended use. Never compromise any safety systems and/or interlocks.

In general, these principles focus on the "Knowledge" characteristic of Personal OD and are actively supported and discussed through the "Leadership" characteristic of Organizational OD.

The Principles are specifically listed in part of the investigation team's form, as is shown in Figure 7. The team fills out this part of the investigation form by answering the question "Did we adhere to the Principle?" with a "Yes," "No," or "Not Applicable" response. A strong Principle receives a "Yes" response. For example, if the employee reports a minor LOC spill immediately, the answer to Principle #5 is "Yes." An example of a weak Principle (one that was not followed) is if the employee decides to substitute a cheaper and subsequently less resistant gasket without confirming its suitability leading to an LOC. In this case, the answer to Principle #3 would be "No."

The team is required to address each of the "No" responses with at least one specific action. The expectation is that all "No" responses will have a specific OD-improving action noted to address and correct the gap. These actions may be combined into a recommendation that is issued with the final incident investigation report. Hence, by monitoring adherence to the Principle, we are measuring an "At Risk" Behavior and can track the Principles to identify and improve the weaker OD-related characteristics.

Initial data analysis from selected incidents over the latter part of 2009 is discussed next. This trending approach indicates that each FMO site and the FMO business will be able to monitor and track poor adherence (lack of OD) and then can address its "weaker Principles" to help improve OD and reduce risk. The Principles trending approach is shown in Figure 8. Since the goal is to focus on weaker principles, those that tend to receive more "No" answers, show up as principles that have the most "No" responses relative to the others. From the 32 incidents noted at this point in Figure 8, it is clear that the weaker principles are "Following Procedures" (Principle #6, at 44%) and "Operating Safely" (Principle #1, at 31%). Do note that these principles are not independent. However, this result indicates that we are struggling to adhere to our safe work practices, an essential element required for effective PSM systems (Figure 3). Assuming that a sufficient LOC incident data set is available, this "trend" indicates that we must re-emphasize across the entire organization the importance of ensuring knowledge of and operating within safe operating limits to help prevent LOC incidents.

4. Summary and Conclusion

The example incident investigation approach described in this paper, proactively measuring and trending an "Operational Discipline" characteristic, may be useful for reducing the number of LOC incidents by improving OD. This paper provides a different perspective on incident data analyses using OD-characteristics, shows that poor adherence to one of the Principles (in particular, "poor OD") can be determined during an LOC incident investigation. More robust trending analysis will provide managers with OD-program focus areas for making OD-characteristic and PSM-related improvements. These OD program improvements should reduce the LOC risk by reducing the frequency of LOC incidents. Therefore, the effectiveness of safety systems and procedures in reducing LOC incidents is dependent on the company's implementation and support for OD-related programs to ensure that the system requirements are rigorously followed day-to-day at all levels of the organization.

5. Future Directions

Work on better data acquisition and analysis of OD characteristics is currently underway; additional work will be performed on existing near miss and unsafe act data already being documented in Cabot FMO's final incident reports.

Note that the frequency and/or consequence risk-reduction approach described in Section 2.3 is the basis for Process Hazards Analysis (PHA) Team evaluations, and helps provide PHA Teams with a framework for proposing practical recommendations that reduce the LOC risk to an acceptable level. However, if poor OD occurs once the process is operating, the actual risk may be higher and unacceptable when compared to the PHA Team's conclusion. Although this will be the subject of a future paper, a brief overview of the PHA Team's process is described here to help elucidate how failure to consider the impact of poor OD increases the risk. First, a PHA team evaluates the frequency and consequences of a LOC event based on a review of hazards of the proposed or existing design of the process. This review includes the all known process safety information: the hazards of the materials, the process design and the equipment design. Then the Team determines potential LOC release scenarios and failures, both without any safeguards (the "worst case scenario") and with existing and/or proposed safeguards. If the risk is deemed unacceptable, the Team must develop additional risk-reduction strategies and propose recommendations for management review and approval. If OD is not considered by the PHA team, then the actual risk may be higher than the calculated risk as some of the safeguards may not function as planned. Consideration of how OD characteristics impact risk to help prevent LOC incidents is therefore an important part of the risk function and should be considered by PHA Teams.

6. References

1 Klein, J. A., "Operational Discipline in the Workplace," Process Safety Progress, Vol. 24, No. 4, p. 228-35, 2005.

2 Klein, J. A. and B. K. Vaughen, "A Revised Program for Operational Discipline," Process Safety Progress, Vol. 27, No. 1, March 2008, pp. 58-65.

3 Vaughen, B. K., K. Lock and T. Floyd, "Improving Operating Discipline Through the Successful Implementation of a Mandated Behavior-Based Safety Program," Process Safety Progress, Early View, Published Online, 27 January, 2010.

4 Lees, F. P., editor, Loss Prevention in the Process Industries (2nd Edition), Elsevier, 1996.

5 Holmstrom, D, et. al., "CSB Investigation of the Explosions and Fire at the BP Texas City Refinery on March 23, 2005," Process Safety Progress, December 2006, Vol. 25, No. 4, pp. 345-349.

6 CSB website on BP Texas City Incident, Completed Investigations, http://www.csb.gov/investigations/investigations.aspx?Type=2&F_All=y

7 BP Website on Texas City Incident, http://www.bp.com/genericarticle.do?categoryId=2012968&contentId=7032902

8 Atherton, J. and F. Gil, Incidents That Define Process Safety, Center for Chemical Process Safety, John Wiley & Sons, 2008.

9 The Baker Panel Report, http://www.bp.com/liveassets/bp_internet/globalbp/globalbp_uk_english/SP/STAGING/local_assets/assets/pdfs/Baker_panel_report.pdf

10 Assenato, G., et. al., "Follow-up of Subjects Who Developed Chloroacne Following TCDD Exposure at Seveso," American Journal of Industrial Medicine, Vol 16, 1989, pp. 119-125.

11 Weiss, J., et. al., Concentrations of dioxins and other organochlorines (PCBs, DDTs, HCHs) in human milk from Seveso, Milan and a Lombardian rural area in Italy: a study performed 25 years after the heavy dioxin exposure in Seveso.

12 Needham. L. L., et. al., "Serum Dioxin Levels in Seveso, Italy, Population in 1976," Teratogenesis, Carcinogenesis, and Mutagenesis 17:225–240 (1997/98)

13 Signorini, S., et. al., "Environmental exposure to dioxin: the Seveso experience," Andrologia 32, 263–270 (2000).

14 Baccarelli, A., and A. C. Pesatori, "Health status and plasma dioxin levels in chloroacne cases 20 years after the Seveso, Italy accident, Clinical and Laboratory Investigations, British Association of Dermatologists, British Journal of Dermatology 2005 152, pp459–465.

15 Willey, R. J., D. C. Hendershot and S. Berger, The Accident in Bhopal: Observations 20 Years Later, Process Safety Progress, September 2007, Vol. 26, No. 3, pp.180-184.

16 Louvar, J. F., Editorial – Bhopal, CCPS and 25 years, Process Safety Progress, December 2009, Vol. 28, No. 4, page 299.

17 US OSHA, "OSHA issues record breaking fines to BP," http://www.osha.gov/dep/bp/bp.html

18 Mobley, R. Keith, Root Cause Failure Analysis, Plant Engineering Maintenance Series published by Newnes, an imprint of Butterworth-Heinemann, 1999.

19 Center for Chemical Process Safety (CCPS), Guidelines for Investigating Chemical Process Incidents, Second Edition, AIChE, 2003.

20 Hendershot, D. C., A History of Process Safety and Loss Prevention in the American Institute of Chemical Engineers, Process Safety Progress, June 2009, Vol 28, No. 2, pp. 105-113.

21 Klein, J. A., "Two Centuries of Process Safety at DuPont," Process Safety Progress, Vol. 28, No. 2, 2009.

22 US OSHA 29 CFR 1910.119, Process Safety Management of Highly Hazardous Chemicals (PSM)

23 Long, L. L., History of Process Safety at OSHA, Process Safety Progress, June 2009, Vol.28, No.2, pp. 128-130.

24 Kleindorfer, P. R., et. al., "Accident Epidemiology and the U.S. Chemical Industry: Accident History and Worst-Case Data from RMP*Info," Risk Analysis, Vol. 23, No 5, 2003, pp. 865-881.

25 Dhara, V. R. and D. Kriebel, "An Exposure-Response Model for Assessing the Long Term Health Effects of The Bhopal Gas Disaster," Disasters, Vol. 17, Number 4, pp. 281- 289.

26 Center for Chemical Process Safety (CCPS), "Process Safety Leading and Lagging Metrics," AIChE Publications, 2008.

Table 1. Matrix Relating the OD Characteristics to the Seveso LOC Incident PSM-related Learnings

Seveso	PSM-related Learning	Organizational				Personal		
	Operational Discipline Characteristic	1 Leadership Focus	2 Employee Involvement	3 Practice Consistent With Procedure	4 Housekeeping	1 Awareness	2 Knowledge	3 Commitment
1	Public control of major hazard installations							
2	Siting of major hazard installations	1				1		
3	Acquisition of companies operating hazardous processes	1						
4	Hazard of ultratoxic substances					1	1	
5	Hazard of undetected exotherms					1	1	
6	Hazard of prolonged holding of reaction mass					1	1	
7	Inherently safer design of chemical processes							
8	Control and protection of chemical reactors			1				
9	Adherence to operating procedures			1				
10	Planning for emergencies	1						
11	Difficulties of decontamination						1	
	Sum	3	0	2	0	4	4	0

Table 2. Matrix Relating the OD Characteristics to the Bhopal LOC Incident PSM-related Learnings

	Operational Discipline Characteristic	Organizational				Personal		
	Bhopal PSM-related Learning	1 Leadership Focus	2 Employee Involvement	3 Practice Consistent With Procedure	4 Housekeeping	1 Awareness	2 Knowledge	3 Commitment
1	Public control of major hazard installations							
2	Siting of and development control at major hazard installations	1						
3	Management of major hazard installations	1						
4	Highly toxic substances						1	
5	Runaway reactions in storage						1	
6	Water hazard in plants					1		
7	Relative hazards of materials in process and in storage						1	
8	Relative priority of safety and production	1						
9	Limitation of inventory in the plant						1	
10	Set pressure of relief valves							1
11	Disabling of protective systems	1		1				1
12	Maintenance of plant equipment and instrumentation				1			1
13	Isolation procedures for maintenance				1			
14	Control of plant and process modifications	1			1		1	
15	Information for authorities and public					1		
16	Planning for emergencies	1				1		
	Sum	6	0	1	3	3	5	3

Table 3. Matrix Relating the OD Characteristics to the BP Texas City LOC Incident PSM-related Recommendations

	Operational Discipline Characteristic	Organizational				Personal		
	BP Texas City PSM-related Recommendation	1 Leadership Focus	2 Employee Involvement	3 Practice Consistent With Procedure	4 Housekeeping	1 Awareness	2 Knowledge	3 Commitment
1	Process Safety Leadership	1						
2	Integrated and Comprehensive Process Safety Management System		1	1	1	1	1	
3	Process Safety Knowledge and Expertise				1	1	1	
4	Process Safety Culture	1	1	1	1			1
5	Clearly Defined Expectations and Accountability for Process Safety	1	1	1	1	1	1	1
6	Support for Line Management	1	1		1			1
7	Leading and Lagging Performance Indicators for Process Safety	1			1	1	1	
8	Process Safety Auditing	1		1	1		1	
9	BP's Board Monitoring	1						
10	Industry Leader	1	1	1	1	1	1	1
	Sum	8	5	5	8	5	6	4

Table 4. Compiled Matrix Summarizing the Relationship Between the OD Characteristics and the LOC Incident PSM-related Learnings

Loss of Containment (LOC)		Number of Learnings	Organizational: 1 Leadership Focus	Organizational: 2 Employee Involvement	Organizational: 3 Practice Consistent With Procedure	Organizational: 4 Housekeeping	Personal: 1 Awareness	Personal: 2 Knowledge	Personal: 3 Commitment		Average OD / LOC
Sum from LOC Incidents		37	17	5	8	11	12	15	7	75	
			41				34				
BP Texas City PSM-related Recommendations	2005	10	8	5	5	8	5	6	4	41	4.1
			26				15				
			63%				37%				
Bhopal PSM-related Learnings	1984	16	6	0	1	3	3	5	3	21	1.3
			10				11				
			48%				52%				
Seveso PSM-related Learnings	1976	11	3	0	2	0	4	4	0	13	1.2
			5				8				
			38%				62%				

Figure 1. A Different LOC Incident Investigation Perspective Using OD

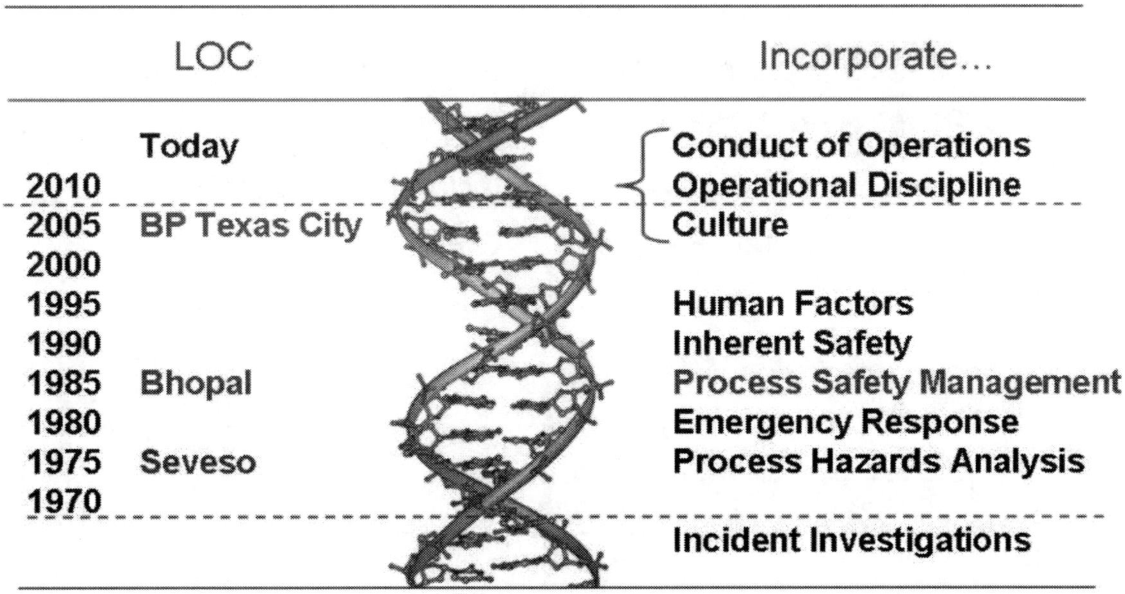

Figure 2. The OD Characteristics Trend from the Three Major LOCs

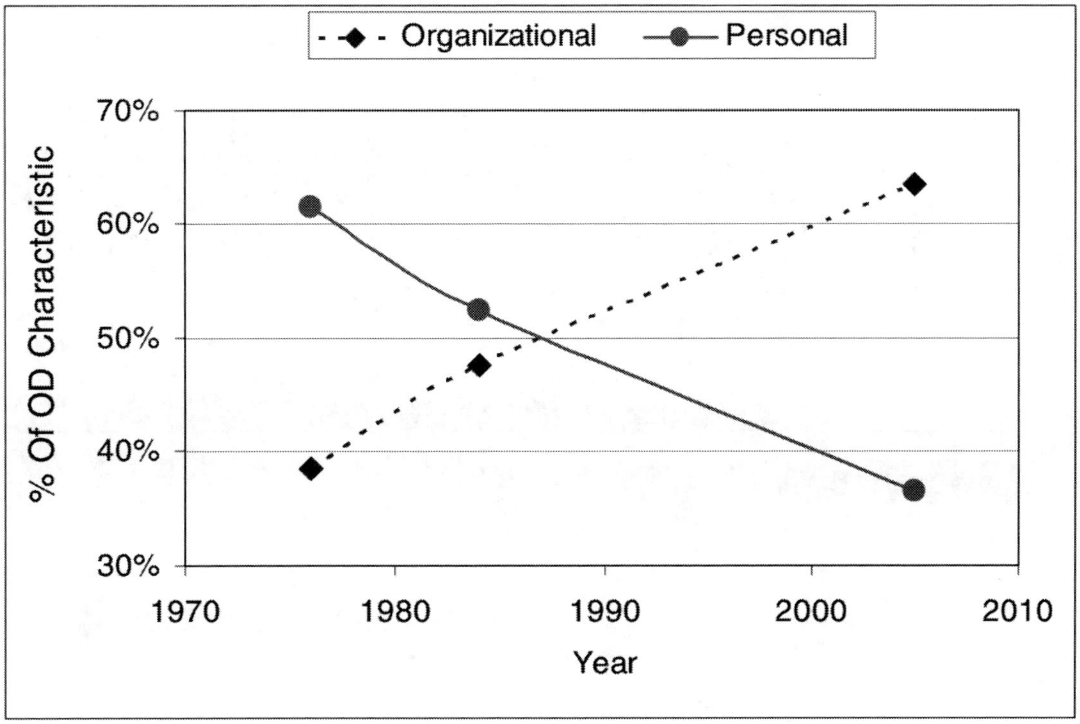

Figure 3. A Link Between the Process Safety Management (PSM) Elements and OD

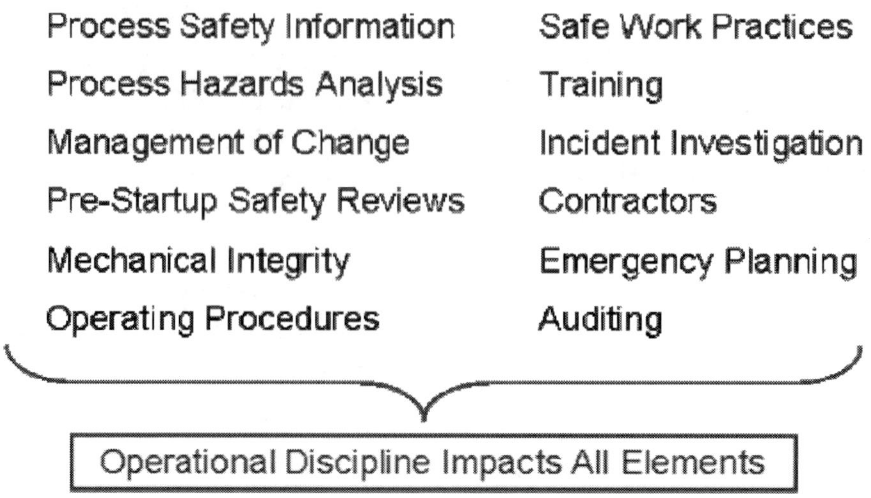

Figure 4. The Impact of Poor OD on the Actual Risk

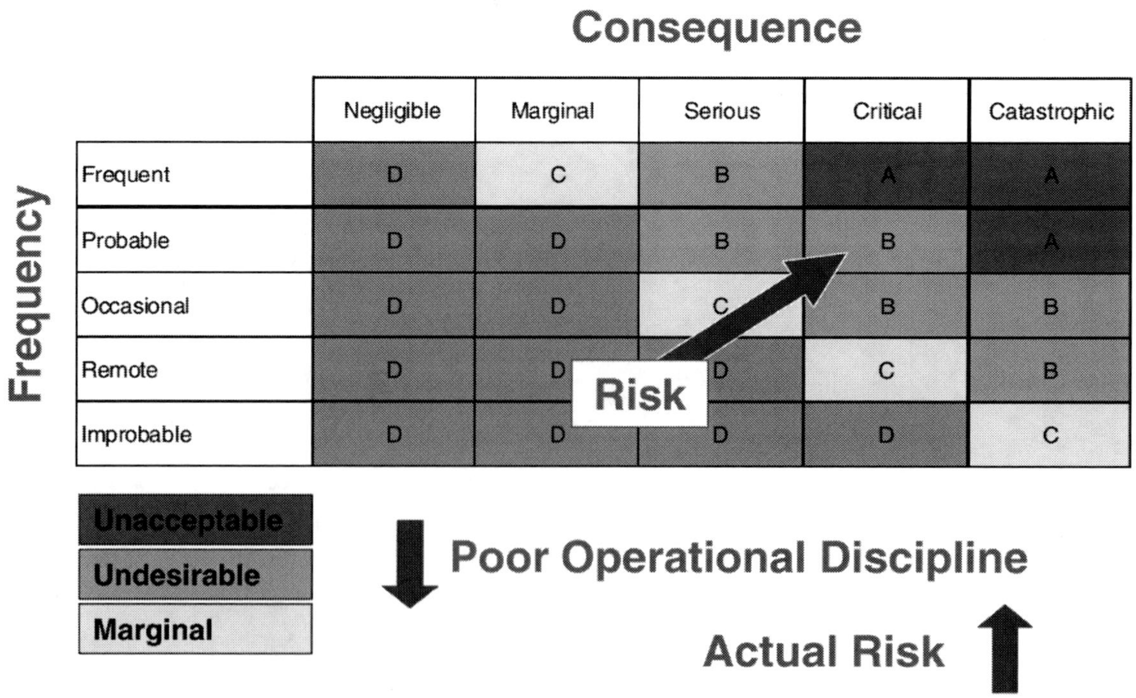

Figure 5. The Effect of OD on the Incident Pyramid

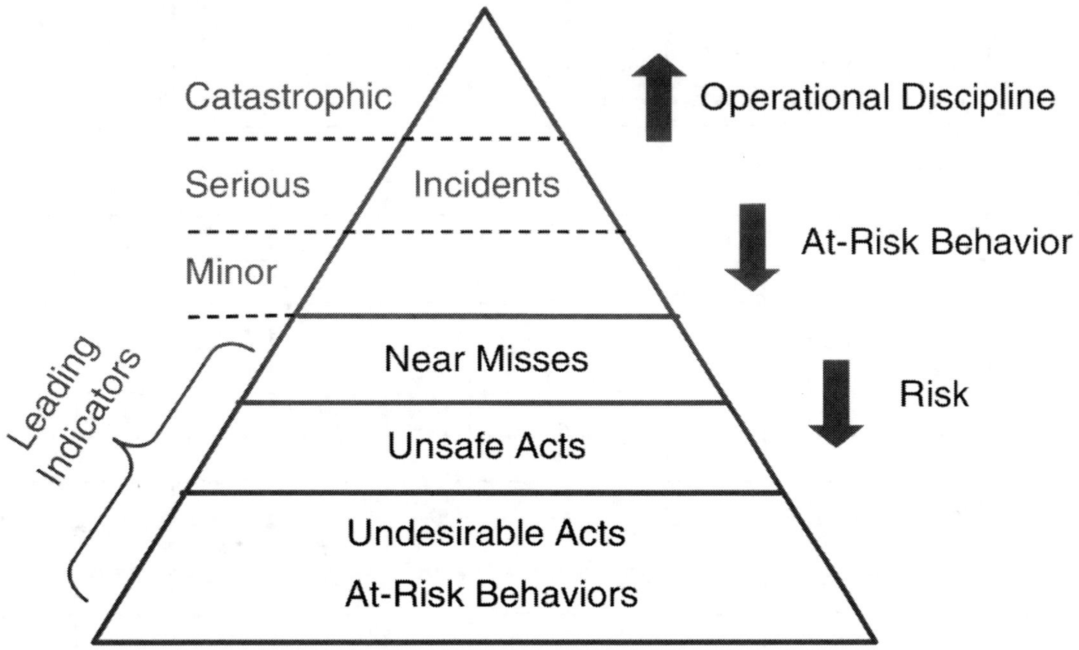

Figure 6. The Link Between the FMO Principles, LOC Incidents and PSM

Incidents

Process Safety Management Element	Principle of Operation							
	Operate Safely	Move to Safety	Seek Help	Ensure Corrections	Report Promptly	Follow Procedures	Address Deviations	Use As Intended
Process Safety Information	X		X				X	
Process Hazards Analysis			X	X		X	X	
Management of Change			X	X		X		
Pre-Startup Safety Reviews			X	X		X		
Mechanical Integrity						X		
Operating Procedures	X					X	X	
Safe Work Practices	X					X		
Training and Performance			X			X		X
Incident Investigation					X	X		
Contractors						X		
Emergency Planning and Response		X				X		
Auditing				X				

Figure 7. The Cabot Incident Investigation Principles Discussion and Action Form

Unrecognized or Unchallenged Hazard	

	Principle of Operation	Adhere to Principle? Yes	n/a	No
1	Always operate equipment and processes within designed safety and environmental limits			
2	Always move to a safe controlled condition OR location when a situation is not understood			
3	etc.			

Shared Learning Discussion from Review of Incident
Discussion:
Action:
Discussion:
Action:

Figure 8. Example of Principles Data Analysis from Incident Investigations

Noted as "Yes" - Following Principle		No - Not Following Principle Need to re-emphasize Principle	
31%	Operate safely	31%	Operate safely
34%	Move to safety	13%	Move to safety
31%	Seek help	13%	Seek help
63%	Ensure corrections	19%	Ensure corrections
78%	Report promptly	9%	Report promptly
34%	Follow procedures	44%	Follow procedures
53%	Address deviations	19%	Address deviations
28%	Use as intended	16%	Use as intended
	Stronger Principles		Weaker Principles
	Prompt reporting		Following procedures
	Ensuring corrections		Operating Safely
	Addressing deviations		

Weaker

OD

Elements

(76c) A Decade of Loss of Containment Lessons – Incident Investigations by the US Chemical Safety Board

Marc G. Sáenz, United States Chemical Safety and Hazard Investigation Board, Washington, DC

The United States Chemical Safety and Hazard Investigation Board (CSB) has investigated about 70 chemical incidents since 1998. Most of the incidents involve loss of containment. Learning from those incidents can prevent similar incidents at other sites. This study reviews the many factors that led to loss of containment and the recommendations to prevent similar incidents. The study relates the factors and recommendations to safety management topics such as the PSM elements. Recognizing those relationships reveals more than a decade of lessons for preventing loss of containment. The study includes all CSB investigations and publications issued to date. The presentation also shows the historical impact of a few incidents to demonstrate the importance of maintaining containment.

See more of this Session: Preventing Loss of Containment Incidents - Beyond Mechanical Integrity
See more of this Group/Topical: Topical 1: Global Congress on Process Safety

Improved Flange Joint Reliability Can Help Prevent Loss of Containment

Tom Sandbrook
DuPont – Fluoroproducts
Parkersburg, WV
Thomas.D.Sandbrook-1@USA.Dupont.com

Mike Emery
DuPont – Protection Technologies
La Place, LA
Michael.B.Emery@USA.Dupont.com

John Ludman
DuPont – Engineering
Wilmington, DE
John.P.Ludman@USA.Dupont.com

Prepared for Presentation at
American Institute of Chemical Engineers
2010 Spring Meeting
6th Global Congress on Process Safety
San Antonio, Texas
March 22-24, 2010

UNPUBLISHED

Improved Flange Joint Reliability Can Help Prevent Loss of Containment

Tom Sandbrook
DuPont – Fluoroproducts
Parkersburg, WV
Thomas.D.Sandbrook-1@USA.Dupont.com

Mike Emery
La Place, LA
DuPont – Protection Technologies

John Ludman
Wilmington, DE
DuPont - Engineering

Abstract

Review of chemical plant process incidents leading to loss of containment shows that Flange Joint Integrity is the single largest contributor of process releases. In response to this data, DuPont has undertaken a holistic approach to improving the reliability of critical process systems. We choose to refer to flange joints as systems, to highlight the complex interactions and functions the various components in a flange joint must perform; failure of an individual element can lead to a Loss of Containment. This paper will review the various components within the flange system, including technical design, development of installation standards, maintenance training, Quality Control, Leak Testing, and Root Cause Failure Analysis.

Historically, attempts at flange joint improvement have examined these components individually. However this approach misses the mark by assuming the flange is a simple device, where in reality, it is a complex interaction of elements, both human and physical. Further, through utilization of Root Cause Failure Analysis, we generate the feedback required to make continuous improvement in Flange Systems which ultimately leads to a reduction of these all too common releases.

Flanged Joint Systems - More Than They Appear

Flanged joint connections (systems) are a critical component of any operating process, for any company. An individual plant may have thousands of these systems in use at any time, performing a critical function- Process Containment- with extremely high reliability, low maintenance, and unfortunately low visibility as an important contributor to the safe operation of the plant. These apparently simple static systems are, in reality, called upon to perform a very complex mechanical function though the dynamic interaction of gasket compression, stud elastic deformation, and flange yielding, while at the same time reacting to the dynamic process conditions, such as pressures and temperatures. In addition to these functions, flanges also are subjected to bending, vibration, and thermal stresses from the attached components. To compound this complexity, flange systems are not easily inspected in service for condition with normal condition assessment techniques. No simple ultrasonic, visual or other methods exist that can reliably predict a flanges continued fitness for service.

Given that flange system condition assessment, without disassembly, is uncertain at best, DuPont has take a position that one of the most effective means of attaining highly reliable flange systems is by ensuring that all the steps required to assemble a properly installed flange system are in place, and that any defects leading to reduction of a flange's ability to perform in the process are being investigated for continuous improvement. This paper will discuss the steps DuPont has initiated to assure that highly reliable flanged joint systems are installed at our plant sites and describe the use of standardized "Why Tree's" to identify and eliminate defects from all flanged assemblies.

The following seven step flowchart describes the major activities, along with the inputs used to accomplish each activity. Correct flange system operation is a combination of technical and human tasks, each requiring clear definition of requirements and responsibilities. While many outside sources of information, such as ASME PCC-1 2000 Guidelines for Bolted Flanged Connections, are used as inputs to the process, none of these are sufficiently robust as to include the entire design-install-operate cycle. Perhaps more importantly these outside guidelines do not include the important function of continuous improvement through root cause failure analysis.

Given that a flange is a complex system, often with very little performance criteria other than it doesn't leak, it is important that the organization clearly define the expected performance specification (i.e., tightness, life cycle, resistance to corrosion etc.). Once this performance expectation is set it is possible to develop the necessary technical, quality, training, and verification standards for use by the organization. Further, once these standards are defined they enable a robust Root Cause Failure Analysis (RCFA) to be conducted for any flanges which fail to meet the expectation. In fact, a RCFA without clearly defined standards to use as the "baseline" is of questionable value. As a point of reference, DuPont has developed internal standards which specifically address each of the seven steps in flange joint make-up. The requirements contained in these standards provide the backbone for the RCFA process used for continuous improvement.

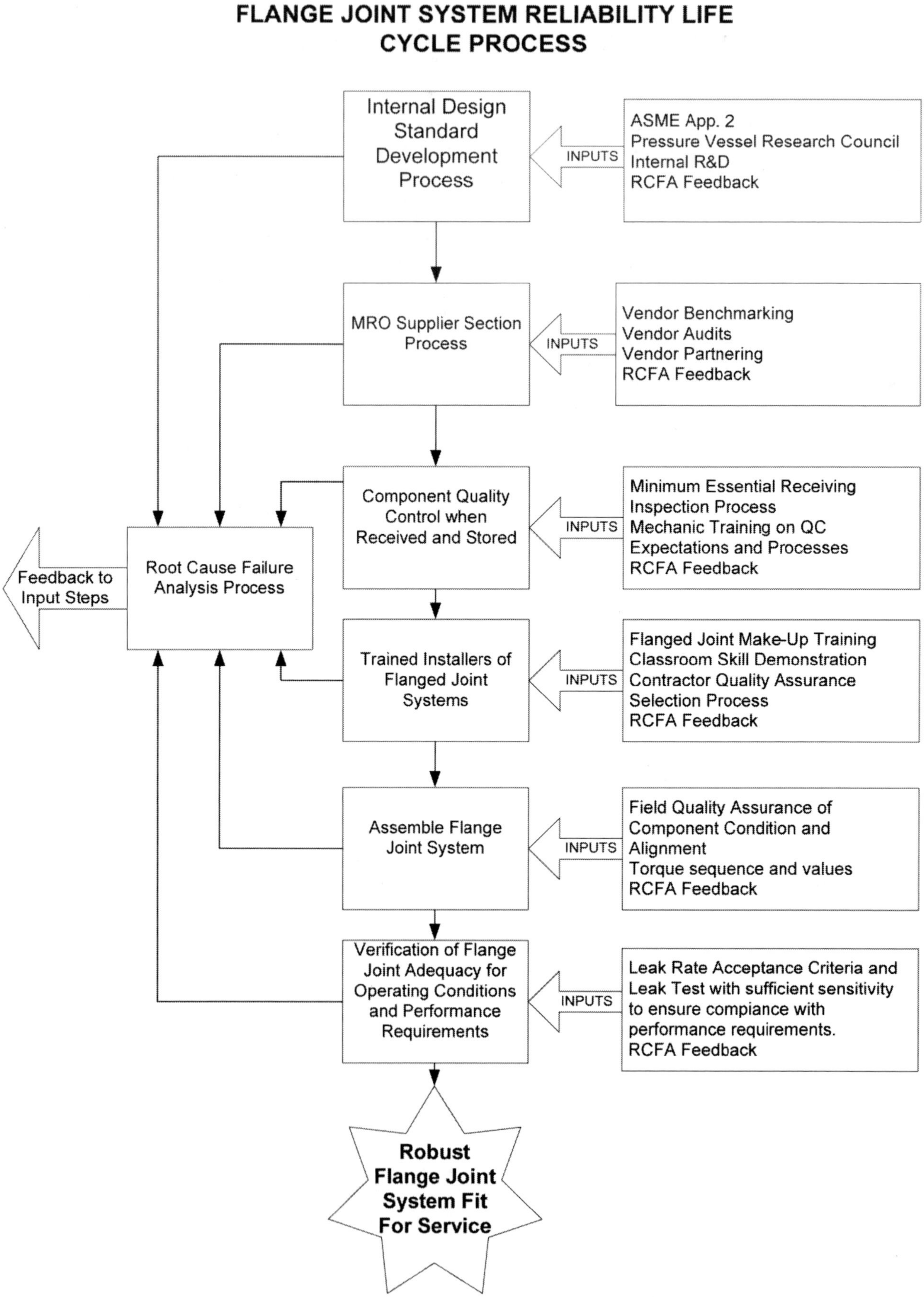
FLANGE JOINT SYSTEM RELIABILITY LIFE CYCLE PROCESS
Internal Design Standard Development Process
INPUTS
ASME App. 2
Pressure Vessel Research Council
Internal R&D
RCFA Feedback
MRO Supplier Section Process
INPUTS
Vendor Benchmarking
Vendor Audits
Vendor Partnering
RCFA Feedback
Component Quality Control when Received and Stored
INPUTS
Minimum Essential Receiving Inspection Process
Mechanic Training on QC Expectations and Processes
RCFA Feedback
Feedback to Input Steps
Root Cause Failure Analysis Process
Trained Installers of Flanged Joint Systems
INPUTS
Flanged Joint Make-Up Training
Classroom Skill Demonstration
Contractor Quality Assurance Selection Process
RCFA Feedback
Assemble Flange Joint System
INPUTS
Field Quality Assurance of Component Condition and Alignment
Torque sequence and values
RCFA Feedback
Verification of Flange Joint Adequacy for Operating Conditions and Performance Requirements
INPUTS
Leak Rate Acceptance Criteria and Leak Test with sufficient sensitivity to ensure compiance with performance requirements.
RCFA Feedback
Robust Flange Joint System Fit For Service

1. Internal Design Standard Development Process

The foundation for achieving a reliably "tight" seal on a flanged assembly is proper design. The bolted flange assembly consists of three primary components; the gasket, fastener, and flange. These components work together as a system to accomplish two primary goals 1) retaining the hydraulic end loads due to pressure and 2) creating a "leak tight" boundary between the process and the environment. Most of the bolted flange connections (BFC's) in the United States chemical industry today, use design criteria from the ASME B31.3[1] piping code or the ASME Boiler and Pressure Vessel Code[2]. Although these standards ensure the components can withstand a given design pressure and temperature, they give little guidance for establishing the tightness of the BFC and provide no consideration for designing the BFC with a design margin against relaxation.

A serious discussion on BFC design should ask the following questions:

1) What level of tightness is required?
2) How do I manage any expected relaxation (thermal or mechanical) of the flange system?

After the required level of tightness (considering any needed margin) is established a proper review of design options becomes possible.

Tightness Considerations

There are many design choices that impact the level of the tightness in a pipe or vessel flange. In general the tightness of the BFC is increased as the preload is increased (provided the mechanical properties of the gasket are not compromised). Unfortunately the preload available to BFC's is often limited. For example, many process piping flanges are specified using design standards (like ASME B16.5[3]). The design criteria of these flanges are primarily based on pressure and temperature considerations, not tightness. The tightness of the BFC can vary dramatically depending on options available within the design standard. Flange type (welding neck vs lap joint), bolting specification (low strength vs high strength), and gasket type all can have an affect on the maximum level of tightness a BFC can obtain. Choosing a robust flange design like the welding neck, and providing higher strength bolting, allows the user to increase the preload resulting in improved tightness.

When selecting a gasket for critical services one should consider performance based gasket tests such as the ROTT[4] (ROom Temperature Tightness Test) or EN 13555[5] , that estimate the level of tightness that can be expected as the gaskets is arebeing compressed and how the tightness decreases as the preload decreases. This information is especially importantwhen selecting a gasket when limited preload is available (i.e. Class 150 lap joint flange). The ASME Pressure Vessel Code (Section VIII-Div 1) has a "Special Working Group on Bolted Flange Joints" working on the development of design rules that consider leakage as related to gasket tightness in the flange design[6]. The premise of the design rules is to determine the preload required to achieve a specified level of tightness of the gasket. This is primarily a function of the design pressure and the characteristics of the gasket (from ROTT test). Once the preload requirements are known, the flange and bolting system is designed to withstand this preload. In cases where a high level of tightness is required it is likely the flange and bolting system will be much more robust (thicker flange and more bolt area) than if no tightness specification was made.

Relaxation Considerations

The flange system design must include provision for managing pre-load relaxation of the fasteners/bolts. The fastener functions as a spring. When tightened it stretches to induce the clamping force on the flanges and gasket. The preload is directly proportional to the amount the fastener is stretched. There are several sources of pre-load relaxation including differential thermal expansion (between fastener and flange), mechanical relaxation of the fasteners over time (possibly due to vibration), and gasket relaxation. As the pre-load is reduced, the tightness of the BFC is reduced. Flange systems that are susceptible to pre-load relaxation need to be designed with sufficient elasticity to effectively compensate for the reduced stretch in the bolts. The designer can either mitigate the extent of relaxation or reduce the amount the preload is reduced for a given relaxation.

Designs for mitigating the extent of relaxation depends on the source. Relaxation induced by gasket creep can be improved through alternative materials. Mechanical relaxation of the fasteners due to vibration can be mitigated by using a locking device (locknut, thread-locking compounds). Relaxation caused by differential thermal expansion can be mitigated by using materials with similar thermal expansion coefficients or keeping the temperatures of the flanges and fasteners close.

The preload loss after assembly can be reduced through addition of spring washers to the fastening system. Properly selected spring washers reduce the spring constant of the fastening system, and as a result the change of gasket load as relaxation occurs is dramatically reduced. Spring washers can also employed to prevent overload of fasteners due to thermal expansion.

Whether the design implements provisions for mitigating the relaxation or the design manages the relaxation through improved elasticity, the possibility of the preload loss due to relaxation needs to be addressed.

2. MRO (Maintenance Repairs and Operation) Materials Supplier Selection Process

The foundation for a robust flanged joint system is quality materials. In today's competitive market place it is incumbent upon the consumer to define the required level of quality, and to hold their suppliers accountable to those standards with an audit process. By following an audit process one can determine if a supplier has:

- Adequate quality systems and is performing adequately against those systems.
- Technical competency to produce and/or supply the specific commodity
- Sufficient process consistency to meet the first two objectives

A business should only deal with suppliers that have been found to be satisfactory through auditing, either at the corporate level or at a local/site level. Auditing provides direct and current information about activities at suppliers' shops or distribution centers.

The audit process should allow for items to be identified as "opportunities for improvement" and defines an acceptable closure process for the audit organization. When arranging an audit, it is important to ensure the supplier's management will be in attendance. Key personnel who can effect change, should it be necessary, must also be present. Suppliers must be open to change if needed or a decision may be required to deal with another supplier.

MRO Supplier audits should be an evergreen process based on audit results and actual supplier performance. The audit process itself may have to change in order to better screen for systemic problems which manifest themselves as defects in supplied components.

Key points regarding the MRO Sourcing Process include

1. Improperly described MRO materials and spare parts allow inadequate or inappropriate supplies to enter the plant.
2. Materials clerks or planners and other personnel who enter parts descriptions into electronic ordering systems require training to ensure the accuracy of their work.
3. Using approved or qualified suppliers provides the best assurance of quality needs being achieved.
4. Auditing suppliers is the only effective way of assessing the quality program of suppliers.
5. Perform receiving inspection with trained personnel
6. Deal with non-conformances following established guidelines.
7. Ensure adequate storage to protect parts before use.
8. Perform field audit before use.

3. Component Quality Control When Received and Stored

Nonconforming MRO materials such as valves, pipe fittings, gaskets, and fasteners may cause process safety incidents, rework, and add to internal costs. MRO material supply chains consists of a series of handoffs between different functional groups, making it difficult for one person at any point in the supply chain to track every issue if solely responsible for reporting or communicating incidents. A system for communicating issues between supplier and user coupled with arbitration by neutral parties who represent commodity procurement and technology issues is fair and beneficial to all.

A user complaint system should be implemented to communicate identified non-conformances back to the supplier to solicit an analysis of the systemic cause. Corrective actions must be communicated back to the end user to build confidence that the problem will not recur.

MRO materials should be stored in an appropriate environment with the each item properly identified, traceable and segregated. A special problem that often happens is "pick errors," which occur when nearly identical items (for example low and high strength fasteners) are stored next to each other and are confused or mixed during stocking or disbursement. Even though storing related items in the same general areas can help keep a storage room organized, plant personnel should always be concerned about storing items immediately next to each other if they are difficult to tell apart.

Defective, degraded or substituted parts identified in the field during the attempted flange assembly step should not only be returned, but rather the underlying causes for the defect investigated.

4. Installation Training

Due to the apparent "simple" design and function of a flanged connection, it is often assumed that anyone with a wrench and some time has the necessary knowledge to properly make-up a flanged connection. It is this belief, that little specialized knowledge is required for this work, is a significant contributor to the less than desired reliability of these important connections. A properly made flange connection is the result of many detailed activities, all of which must work together when called upon, to contain the process fluid. The only way to properly convey this complexity is through a rigorous training program, not only on how to properly make a connection, but also why all the steps- design, material selection, parts storage, installation quality control, verification, and Root Cause Failure Analysis-are important for the successful installation of a flanged connection. Obviously, operators and mechanics, the people normally doing this work require this training, but equally important, and often overlooked, are the supporting staff, such as supervisors and engineers, those who set the expectations on the importance of robust (i.e. Leak Free) flanges, should also be fully trained. It is also important that the people who make-up the flange also participate in the verification step, the final leak test, as a means of immediate feedback on the results of their work

The need for workforce trained on Flange Joint Make-Up has never been more important. Today's equipment is more complex, requiring a greater degree of system knowledge than in the past. Leak testing criteria and fugitive emission monitoring requirements are increasing in sensitivity and miniscule leaks are not tolerated, nor should they be. Some of the available standards for this important task are available, such as ASME Standard PCC-1 "Guidelines for Pressure Boundary Bolted Flange Joint Assembly" give recommendations for procedure qualification. The Materials Testing Institute (MTI) has a created a standardized joint assembly procedure (based on ASME PCC-1-2001), which includes a demonstration test demonstrating that the mechanic can take the bolts to the target stress.

Qualification

Proper assembly of a bolted flange joint requires an understanding of the procedures. Demonstration of these skills may be required to qualify the mechanic. Qualification may include an understanding of selecting the proper torque value and required lubrication, knowledge of flange inspection criteria, gasket and fastener identification, and demonstration that that they can tighten bolts to a specified procedure and achieve a target bolt stress.

ASME Standard PCC-1 "Guidelines for Pressure Boundary Bolted Flange Joint Assembly" gives recommendations for procedure qualification. The Materials Testing Institute (MTI) has a created a standardized joint assembly procedure (based on ASME PCC-1-2001, which includes a demonstration test showing that the mechanic can take the bolts to the target stress. MTI Report R-5 can be used by Sites to qualify their mechanics.

Develop Competency Evaluation System
The number of hours of training or type of training is not as important as the ability of the employee to demonstrate how to independently perform the task properly. An evaluation of the employee's competency to perform the task is mandatory. The evaluation of the employee's competency is best done by his or her performance of the task with a knowledgeable person watching. The competency evaluation can be done using a simulated task or can be done by watching the person do a real task. For tasks that are difficult to simulate and are done only rarely, alternate means, such as detailed interview, might be appropriate.

5. Flange Assembly Process and Quality Control

Reliable "leak tight" installation of a bolted flange connection requires the maintenance organization to have systems in place to control the assembly process. Training requirements, material receiving inspections, calibration of assembly tools, and the establishment of detailed assembly procedures (including target preload) are all essential. A recent study by Garlock Sealing Technologies illustrates the importance of the installation process. Their study revealed 82 percent of the gaskets returned for analysis (after the joint leaked) showed signs of problems with the gasket compression generated during installation[7]. Gasket compression issues can be related to non-uniform loading, inadequate preload or excessive preload. In each case, these problems can be mitigated by controlling the flange assembly process.

A quality control system ensures processes and procedures are followed, correct materials are used, and only qualified personnel perform the work. In the case of the flange assembly process, a check list referencing detailed procedures can be an effective method. The check list ensures that all aspects of the assembly procedure have been followed and there is traceability of the work processes. When defects are identified, the underlying systemic causes should be investigated and remedied (in addition to rectifying the problem).

Flange assembly procedures need to include mechanic qualification requirements, inspection requirements, tightening sequence (including provisions for re-torque), and target preload. A brief description of each follows.

Qualification Requirements

ASME Standard PCC-1 "Guideline for Pressure Boundary Bolted Flange Joint Assembly"[8] gives recommendations for procedure and personnel qualification. Training and subsequent qualification should include a combination of classroom and field demonstration. The content of the training should include all key aspects of the procedure.

Inspection Requirements

There are two main types of inspections that need to be performed during assembly of the bolted flange connection; 1) materials verification (gasket, fastener, lubricant) 2) condition assessment of flange system.

Materials verification can be critical to the installation. The installation of low strength fasteners when high strength fasteners are specified, will likely result in the low strength fasteners being stretched beyond the elastic limit. Likewise installation of the wrong gasket material can result in a material compatibility issue. In both cases a large scale leak can result.

Inspection by the end user needs to occur in two phases; : receipt inspection and condition assessment. Receipt inspection is important to ensure that procured materials are in fact adequate to support future work. Inspection at the time of assembly ensures that an unsafe condition doesn't result from use of inadequate components. So it is critical that the people assembling the flange system are familiar with the materials required and can positively identify materials that do not meet the required specification or are degraded.

Condition assessment of a flange installation ensures the installation still meet the design basis assumptions. Tolerances for flange surface finish, flange thickness, thread engagement, thread condition and alignment are given by standards like ASME B16.5.and ASME PCC-1 and should be referenced by the flange assembly procedure. These observations should be noted and corrected as needed.

Tightening Sequence

Uniform compression of the gasket is essential to long term tightness. Since a majority of flanged connections are tightened 1 or 2 bolts at a time it is not always possible to apply uniform loading. As each bolt is tightened and the section of flange is brought together, another section of flange and accompanying bolt loosens, an effect known as "cross-talk". In order to manage this "cross-talk", the industry has established a best practice for tightening bolts, which includes tightening the bolts using a star-bolting pattern in one-third increments followed by a rotational pattern until no additional nut rotation occurs. The sequence in which bolts or studs are tightened has a substantial effect on the distribution of the gasket compression[7]. In addition to proper torque sequence the tightening sequence should include provisions for re-torque to accommodate any thermal, mechanical or gasket relaxation

Target Preload

There are several methods currently used today to establish target preloads. In many cases a "rule of thumb" approach is applied in which the bolts are stressed to a given percentage of yield. Slightly more sophisticated approaches include consideration of gasket stress in addition to the bolt stress. In this method the preload ensures a minimum gasket stress (given from gasket manufacturer) is provided to achieve a specified level of tightness. In some cases the preload required to achieve this level of tightness can result in excessive flange stress and deflection, especially in lap joint style flanges or fragile flanges (glass-lined steel, reinforced plastic)

The optimal approach is to determine the maximum preload the system can handle without compromising the flange, gasket or fastener. This maximum preload provides the tightest seal and the largest margin against relaxation. DuPont has established torque values, based on this criteria, for Class 150 and Class 300 Welding Neck and Lap Joint Flanges. Similar guidance needs to be established for ASME Pressure Vessel Flanges. As mentioned earlier the ASME Boiler and Pressure Vessel Code does not provide guidance for recommended preload.

6. Verification of Flange Joint Integrity

The final element to consider prior to placing a system in service is the Quality Assurance of the completed flange joint assembly. Regardless of the care taken in the previous steps one can not be certain the design intent will be met without a leak test. Ideally, flange joint systems in process equipment and piping would be free of leaks. In reality, many types of joints are permitted some degree of leakage. An allowable leak rate is a rate that allows safe, legal and efficient operation. The determination of an acceptance criterion for the quality assurance of the assembled connection needs to be based on the hazards of the process. This leak rate acceptance criterion should be documented as part of the plant's overall MIQA program. It is important to note that while the relevant codes (notably ASME Section V Article 10- Leak Testing) refer to leak testing; none of them provide specific guidance on acceptance criteria. Further, these codes typically apply to new construction and repair work, they are not intended for routine maintenance of flange joint connections. It is up to the Owner/User to define the method, test medium, test pressure and acceptance criteria based on the unique operating conditions, and hazards, of their process. Of course, the people performing the leak testing need to be trained and certified in the method being used for the final acceptance QA.

While there are many different leak test methods, they vary widely in sensitivity and complexity. It is important that the leak test method selected is sufficiently sensitive as to detect your previously determined leak acceptance criteria, but not excessively complex as to make them impractical in the field. By comparing costs versus capabilities, it is possible to accurately assess a flange's leak rate against your previously determined permissible values to permit a go/no-go decision to be made. Advances in instrumentation sensitivity and reduction in equipment pricing, has made the traditionally "advanced" techniques very practical for use in most plant applications. In actuality, it is common that several leak tests will be used when testing systems to gain advantage of each method at the appropriate time in the flange joint's Quality Assurance testing.

7. Defect Elimination through Root Cause Failure Analysis

There are a few commonly held beliefs which may stand in the way of defect elimination from bolted flanged joints including:

- Flanges are common, simple components which do not typically merit engineering attention.
- Flange leaks, although typically small, are common and inevitable.

Neither of these beliefs is well founded:

- Bolted flanged connections are complex systems involving a host of esoteric design issues
- By identifying the causes of the leaks and addressing them....instead of accepting them....plant performance can be improved over time. Many defects, such as many design defects, can be identified and eliminated permanently. Other defects, such as those introduced by human errors, can be greatly reduced...if not eliminated.

Root cause failure analysis is a widely accepted approach to defect elimination which can be applied to flange leaks. Due to the complexity of bolted flanged joints some investigation teams may need to include resources with specific knowledge of flange design.

Careful collection of evidence before and during flange disassembly must be conducted. Otherwise valuable clues may be lost forever, and later failure diagnosis may be ill-founded. A checklist of desired observations can be helpful in ensuring pertinent observations are made.

<u>Potential Failure Causes for Consideration:</u>

<u>Design issues</u>

- Problems can occur even if standard ASME B16.5[3] flanges are used. For example, in the case of Class 150 lap joint flanges used with spiral wound gaskets.
- Adequacy of custom designed flanges is much less certain.
    - Stipulation of target bolt loads which effectively create a seal while not distorting the flanges, must be considered a necessary product of the design effort.
    - Design may need to be reviewed and target load calculated, if evidence of appropriate review (i.e., calculations regarding engineering basis) do not exist

<u>Component problems</u>

- Are the flanges deformed or degraded?
- Is the gasket the wrong type, material or dimensions?
    - Adequacy assessment needs to consider mechanical properties as well as process compatibility
- Are the fasteners the right type?
- Are the fasteners in good condition?
    - Integrity to withstand load
    - ability to predictably generate load (normal friction)

<u>Assembly problems</u>

- Flanges out of alignment
- Fasteners under, over or unevenly loaded
- Gasket installed out of position

<u>Exceeding Design Capabilities</u>

- Pressure
- Temperature

- Externally applied loads excessive
    - Thermal expansion
    - Supplemental weights and moments

The ultimate goal of the analysis being the generation of systemic remedies such as institutionalization of systems and tools to:

- control quality of incoming parts
- ensure adequacy of assembly
- ensure adequate design of new flanges

The systemic upgrades will not only prevent recurrence of leakage from the flange under review, but from a host of others similarly installed

References

[1] ASME Code for Pressure Piping ASME B31.3-2008
[2] ASME Boiler and Pressure Vessel Code, 2007
[3] Pipe Flanges and Flange Fittings ASME B16.5-2009
[4] WRC Bulletin 427
[5] BS EN 13555:2004 Flanges and their Joints. "Gasket Parameters and Test Procedures relevant to the Design Rules for Gasketed Circular Flange Connections"
[6] Robert Mikitka, PVP-Vol 419 Pressure Vessel and Piping Codes and Standards-ASME 2001, Alternative ASME Rules for Bolted Flanged Joints Based on Tightness
[7] Jim Drago P.E., Jim Heffron, and Dave Burgess, Pumps and Systems, "Proper Installation of a Bolted Joint Gasket", May 2008
[8] ASME PCC-1 2000 Guidelines for Pressure Boundary Bolted Flange Joint Assembly

The Hazards of Thermal Expansion

Juan C. Ramirez,
Exponent Failure Analysis Associates,
1011 Warrenville Road, Lisle, IL 60532,
(630)-743-7716, jramirez@exponent.com

Russell A. Ogle, Andrew R. Carpenter, Delmar "Trey" Morrison
Exponent Failure Analysis Associates,
1011 Warrenville Road, Lisle, IL 60532,
rogle@exponent.com, acarpenter@exponent.com, tmorrison@exponent.com

Prepared for Presentation at
American Institute of Chemical Engineers
2010 Spring Meeting
6th Global Congress on Process Safety
San Antonio, Texas
March 22-24, 2010

UNPUBLISHED

The Hazards of Thermal Expansion

Juan C. Ramirez,
Exponent Failure Analysis Associates,
1011 Warrenville Road, Lisle, IL 60532,
(630)-743-7716, jramirez@exponent.com

Russell A. Ogle, Andrew R. Carpenter, Delmar "Trey" Morrison
Exponent Failure Analysis Associates,
1011 Warrenville Road, Lisle, IL 60532,
rogle@exponent.com, acarpenter@exponent.com, tmorrison@exponent.com

Keywords: thermal expansion, liquid, incompressible, expansion coefficient

Abstract

We examine the sometimes overlooked hazards posed by thermal expansion of trapped liquids in piping systems. Trapped liquids can generate large pressure increases with only modest temperature changes. This phenomenon is well understood by the process safety community. However, it is also a phenomenon that is easily overlooked or dismissed as being of small consequence.

This paper discusses a simple model for estimating the overpressure hazard from thermal expansion and illustrates its use with some typical liquids encountered in heat transfer systems. The control of these overpressure hazards is discussed in terms of applicable safety guidelines and standards. We then illustrate the hazard of heating trapped liquids with two case studies; one was a heating system and the other was a refrigeration system. We show that the basic cause of these catastrophic failures could have been avoided by following applicable safety standards in conjunction with a process hazard analysis during the design phase. Our findings underscore two important design guidelines: 1) install a pressure relief device wherever liquid can be trapped and subjected to heating; and 2) verify that each component of the piping system that can be isolated from pressure relief satisfies the design pressure.

1. Introduction

The heating of trapped liquids can lead to the catastrophic failure of piping or process equipment. Although the hazard is well-known, it can be easily overlooked in the absence of a systematic process hazard analysis. It is a relevant hazard scenario for process systems in which the system temperature is different from the ambient temperature or where ambient temperature changes can have significant effects. Thus it can be an especially significant hazard in process systems involving either heating or refrigeration. A loss of containment of process fluid caused by thermal expansion overpressure may seem like a *de minimis* risk when compared to hazard scenarios like fire exposure or runaway chemical reactions. But the loss of containment of a process fluid can lead to a catastrophic loss if it is a high hazard material. Examples of this situation include a combustible liquid heated near or above its flashpoint, a liquid heated above its normal boiling point, or a volatile toxic liquid. These catastrophic incidents can cause flash fires, explosions, and toxic releases with all the related casualties and property damages.

This paper is divided into five sections. First we present a thermodynamic model for evaluating the potential magnitude of the overpressure hazard posed by heating a trapped process liquid. Next we describe safeguards for controlling the overpressure hazard and identify several design standards and guidelines that document good engineering practices for implementing these safeguards. This is followed by a presentation of two case studies of explosions caused by heating trapped liquids. Finally, we offer some lessons learned that may assist the engineer in performing a process hazard analysis on systems that involve the heating of process liquids.

2. Thermal Expansion Hazards

Liquids contained within process vessels and piping will normally expand when heated. The expansion will damage pipes and vessels if the pipe or vessel is filled completely with fluid and the liquid is blocked in. Therefore, trapped liquid subjected to heating is an example of hazardous energy storage. This hazard is specifically indicated in the safety literature as the potential for excess pressure to develop due to thermal expansion or vaporization of trapped liquid [1,2].

The thermal expansion of liquids is a phenomenon with which we are all exposed due to its presence in everyday life. For instance, the mercury thermometer's operation is based on this principle. Similarly, a common recommendation for automobile drivers to help save fuel is not to overfill the tank, because this might cause the fuel to overflow later due to thermal expansion [3]. The potential for expansion of liquid propane, for example in 20-pound gas grill cylinders, has also resulted in the legal requirement for overfill protection devices to limit the likelihood for cylinder releases. The phenomenon of thermal expansion applies to fluids that are liquids at normal conditions as well as liquefied gases.

From the thermodynamics of fluids in single-phase regions, pressure, P and temperature, T are independent and we can think of the specific volume, v as being a function of these two, $v = v(T, p)$. The differential of such a function is

$$dv = \left(\frac{\partial v}{\partial T}\right)_P dT + \left(\frac{\partial v}{\partial P}\right)_T dP \tag{1}$$

Two thermodynamic properties related to the partial derivatives appearing in the differential are the coefficient of thermal expansion β

$$\beta = \frac{1}{v}\left(\frac{\partial v}{\partial T}\right)_P \tag{2}$$

and the isothermal compressibility coefficient κ

$$\kappa = -\frac{1}{v}\left(\frac{\partial v}{\partial P}\right)_T \tag{3}$$

The unit for β is the reciprocal of that for temperature and the unit for κ is the reciprocal of that for pressure. The volume expansivity is an indication of the change in volume that occurs when the temperature changes while pressure remains constant. The isothermal compressibility is an indication of the change in volume that takes place when pressure changes while temperature remains constant [4].

For the liquid phase, the constant temperature lines on a $P - v$ diagram and the constant pressure lines on a $T - v$ diagram are very steep and closely spaced, thus both $(\partial v/\partial T)_P$ and $(\partial v/\partial P)_T$ and hence both β and κ are small. This characteristic behavior of liquids – far from the critical point – leads to the idealization of the incompressible fluid, for which both β and κ are zero. For real liquids, β and κ are weak functions of temperature and pressure, so for small changes in P and T little error is introduced if we regard them as constant [5].

Note that, for constant volume (∂v = 0, as in a rigid container or pipe) equations (1) through (3) can be combined and integrated to yield:

$$\Delta P = \frac{\beta}{\kappa}\,\Delta T \tag{4}$$

where ΔP is the pressure increase induced by the temperature increase ΔT. Thus, Equation (4) can be used to estimate the pressure increase with temperature of a liquid when trapped inside a fixed volume, rigid container. This calculation is only an estimate since, at smallΔT, real piping and pressure vessels will undergo elastic deformation which will relieve some of the overpressure caused by thermal expansion. As ΔT increases, however, the overpressure can easily reach the burst pressure of the component. The burst pressure for an individual vessel or component can be estimated using various methods, which won't be covered here. Calculations and tests are described in the pressure vessel literature and the ASME pressure vessel design code (e.g., see Table 2).

Compressibility and expansion coefficients are tabulated for some fluids, but most often these parameters must be calculated from thermodynamic state (P, v, T) diagrams. Figure 1 is a schematic representation of a typical pressure (P) – enthalpy (h) diagram focusing on the region left of saturation dome, the compressed (or subcooled) liquid region.

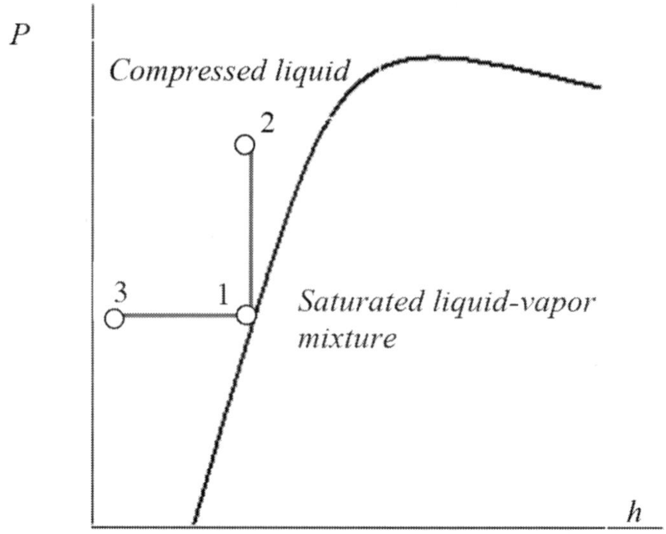

Fig 1. Graphical estimation of isothermal compressibility and thermal expansion coefficients

Under the assumption that β and κ are approximately constant, the partial derivatives in equations (2) and (3) can be estimated with finite differences. Note that in Figure 1, states 1 and 2 are at the same temperature, and states 1 and 3 are at the same pressure. Therefore:

$$\kappa \approx \frac{-1}{v_1}\left(\frac{v_1 - v_2}{P_1 - P_2}\right) \tag{5}$$

and

$$\beta \approx \frac{1}{v_1}\left(\frac{v_1 - v_3}{T_1 - T_3}\right) \tag{6}$$

where the subscript refers to the state at which the thermodynamic property is evaluated.
Table 1 summarizes the thermal expansion and isothermal compressibility coefficients for several typical process fluids or refrigerants.

Table 1. Isothermal compressibility and thermal expansion coefficients for select liquids

Liquid	ΔT range, °C	ΔP range, bar gauge	β, °C^{-1}	κ, bar^{-1}	β/κ, bar/°C	Reference
Water	20 to 80	0-2000	$4.5 \cdot 10^{-4}$	$3.64 \cdot 10^{-5}$	12.4	6
Fuel oil	27 to 44	28 to 165	N/A	N/A	9.3	7
Silicone oil	25 to 150	0 to 55	10^{-3}	$1.5 \cdot 10^{-4}$	6.7	8
Carbon dioxide	-40 to -18	28 to 69	$4.0 \cdot 10^{-3}$	$4 \cdot 10^{-4}$	100	9
Propane	-183 to 27	10 to 30	$1.56 \cdot 10^{-3}$	$6.18 \cdot 10^{-4}$	2.5	10
R-134a	-60 to 0	3 to 40	$2 \cdot 10^{-3}$	$3.09 \cdot 10^{-4}$	6.6	9
Ammonia	0 to 60	26 to 50	$2.47 \cdot 10^{-3}$	$3.7 \cdot 10^{-4}$	6.8	9

Figure 2 shows the pressure increase as a function of the temperature increase in a rigid container for the 7 liquids considered. The slope of each line is the ratio β/κ.

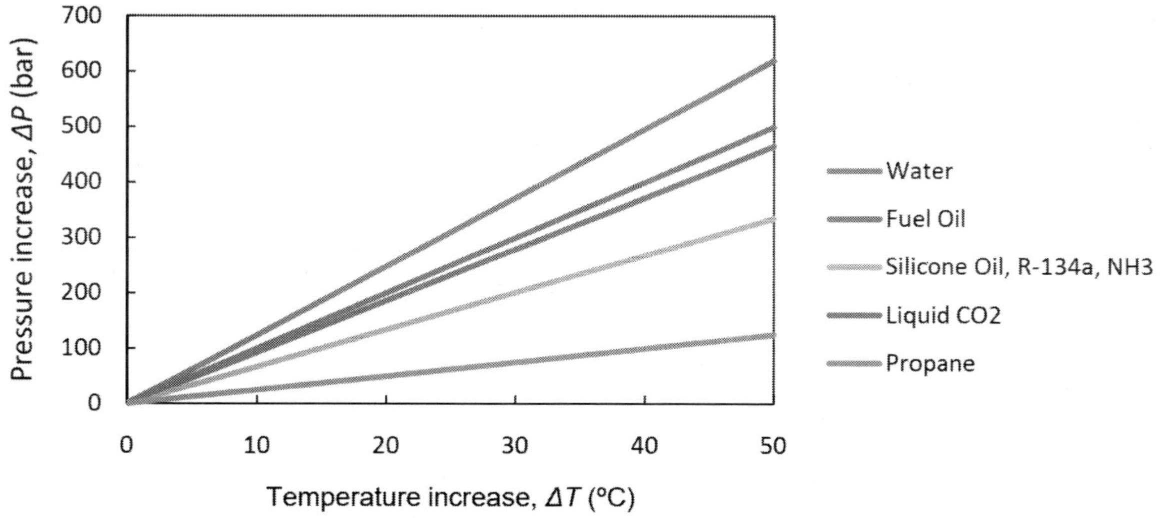

Fig 2. Pressure increase in a rigid container vs liquid temperature increase

3. Thermal Expansion Safeguards

The hazard presented by the heating of a trapped liquid is overpressure. If the overpressure generated by thermal expansion exceeds the burst pressure of its container, then a release of the pressurized liquid will occur. The significance of a loss of containment accident depends on the magnitude of its consequence. If the process fluid is a high hazard material, then appropriate safeguards should be implemented. A high hazard material as considered here poses a significant hazard to health, safety, and surrounding equipment if suddenly released due to vessel failure. Depending upon the mode of failure for the vessel (e.g., leak versus catastrophic/fragmenting failure), the nature of the fluid (e.g., flammable, above its boiling point, and/or toxic), and the surroundings (e.g., inside an occupied building or adjacent to other high hazard process units), the consequences of overpressure-induced vessel failure can be extreme.

The discipline of system safety suggests that hazard control is best considered in terms of a hierarchy of safeguards. In order of decreasing effectiveness, the system safety hierarchy is

- Eliminate the hazard
- Use passive engineering controls to manage the hazard
- Use active engineering controls to manage the hazard
- Manage the hazard with warnings and procedures

Depending on the severity of the consequences, it may be best to manage an overpressure hazard with multiple safeguards.

There are several design standards and guidelines that contain requirements for overpressure protection. These documents tend to emphasize two strategies for hazard control, pressure containment (passive engineering control) and pressure relief (active engineering control). In essence, the recommended practice is this: if a liquid can be trapped and heated in a section of piping or within a container, then the designer must either provide a means for pressure relief or he must design that component to withstand the greatest pressure that can be developed. Table 2 is a partial listing of references for recommended practices.

Table 2. Recommended practices for overpressure protection from thermal expansion

Component	Design Standard or Guideline
Pressure vessels	ASME *Boiler and Pressure Vessel Code*
Process piping	ASME *Code for Pressure Piping B31.3 Process Piping*
Refrigeration piping	ASME *Code for Pressure Piping B31.5 Refrigeration Piping*
Refrigeration systems	ASHRAE 15 *Safety Standard for Refrigeration Systems*
Pressure relief design requirements	CCPS *Guidelines for Pressure Relief and Effluent Handling Systems* API 521 *Guide for Pressure-Relieving and Depressuring Systems*
Propane systems	NFPA 58 *Standard for the Storage and Handling of Liquefied Petroleum Gases*
Ammonia refrigeration systems	IIAR 2 *Equipment, Design, and Installation of Ammonia Mechanical Refrigerating Systems* IIAR 116 *Avoiding Component Failure in Industrial Refrigeration Systems Caused by Abnormal Pressure or Shock*
Carbon dioxide systems	CGA G-6 *Carbon Dioxide* CGA G-6.1 *Standard for Insulated Liquid Carbon Dioxide Systems at Consumer Sites*

4. Case Studies

In this section we present synopses two case studies that have resulted in serious injuries and/or significant business interruption. The first case involved an industrial oven (heating system) which used a heat transfer oil and the second case involved a custom-made freeze dryer for a pharmaceutical application in which nitrogen vapor cooled a stream of silicone oil in a heat exchanger (refrigeration system).

4.1 Case study #1: Industrial Oven

An industrial oven for a food processing application used a heat transfer oil (HTO) as the heat source. The oven was comprised of several contiguous zones and each zone was connected to a plant-wide HTO network via supply and return pipes. Figure 3 shows a schematic representation of this system.

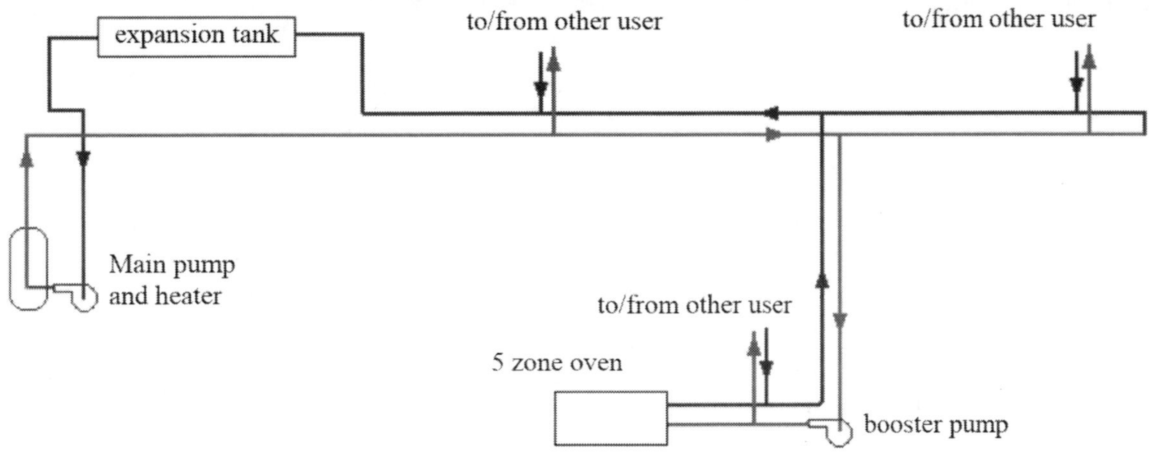

Fig. 3. Schematic representation of plant-wide heat transfer oil system

A pump circulates the thermal fluid through a natural gas fired heater and on to the rest of the thermal fluid system. The heated fluid is distributed across the plant to several users, one of which is the 5-zone oven. After circulating through the oven and other users, the thermal fluid is returned to the heaters, passing through an expansion tank along the way.

The temperature of each zone is controlled by independent flow control valves for each zone. It is possible to operate the oven with each zone operating at different temperatures, or while some zones are off. Manual isolation valves were installed on both the upstream and downstream side of the control valves to allow for maintenance of the valves. Additionally, the facility installed manual isolation valves at the supply and return of each oven zone, as well as a drain valve to drain the HTO from individual zones.

One the day prior to the incident, zones 2 and 3 were manually isolated from the oil network by a contractor to repair a minor leak. The oven was started the following day, and the pump was activated. However, the manual isolation valves were not reopened following the previous day's maintenance activities. The operator was unaware that the piping for zones 2 and 3 contained trapped liquid. The temperature in three of the five zones increased as he expected, but temperatures in the isolated zones rose at a much slower rate than normal.

The operator determined that the zones must still be isolated and dispatched a separate operator to open the isolation valves for the two zones. Immediately after the isolation valves were opened, the primary operator observed a pool of HTO flowing out from one of the previously isolated zones. While the fluid was trapped, the heat from the adjacent zones heated the HTO resulting in thermal expansion of the oil which caused a rupture in the internal piping. The fresh, hot oil spilled out from the rupture and eventually ignited resulting in an explosion and fire which caused minor injuries to facility personnel, structural damage to the facility and a significant interruption to operations. During the accident investigation it was determined that the ruptured piping component had a lower burst pressure than the surrounding portions of the piping system. The weakest point in the piping system happened to be in the oven, thus the high consequence of the release.

4.2 Case study #2: Cryogenic refrigeration system

Lyophilization, also referred to as freeze-drying, is a process of removing moisture from a pharmaceutical product by first freezing it and then by subjecting it to a vacuum. This process promotes the direct phase change from solid ice to water vapor without a transition through a liquid phase - a process known as sublimation. In a conventional drying process, a product is heated at atmospheric pressure, requiring much higher temperatures, to promote this phase change. Lyophilization benefits pharmaceuticals by increasing product shelf life without heating which can have a deleterious impact on product quality and/or performance.

The subject freeze-dryer (Figure 4) was composed of a process chamber, a condensing unit, vacuum pumps, and a utility skid. The utility skid cooled silicone oil that was used as a heat transfer fluid with liquid nitrogen. The cold silicone oil was circulated through a heat exchanger in the process chamber to chill the product. The silicone oil was also used to cool the condenser. After the product was chilled, the vacuum pumps would lower the pressure in the process chamber causing water to sublime as vapor from the product. This vapor was then deposited on plates in the condenser. Electric immersion heaters were also installed in the silicone oil flow circuit for additional temperature control.

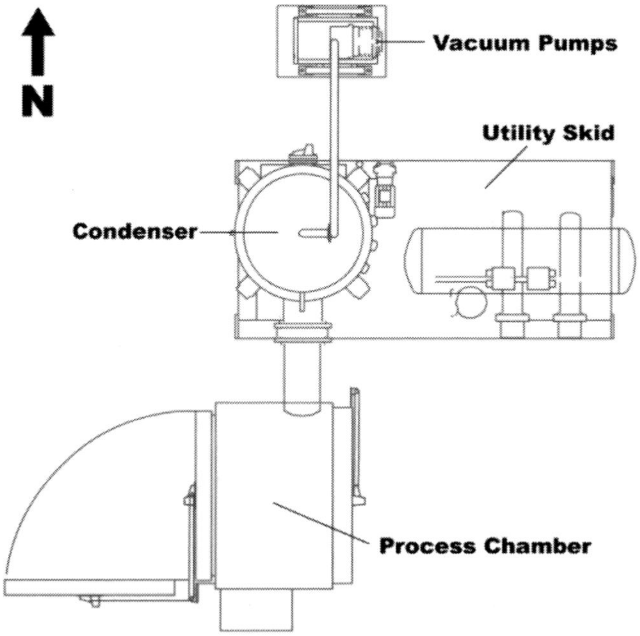

Figure 4. Major components of subject freeze dryer

One the day if the incident, a power outage occurred on the freeze-dryer's control system power supply. As a result of the loss of power a number of key sensors and control elements were unavailable to the programmable logic controller. In response, the control system managed a chain of events that led to the freeze-up of the oil-nitrogen heat exchanger and the call for heat in the electric heater. Figure 5 shows a schematic representation of the shelf cooling system.

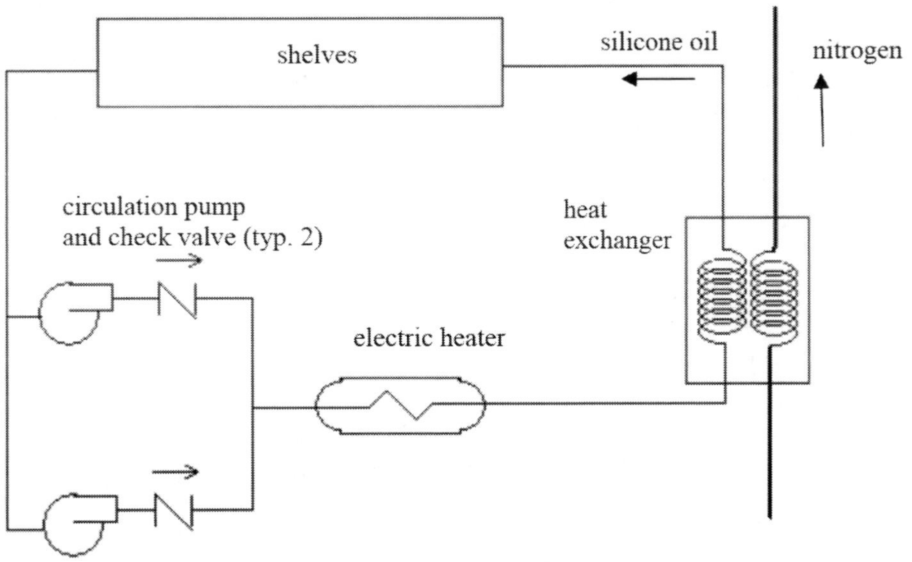

Figure 5. Schematic representation of silicone oil circuit

The silicone oil inside the heat exchanger froze, forming a solid plug. There was liquid silicone oil trapped between the frozen oil in the heat exchanger and the check valves. The control system initiated a call for heat which resulted in the thermal expansion of the trapped liquid oil. This resulted in the catastrophic rupture of the housing for the electric heater. The ejected oil ignited causing a flash fire and serious injury to an operator. Subsequent to the accident it was discovered that the housing for the electric heater had been defectively fabricated, causing it to rupture at a pressure significantly less than the burst pressure of the piping system.

5. Recommendations

Based on observations from these case studies and from similar accident investigations, we offer some lessons learned that may assist the engineer in performing a process hazard analysis on systems that involve the heating of process liquids.

5.1 Understand the limitations of the pressure relief system

A process system typically consists of a network of vessels and piping. Pressure relief devices are incorporated into the design with specific hazard scenarios in mind. It is essential that the design basis for pressure relief is thoroughly understood and documented. As process changes are implemented over the lifetime of the process system, the impact of the change on the pressure relief system must be considered (via the management of change program). For example, most indirect heating systems circulate heat transfer oil through an expansion tank. Pressure relief for the entire flow circuit is often provided only at the expansion tank [12]. Care must be taken to ensure that any process modification does not present the opportunity to trap liquid anywhere in the flow circuit.

5.2 Guard against blockages in the flow path

Flow blockages come in all shapes and sizes. Once a process liquid has become isolated from the pressure relief system, it becomes a potential victim of uncontrolled thermal expansion. Manual block valves, control valves that fail closed, and check valves are obvious forms of flow blockage in a piping circuit. However, other blockages may not be so obvious. For example, in refrigeration systems with secondary coolants it may be possible to freeze a heat exchanger to the extent that it completely blocks the flow of the secondary coolant. The potential for non-obvious hazard scenarios underscores the importance of conducting a systematic process hazards analysis.

5.3 Know/verify the design pressure of each piping component

Just like a chain will fail at its weakest link, so too in any given piping circuit, the component with the lowest design pressure will be the first component to fail under overpressure. The lowest design pressure for a component will usually dictate the design pressure for the system. The setpoint for the pressure relief system should be based on protecting the component with the lowest design pressure.

5.4 Beware of defectively fabricated piping components

Defectively fabricated piping components may have failure pressures far less than expected, presenting the possibility of premature failure under abnormal operating conditions. A mechanical integrity program is an essential safeguard for overpressure protection, and one of its objectives should be to verify that the process piping and vessels have been fabricated in accordance with the design documents.

6. Conclusion

The heating of a trapped liquid is a well-recognized hazard scenario, but it is sometimes overlooked. This seemingly minor source of overpressure can actually result in catastrophic accident. The two primary strategies for protecting against this overpressure hazard are pressure containment and pressure relief. There are numerous design standards and guidelines that document good engineering practices for implementing these strategies. Recommendations were presented to assist in the process hazard analysis effort.

7. References

1. Center for Chemical Process Safety, *Guidelines for Hazard Evaluation Procedures, Second Edition with Worked Examples*; New York: AIChE, 1992.
2. Crowl, D., and Louvar, J. *Chemical Process Safety; Fundamentals with Applications, 2nd edition*. Englewood Cliffs, NJ: Prentice Hall, 2001.

3. Cengel, Y. and Boles. *Thermodynamics (An Engineering Approach), 6th edition.* McGraw-Hill, 2006
4. Moran, M. and Shaprio, H. *Fundamentals of Engineering Thermodynamics, 6th edition,* Hoboken, NJ: Wiley, 2008
5. Smith, J., Van Ness, H., and Abbott, M. *Introduction to Chemical Engineering Thermodynamics, 5th edition.* New York: McGraw-Hill, 1996.
6. Dean, J. (Ed.), *Lange's Handbook of Chemistry, 11th edition.* New York: McGraw-Hill, 1973
7. 1960 Crane's Catalog, p. 14
8. Clearco, *Compressibility of Clearco Pure Silicone 20CST*, and *Product Information Sheet*, www.clearcoproducts.com.
9. ASHRAE, 1997 *ASHRAE Handbook: Fundamentals*. Atlanta: ASHRAE, 1997
10. ASTM, *Standard Specification for Propane Thermophysical Property Tables D4362-08*, West Cronshohocken, PA, 2009.
11. Oetinger, Jim, *Prevent Fires in Thermal Fluid.* www.paratherm.com
12. Singh, J. *Heat Transfer Fluids and Systems for Process and Energy Applications*. Boca Raton: CRC Press, 1985.

Using Real Time Process Models to Detect Loss of Containment and Mitigate Consequences

Harri K. Kytömaa
Exponent, Inc.
9 Strathmore Road
Natick, MA 01760
(508) 652-8572
hkytomaa@exponent.com

Timothy J. Myers, Alfonso F. Ibarreta, and Nicolas F. Ponchaut
Exponent, Inc.
9 Strathmore Road
Natick, MA 01760
(508) 652-8500

Prepared for Presentation at
American Institute of Chemical Engineers
2010 Spring Meeting
6th Global Congress on Process Safety
San Antonio, Texas
March 22-24, 2010

UNPUBLISHED

Using Real Time Process Models to Detect Loss of Containment and Mitigate Consequences

Harri K. Kytömaa
Exponent, Inc.
9 Strathmore Road
Natick, MA 01760
(508) 652-8572
hkytomaa@exponent.com

Timothy J. Myers, Alfonso F. Ibarreta, and Nicolas F. Ponchaut
Exponent, Inc.
9 Strathmore Road
Natick, MA 01760
(508) 652-8500

Keywords: loss of containment (LOC), process monitoring, process safety information (PSI), process design, equipment design, alarms, emergency response, early response, process upset, data analysis, overflow, leak

Abstract

Review of process data from loss of containment incidents reveals that in some cases the loss of containment could have been identified well before plant staff became aware of the failure. If operators become aware of a loss of containment in its early stages, they can potentially mitigate the consequences of the event. Automated real time analysis of fundamental process safety information (PSI) variables can detect some loss of containment (LOC) incidents in their early stages. In this presentation we provide examples of loss of containment events that could have been detected by analyzing mass and energy balances calculated from fundamental process variables. These incidents include:

1) Overflow of a petroleum storage tank in the 2005 Buncefield incident
2) Leak in a hydrocracker preheater
3) Leak of heat transfer fluid into a fixed bed catalyst reactor

In these incidents, early detection of the loss of containment would have allowed operators to take actions to prevent or significantly reduce the eventual consequences of the event.

1. Introduction

In some cases, catastrophic accidents that result from a loss of containment (LOC) do not occur instantaneously, but are preceded by a process upset and sometimes a prolonged period of release. The review of fundamental process safety information (PSI) from these LOC incidents reveals that, in these cases, the loss of containment could have been identified well before plant staff became aware of the failure.

If operators are alerted to a loss of containment in the early stages of a failure, they could potentially mitigate the consequences of the event. Automated real time analysis of fundamental process variables can detect some loss of containment incidents in their early stages. In this presentation we provide examples of loss of containment events that could have been detected by analyzing fundamental process variables and performing checks based upon conservation of mass or energy.

2. Overflow of a Petroleum Storage Tank in the 2005 Buncefield Incident

At approximately 6:00 am on Sunday 11 December 2005, an explosion and fire occurred at the Buncefield fuel terminal. The incident caused significant damage to the facility and the surrounding area, but no fatalities. The fuel terminal received fuel from multiple pipelines, stored fuel, and distributed fuel to outgoing pipelines and tanker trucks. The initial explosion was caused by the overfilling of Tank 912, which caused approximately 300 m^3 of premium gasoline to spill from roof top vents on the tank; creating a vapor cloud over a large portion of the facility [1]. Figure 1 shows the recorded liquid level for Tank 912 in the hours preceding the incident.

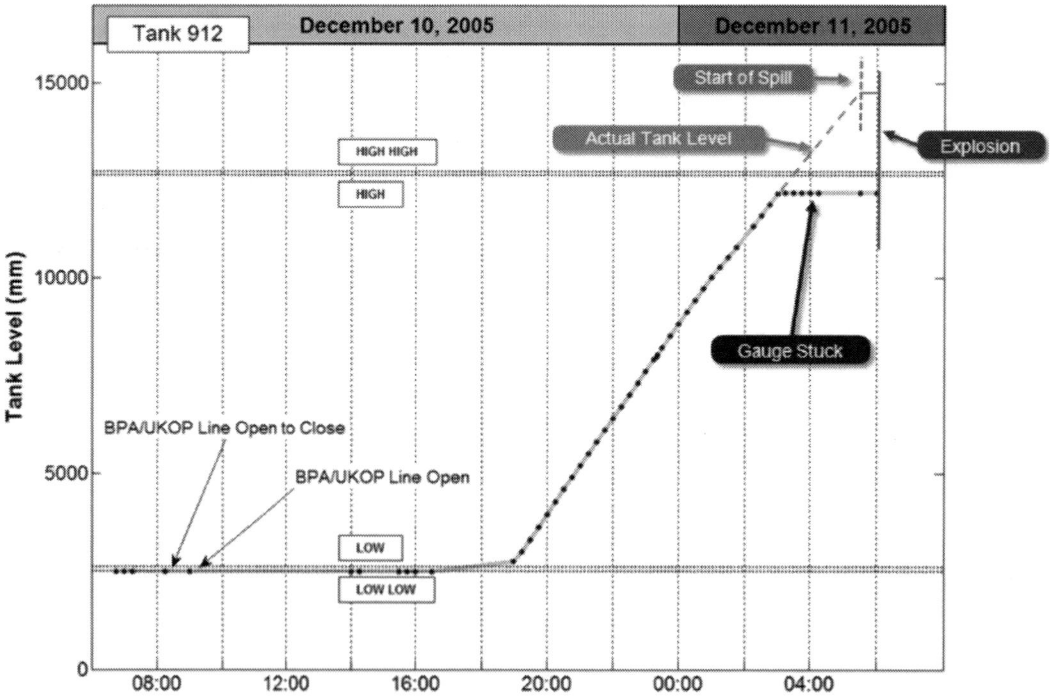

Figure 1 Liquid level in Tank 912 in the hours prior to the explosion.

Earlier that morning at approximately 3:05 am, while Tank 912 was being filled from a pipeline, the level gauge became stuck and failed to record the increasing tank level after that time. Because the level gauge was stuck, the tank level monitoring system used by operators at the facility reported that the level of fuel in the tank was no longer rising, even though the tank continued to fill.

The tank liquid level alarms, designed to activate and alert operators before the tank overflows, did not activate because the gauge they relied on was stuck. Additionally, an ultimate high level alarm and overfill prevention system, designed to shutdown the pipeline filling the tank before the tank overflows, did not operate.

2.2 Post-incident Detection of Stuck Gauges

The Buncefield fuel terminal computer system contained a record of the liquid levels of all tanks as a function of time for the previous six months. Post-incident analysis of the historical data indicated that some of the Buncefield level gauges often became "stuck" as the tank level continued to change. During these sticking events the level gauge displacer became submerged in fuel when the tank was filling or remained hanging in air when the tank was draining.

Prior stuck points can be identified in the historical data. During a sticking event while filling, the level gauge displacer stops rising even though the tank is still filling. When the gauge eventually unsticks, the displacer rapidly returns to the actual liquid surface, and once again accurately measures the tank level. The gauge often returns to a level that is consistent with continued filling of the tank during the period that the gauge is stuck. The rapid change in reported tank level that occurs when the gauge unsticks typically corresponds to filling (or emptying) rates that are not possible in practice. The apparent flow rate can therefore be used as a simple means to identify the occurrence of a past sticking event.

A sample of multiple sticking events for Tank 912 in a single day is shown in Figure 2. Analysis of the data clearly shows that on this particular morning the recorded gauge level did not follow the true liquid level for much of the time.

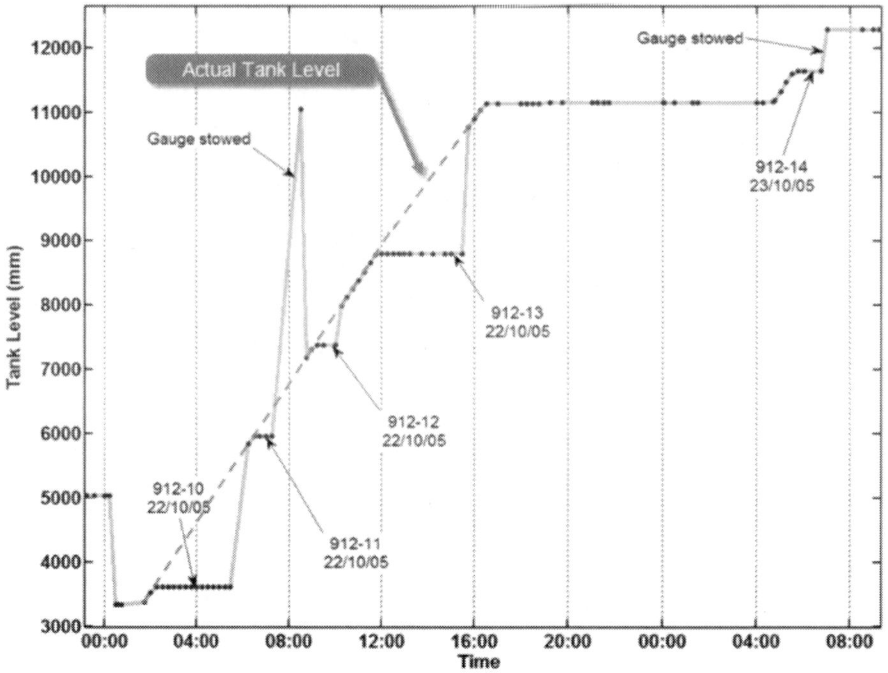

Figure 2 Plot of Tank 912 liquid level in October 20, 2005, showing multiple stuck points.

2.3 Early Detection Strategy – Mass Balance

Level gauge sticking events could have been identified by performing a mass balance and comparing the flowrate of fuel in pipelines feeding and exiting the facility to the rate of fuel entering and leaving all tanks. In its simplest form, this can consist of providing the operator with a display of both the tank filling (or emptying) rates and pipeline flowrates. Such displayed information can empower the controller to recognize an inconsistency between a stationary (stuck) gauge and continued pipeline flow.

It is our understanding that the pipeline used to fill the tank at the time of the incident was controlled by another facility and the tank facility therefore did not have real-time readings of the pipeline flow rate. Installation of pipeline flow meters and associated displays in the control room or importing flowrate data from the pipeline to the control room would allow real time mass balances to be performed on each tank to determine when the tank level is not changing in a manner consistent with flow of fuel into and out of the facility. An alarm could be programmed into the software to alert operators of an inconsistency between tank filling (or emptying) rates and the pipeline flowrates. This approach would not only detect stuck gauges, but could also identify other losses of containment due to leaks in tanks, pipes, or other equipment. It should be pointed out that such a function existed in principle but relied upon a comparison between tank level changes based on the level gauge, and an operator input of the expected status of the tank (filling, emptying or stationary).

3. Leak in a Hydrocracker Preheater

An explosion occurred at an oil refinery, causing one fatality. The incident originated in a hydrocracker heating furnace. The hydrocracker was being used to produce high-octane gasoline from secondary crude products.

On the morning of the incident, a worker reported the presence of smoke coming out of the hydrocracker furnace stack. At the same time, the hydrocracker operator saw that the hydrocracker furnace stack excess oxygen value was dropping. As a result, the operator instructed another worker to increase the air supply to the furnace. As workers were attempting to fix the under-ventilated burner problem, an explosion occurred.

The explosion occurred because of the sudden rupture of the crude feed tube in one of two hydrocracker heaters. This break caused the heated crude and air to mix and subsequently ignite. Tube failure was attributed to deterioration due to local overheating.

Analysis of the data showed that prior to the catastrophic failure, the crude had been leaking and burning, causing the furnace to become starved of oxygen. Exponent's analysis of the process efficiency clearly indicated that crude was leaking and burning almost two hours prior to the event.

3.1 Heat Balance and Process Efficiency

When the process heater used to heat combustible liquids sustained a leak, combustion of the leaking product added additional heat. In such a situation, a heat balance for the process heater can reveal this new heat source. An estimate of the heat balance is provided by the energy efficiency of the process. This can be quantified by dividing the energy absorbed by the feed by the total energy released by the combustion of fuel in the furnace:

$$\text{Process Efficiency} = \frac{\dot{m}_{feed} \cdot C_p \cdot \Delta T_{feed}}{\dot{m}_{fuel} \cdot \Delta H_{comb}},$$

where $\dot{m}$ represents the mass flowrates, C_p is the specific heat of the feed, and ΔH_{comb} is the heat of combustion of the fuel. The energy absorbed by the feed is calculated from the feed's temperature difference across the furnace, the specific heat capacity of the feed, and the feed flow rate. The energy released by the combustion of the fuel is calculated by multiplying the specific heat of combustion of the fuel by its mass flow rate.

During normal operation the process efficiency is well below 1 and remains relatively constant. Plotting of the process efficiency factor in the day prior to the incident (shown in Figure 3 and Figure 4) shows how the factor remains below 0.55 during normal operation.

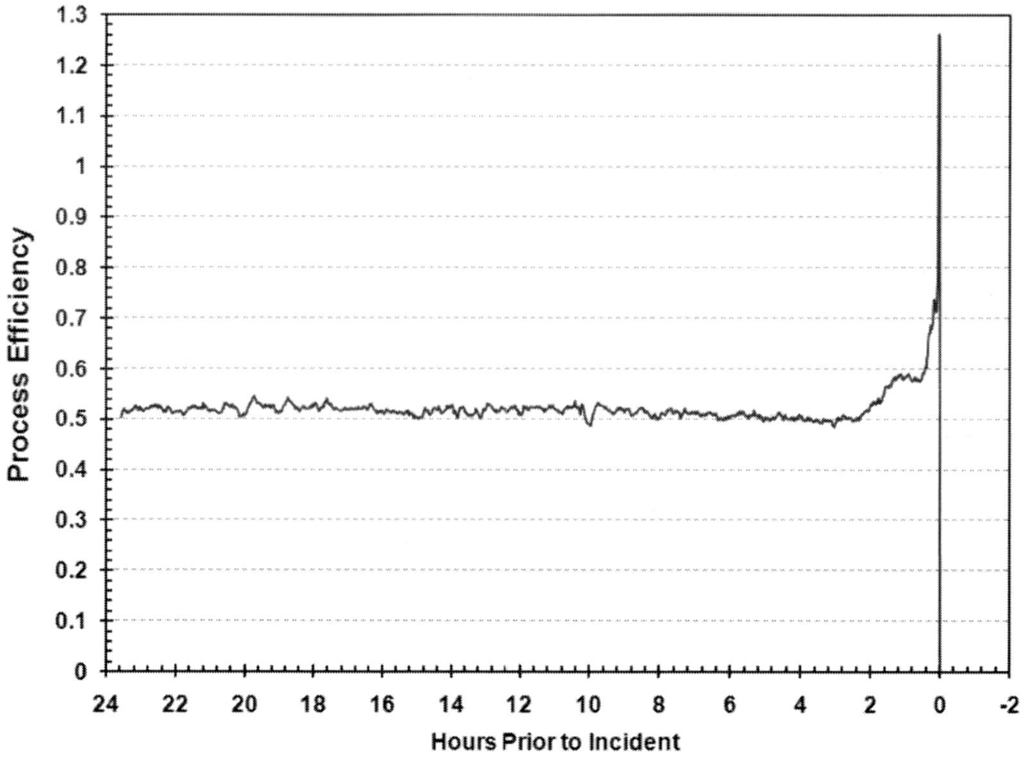

Figure 3 Process efficiency in the day prior to the incident.

Figure 4 shows how the process efficiency began to rise above normal operating conditions about 1-½ to 2 hours prior to the incident and eventually peaked at 1.25 immediately before the incident. The reason for this increase was the leakage of crude which burned with the excess air and contributed to the heat input into the process fluid, and decreased the flow rate of crude leaving the heater. The leak therefore added excess energy to the process.

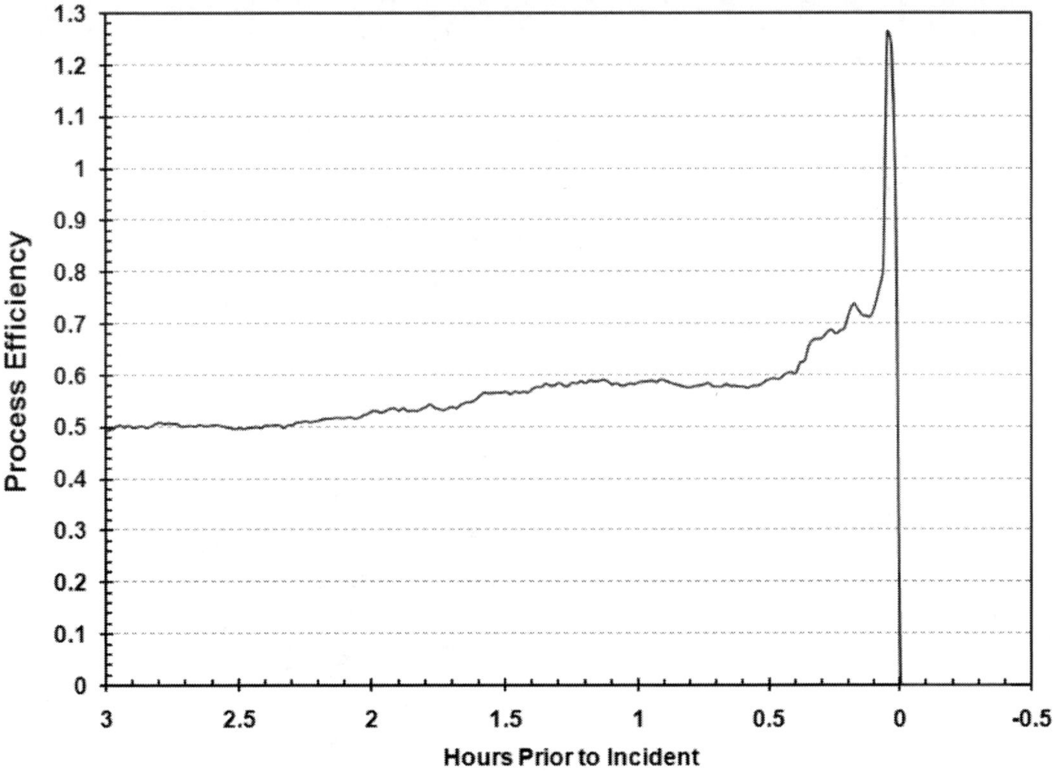

Figure 4 Process efficiency in the hours prior to the incident.

3.2 Early Detection Strategies

Operators could have been alerted to the energy imbalance almost two hours prior to the catastrophic event if the overall process efficiency had been calculated and monitored in real time. If this variable had been calculated by the software and appropriate alarms had been set, the operators would have been immediately alerted and the process could have been aborted prior to the catastrophic failure.

4. Leak of Heat Transfer Fluid into a Fixed Bed Catalyst Reactor

A failure occurred in a fixed bed catalyst reactor that was similar in construction to a vertical shell-and-tube heat exchanger. Tubes of the reactor were filled with a mixture of catalyst and inert packing material and an organic heat exchange fluid was circulated on the shell side to control the temperature of the reactor. A schematic of the reactor is shown in Figure 5.

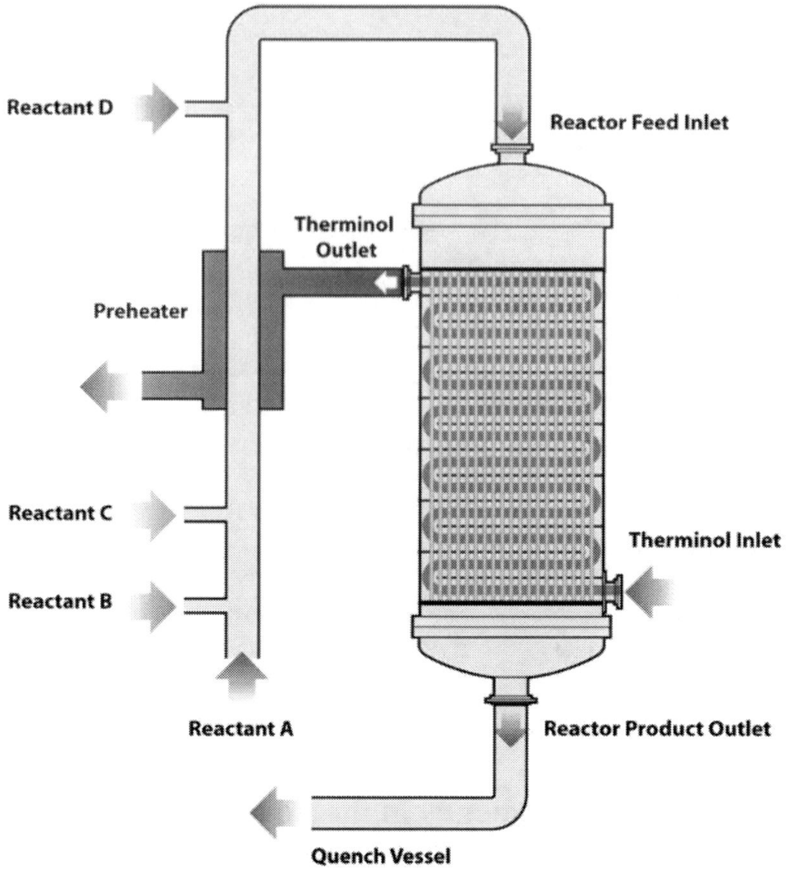

Figure 5 Schematic of process.

On the day of the incident, over-temperature conditions occurred during repeat start-up attempts of the reactor, causing multiple reactor shutdowns. Eventually, the reactor was disassembled and inspection revealed melting of the tubesheet, damage to the tube-to-tubesheet connections, and heat transfer fluid on the tube side of the reactor. Similar problems were experienced in subsequent startups after repair of the reactor, and eventually the reactor was replaced.

An investigation of the reactor failure found that an initially small breach of a tube caused heat transfer fluid to leak from the shell-side to the tube-side. An exothermic reaction of the heat transfer fluid with the catalyst resulted in localized heating of the tubes. The high temperatures further degraded the tube-to-tubesheet connections, allowing even more heat transfer fluid to enter the tube side of the reactor. Although the reactor automatically shut down when local temperatures reached a pre-set maximum, operators did not identify the cause of the temperature spikes and continually attempted to re-start the reactor, leading to further damage.

Analysis of the temperatures observed prior to the reactor failure shows large temperature excursions, which provide indication that heat transfer fluid was entering the reactor.

3.1 Process monitoring and upset

As part of the process monitoring system, thermocouple probes were placed inside some of the reactor tubes. Each probe is made up of twenty thermocouples that record temperatures along the length of the reactor.

During normal operation temperatures in the reactor were relatively uniform and there was only a small temperature difference between the average recorded temperatures (temperature averaged over the length of tube) of the coldest tube and of the warmest tube. A plot of the average tube side and shell side temperatures versus distance (Figure 6) shows that on average at the top of the tubes, the heat transfer fluid is warmer than the reactants and is actually heating reactants while over the remainder of the tubes the heat transfer fluid cools the reactant/product stream.

The process flowrates and tube catalyst distributions were designed to maintain a fairly constant temperature of the reactant through the reactor. At the tube entrance, the catalyst concentration was relatively low so as to achieve a lower reaction rate with the higher concentration of reactants. Towards the end of the reactor tubes, the catalyst concentration was designed to be higher in order to maintain the reaction rate as the reactant concentration decreased.

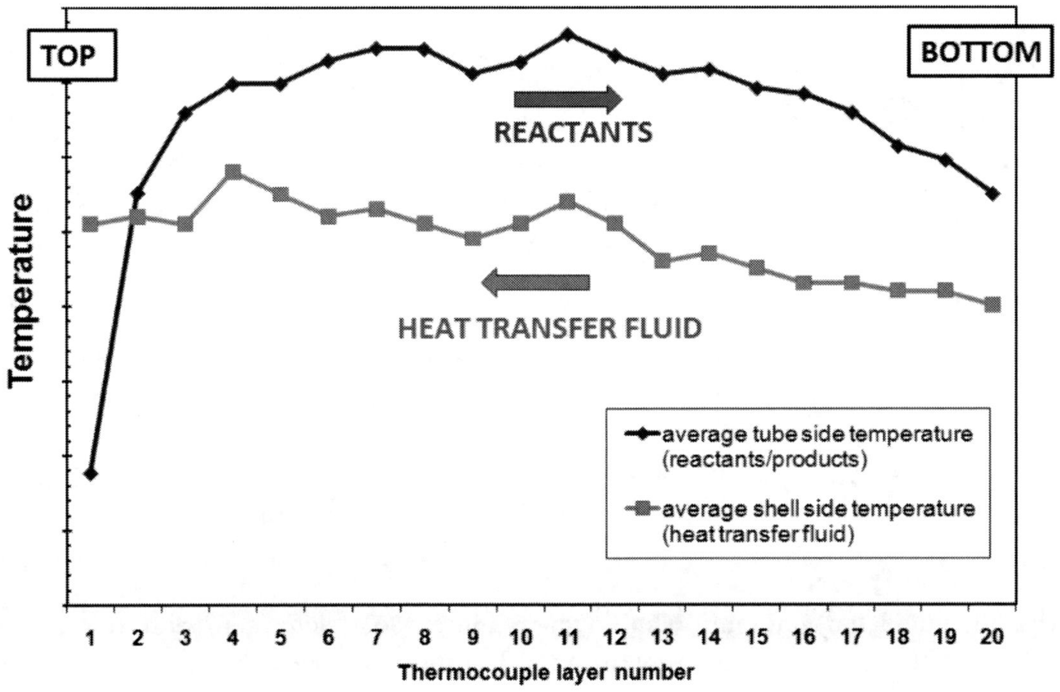

Figure 6 Average tube side and shell side reactor temperatures vs. axial length of reactor. Layer 1 is near top of the reactor and layer 20 is near bottom of the reactor.

The reactor control system was originally designed to cut the feed of reactants to the reactor if a recorded temperature at any location within the reactor tubes exceeded maximum values in order to limit reactor temperatures below the reactor vessel design temperatures. A review of the process data after the incident showed that during the initial process upset, temperatures in a small number of tubes rose towards the outlet of the reactor. This temperature rise had automatically shut down the reactor, as intended, but operators continued to attempt to operate the reactor, unaware of the mechanical failure at the tube-to-tubesheet connections.

Figure 7 shows the temperature profile of one of the affected tubes prior to and immediately after one of the initial emergency shutdowns. Only 9 minutes elapsed between the two temperature readings. The later temperature profile shows a high temperature at the outlet of the reactor, which is not consistent with normal operation temperature profiles, where the temperature maximum is reached near the center of the reactor,

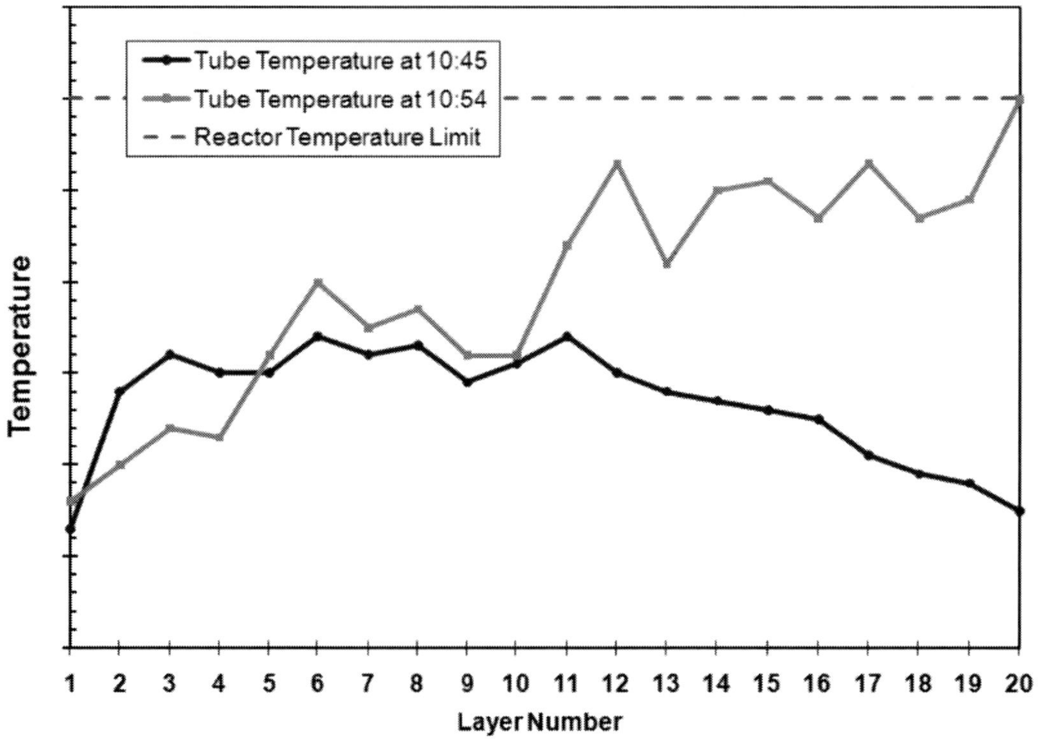

Figure 7 Temperature profile in the reactor tube before and at time of emergency shutdown at 10:54 on August 11.

Operators attributed the automatic shutdown to increased reactor temperatures caused by too much flow of reactants. They attempted multiple times to restart the reactor at lower reactant flowrates, but the reactor kept shutting down due to temperature limit alarms. As the restarts continued, the automatic shutdowns occurred at lower and lower reactant flowrates until the reactor was finally taken offline.

Post-incident testing was performed in a research reactor where process temperatures were examined under different feed conditions. The testing showed that under nominal conditions, temperature profiles remained stable over a range of reactant flowrates. Figure 8 shows the temperature profiles observed in a test reactor with varying reactant flowrates and with injection of the heat transfer fluid into the reactant stream. The test data indicate that the temperature profile only changes significantly when heat transfer fluid is introduced into the reactant stream. The tube temperature spikes observed during the initial process upset (Figure 7) were therefore not achievable with the reactant flowrates employed at the facility, and the automatic shutdowns were therefore symptoms of a leak of heat transfer fluid into the reactor.

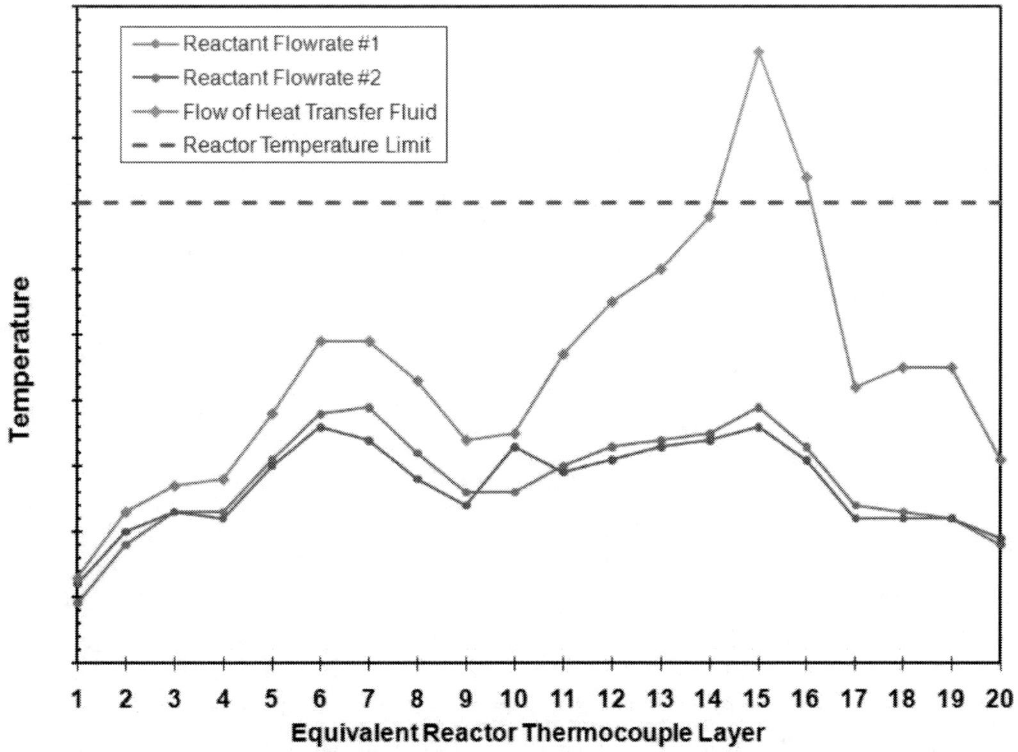

Figure 8 Test reactor temperature profiles with varying reactant flowrates and injection of heat transfer fluid.

4.2 Early Detection Strategies

In this case the reactor shutdown strategy worked as intended, automatically shutting down the reactor when temperatures rose above a preset maximum. The operators, however, did not diagnose or recognize the process upset condition. They incorrectly associated the high temperatures near the outlet of the reactor with excess reactant being fed to the reactor, and were unaware of the possibility that heat transfer fluid could be leaking and reacting. Unfortunately, the process designers had not identified this upset condition.

In addition to providing maximum allowable temperatures, a heat balance calculation could have been used to compare reactor and heat transfer fluid temperatures to the expected temperatures for a given feed rate. The heat released during reaction would have been accounted for in the temperature increase of the reactant stream and heat transfer fluid across the reactor. For conditions where most of the reactants are consumed and the total specific heat of the reactants and products remains approximately constant, a process efficiency variable can be defined as:

$$\text{Process Efficiency} = \frac{\text{total increase of sensible enthalpy}}{\text{heat released during reaction}} \;,$$

$$\text{Process Efficiency} = \frac{\dot{m}_{HTF} \cdot C_{p,HTF} \cdot \Delta T_{HTF} + \dot{m}_{react} \cdot C_{p,react} \cdot \Delta T_{react}}{\dot{m}_{react} \cdot \Delta H_{reaction}} \;,$$

where $\dot{m}$ represents the mass flowrates – with *HTF* standing for heat transfer fluid and *react* the reactants; C_p is the specific heat of the fluid; ΔT the temperature increase between the inlet and outlet; and $\Delta H_{reaction}$ is the heat of reaction of the process stream. Under perfect conditions (no heat loss) and with complete reaction of the reactant stream, this process efficiency should be equal to 1. The monitoring of this process variable could therefore identify times when excess energy is released due to a process upset, such as a heat transfer fluid leak. As the catalyst degrades and a lesser fraction of the reactants react, the above process efficiency will drop and reflect the condition of the catalyst.

5. Monitoring and Early Detection

In the three incidents discussed in this paper, the failures were not instantaneous, but occurred over a matter of hours. The final catastrophic event in each of these cases could have been prevented or mitigated by early detection based upon the real time analysis of fundamental process variables using mass and energy balances. A better understanding and communication of fundamental and critical process safety information (such as process mass balance or heat balance) to the operators can provide a real-time process alarm and potentially allow them to diagnose the problem and respond earlier. Early detection of the loss of containment in all these cases would have allowed operators to take actions to prevent or significantly reduce the eventual consequences of the event.

Real-time analysis of the process data can be a useful tool to identify process failures during their incipient stage. It is important to know what the process parameter boundaries are in terms of flowrates, local temperatures and efficiencies. The PSI-based early detection strategies proposed in this paper require a fundamental understanding of the process operating limits, as well as improved capture and analysis of available data. Any inconsistencies or unphysical values should be treated as an indication of a process upset that must be immediately addressed. Automated analysis and alarms can be configured to warn operators of these process deviations.

6. References

[1] Buncefield Major Incident Investigation Board Health "The Buncefield Incident - 11 December 2005: The final report of the Major Incident Investigation" Safety Executive (HSE) Volumes 1, 2 & 3 (2008) http://www.buncefieldinvestigation.gov.uk/reports/index.htm#final

Lessons Learned from Real World Application of the Bow-tie Method

Steve Lewis
Risktec Solutions Limited
Warrington, UK
email steve.lewis@risktec.co.uk

Kris Smith
Risktec Solutions Inc.
Houston, TX
email kris.smith@risktec.com

Prepared for Presentation at
American Institute of Chemical Engineers
2010 Spring Meeting
6th Global Congress on Process Safety
San Antonio, Texas
March 22-24, 2010

UNPUBLISHED

Lessons Learned from Real World Application of the Bow-tie Method

Steve Lewis
Risktec Solutions Limited
Warrington, UK
email steve.lewis@risktec.co.uk

Kris Smith
Risktec Solutions Inc.
Houston, TX
email kris.smith@risktec.com

Keywords: Bow-tie, hazard assessment, lessons learned

Abstract

The benefits of using bow-tie diagrams for risk management have been realized by organizations world-wide across a variety of business sectors. Also known as barrier diagrams, they provide a readily understandable visualization of the relationships between the causes of business upsets, the escalation of such events to a range of possible outcomes, the controls preventing the event from occurring and the preparedness measures in place to limit the consequences.

More importantly, the preventive and mitigating measures are linked to tasks, procedures, responsible individuals and competencies. This demonstrates the crucial connection between risk controls (whether hardware or human intervention) and the management system for assuring their ongoing effectiveness.

This paper draws on Risktec's unparalleled experience in applying the bow-tie methodology and is intended to be of interest to those who are new to the technique and experienced users alike. It summarizes the history of the bow-tie method, gives an overview of how to apply it and describes in detail its practical uses and benefits as well as potential pitfalls and guidelines for success.

1. Introduction

The exact origins of the bow-tie methodology are a little hazy. The earliest mention appears to be an adaptation from the ICI plc Hazan Course Notes 1979, presented by The University of Queensland, Australia[1].

Undoubtedly, the Royal Dutch/Shell Group was the first major company to integrate fully the total bow-tie methodology into its business practices[2,3,4] and is credited with developing the technique which is widely in use today. The primary motivation was to seek assurance that fit-for-purpose risk controls were consistently in place throughout all operations world-wide.

Use of bow-ties has spread between companies, industries, countries and from industry to regulator. Examples of bow-tie analysis have been published by the UK defense industry[5], the French government[6], the UK Health and Safety Executive[7,8,9a,9b], an Australian State Regulator[10], the Land Transport Safety Authority of New Zealand[11a], petroleum industry international associations[11b,11c] and international standards[11d], the European aviation industry[12a] and US Federal Aviation Authority[12b], and in the banking industry[12c].

This paper aims to demonstrate, through reference to actual case studies and examples, the practical uses and benefits of this versatile tool which can be used to qualitatively assess and demonstrate control of all types of risk in many industries and business sectors.

2. Bow-tie Method

The bow-tie method provides a readily understood visualization of the relationships between the causes of business upsets, the escalation of such events, the controls preventing the event from occurring and the preparedness measures in place to limit the business impact (Figure 1).

Figure 1 - The Bow-tie

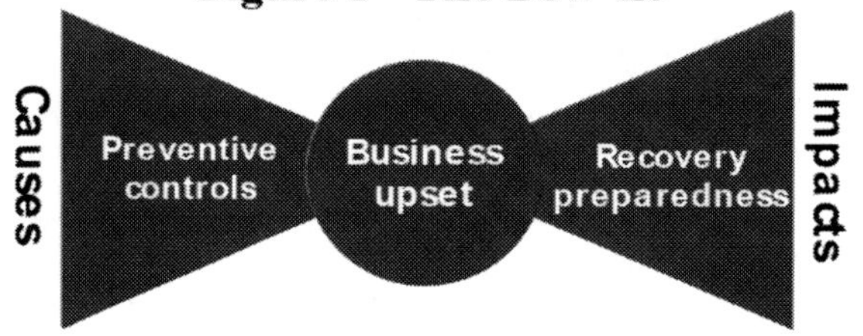

The method for building a bow-tie diagram is well-documented[2,3,13,14,15,16,17], hence it is only covered briefly here.

In its most common use, the ultimate aim is to demonstrate control of health, safety and environmental (HSE) hazards; it is therefore necessary, firstly, to identify those hazards requiring bow-tie analysis. Most companies involved in hazardous activities have an HSE management system[18,19,20] within which there will be formal procedures and/or guidance for identification of potential hazards and assessment of risks. Similarly, other companies have systems and standards for management of commercial, security, business continuity and corporate governance issues[21,22], to which the bow-tie method is equally applicable.

Once hazards have been identified, the bow-tie method can be applied to further assess risks and provide a framework for demonstrating their effective control. Typically bow-ties are developed by asking a structured set of questions which build up the diagram step-by-step (Figure 2).

Figure 2 - Building the Bow-tie

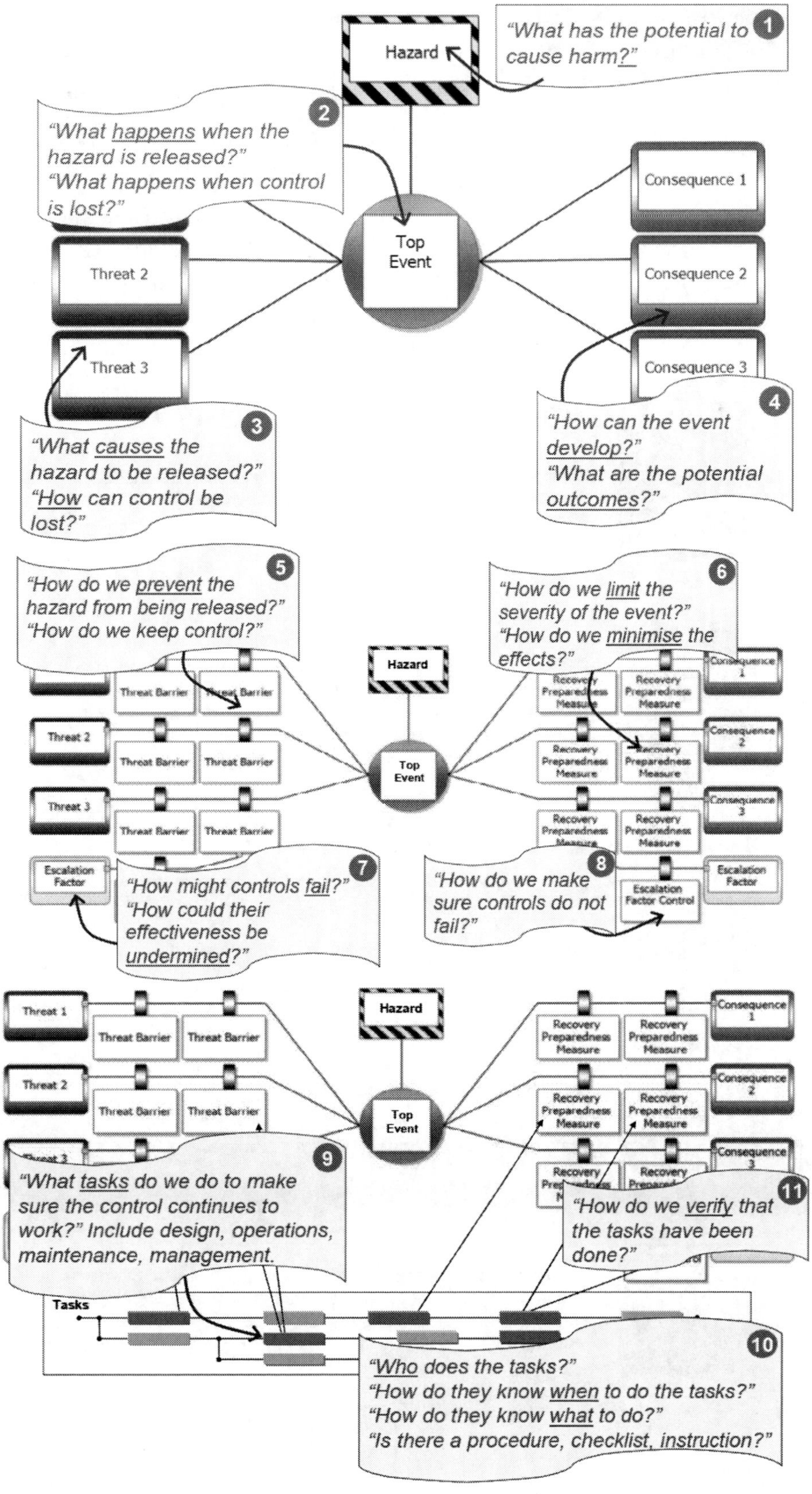

2.1 How to Apply the Technique

Developing the bow-tie diagrams and critical tasks should be carried out in a structured manner in order to obtain quality information and best represent the actual risk control arrangements. Figure 3 summarizes an effective bow-tie building process, which has been developed and refined through experience with a variety of companies, industries and work groups.

Facilitated workshops involving people who are regularly confronted with the risks have proven to be the most effective way of identifying real controls and capturing past incidents and current practice. Openness is an essential ingredient during these sessions if any weaknesses in controls are going to be uncovered. To encourage free discussion, the workshop needs to be run in an honest and engaging fashion, and, like HAZOP study for example, an independent facilitator can often help to create such an environment.

Figure 3 - Practical Application of the Bow-tie Technique

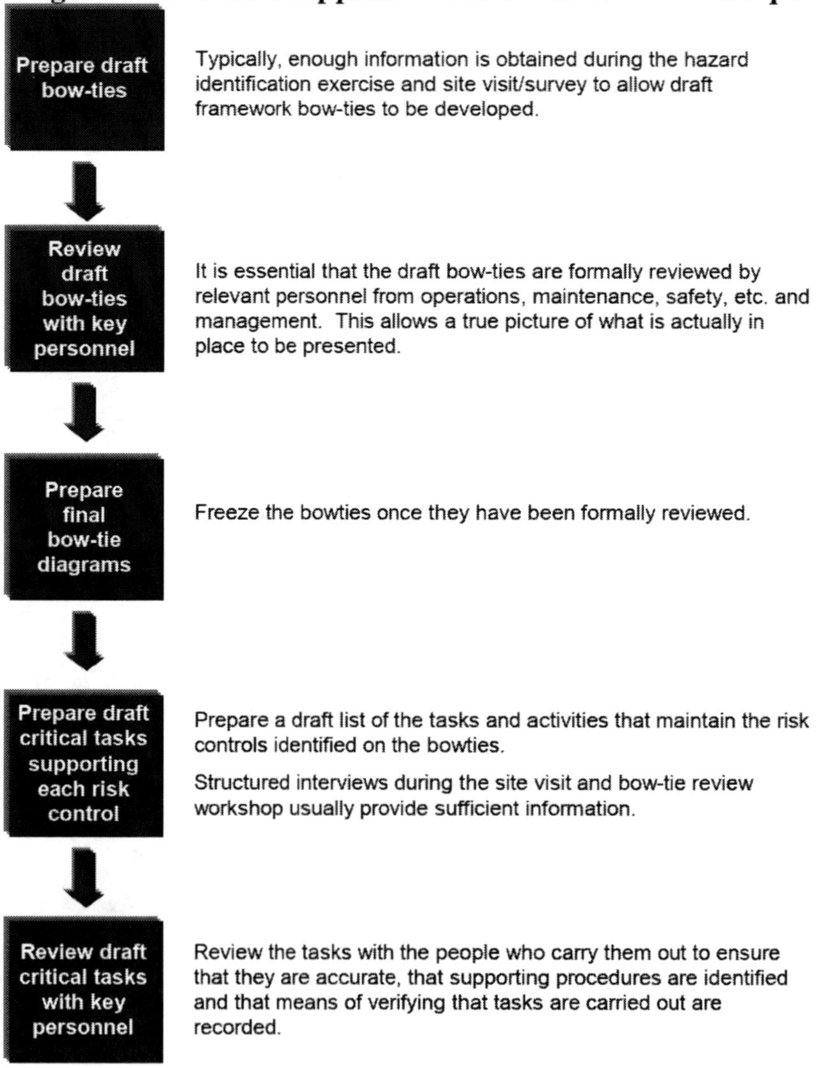

2.2 *Practical Uses*

The bow-tie technique is incredibly versatile and has proven to be successful in many applications.

2.2.1 Logical Structured Approach

The structured approach of the bow-tie forces an assessment of how effectively all initial causes are being controlled and how well-prepared the organization is to recover should things start to go wrong. This logical approach often identifies gaps and issues that are missed by other techniques.

2.2.2 Complete Risk Management

Risk assessments can have a tendency to concentrate on the level of risk only, rather than considering all aspects of the management of risk. The bow-tie method however highlights the direct link between the controls and elements of the management system (Figure 4). As such, it covers far further than HAZOP study, quantitative risk assessment (QRA), etc.

2.2.3 Demonstration

Due to their origin, bow-ties are most commonly used where there is a requirement to demonstrate that hazards are being controlled, and particularly where there is a need to illustrate the direct link between the controls and elements of the management system (Figure 4). For example, bow-ties have been used successfully in safety reports produced for compliance with the UK onshore chemical industry Control of Major Accident Hazard (COMAH) Regulations[23,24]. Indeed, the Competent Authority has stated that bow-ties are an acceptable means of demonstrating a systematic approach to hazard assessment[25].

The COMAH Regulations require that *"The demonstration should provide a clear link between the various accident scenarios identified and the measures which are in place to defend against them. The safety report should demonstrate how the necessary measures will prevent foreseeable failures which could lead to major accidents. There should also be a clear link to the SMSs which keep the necessary measures in place."* [26].

There are other means by which such a link can be demonstrated (e.g. tables) but the bow-tie provides the clearest graphical illustration which is easy to understand.

In Abu Dhabi, the ADNOC HSEMS Guidelines require *"...a demonstrable link between each significant risk and a set of controls or risk reduction measures."*[27] and the associated Code of Practice on Control of Major Accident Hazards (COMAH)[28] outlines the bow-tie methodology as the means to achieve this.

Figure 4 - Link with Management System

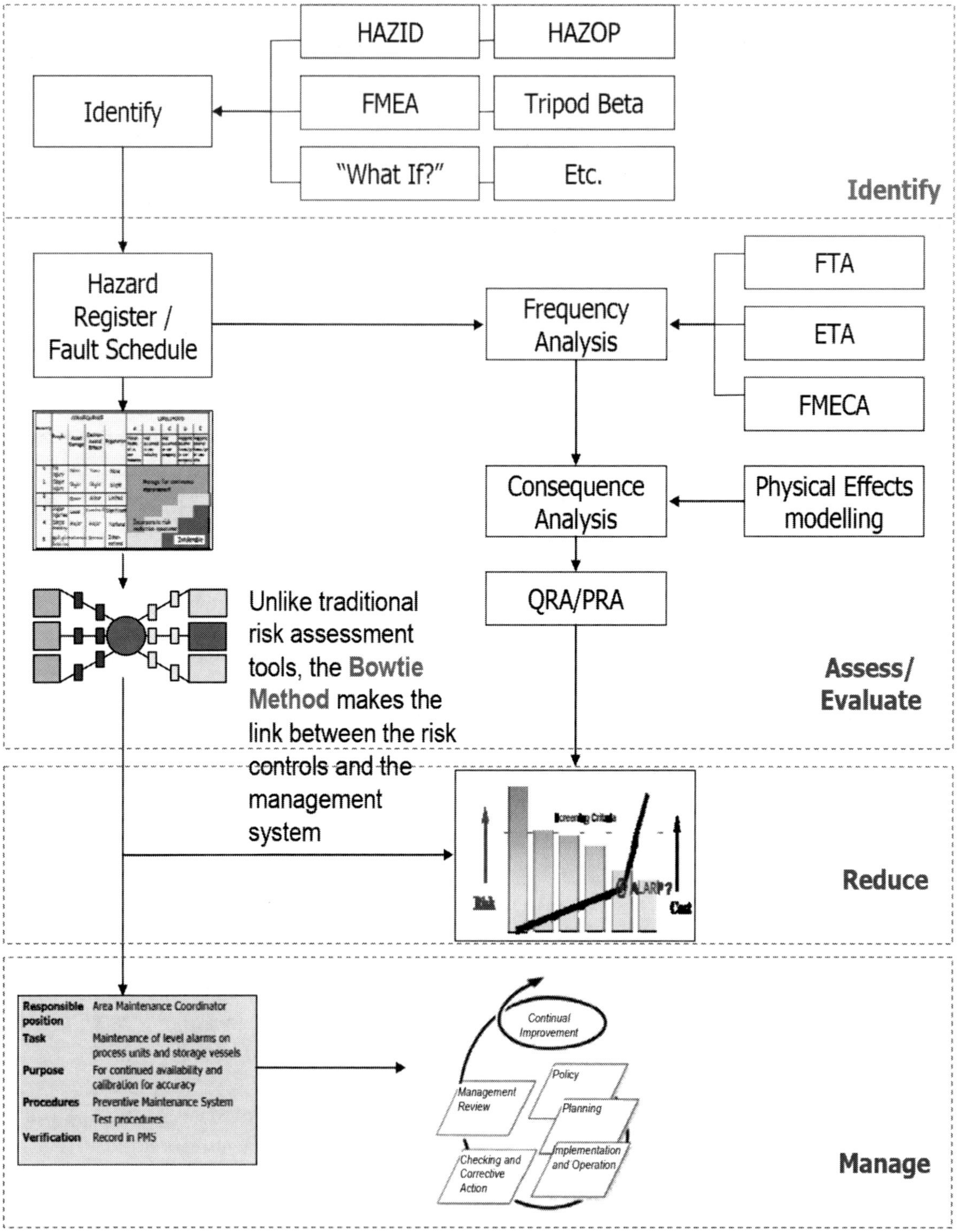

2.2.4 Communication

In its simplest, graphical form, the diagram can be understood by personnel at all levels of an organization, including those who are not connected with the day-to-day business operation being assessed. The bow-tie lends itself to being displayed on posters highlighting key risk control issues. Pocket books and leaflets have also been produced for dissemination of the risk management message. There are examples of bow-ties being included in web-based HSE Cases as part of on-line training and information systems.

It is not always necessary to use sophisticated publishing techniques to get the most out of the bow-tie method. Talking through the components of a particular scenario (causes, consequences, controls, tasks) whilst simultaneously sketching a bow-tie layer by layer can clearly illustrate how the hazard is managed effectively, particularly to those who may not be familiar with the details of the operation. This approach has been used successfully during safety workshops, pre-job planning meetings, discussions with regulators and interested members of the public.

Bow-ties provide a robust, comprehensive yet simple means of 'rolling out' the main points from a risk assessment exercise or HSE Case. Workshops involving workforce teams can be taken through the bow-ties step by step, in order that they understand the significant risks associated with their place of work, the measures taken to manage these risks and the importance of their individual role in preventing and mitigating hazardous events.

Once personnel are comfortable with the bow-tie technique, building a bow-tie becomes a viable alternative to traditional hazard assessment tools such as checklists and prompts. The bow-tie structure provides an excellent framework for 'brainstorming' sessions.

2.2.5 Critical Systems

Critical hardware systems can be defined as those which cause, prevent, detect, control or mitigate a hazardous event. The nature of a bow-tie diagram means that these critical systems are clearly illustrated along the threat and consequence branches, providing a ready means of identifying systems which are critical to ensuring ongoing asset integrity. Critical systems can be linked to defined performance standards and means of verification.

2.2.6 Organizational Improvements

It is possible to use bow-ties in conjunction with organizational and cultural survey techniques such as Tripod Delta[29] to highlight the branches of the bow-tie where organizational control is weak, i.e. controls are not effective (Figure 5). Recognition that organizational failures are the main cause of accidents is the principle behind Tripod Delta, which is a questionnaire-based survey used to build a picture of an organization's strengths and weaknesses against eleven basic risk factors. This enables proactive, sustainable strategies for reducing and managing risk to be focused on weak spots, for example a higher level of inspection or auditing.

Figure 5 - Link to Basic Risk Factors

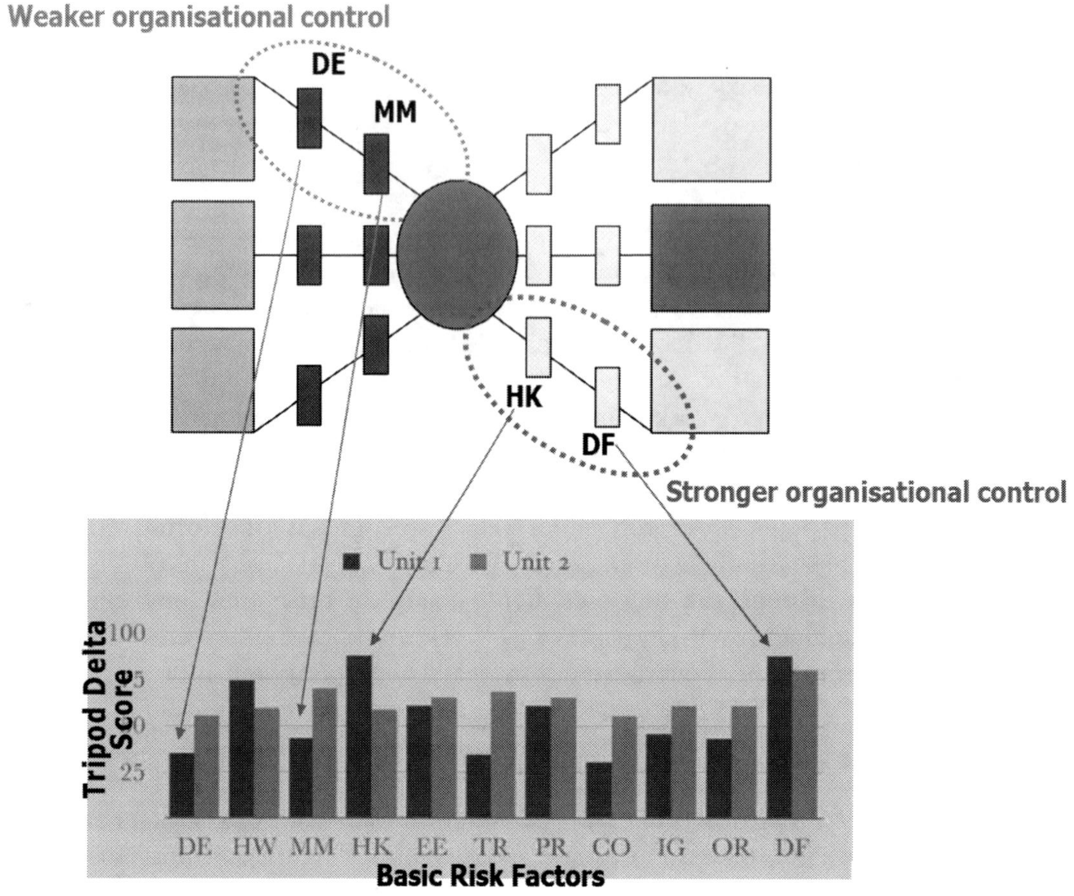

Similarly, bow-ties have also been used to ensure that no critical controls "fall through the cracks" after a company re-organization, merger or acquisition.

Also, the bow-tie can be linked to incident investigation techniques such as Tripod Beta[30] which identify the organizational failures that enabled risk controls (defenses) to fail (Figure 6).

Figure 6 - Link to Incident Investigation

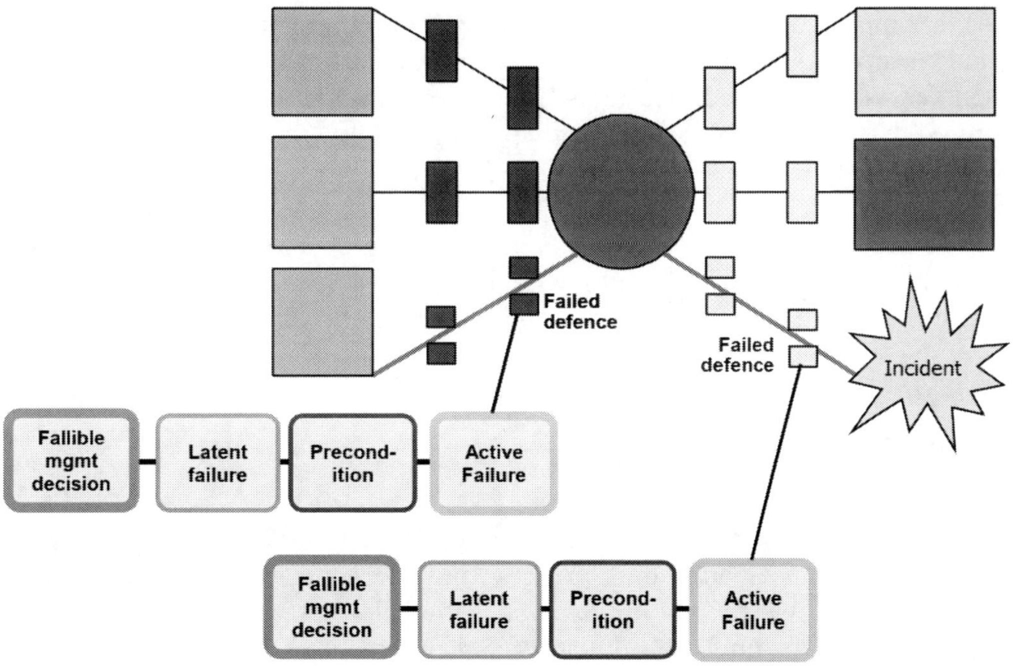

2.2.7 Specific Risks

Bow-ties are of value not only to meet formal regulatory or corporate standards but for undertaking thorough analysis of specific risks (e.g. non-routine activities or particularly problematic areas) to provide reassurance that everything that can reasonably be done to reduce and manage risk is actioned.

2.2.8 Procedures and Competence

A completed bow-tie analysis includes a list of critical tasks undertaken to ensure ongoing integrity of the risk controls. As a minimum, the tasks need to be reviewed with the individuals responsible for carrying them out, but they also lend themselves to being incorporated into procedures, work instructions, individuals' job descriptions, etc. A thorough bow-tie analysis can also detail, for each task, the input procedures required to undertake the task correctly and any records which verify completion of the task, as well as competence requirements for the role responsible for undertaking the task. The tasks can therefore be used to verify the adequacy of a company's competence assurance system; the competencies defined for an individual or role should align with the bow-tie controls. In this fashion, bow-ties have also been used to manage handover and new-starter responsibilities.

2.2.9 Layer of Protection Analysis

Bow-tie diagrams have also been used for conducting Layer of Protection Analysis (LOPA), a simplified risk assessment tool for determining if sufficient protection is in place and quantifying the residual risk where it is insufficient[31]. Well-constructed bow-ties, together with clearly defined rules for applying failure data for hardware and human intervention barriers, enable conservative estimates of residual risk to be compared against risk tolerability criteria. The approach tends to work well during design phases.

It is also possible to make direct links between the engineered safeguards on the bow-ties and specific work orders in a preventive maintenance scheduling and tracking system. In this way, critical hardware systems essential for ongoing risk management are prioritized.

2.3 Benefits

The bow-tie method has three main benefits:

1. Clear communication and improved understanding - Visually illustrating the hazard, its causes and consequences, and the controls to minimize the risk, the bow-tie can be readily understood at all levels, from senior managers and operations personnel, to regulators and members of the public. Bow-ties keep sight of the big picture and can capture the sequence of events as well as previous incidents.
2. Greater ownership - Recognizing that effective risk management is only possible if people are assigned responsibilities for controls. Bow-tie workshops encourage participation and stimulate communication between key stakeholders, whether from the company, contractors or external parties, who all have a role to play in managing risk and yet may not be involved in more traditional techniques. When people feel involved they tend to 'buy-in' to the process. When action is taken based on what they say, people will take ownership. Bow-ties should especially be considered where lack of ownership of process safety by all levels of operations personnel may be an issue.
3. Efficiency gains – Realizing efficiency improvements through a number of different ways, for example: the method is less labor intensive than many other traditional techniques; it identifies where resources should be focused for risk reduction (i.e. prevention or mitigation); it can reduce the volume of safety analysis – it is true that a picture paints a thousand words; it can lead to a potential reduction in unnecessary/lower importance barriers (where fully justified); it helps to target maintenance, inspection and testing activities on critical hardware barriers; and it provides a 'corporate memory' to avoid reinventing the wheel every few years.

The approach also has a number of other benefits:

- 'Future proof' risk management - Illustrating not only what controls are currently in place today, but, through the use of critical tasks, why they will still be there tomorrow.
- Fit-for-purpose management system - Linking the elements of the organization's management system to specific controls to show how it ensures the ongoing management of risk. This avoids the development of over-burdensome management systems and unnecessary procedures, by focusing on those procedures required to support risk control.
- Practical approach - Focusing on risk management by people on a day-to-day basis, rather than analytical studies by technical risk specialists. All too often, risk analysis can become progressively more complex leading to 'analysis paralysis' which overwhelms the need to take positive action.
- Workforce involvement - Risk management is the responsibility of line managers and their people; all staff can see why what they do is critical for risk control.
- Logical structured approach - Considering all aspects of the management of risk, from initial cause to final consequence in a sequential manner. This logical approach often identifies gaps and issues that are missed by other techniques.
- Auditable trail - The diagrams and critical task lists provide a protocol around which auditing by internal departments or regulators can focus on what people are actually doing rather than the condition of physical systems.
- International application – The graphical-based approach is easy to implement with multi-national teams where language difficulties may otherwise hinder progress.
- All risks – The technique is not limited to assessment of HSE risks. Bow-ties have been developed for demonstrating management of security, information technology, business interruption and project risks. The possibilities are endless.
- 'Living case for safety' – Comprehensive bow-tie assessments can be captured in a relational database which supports ready and wide access across an organization, and enables easier periodic updating. Links to current safety-critical procedures help to maintain any case for safety as a live tool rather than a document for risk practitioners that sits on the shelf..

The bow-tie illustrates not only what controls are in place today, but why they will still be there tomorrow

2.4 *Limitations*

Of course, bow-ties are not the panacea for all risk management problems. If you want to quantify your level of risk in absolute terms then the bow-tie method will not help directly. If you want to model complex inter-relationships between your risk controls, there are better ways than using bow-ties. If you want to identify individual safeguards for every line of every section of every unit of your process facility, then HAZOP study is the solution. But if you want to remove the mystique of risk management and obtain insights into your risk controls that are easy to understand and easy to communicate, and at the same time realize some efficiency gains, there is no better method than bow-ties.

3. Example Case Studies

3.1 *Public Meetings*

In one oil and gas industry case where onshore sour gas wells are periodically drilled close to third party land, the operating company has pioneered the use of bow-ties to illustrate to the regulator and members of the public that the hazards associated with the operation are recognized, understood and well managed, both from a preventive point of view and for preparedness in the event of an emergency. Simply drawing bow-ties freehand during public meetings helped considerably in putting across the message that the company is in control of the hazards and the risks are minimized.

3.2 *Adequacy of Controls*

The bow-tie has been used successfully as a means of assessing the adequacy of controls and identifying areas for risk reduction for a rail transport network[11]. A series of stakeholder workshops employed the bow-tie method to test the robustness and number of existing safeguards and identify improvements. For each risk control, critical operating parameters were identified and links were made to rail operating procedures, maintenance systems and international standards. Actions were identified to strengthen particular controls.

3.3 *Auditing*

As an example of how the technique lends itself to use during audits, bow-tie controls and critical tasks have been used as the basis for pre-start up inspections of land drilling rig activities, to make sure that the drilling contractor personnel were fully aware of their responsibilities in managing hazards and that all controls were in place. The same audit also allowed the contracting oil company to assure itself that the operation was fit to continue.

3.4 *Dissemination of Information*

Use of bow-ties enables companies to streamline their risk management documentation, as the amount of information which can be included on a single bow-tie diagram would require several pages if presented in text or tabular format. A company operating a number of oil and gas production installations, together with the associated exploration, infrastructure and transport facilities, found that its HSE Case documents, which originally ran to several volumes per asset, could present the information at the same level of detail but in slimline, single volume versions using bow-tie diagrams. Transferral of this information onto the company intranet with electronic links between bow-tie controls and relevant responsibilities, procedures, instructions, etc. allowed online HSE Cases to be readily accessible to all.

Use of web-based bow-ties[32] has enabled one organization to ensure that up-to-date, consistent information on HSE critical roles and responsibilities is managed effectively, aligned with business processes and disseminated to individuals, communities, disciplines and projects. In this way the HSE management system is 'operationalized', enabling it to serve as a dynamic corporate HSE memory.

3.5 *Specific Risk Assessment*

Bow-ties have been used for risk assessment of a driver-to-signaler train radio communication system on behalf of the rail network management company. The graphical representation enabled easy understanding of the relationships between causes of unwanted events and their controls. The assessment systematically identified procedural controls as well as functional and integrity requirements for engineered measures and established issues for further assessment, information or action where the effectiveness of a control was questioned.

3.6 *Periodic Process Safety Review*

Many regulatory regimes and corporate standards have a requirement for 'periodic' or 'thorough' review of a safety case to ensure that it remains valid. At first sight, the potentially wide ranging nature of periodic reviews can appear daunting and onerous, with the potential to place a heavy burden on resources. But a well-managed periodic review provides a great opportunity to strengthen the connection between the safety case and the real world – recognizing that the case for safety is based on real plant, processes and, arguably most importantly, people.

A large nuclear site operator took the opportunity to go beyond the traditional desktop study of the safety case documentation by safety personnel. Bow-tie workshops were arranged to involve operational personnel, whose detailed knowledge of the safety case was limited, but who had an excellent first-hand perspective of its practical implementation and direct experience of incidents. The approach improved understanding of both safety and operations personnel, resulting in a safety case that better reflected operations and operations that better reflected safety requirements.

3.7 *HSE Critical Systems*

A large oil company has chosen to use the bow-tie technique to identify hardware systems and operating and maintenance procedures which are critical to the safeguarding of asset integrity, i.e. those which are 'claimed' as controls on the bow-tie diagrams. These HSE-critical systems and procedures are subject to a transparent inspection, examination and testing program which includes verification by independent third parties.

3.8 *Enterprise Wide Risk Management*

A major natural resources organization has applied the bow-tie method to map its company-wide corporate risk management strategy, covering all risks including quality, financial, business, political, environmental, information technology, human resources, design and new technology.

At a medium-sized oil company the chief executive uses a 'risk dashboard' which comprises a single bow-tie encompassing all risks, with color-coded controls highlighting priorities for improvement, together with responsibilities for action.

If you want to remove the mystique of risk management, there is no better method than bow-ties

4. Tips for Success

There are several common pitfalls which can be encountered when applying the bow-tie method; the key ones are discussed here together with ways of avoiding them.

- Avoid working in a vacuum - Operational experience or bow-tie experience alone gives sub-standard results; a combination of the two is essential.
- Involve people – Use people in building bow-ties, reviewing tasks and identifying areas for improvement. Whilst a first pass bow-tie can be developed as a desk top study, quality can only be assured by involving competent people who know how activities are in reality carried out in the workplace and have an in-depth understanding of the plant, the operation or design. This ensures that risk controls and supporting tasks accurately reflect actual practice rather than the preconceptions of senior management or the risk department. However, to arrive quickly at a truly representative bow-tie structure and avoid getting stuck in the detail, it is accepted best practice to use independent facilitation to solicit the input from the workforce. The facilitator should have practical experience of building bow-ties. The subjective nature of the method means that different groups of people may produce different bow-ties for the same event. However,

this is of secondary importance; the main point is that the assessment should be complete and the key risk controls and tasks captured.

- Pitch at the correct level of detail - Too high and the bow-tie is meaningless; too low and the exercise is labor intensive.
- Make sure the control responsibilities don't all finish up at the manager level - Care needs to be taken when setting the level of detail. Controls should be independent and self-explanatory. Tasks need to be meaningful and assigned at a level where their completion can be verified. Typically when building bow-ties for assessing major hazards (e.g. multiple fatalities, massive environmental impact, extensive asset loss, international reputation damage), tasks supporting the controls should be pitched at the supervisor or team leader level.
- Balance the information between the bow-ties and the tasks - Depending on the intended use of the bow-ties, it may be appropriate to keep the diagrams simple, with few words, and present any detailed information in the tasks. This approach can be appropriate when using bow-ties to represent the management of major business risks across the company for example. Alternatively, it is sometimes beneficial to keep the detail in the diagram, e.g. where a bow-tie is to be used for training purposes during a pre-job safety meeting.
- Prioritize effort - Don't get carried away. The bow-tie method is a flexible, generic tool that can be applied to any type of risk at any level. It can equally be used for assessing lower consequence 'workplace' hazards (e.g. slips, trips, and falls) as well as for assessing major business risks. It provides useful information for pre-job safety meetings, for example[33,34]. When using the technique at this level, it may be prudent to select representative workplace hazards with care and prioritize resources at those which are most common, are of greatest concern or are unique, otherwise significant resources could be demanded and the method called into question.
- ALARP demonstration - Bow-ties are an appropriate tool for qualitative demonstration that risk is managed to a level which is As Low As Reasonably Practicable (ALARP). However, avoid barrier counting where possible. Instead, in addition to the controls currently in place, the team must ask "what additional, practical controls can we implement?" There is one school of thought that advocates setting numerical risk acceptance criteria for bow-ties, (e.g. there must be at least three independent controls for every threat). The danger with this approach is that the assessment becomes a 'control counting' exercise, with dependent controls artificially represented as separate control measures in order to meet the criteria and with the assessment stopping once the requisite number of controls has been confirmed. It is better to use the cumulative experience of the bow-tie building team to review the completeness of the assessment as a whole and confirm the number, suitability, quality and effectiveness of the controls and supporting critical tasks. The important question must always be asked "is there anything more we can reasonably do?"

- Verify control measures and tasks - Follow-up with an audit. Depending on the make-up of the group who develop the bow-tie and the expertise of the facilitation, there can be a danger that the diagram represents only a single opinion, has serious omissions or does not represent what actually happens in the 'real life' situation. A follow up audit or inspection helps to ensure the credibility of the bow-tie and the completeness of the management arrangements. It verifies that the controls are actually in place and the critical tasks ensuring control effectiveness are being carried out. This is particularly useful when bow-ties have been developed for a new project, when there may be limited information available, when procedures are still to be developed and roles are not yet assigned. The verification audit can be carried out as part of pre-start up activities.

- Software helps but don't get hung up on it - The true benefits from the bow-tie process are largely independent of the means by which the bow-tie is constructed, e.g. by hand or electronically[35]. A number of software tools are available to construct bow-ties and manage the information behind the diagram. It should not be forgotten however that many of the benefits of the approach are associated with the actual implementation of the process and involvement of the workforce, which is often easiest to achieve using hand drawn bow-ties in a brainstorming, workshop setting. Software is ideal for speeding up the reproduction of bow-tie diagrams and organizing the information for future interrogation, retrieval and update.

- Use the method to its full potential - As use of bow-ties has become more widespread, partial assessment has become more common, with the analysis ceasing once the diagram is constructed. While this gives a graphical demonstration of risk control, it provides no more information than other risk assessment tools which illustrate the controls in place at that moment in time, e.g. HAZOP, What-If. In other words the bow-tie diagram on its own is just a 'snap shot' of the current risk control arrangements. What is missing is the direct and visible link between the controls as they are today and the procedures and people responsible for ensuring they will continue to be effective tomorrow. This understanding is only achieved by identifying and documenting the critical tasks and/or critical procedures which are crucial for ongoing functioning of the controls.

- Keep the end objective in mind - It can be easy to get carried away when using the bow-tie method. For example, for an operating plant the key controls are in making sure that the installed equipment keeps working properly, not in assessing the quality control of the design process from many years before. Similarly, there may be little point in reproducing the large number of individual steps taken to control an event if they are already well documented in a work instruction or manual. Reference to the external document is usually sufficient.

- Quantification - use the right tool for the job - The bow-tie is sometimes described as "like a fault-tree on the left hand side with an event-tree on the right". Some risk assessors interpret this as an opportunity to try to quantify the risk level, but the bow-tie is wholly qualitative, without any of the complex logic of fault and event trees. While a well-constructed bow-tie can be used to support a quantitative LOPA approach, most attempts to quantify the risk tend to miss the main point of the bow-tie, which is to identify how the management system provides assurance that risks will continue to be properly managed in the future.

The method helps to ensure that risks are managed rather than just analyzed

5. Conclusion

This paper presents the lessons learned from application of the bow-tie method across a number of business sectors. Our experience has shown that the bow-tie is ideal for structured assessment and communication of risks, clearly demonstrates the link between control measures and management system arrangements and can be used to qualitatively assess and demonstrate control of all types of risk.

6. References

[1] University of Queensland, Australia, Minerals Industry Safety and Health Centre, National Minerals Industry Safety and Health Risk Assessment Guide. www.mishc.uq.edu.au/NMIRAG/NMISHRAG_Chapter4_4.1.5.asp

[2] Primrose, M.J., Bentley, P.D., van der Graaf, G.C., Sykes, R.M. Shell International Exploration and Production B.V., "The HSE Management System in Practice-Implementation," SPE 35826, 1996.

[3] Primrose, M.J., Bentley, P.D., van der Graaf, G.C. Shell International Exploration and Production B.V., "Thesis – Keeping the Management System "Live" and Reaching the Workforce," SPE 336034, 1996.

[4] Gower-Jones, A.D., van der Graaf, G.C., Milne, D.J. Shell International Exploration and Production B.V., "Application of Hazard and Effects Management Tools and Links to the HSE Case," SPE 36031, 1996.

[5] Gifford, M.J., Gilbert, S.M. Atkins Defence Systems, Barnes, I. Ministry of Defence, "The Use of Bow-tie Analysis in OME Safety Cases," ESAS 03, 2003.

[6] Couronneau, J.C., Tripathi, A, Transoft International, "Implementation of the New Approach of Risk Analysis in France," 41st International Petroleum Conference, Bratislava, 2003.

[7] Marine Risk Assessment, Prepared by Det Norske Veritas Ltd. for the Health and Safety Executive, Offshore Technology Report 2001/063, 2002.

[8] "Application of QRA in Operational Safety Issues," Prepared by Det Norske Veritas Ltd. for the Health and Safety Executive, Research Report 025, 2002.

[9a] "Lines of Defence/Layers of Protection Analysis in the COMAH Context," Prepared by AMEY Vectra Ltd. for the Health and Safety Executive. www.hse.gov.uk/research.

[9b] Guidance on Risk Assessment for Offshore Installations, UK HSE Offshore Information Sheet No. 3/2006.

[10] Victorian WorkCover Authority, Major Hazards Division, Major Hazards Facilities Regulations Guidance Note 14, Safety Assessment under the Occupational Health and Safety (Major Hazard Facilities) Regulations, 2002. mhunit@workcover.vic.gov.au.

[11a] Review of the South Island Rail Coal Route, by Kellogg Brown and Root for the Land Transport Safety Authority of New Zealand, MET351-X-REP-001, Rev 1, June 2004. www.ltsa.govt.nz/rail/coal-route.

[11b] Asset Integrity – the Key to Managing Major Incident Risks, The International Association of Oil and Gas Producers (OGP), Report No. 415, December 2008.

[11c] Petroleum and Natural Gas Industries – Offshore Production Installations – "Guidelines on Tools and Techniques for Hazard Identification and Risk Assessment," ISO 17776, 2000.

[11d] Health, Safety and Environment Case Guideline for Mobile Offshore Drilling Units, International Association of Drilling Contractors, Issue 3.2, October 2006.

[12a] Safety Management System and Safety Culture Working Group (SMS WG) – Guidance on Hazards Identification, ECAST European Strategic Safety Initiative, March 2009.

[12b] Bow-Tie Analysis, https://www2.hf.faa.gov/workbenchtools/default.aspx?rPage=Tooldetails&subCatId=43&toolID=21

[12c] Safety First – Scenario Analysis under Basel II, McConnell & Davies, April 2006.

[13] Shell International Exploration and Production, DHSE – Developing the HSE Case Training Course.

[14] "Demonstrating How Hazards Are Being Properly Managed," RISKworld, The Newsletter of Risktec Solutions Ltd, Issue 1, 2002.

[15] Lidstone, A., EQE International Ltd. THESIS: "The Health Environment and Safety Information System – Keeping the Management System 'Live' and Reaching the Workforce," IChemE Hazards XIV, Cost Effective Safety, 1998.

[16] Kandola, B., Sullivan, M, ABS Consulting, "The Use of Bow-tie Model and THESIS in the Control of Major Accident Hazards," FABIG Technical Meeting, October 2003.

[17] Trbojevic, Dr. V.M., EQE International Ltd., "The Use of Risk Assessment to Improve Safety Management Systems in Ports," Journal of The Dock and Harbour Authority, Volume 79, Nos. 889, 890, 891, 892, 1999.

[18] BS EN ISO 14001, Environmental Management Systems, 1996.

[19] BSI-OHSAS 18001, Occupational Health and Safety Management Systems – Specification, 1999.

[20] E&P Forum, "Guidelines for the Development and Application of Health, Safety and Environmental Management Systems," Report No. 6.36/210, July 1994.

[21] BSI PAS 56:2003, "Guide to Business Continuity Management," 2003.

[22] The UK Financial Reporting Council Combined Code on Corporate Governance, 2003.

[23] "The Control of Major Accident Hazards Regulations (COMAH)," Statutory Instrument No. 743, 1999.

[24] Considine, M. BP, "Demonstration in COMAH Safety Reports," 26th FABIG Technical Meeting, April 2002.

[25] Joint Guidance Prepared at a Workshop of Representatives of the COMAH Competent Authority and the Chemical Industries Association (CIA), Issue 1.1, March 2002. www.hse.gov.uk/hid/land/comah2.

[26] "A Guide to the Control of Major Accident Hazard Regulations 1999," HSE Books, L111 1999.

[27] ADNOC "Health, Safety & Environment Management System Guidelines," Version 2.0, January 2002.

[28] ADNOC "Code of Practice on Control of Major Accident Hazards (COMAH)," ADNOC-COPV5-01, May 2004.

[29] Tripod Delta. www.tripodsolutions.net.

[30] Tripod Beta Incident Analysis. www.tripodsolutions.net.

[31] Dowell, A.M. III "Layer of Protection Analysis for Determining Safety Integrity Level," ISA Transactions 7 155-165, 1998.

[32] Amadi, A., Engelmann, G., de Gier ,W., Goler, W, Heckman, J., Row, Z., Twilhaar, R, Schwartz, D., Wilne, T. Shell Deepwater Services and EP Projects, Five Winds International, "Taking the Next Step in Managing HSE Critical Activities & Hazards." www.checkmatesoftware.info.

[33] Canadian Petroleum Products Institute Western Region Guidance Document, "Working Alone Safely," May 2001. www.cppi.ca/publ.

[34] "Step Change in Safety, Task Risk Assessment Guide," 2003. www.stepchangeinsafety.net.

[35] BowTieXP, "The Next Generation Bow-tie Methodology Tool," 2009. www.bowtiexp.com.

High Integrity Protective System Design Using a Risk-Based Approach

Robert J. Stack
The Dow Chemical Company
Hillsdale, MI 49242
stackrj@dow.com

Dow Co-Authors:
John Armstrong [Houston, TX]
Don Eure [Plaquemine, LA]
Ron Johnson [Midland, MI]
Scott Tipler [Midland, MI]
Tim Wagner [Midland, MI]
Sofka Werkmeister [Freeport, TX]

Prepared for Presentation at
American Institute of Chemical Engineers
2010 Spring Meeting
6th Global Congress on Process Safety
San Antonio, Texas
March 22-24, 2010

UNPUBLISHED

KEYWORDS: HIPS, Integrity, Protection, System, LOPA, Consequence, Relief

ABSTRACT

In 2007 API 521/ISO 23251 published guidance on the use of High Integrity Protective Systems (HIPS). The Dow Chemical Company (Dow) updated its internal work process to apply Dow's risk based work processes to HIPS design, application and evaluation. Rather than just focusing on the vessel related consequences of substituting a HIPS for a pressure relief device (PRD), the work process calls for stepping back and looking at the overall scenario and all the potential consequence outcomes. Using a risk based methodology requires looking at a much wider consequence potential than what Dow has traditionally focused upon for the design of relief systems.

By using Layer Of Protection Analysis (LOPA) to evaluate the consequences of the overpressure scenarios and determine the consequence severity factor or LOPA Target Factor for each scenario, the wider consequence potential is covered. The High Integrity Protection System (HIPS) safety integrity level (SIL) is determined by the highest risk scenario, which is typically the scenario with the greatest LOPA consequence severity factor. The resulting SIL level for the HIPS is adjusted to close the HIPS scenario risk gap. Other applicable LOPA independent Protection Layers (IPLs) are included in the LOPA risk evaluation. The HIPS design is verified by calculation to meet the required SIL level.

There are pressure relief device (PRD) services for which PRDs can't deliver the over-pressure protection that Dow needs and expects. The updated work process also identifies key opportunities for using a HIPS when: a conventional PRD is not practical or possible, a conventional PRD will not be reliable or a conventional PRD will work but will result in high treatment cost. Each of these three opportunities is discussed.

The overall result is an efficient process that links together the existing work processes for conventional relief design, Layers of Protection Analysis (LOPA) and Safety Instrumented Systems (SIS). This produces a risk based method for applying and designing protection layers into a HIPS. This paper gives an overview of this work process with some application examples and describes how this work process is used to improve safety while reducing risks and potentially reducing costs.

1. INTRODUCTION

Conventional pressure relief devices provide highly reliable and cost-effective protection for most pressure containing equipment. But there are applications where conventional relief devices are not reliable or practical. In these scenarios a High Integrity Protection System (HIPS) can be used.

So what is "HIPS?" In the simplest sense it is an "instrumented protection system" that is used in lieu of a relief device to mitigate a single overpressure scenario. A HIPS normally is comprised of sensors detecting the development of an overpressure situation, final element(s) to interrupt the pressure increase and a logic solver to tie the sensors and final element(s) together.

Typically, HIPS reduces the probability of a specific overpressure scenario to a point where the specific scenario meets some predefined risk. This normally will allow the relief designer to specify a smaller relief system to address the remaining scenarios. There are rare occasions where a HIPS totally replaces a relief device.

The HIPS concept is defined in API 521 (5th edition – 2007) Annex E, along with DIN V 19250 (DIN EN 51508). It is also important to recognize that HIPS applies to "underpressure" scenarios as well as "overpressure" scenarios.

A HIPS system is typically used for one of the following primary reasons:

1. A conventional pressure relief device is "not possible or practical"
2. A conventional pressure relief device is "not reliable"
3. Installing a conventional pressure relief device requires "excessive cost for treatment"

Let's look at these individually, in more depth:

NOT PRACTICAL OR POSSIBLE:
A conventional relief device may not be practical or possible if the relief device needs to be overly large. This often occurs when designing a relief device for:

- reactive chemical relief scenarios
- scenarios involving small equipment, such as in research and market development facilities
- relief scenarios protecting against excessive temperature, such as over-firing a cracker or heat transfer furnace

NOT RELIABLE:
A conventional pressure relief device may not be reliable when used in applications:

- that may cause plugging, such monomer service
- that are involved with overly corrosive or erosive materials
- with fluids that tend to freeze, or cure in the relief system

EXCESSIVE COST:
When relief valves are routed to a flare or scrubber system, the real cost is often much higher than that for just the relief devices. That is especially true for new installations and for existing systems that are already at their maximum capacity.

- The incremental load from a relief device to an existing flare system or scrubber system might be too much for the system to handle, resulting in a costly capacity expansion project.
- For new flare system or scrubbers, the initial cost can often be greatly reduced by selectively using HIPS to eliminate high loads from a limited number of relief devices. For example, plant-wide loss of cooling is often the controlling scenario that determines the size of a flare system. For distillation columns, the pressure relief load due to loss of cooling is often much higher than that for the other remaining scenarios.

- The flow from new relief devices into an existing flare header might create excessive backpressure on some of the other safety valves already in that system. Upgrading the existing safety valves, or the flare header, could be very costly.

2. IMPLEMENTING HIPS

So now that we know why HIPS are used, let's discuss how they are implemented. A relief device's main purpose is to prevent catastrophic failure of a pressurized vessel that may create destructive shrapnel and/or destructive pressure waves. And since a HIPS is replacing a relief device for specific relief design scenario, one might simplistically assume that the HIPS needs to be designed to the same integrity as the relief device. A typical relief device has a failure probability similar to a Safety Instrumented System (SIS) that is designed to a Safety Integrity Level (SIL) between SIL-2 and SIL-3.

Traditional HIPS Work Processes have used a **prescribed solution** to accomplish this task. For example: "a HIPS shall be a SIS with integrity of at least SIL 2.5." Some institutions even mandate a Safety Integrity Level of "3" as a minimum. This rule based solution is:

- simple to implement
- safely covers most cases because it is conservative by nature
- consistent with the philosophy that a relief device simply prevents a vessel from failing due to over pressure
- needs only one HIPS PFD calculation for the standard solution
- consistent with other rule bases standards, such as NFPA

But this prescribed solution is based on the concept of simply substituting a HIPS for a PSV. By stepping back and looking at the overall scenario, Dow is able to consider the "consequence" or "risk" of over pressurization. (Remember that a relief device or its equivalent SIS used in a HIPS has a probability of failure on demand.) Vessel fragmentation and the resulting pressure waves are certainly important, but what if the **contents** of the vessel create a greater hazard? This may be a very serious issue if toxic or flammable materials are involved. Shouldn't the population density in the vicinity of the potential accident be an important design consideration? In some situations can't the release of materials be an even more important concern than vessel fragmentation?

Let's look at some examples of varying consequence:

1. If a hot water tank over pressurizes, the resulting shrapnel or pressure wave could certainly create potential fatalities
2. When an extruder feed line containing a heat exchanger over pressurizes, the only concern is some hot polymer oozing out of ruptured flange gaskets (which are the weak spot in the line)
3. If a chlorine sphere over pressurizes and ruptures, there could potentially be multiple fatalities in both the production facility and in the community at large

IEC 61511 defines rules for implementing Safety Instrumented Systems in the process sector. These rules center around the concept of first defining the hazard, followed by defining the consequence of the hazard. As the severity of the consequence increases, so does the required safety system design integrity or reliability. So why hasn't the consequence of over pressurization ever been considered in HIPS design?

Dow decided to use the IEC 61511 philosophy in HIPS design. Since Dow already uses LOPA for general process hazard quantification, scenario likelihood, and protection layer identification, it was a natural evolution to use LOPA to assess the HIPS scenario for:

- the consequence of over pressure
- the likelihood of over pressure
- the necessary layers of protection to reduce the likelihood of over pressure

These protection layers – as per the LOPA Work Process - are not limited to Safety Instrumented Systems, but can also consist of:

- Basic Process Control actions
- Operator Response to Alarms
- Management Systems
- Other Safety Related Protection Systems

This higher level of analysis is more work than simply substituting a HIPS for a PSV. However it provides the opportunity to incorporate the existing protection layers and adjust the protection to fit the consequences. Therefore, Dow has no rule that requires a HIPS to be a SIL-2 or SIL-3 SIS. Instead, LOPA is used to determine the integrity level of any required SIS beyond SIL-1. The only rule regarding Safety Instrumented Systems is that the mitigation scheme must include a minimum of one SIL-1 SIS. But LOPA could also identify the need for multiple SISs or multiple HIPSs to manage severe risks.

3. WHAT API 521 SAYS

A very interesting idea and argument follows, but does it comply with API 521?

API 521 section E.4.2 says:

> *In accordance with ISA S84.01, a necessary step in safety instrumented system design is to* ***set a safety integrity level (SIL)*** *or availability value target for system design. The system is assigned as a SIL-1, SIL-2 or SIL-3 system, with SIL-3 being the most robust and most reliable and SIL-1 being the least.*

API 521 doesn't require a HIPS to be a SIL-2 or SIL-3.

API 521 section E.4.2 continues to say:

The determination of target SIL *for a given system is dependent upon:*

- *the risk associated with the hazard that the system is protecting against,*
- *i.e. the likelihood of the initiating and contributing events*
- *the magnitude of the consequences*
- *the credit that can be taken for other safeguards*

This sounds exactly like the philosophy that is the foundation behind LOPA. A closer look at this requirement reveals:

- The "*risk associated with the hazard*" is like the LOPA "scenario"
- The "*likelihood of initiating events*" is like the LOPA "Initiating Event Freq."
- The "*magnitude of the consequences*" is like the LOPA "Severity level" which is used to assign a LOPA target factor
- The "*credit taken for other safeguards*" is like non-SIS LOPA "protection layers"
- The "*determination SIL target*" is what you do for SISs defined in LOPA

So "YES," LOPA can be used (provided corporate risk criteria has already been defined).

API 521 section E.4.4 HIPS availability analysis says:

The purpose of the HIPS availability analysis is to evaluate the system performance of the proposed configuration. The availability analysis should utilize standard techniques, such as fault-tree analysis (see ISA TR-84.02).

Dow does this availability or PFD analysis for every HIPS system.

4. DOW'S HIPS WORK PROCESS

But implementing a HIPS at Dow is not an insignificant task. It is often more complex than designing and installing a relief device. There are many requirements that must be met. They are summarized here, and explained in further detail below:

1. Determine if HIPS is a better alternative
2. Gatekeeper review
3. Perform LOPA to determine SIL level
4. HIPS implementation using existing SIS work process
5. Follow Pressure Relief Design Work Process
6. Final review & approval

This Work Process includes the following as a minimum:

1. Determine if HIPS is a better alternative
   - Relief design, process design engineer or process safety review discovers the need for a HIPS
   - Defined HIPS application criteria must be met
2. Gatekeeper Review
   - Ensure HIPS meets application criteria
   - Provide guidance for following HIPS work process

3. Perform LOPA on HIPS scenario
    - Identify and verify applicable independent protection layers (IPLs)
    - Determine the required SIL for HIPS SIS (minimum SIL-1 required)
4. Follow Dow's SIS Work Process for identification and implementation of any Safety Instrumented Systems (SIS)
    - All SIS and LOPA work processes apply
    - Special HIPS identifier in SIS, LOPA, and relief design records
5. Follow Dow's Pressure Relief Design Work Process, which includes:
    - Responsibilities,
    - Design requirements,
    - Approvals
    - Integrity Management,
    - Flow chart of the overall relief design process.
6. Final Review and Approval
    - Stakeholder review and approval
    - HIPS registration and documentation archived
    - Note: Dow minimum design requirements must be followed:
        - Design must include at minimum a SIL1 SIS, regardless of how benign the consequence is determined to be (in accordance with API 521),
        - A qualified Relief Designer and reviewer has to be involved in the overall HIPS design,
        - The HIPS registration process is used for tracking and auditing purposes.

The benefit of using LOPA to quantify both the risk and the associated protection layers is that you have more flexibility to better meet the actual risk reduction needs. But these benefits are not free. The cost is that you must:

- define the LOPA overpressure scenario(s),
- calculate the LOPA target factor(s),
- design and verify the independent protection layers,
- demonstrate by PFD calculation per API 521 E4.4 that each HIPS meets the required SIL level,
- Audit all aspects of the study, design, and protection layers.

However, using LOPA is the normal work process required in Dow for new equipment – much of this work was completed in parallel with the Relief Design process.

DOCUMENTATION:
The requirements for Dow's SIS Work Process and for Dow's Pressure Relief Design Work Process must both be met.
The SIS work process includes documentation of the scenario and mitigation strategy in LOPA, a SIS functional specification, verification, validation, periodic proof testing, and audits/assessments required during the SIS's life cycle. Lastly, all engineering calculations and assumptions must be documented.

The Pressure Relief Design work process requires that a mechanical integrity equipment file be created for each relief device. Relief device documentation includes a description of the design scenario, relief design calculations, inlet/outlet piping pressure drop calculations, and a specification record. The documentation for the HIPS in included in this file. The LOPA scenario number is included for cross reference purposes.

For the rare case where the relief device is eliminated by the use of a HIPS system, UG-140 (previously known as Code Case 2211) verification and documentation is required. The UG-140 documentation is placed in the mechanical integrity equipment file for the process containment equipment.

APPROVALS:
Dow's Process Safety Technology Leader (PSTL) is the HIPS Gatekeeper for his/her business. This is a technical role and the PSTL must give the final approval for:

1. the application of HIPS
2. the Hazard Assessment and Evaluation (documented in LOPA)
3. the required SIL for the SIS component of a HIPS

<u>Why does Dow have a gatekeeper?</u> To ensure control of the work process to:

1. avoid misapplications of HIPS by ensuring the HIPS system fits prescribed applications
2. avoid unnecessary cost and waste of resources for incorrect applications
3. ensure the Process Safety Technology Leader is drawn into process
4. gain widespread and mutual approval of the LOPA work
5. conform with Dow's stated preference to use conventional relief devices
6. reinforce that HIPS is not intended to make relief devices "go away", rather it is to mitigate a specific relief scenario in accordance with Dow's risk criteria

Additional approvals are needed from:

1. the Production Leader
2. the Business Technology Center
3. a "qualified" relief reviewer
4. all the normal approvals required for a Safety Instrumented System

5. HIPS VERSUS SIS

Dow has historically performed risk assessments for Process Safety separately from our Relief Design process. They are separate processes performed by separate groups using separate acceptance criteria. This new HIPS work process presented a new paradigm and one important clarification had to be made: defining the distinction between when to use HIPS and when it is just a SIS:

- If the size of the relief device is affected, it is a HIPS application: elimination of a relief device sizing scenario, or making a relief device smaller
- If the frequency or probability of a relief device opening is affected, then it is purely a SIS/LOPA application. (In this case, the relief device is fully sized for all scenarios.)

  Note: Others may consider this a HIPS application.

So a HIPS system must include a SIS (for overpressure protection), but a SIS is not always part of a HIPS system. HIPS and SIS requirements are amazingly similar.

Table 1: Summary of HIPS versus SIS requirements

Requirement	HIPS	SIS
• LOPA Risk Assessment		
• Documentation		
• Registration in Relief files		
• Approvals		
• Follow the Work Process[1]		
• Testing		

The SIS work process used within the HIPS work process is a normal Dow work process. The PFD of the SIS must be calculated and verified to be adequate for HIPS application and/or required SIL level.

6. Common Examples Of Beneficial HIPS Application:

Figure 1. conventional PRD is not practical or possible to use

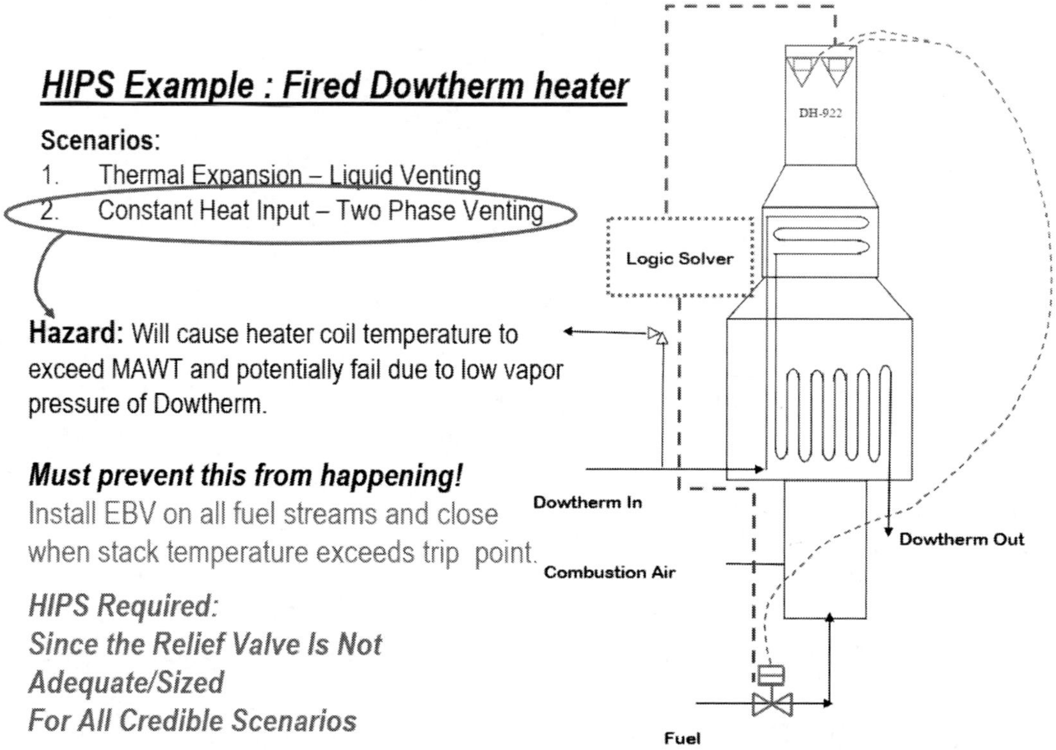

If the heater coil ruptures from high temperature, the sudden release of hot Dowtherm into the heater combustion chamber would cause a "vapor explosion" type scenario which could not be adequately vented by a conventional PRD.

Figure 2. conventional PRD will not be reliable example

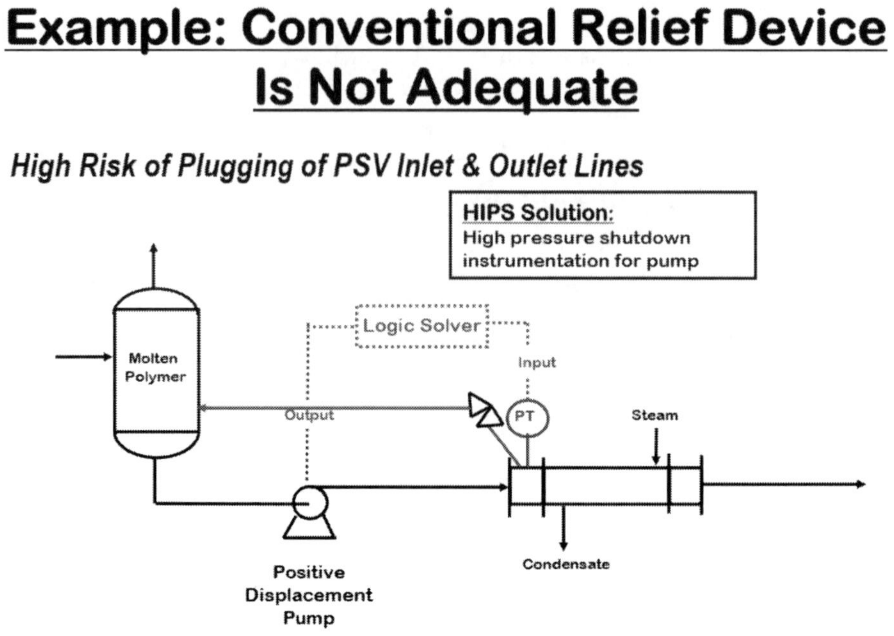

Figure 3. conventional PRD will work but will result in high treatment cost example

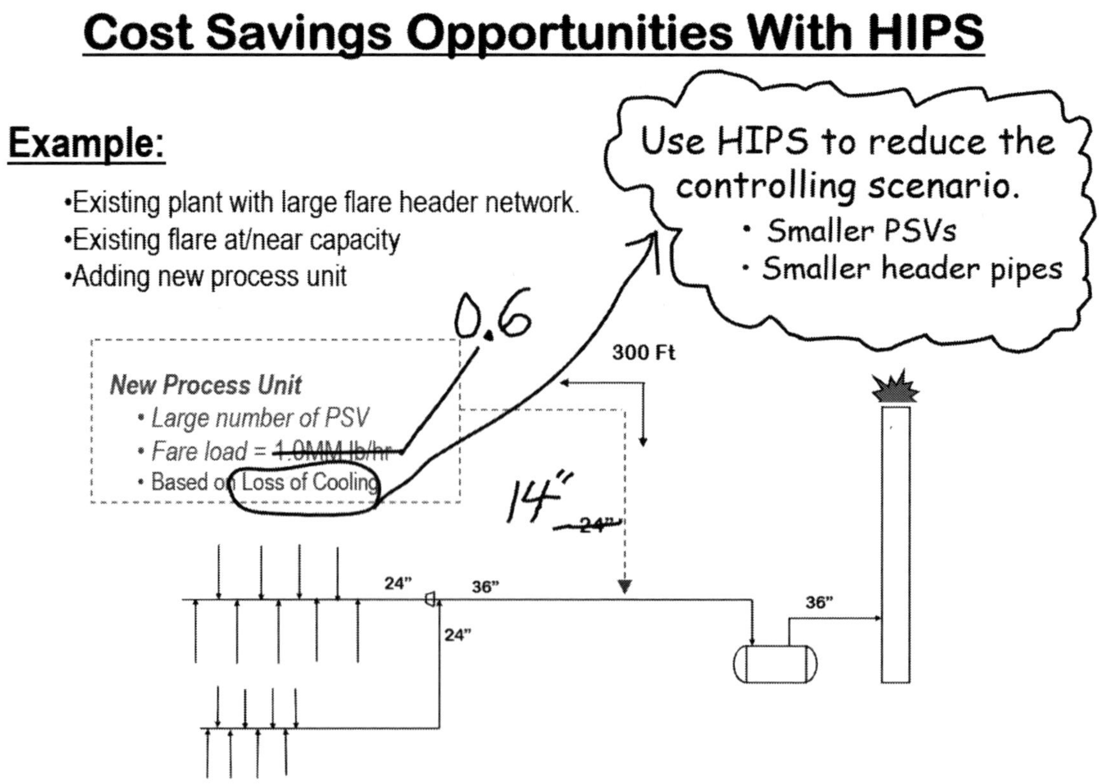

Figure 4. conventional PRD will work but will result in high treatment cost example

Cost Savings Opportunities With HIPS

Equip. Tag / Scenario	Relief flowrate (lb/hr)
C-1000	
N2 control valve failure	2000
Fire	50,000
Loss of Cooling	170,000
C-2000	
Reboiler tube Failure	10,000
N2 control valve failure	2,000
Fire	40,000
Loss of Cooling	150,000
C-3000	
N2 control valve failure	2,000
Fire	55,000
Loss of Cooling	160,000
C-4000	
Reboiler tube Failure	12,000
N2 control valve failure	2,000
Fire	45,000
Loss of Cooling	120,000

Use HIPS to defend against the loss of cooling scenario for these 4 columns.

- ✓ Reduces flare load by 410,000 lb/hr.
- ✓ Requires 4 HIPS....shutoff stm to reb.
- ✓ Smaller PSVs - sized for fire scenario

✓ Big cost savings !!

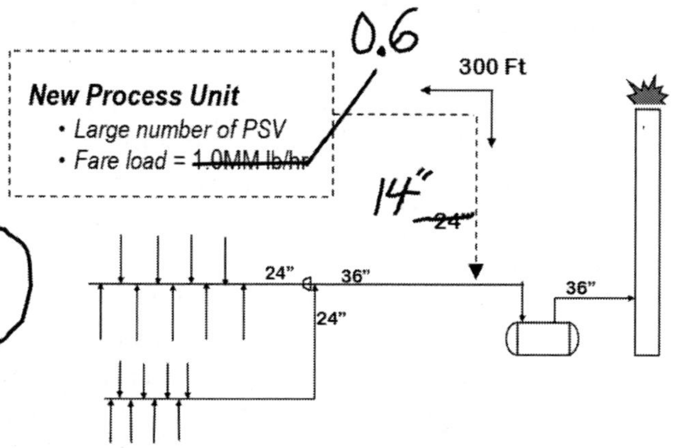

Note that most HIPS applications only mitigate specific relief scenarios, not all relief scenarios. HIPS process trip inputs can be pressure, temperature, flow, level or concentration, etc. depending on the relief scenario. Multiple process inputs are typically needed to design a higher SIL HIPS system. Process equipment can have multiple HIPS systems for multiple HIPS scenarios.

7. SUMMARY:

The consequence of a relief device failing is not limited to vessel failure and should also include the potential consequences of the specific material released. LOPA is an appropriate tool to quantify the consequences and risks of a failed relief device. It is a standard risk assessment tool at Dow. API 521 does not mandate any specific SIL level for a SIS used in a HIPS, but does state that the SIL determination is dependent upon "*credit taken for other safeguards*". A risk based approach provides more flexibility to better meet the actual risk reduction needs and allows for improved safety while reducing risks, and also potentially reducing costs.

8. References:

1. AANSI/API STANDARD 521 FIFTH EDITION, JANUARY 2007 ISO 23251, Petroleum and natural gas industries Pressure-relieving and depressuring systems
2. Anderson Greenwood Crosby Technical Seminar Manual - 4.0 Department Of Transportation Code Requirements - Gas Transmission And Distribution Piping Systems 4.2.4, 4.2.5 (2001)
3. API Recommended Practice 520: Sizing, Selection, and Installation of Pressure-Relieving Devices in Refineries
4. Sizing, Selection, and Installation of Pressure-Relieving Devices in Refineries," American Petroleum Institute (API), API Recommended Practice 520, July 1990
5. API 617 API Std 617 Axial and Centrifugal Compressors and Expanders -2003
6. API 618 API Std 618: 2007 Reciprocating compressors for the petroleum, chemical and gas industry services. - 5th edition
7. API 619 API Std 619: 2004 Rotary-Type Positive-Displacement Compressors for Petroleum, Petrochemical, and Natural Gas Industries, Fourth Edition
8. API 672 API STD 672 Packaged, Integrally Geared Centrifugal Air Compressors for Petroleum, Chemical, and Gas Industry Services ISO/DIS 10442
9. API 675 Positive Displacement Pumps - Controlled Volume 1994
10. API STD 674 Positive Displacement Pumps - Reciprocating 1995
11. API STD 676 Positive Displacement Pumps - Rotary 1994
12. Application of Safety Instrumented Systems for the Process Industries, ISA Standard, ISA-S84.01-1996 and S84.01-2004
13. ASME Code Case 2211-1 Pressure Vessels With Overpressure Protection By System Design, Section VIII Division 1 and 2
14. Deutsches Institute für Normung (DIN) – DIN 19250, "Fundamental safety aspects to be considered for measurement and control equipment," 2008
15. DOT Guidelines 4.2.4. 4.2.5
16. DOT Pipeline Standard
17. Guide for Pressure-Relieving and Depressurizing Systems, High Integrity Protective Systems (HIPS)", American Petroleum Institute (API), API Standard 521, Jan. 2007
18. Guidelines for Pressure Relief and Effluent Handling Systems, Sec. 5.9, CCPS, 1998
19. International Electrotechnical Commission (IEC) - IEC 61511-1, Functional Safety: Safety Instrumented Systems for the Process Industry Sector, 2002
20. International Electrotechnical Commission (IEC) IEC 61508, Functional Safety of electrical/electronic/programmable electronic Safety Related Systems, 7 Parts, 1998 – 2000
21. ISO 4126-5 Controlled Safety Pressure Relief Systems
22. VDI/VDE 2180 Safety device of plants of process engineering with means of the process instrumentation (PLT) introduction, terms, conception (April 2007)

Utilizing Integrated Risk Assessment to Maximize Asset Integrity Management

Steve Soos IV
Meridium Inc, Roanoke, US
ssoos@meridium.com

Ramanathan B T Viswanathan
Meridium International, Dubai
rviswanathan@meridium.com

Prepared for Presentation at
American Institute of Chemical Engineers
2010 Spring Meeting
6th Global Congress on Process Safety
San Antonio, Texas
March 22-24, 2010

UNPUBLISHED

Utilizing Integrated Risk Assessment to Maximize Asset Integrity Management

Steve Soos IV
Meridium Inc, Roanoke, US
ssoos@meridium.com

Ramanathan B T Viswanathan
Meridium International, Dubai
rviswanathan@meridium.com

Keywords: Risk Management, RCM, RBI, SIS, HAZOP, EAM

Abstract

The advent of various safety standards like ANSI /ISA 84, IEC 61511, API 14C, API 580 / 581, SAE JA1011, IEC 60330 and PSM guidelines has been a good impetus in improving the availability and reliability of safety systems and processes with an objective to meet the requirements defined by OSHA, EPA, HSE and other governmental groups around the world.

Hazards analysis and risk assessments have become standard operating procedures in process plants over the past decade. Many organizations are carrying out more detailed asset specific risk assessments as part of Enterprise Asset Management programs. These risk assessments produce risk rankings and recommendations to maintenance / QA programs to mitigate these risks.

This paper addresses the interaction between various risk assessment processes like Risk Based Inspection (RBI), Reliability Centered Maintenance (RCM), Safety Integrity Level (SIL) analysis and movement toward developing a comprehensive risk management strategy. This paper envisages an asset integrity management approach to maintenance / QA programs wherein the integration of these analytical processes will result in a unified risk ranking. The unified risk ranking of assets can enhance safety, improve availability, reduce maintenance costs, and result in better operability.

1. Introduction

Over the years many new developments have taken place in the field of maintenance engineering and asset management and billions of dollars are spent on equipment maintenance around the world. Even though plant maintenance has undergone impressive progress in maintaining the equipment, still some challenges like cost, complexity, performance, environmental impact and safety implications in balancing the operating margins and profitability are yet to be addressed in achieving an effective asset management program.

Many major companies have adopted asset integrity programs of staged reviews throughout the plant life, to confirm all health, safety and environmental aspects have been identified and that solutions are correctly actioned in a timely manner as part of enterprise-wide asset management. The risk informed asset specific maintenance strategies are developed to address the equipment maintenance (such as in Reliability Centered Maintenance, Risk Based Inspection and Safety Instrumented System life-cycle management) which provides a comprehensive method of organizing an efficient maintenance program. Although these analytical methods provide function / asset specific maintenance strategies by evaluating the risk exposure, the balancing of risk from an overall asset management perspective is always a challenge. An asset integrity program without balancing the risks can lead to incorrect strategies which can increase the risk exposure or maintenance costs. This paper specifically addresses balancing of risks to develop effective asset performance management strategies which can result in critical success factors such as safety, product quality, better operability, profitability and reliable processes.

2. Risk Management and Balancing of Risks

The history of tragic accidents in the process industries (chemical, petrochemical, oil & gas, etc) in the late 70's and 80's has led to the development of various international safety and environmental standards to address all facets of process plants viz. mechanical integrity (API-570, API-580 / 581), reliability in design and maintenance engineering (SAE JA1011, IEC 60330), safety / protection systems (ISA-84-2004, IEC 61508, IEC 61511), etc. Many government organizations have stepped in to define their national health and safety standards for process industries like OSHA 29CFR 1910.119, the Seveso Directive, and UK HSE guidelines, in order to improve the safety and reliability of the processes by knowing the inherent process risk. These standards and guidelines have triggered the process of hazard identification and risk assessment for the industries in identifying the potential threats / process hazards and their impact on process safety / plant integrity.

2.1 Process Hazard Analysis

Hazard is defined as an inherent physical or chemical characteristic that is intrinsic to a material, or its conditions of use which has the potential for an accident resulting in undesirable consequences like fire, explosion or toxic release impacting the health, safety, environment and economy. Process Hazard Analysis (PHA) uses a systemized approach to identify and analyze hazards in a process. Process Hazard Analysis is a key requirement of the OSHA regulation "Process Safety Management (PSM) of Highly Hazardous Chemicals" (29 CFR §1910.119) and the USEPA's rule, "Risk Management Programs (RMP) for Chemical Accidental Release Prevention," (40 CFR Part 68). Process plants must identify the hazard scenarios in the process and the associated consequences / impacts on health, safety and environment using any of the following PHA methods:

1. What-If
2. Checklist
3. What-If/Checklist
4. Hazard and Operability Study (HAZOP)
5. Failure Mode and Effects Analysis (FMEA)
6. Fault Tree Analysis (FTA)

Hazard and Operability analysis (HAZOP) is the most common and widely used PHA method. It is a structured and systematic examination of a planned or existing process or operation in order to identify and evaluate problems that may represent risks to people, the environment or equipment, or efficient operation. HAZOP is a qualitative technique based on guidewords and is carried out by a multi-disciplinary team (HAZOP team) during a set of meetings. This paper discusses the risk-ranked HAZOP study (refer to Figure 1) on a process system that requires documents such as Process Flow Diagrams (PFDs), Piping and Instrumentation Diagrams (P&IDs), layout diagrams, Material Safety Data Sheets (MSDS), provisional operating instructions, heat and material balances, equipment data sheets and start-up and emergency shut-down procedures and involves systematic examination as follows:

1. Identify the process system
2. Divide system into subsystem / sections
3. Choose a node (specific location in the process) in the subsystem / section
4. Define the design basis and process parameters
5. Define the deviations by applying the guidewords to parameter
6. Identify the causes / initiating events for each deviation and the associated scenario
7. Determine the consequence for each cause
8. Determine the enabling conditions / consequence modifiers for each consequence
9. Evaluate cause and consequence pair to determine the unmitigated risk value
10. Define the existing safeguards for mitigation or prevention of hazard
11. Evaluate the safeguards to determine mitigated risk value
12. Identify recommendations to achieve the target safety level or risk value (viz. additional layers of protection, increased system testing, procedural changes, changes to design)

Figure 1. HAZOP Workflow

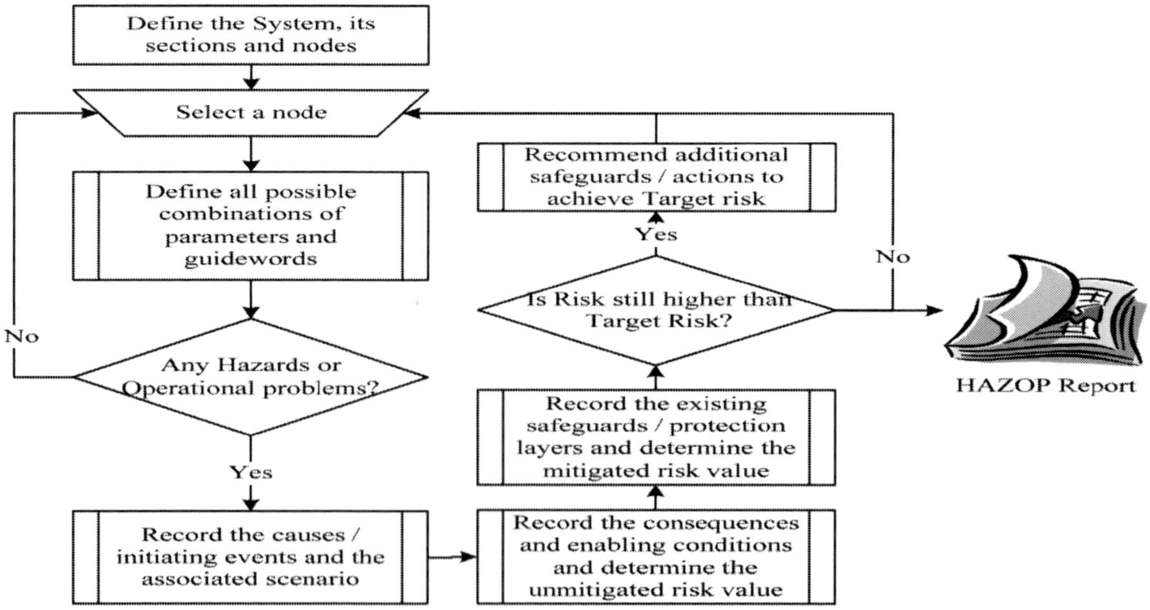

The above method helps in prioritizing the recommendations and to determine the implementation plan as part of planning and work management when integrated in an enterprise-wide asset management system. Also, this forms a basic framework for further asset specific analysis like Layer of Protection Analysis (LOPA) for Safety Integrity Level (SIL) determination and RCM-FMEA studies.

2.1.1 HAZOP Case Study

The sample HAZOP study for a chemical reactor - refer Figure 2 and Table 1 for details.

Figure 2. Case Study of Ethylene Reactor Process

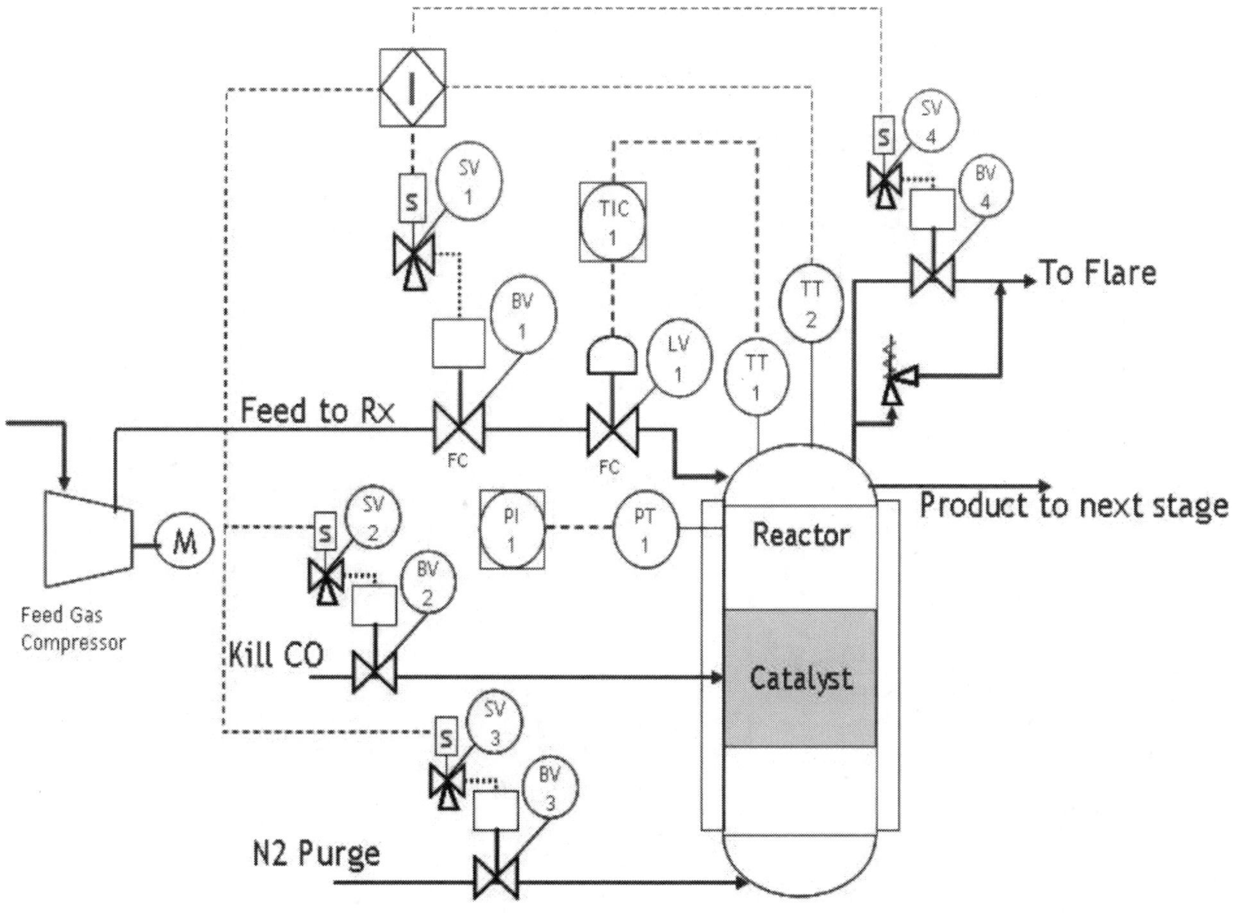

Table 1. Risk Ranked HAZOP Analysis

Site	**Petrochemical Co, USA**				
HAZOP#	**113**	**HAZOP title**	**Ethylene Process**	**Sheet**	**1 of 4**
Node #	**01**	**P&ID / Drwg #**	**P&ID/03003/PR-01.1**	**Date**	**28/10/2008**
Node #	**Ethylene Reactor, PR-01**			**Rev**	**01**

Deviation	Cause / IE	Scenario	Consequence	Enabling condition	Unmitigated Risk			Safeguards	Mitigated Risk			Recommendation
					S	L	RV		S	L	RV	
High Pressure	Feed valve fail open	Excess feed to Reactor causing Overpressure in Vessel	Loss of containment leading to release of flammable material to environment	Wall thinning due to corrosion / Hydride attack	500	0.3	150	1. PSV 2. Hardwire Alarm	500	0.1	50	1. Overpress SIF 2. Inspection Program
			Possible fire and exposure to operators	Presence of ignition source and operators	1000	0.3	300	1. PSV 2. Deluge system	100	0.1	10	
			Possible Explosion and affect of plant personnel	Presence of ignition source and failure of fire protection systems	1000	0.1	100	1. PSV 2. Entry restricted	500	0.1	50	
	Run-away reaction	Overpressure in Rx. Vessel	Same as above	-do-	1000	0.3	**300**	1. PSV 2. BPCS 3. Hardwire Alarm	1000	0.1	**100**	1. Overpress SIF 2. Reaction killing system
	High Inlet pressure due to compressor malfunction	Overpressure in Rx. Vessel	Same as above	All of the above and Failure of Rx inlet pressure control	500	0.3	150	1. PSV 2. Hardwire Alarm	500	0.1	50	1. Overpress SIF
	BPCS Failure	Overpressure in Rx. Vessel due to large overshoot from the set press.	Same as above	-do-	500	0.3	150	1. PSV 2. Hardwire Alarm	500	0.1	50	1. Overpress SIF

The HAZOP team selected the Ethylene Reactor system for study, and on review of the P&ID the first node was the Reactor Vessel (PR-01). Possible deviations and causes were then analyzed. "High Pressure" was selected as the first deviation and the team assessed the causes, consequences, safeguards and risk for each scenario. The unmitigated risk before safeguards and mitigated risk with safeguards are calculated using the 5x5 risk matrix as shown below to arrive at the Risk Value (RV).

Figure 3. HAZOP Risk Matrix

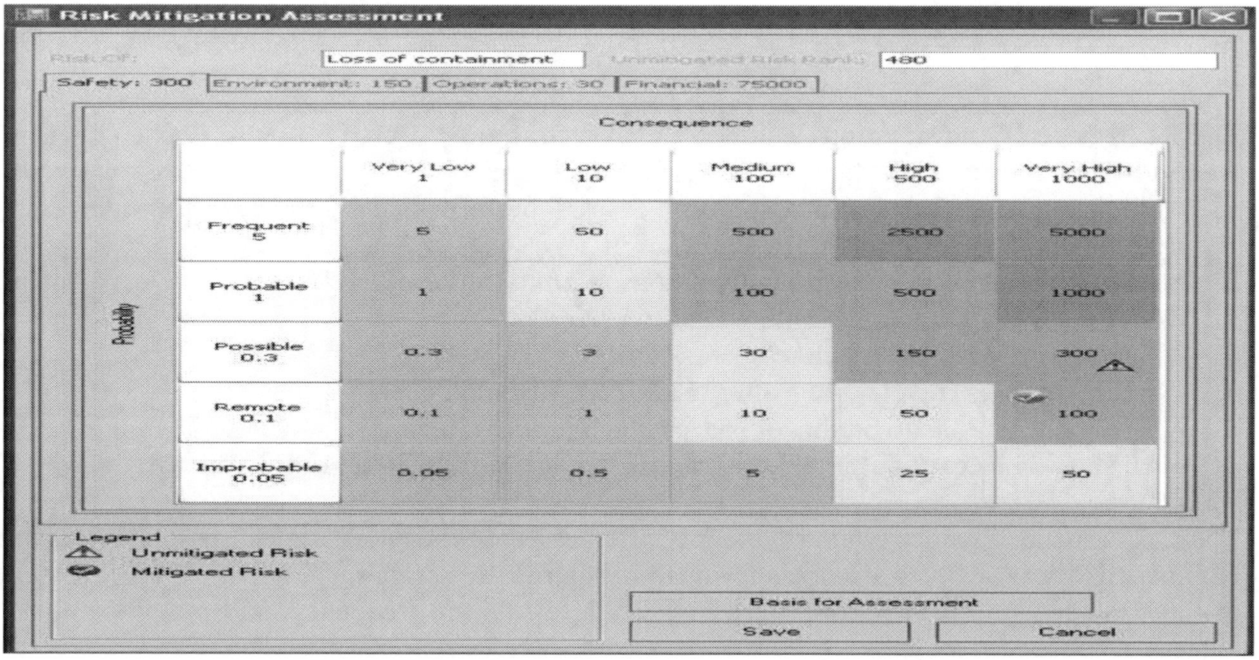

The HAZOP study indicates that the highest risk related to a high pressure deviation is due to a runaway reaction which leads to a risk value of '100' (refer to Figure 3). An overpressure protection function (SIF) is recommended to vent the reactor vessel in case of overpressure, as well as a "kill system" to stop the reaction, to achieve the target limits for safety, environment and economy.

In the HAZOP analysis many deviations, their causes, consequences, and safeguards were analyzed along with the associated risks.

2.2 Reliability Centered Maintenance (RCM)

"If you want to change the way people act... you have to change the way that they think"
– John Moubray

The term "Reliability Centered Maintenance" appeared for the first time as the title of a report prepared by United Airlines, on the processes used by the civil aviation industry to prepare maintenance programs for aircraft. In 1999 the Society of Automotive Engineers, Inc. (SAE) issued their Evaluation Criteria for RCM processes – SAE J 1011.

Reliability centered maintenance is a systemic process used to determine the maintenance strategy that has to be accomplished to ensure that any asset (equipment) is able to continuously meet its designed functions in its current operating context. RCM is concerned more with maintaining system function as opposed to maintaining individual component function. Thereby it leads to maintenance programs that focus Preventive Maintenance (PM) on specific failure modes likely to occur and is not equipment oriented, but rather is driven by its functions and risks. The functional failure consequences and risks (safety, environment and economic) define the maintenance priorities for a PM program.

Nowlan and Heap defined RCM as "A logical discipline for developing a scheduled-maintenance program that will realize the inherent reliability levels of complex equipment at minimum cost".

The RCM seven basic questions outlined in **SAE Standard JA1011 Evaluation Criteria for Reliability Centered Maintenance** (RCM) Processes are:

1. What are the functions and associated desired performance of the (**asset/system**) in its present operating context (**Functions**)?
2. In what ways can it fail to fulfil its functions (**Functional Failures**)?
3. What causes each functional failure (**Failure Modes**)?
4. What happens when each failure occurs (**Failure Effects**)?
5. In what way does each failure matter (**Failure Consequences**)?
6. What should be done to predict or prevent each failure (**Recommendations**)?
7. What should be done if a suitable RCM task cannot be found (**Default Actions**)?

The risk associated with the failure effect is often assessed using a Risk Matrix or Risk Priority Number (RPN). A RPN is a process for quantifying risk by assigning a numerical score for each of three elements: criticality (or severity of a failure), failure frequency, and the probability of detection.

RCM studies are usually conducted by the maintenance and reliability groups. The main objective is to develop appropriate maintenance plans based on probability and consequence of the analyzed failure modes.

2.2.1 RCM Case Study

In reference to the above case study on the Ethylene Reactor (refer to Figure 2), this system comprises various RCM components (sub-systems) such as the feed gas compressor, etc which are to be analysed for their functional failures, failure mode and effects along with the associated risk to determine the appropriate maintenance tasks / plans for these assets.

Table 2. RCM Analysis

RCM#	011	RCM title	Ethylene Feed gas Compressor	Sheet	1 of 8
Site	Petrochemical Co, USA	P&ID / Drwg #	P&ID/03003/PR-01.1	Date	05/12/2008

System	Sub-system / Asset ID	Function ID	Function	Functional Failure	Failure Mode	Failure Effect	Criticality / Risk score
Ethylene Reactor	Feed gas compressor (FC-001)	C1	Supply compressed feed gas to ethylene reactor	Fail to supply feed gas at a pressure of 100 bar to ethylene reactor.	Compressor last stage seal failure	External leak leading to fire	B (150)

Figure 4. RCM Risk Matrix

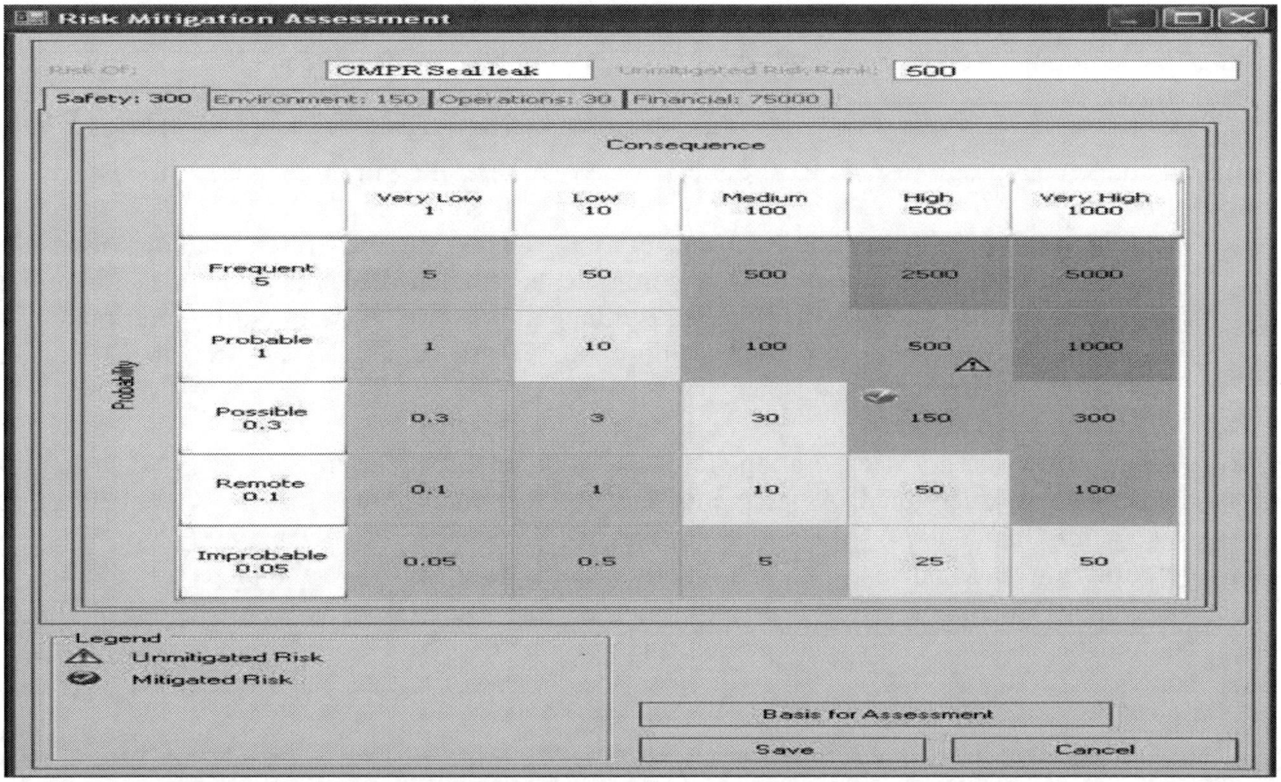

In the feed gas compressor RCM analysis (refer to Table 2), failure of the compressor to supply feed gas at the required pressure was analysed and the predominant failure mode was found to be "seal failure" which can result in an external leak and may lead to fire. The unmitigated risk (refer to Figure 4) was assigned 'High' (500) from safety / environmental aspects after considering the design aspects and material of the seals. The recommended maintenance tasks / actions to mitigate this failure will bring the risk down to a value of '150' (still 'High Risk').

In full RCM analysis of the feed gas compressor, there are multiple functions / failures / failure modes and effects with associated risks analysed to strategise the appropriate maintenance tasks / actions to mitigate the failure risks.

2.3 Risk Based Inspection

A Risk Based Inspection (**RBI**) process is primarily focussed on maintaining the mechanical integrity of pressure equipment items and minimizing the risk of loss of containment due to deterioration. This is normally accomplished by inspection and testing programs which are established to detect and evaluate deterioration due to in-service operation and may widely range from reactive programs to proactive programs (time-based or calendar-based). RBI, as a risk-based approach, focuses on static equipment and associated deterioration mechanisms representing the most risk to the facility. Mitigation thereby provides a better linkage between the mechanisms that lead to the failure (loss of containment) and the inspection approaches that will effectively reduce the associated risks.

API RP-580 is the guideline specifically targeted at the application of RBI in the hydrocarbon and chemical process industry. RBI is focussed on a systematic determination of relative risks to rank the equipment or components so as to focus the risk management efforts on the higher ranked risks.

Risk management programs viz. RBI; generally involve the following four-phase process:

1. Identify Specific Failure Modes or Deterioration Mechanisms - Different Failure Modes that apply to individual equipment items or components are identified. Failure modes are defined as the different mechanisms that may cause the equipment or component to deteriorate over time and eventually lead to failure. Failure Modes are sometimes referred to as Damage Types or Degradation Mechanisms.

2. Assess Risk - The deterioration rate and equipment tolerance to each deterioration mechanism is ascertained. Assessments can be Quantitative, Qualitative, or a blend of the two (semi-Quantitative). Quantitative assessments results in a Damage Factor for each deterioration mechanism. The Cumulative Damage Factor is obtained by adding the individual Damage Factors and is used to assess risk. Qualitative assessments depend on the equipment history, plant experience, and largely the expertise of the analyst performing the study.

3. Identify Risk Mitigation Alternatives – High risk equipment is examined and different risk mitigation alternatives are investigated. These alternatives might include a more sensitive Inspection Plan, Equipment Re-design, Equipment replacement, Equipment repair, or in some extreme cases a Change in Process.

4. Develop an Action Plan – An action plan is established to carry out the recommended actions that have been identified in phase three.

RBI alone cannot adequately cover all aspects of risk management. It should complement existing processes like PHA or HAZOP and RCM, but not serve as a substitute for them. RBI compliments the PHA by focussing on the mechanical integrity related deterioration mechanisms and risk management through inspection. For instance, an RBI assessment may define how inspection activities could mitigate the risk associated with loss of containment for a piece of equipment identified in HAZOP. However, when the same equipment neared the end of its life, RCM could define an "end of life" strategy involving equipment replacement or repair. RCM analyses are applied at a system level, and whenever a Loss of Containment function is subjected to an RCM analysis, and inspection activities are planned to mitigate the risk associated with the loss of containment, the RBI Analysis results could be used to complement the Equipment Plan.

2.3.1 RBI Case Study

The RBI study can include a review of the output from a HAZOP that has been conducted on the process unit being evaluated. In this case consider the HAZOP conducted on the Ethylene Reactor (refer to Figure 2 and Table 1). Hazards identified in HAZOP can be specifically addressed in the RBI analysis.

Table 3. RBI Analysis

Risk Based Inspection (RBI) Worksheet			
Site	**Petrochemical Co, USA**	**Sheet No.**	3 of 8
Units	**EO Process Unit**	**Date**	09/11/08
Asset	**Ethylene Reactor, PR-01**	**Ref Drawing:**	P&ID/03003/PR-01.1
RBI Component	**Reactor Vessel, PR-01**	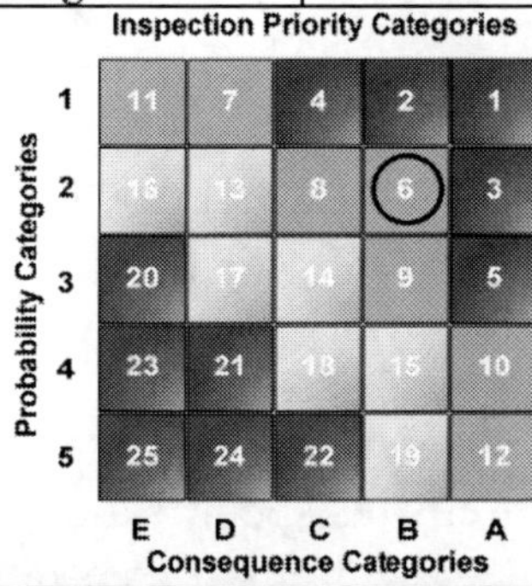	
Rep. Fluid	**EO**		
Rolled Up COF	**C**		
Rolled Up POF	**3**		
Overall Risk Ranking	**Medium (14)**		

Degradation Mechanism	**Probability of Failure (POF)**	**Consequence of Failure (COF)**	**Risk Ranking**	**Inspection Priority**
Internal Corrosion	4	B	Medium	15
External Corrosion	4	B	Medium	15
Erosion	4	C	Medium	18
Ext Chloride SCC	3	B	Medium High	9

RBI analyses are typically performed at a level of detail below that of the overall asset. This is because potential degradation mechanisms, rates of degradation, material resistance, consequences, etc are often different for the various portions of the equipment. These various portions are referred to as RBI Components. The RBI analysis data sheet above shows the four potential degradation mechanisms that were evaluated for the Ethylene Reactor (pressure vessel). In the case above, past inspection data revealed that External Chloride Stress Corrosion Cracking (SCC) and in turn, its Probability of Failure rating is driving the overall risk ranking of the reactor vessel. Once the RBI Analyses are completed for the other relevant portions of reactor vessel (viz. head, cylindrical shell, internals), an overall inspection strategy for the asset can be effectively established to address the various degradation mechanisms and prevent unplanned failures.

2.4 Safety Instrument System Life-Cycle Management

Safety Instrument System life-cycle management is the methodology to manage from conception, design and engineering, operation and maintenance, until decommissioning of those instrumented safety systems which perform the functions of protection or mitigation of process / critical equipment, from undesired consequences / failures.

SIS Life-Cycle Management was formalized by (ISA) when it published the ANSI/ISA 84-01-1996 'Application of Safety Instrumented Systems for Process Industries'. This standard was developed to provide a standard process for assessing the requirements for applying a safety instrumented system, designing the SIS and maintaining it. ISA superseded this standard in 2004 by the adoption the International Electro-technical Commission (IEC) 61511 Functional Safety – Safety Instrumented Systems for the Process Industry Sector which is actually a subset of IEC 61508 'Functional safety of electrical/electronic/programmable electronic safety-related systems'.
The new ANSI/ISA-84.00.01 2004 standard describes and defines the requirements for each of the following SIS Life-Cycle phases:

1. Hazards and risk assessment
2. Allocation of safety functions to protection layers
3. Safety requirements specification for the safety instrumented system
4. Design and engineering of the safety instrumented system
5. Installation, commissioning and validation
6. Operation and maintenance
7. Modifications
8. Decommissioning
9. Verification
10. Management of functional safety and functional safety assessment and auditing
11. Safety life-cycle structure and planning

The allocation of safety functions to protective layers is the phase that defines the process of assessing the risk and assigning a Safety Integrity Level (SIL) to a Safety Instrumented Function (SIF). The Safety Integrity Level is a measure of the safety margin or reliability of the instruments in the safety loop and is expressed as average probability of failure of the instrument loop.

A Safety Instrumented Function (SIF) is a safety function with a specific SIL that mitigates or prevents a consequence of a hazard / failure. ISA/IEC has defined the specific ranges (PFDavg) of safety integrity for each SIL. Table 4 represents the SIL levels for low demand operations.

Table 4. SIL Levels for Low Demand Mode

Low Demand Mode of Operation			
Safety Integrity Level (SIL)	Target average probability of failure on demand (PFD Avg)	% Availability	Risk Reduction Factor (RRF)
4	$\geq 1x10^{-5}$ to $< 1x10^{-4}$	> 99.99%	> 10,000 to ≤100,000
3	$\geq 1x10^{-4}$ to $< 1x10^{-3}$	99.9% to 99.99%	> 1000 to ≤10,000
2	$\geq 1x10^{-3}$ to $< 1x10^{-2}$	99% to 99.9%	> 100 to ≤1000
1	$\geq 1x10^{-2}$ to $< 1x10^{-1}$	90% to 99%	> 10 to ≤100

2.4.1 SIL Assessment

A SIL Assessment is conducted to determine the appropriate assignment of the SIL for each SIF. The standard allows for both qualitative and quantitative methods for selection of the SIL. Risk Matrix, Risk Graph, Layer of Protection Analysis (LOPA) and Fault Tree Analysis (FTA) are all acceptable methods. Risk Matrix (refer to Figure 5) is the most popular risk (SIL) assessment method used in the process industries. An organization will define the risk matrix and add an assignment of a SIL to all the risk levels that require an additional safeguard to achieve an acceptable or tolerable risk.

Figure 5. SIL Assessment Risk Matrix

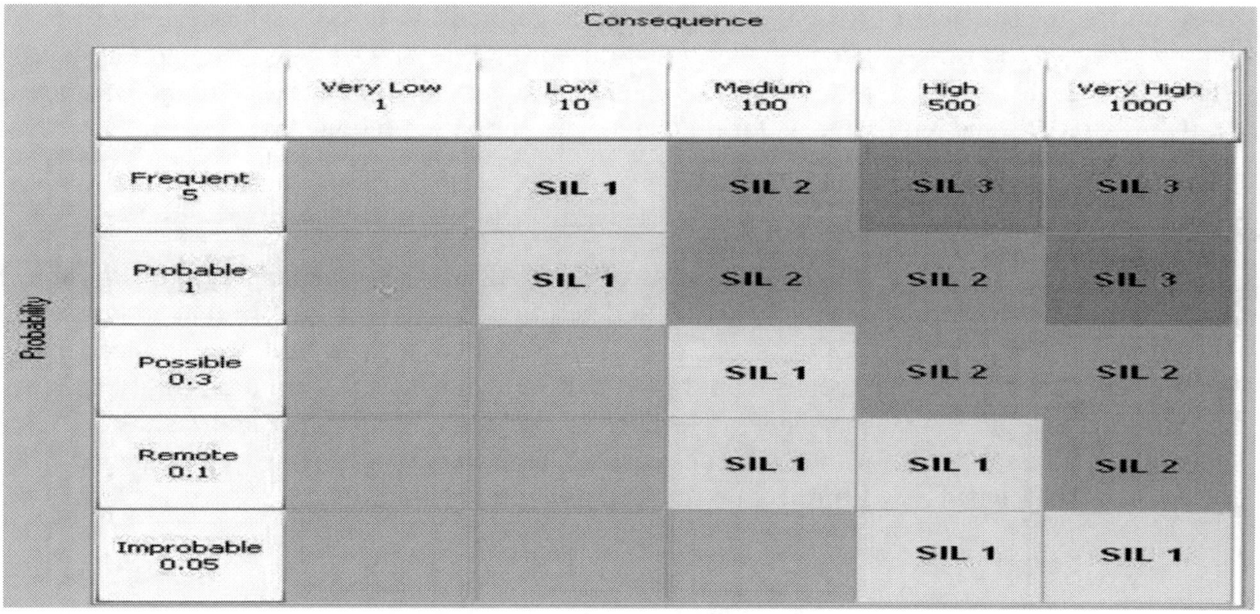

In Layer of Protection Analysis (LOPA), the hazard and its cause frequency and its consequences are determined and assigned a quantitative value as defined in their corporate Risk Matrix. Then the existing safeguards / layers of protection (like process controller, process alarms, relief devices, etc) for this event are determined and are assigned a failure probability to calculate the value of mitigated risk and remaining unmitigated risk. The unmitigated risk or the gap specifies the required safety integrity level for the SIF (refer to Figure 6).

Figure 6. Layer of Protection Analysis

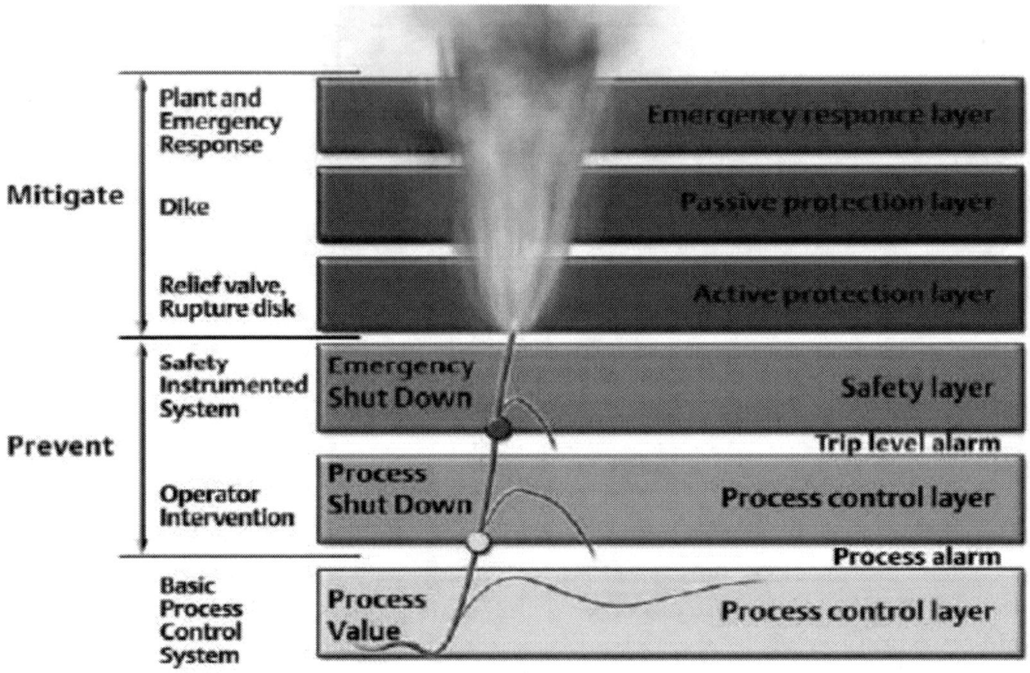

Once a SIL has been selected for a SIF, an SIS will be designed to take the process to a safe state based on the measurement of the initiating event as defined by the SIF. The design of the SIS will be analyzed to insure it can meet the Probability of Failure on Demand Average for that SIL. Higher SIL levels tend to cost more to implement and maintain because of the increase in availability targets and requirements for redundancy of the SIS components.

2.2.2 SIL Assessment Case Study

This SIL Assessment study utilized the results of the HAZOP risk assessment (refer to Table 1) for the possible release of the ethylene oxide to determine the SIL for the SIF that addresses that risk. If the Risk Matrix were not the same, the SIL Assessment team would need to reconcile the differences. Even less advantageous would be if the SIL Assessment team didn't utilize the risk assessment from the HAZOP or have a HAZOP to reference in an integrated fashion. Table 5 shows the SIL assessment for the ethylene reactor where SIF-2 and SIF-3 were assigned SIL 2 and SIF-1 was assigned SIL 1 (protection for feed gas compressor).

Table 5. SIL Assessment

SIL Assessment					
Site	Petrochemical Co, USA		Sheet No.	2 of 7	
Units	EO Process Unit		Date	09/11/08	
System	Ethylene Reactor, PR-01		Ref Drawing:	P&ID/03003/PR-01.1	
SIF	**SIF Description**	**Initiating Event**	**Hazardous Event**	**Consequence**	**Assigned SIL**
SIF - 2	PR-1 High pressure shutdown	Feed valve failure / run-away reaction	Overpressure of PR-1	Fire and/or explosion	2
SIF - 3	PR-1 High temperature shutdown	Exothermic reaction	Overpressure of PR-1	Fire and/or explosion	2
SIF - 4	PR-1 Inlet pressure high trip	High pressure due to compressor failure	Possible overpressure leading to inlet line rupture.	Release of EO and/or fire.	1

2.5 Assessed Asset Integrity Level Approach

The assessed asset integrity level is the process by which risk informed methodologies (HAZOP, RCM, RBI, SIS) and deterministic / performance based information (Event history and reliability analytics) are combined to establish appropriate levels of programmatic controls for systems, equipment and components in order to provide necessary assurance that these items will operate safely and activities are accomplished as prescribed.

The management and utilization of data for this process can be resolved if all these data are managed in one comprehensive system or database. The prime requirement is to have the plant equipment / assets identified in the HAZOP be defined in an equipment database / functional location in the Computerized Maintenance Management System (CMMS) / Enterprise Resource Planning (ERP) [SAP / Maximo] systems and be integrated with the risk assessment and reliability tools as part of an Enterprise Asset Performance Management (EAM) framework. The risk assessment definitions need to be uniform over all the methodologies (HAZOP / RCM / RBI / SIS), so that information or outcomes from one methodology can be mapped to other in order to achieve the risk balancing. This improves consistency of risk assessment information between these methodologies as well, and serves as an aid in compliance auditing and effective recommendation management. For example, HAZOP results can be complimented with RBI / RCM analysis outcomes to balance the risks and later can provide more qualitative support to the decision making process of risk management. Similarly, information from a HAZOP study can be directly mapped into a SIF during the process of SIL assignment which can reduce the effort of SIL assessment to a larger extent. Moreover, this process aids in management of mechanical integrity requirements for OSHA 1910 / PSM practices.

In the above case study of an ethylene reactor, an RBI study is conducted on the reactor vessel. The RBI study assesses the risk of loss of containment at 'Medium Risk' based on the past inspection information as well as the current inspection strategy being adopted. The Hazards Analysis set the risk of loss of containment at 'Medium High'. Based on HAZOP the SIL level for the SIF is set to SIL 2. This result does not pose a safety concern, but it does lead to more capital and maintenance cost to maintain the SIL 2 function. Also SIL 2 mandates stringent requirements on the frequency of functional tests / proof tests for the instrument loop which can lead to further burdens on the operations. If the results of the RBI study could be fed back to the hazards analysis team perhaps the risk in HAZOP would be reduced, which may lead to a lower SIL (i.e. SIL1), in turn saving time, money and posing a lower burden on operations.

Another scenario is considering the RCM study conducted on the feed compressor. The RCM study assesses all the functional failure modes and related consequences. The RCM study concluded that there is a significant risk (Medium-High) due to a moderate probability of a high consequence occurring. The Hazards Analysis set the probability of this consequence occurring at Low. The SIL assessed for the related SIF in this case was set to a SIL 1, considering the HAZOP output. In this case there could be a gap in the design of the SIF due to the lower-than-expected frequency of occurrence of compressor failure. This would have exposed the process / asset to an unmitigated risk leading to unreliable operation if the risk from the RCM study was not considered for Safety Instrumented Function design.

The Figure 7 below defines the logic and workflow of the assessed asset integrity level process in determining the appropriate level of program controls to be applied to plant assets. This program has three graded levels of program control such as 'Category-A', 'Category-B' and 'Category-C'.

Figure-7: Assessed Asset Integrity Level process workflow

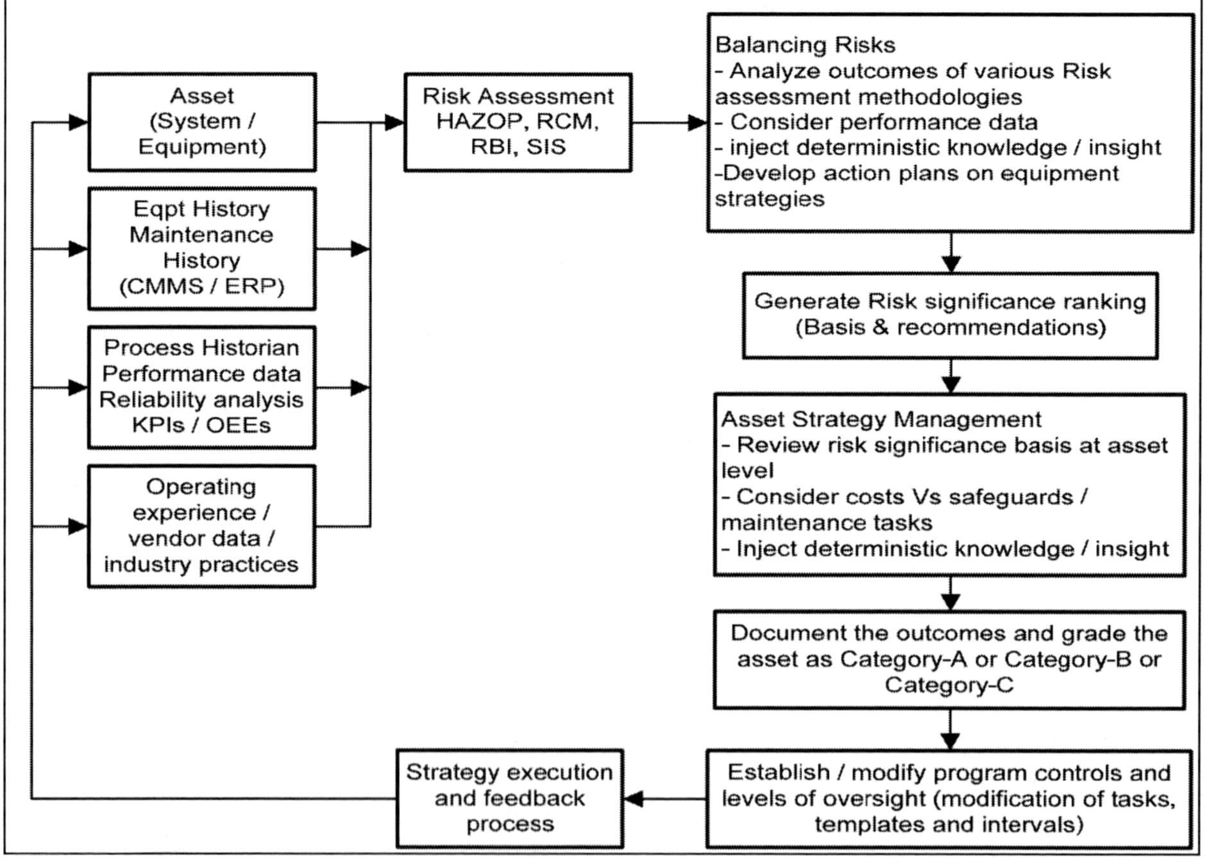

'Category-A' is for high integrity program controls that can be applied to systems / equipments determined to have 'High Risk' significance after balancing the risks. These programmatic controls are in full compliance with the HSE requirements and standards that they endorse. These controls shall be prescribed in implementing procedures and are afforded multi-tiered levels of oversight consisting of focused independent oversight in the form of audits, performance monitoring, risk / reliability assessment, evaluation, inspection and testing, as appropriate. These systems / equipment shall remain in this category, regardless of performance measures, due to their high risk significance. The test/inspection intervals and methodologies as well as PM tasks are to be strictly adhered to as recommended by the appropriate risk management methodology without any deviation.

'Category-B' is for system specific program controls that can be applied to systems / equipments determined to have 'Medium Risk' significance after balancing the risks. These are lower levels of control and oversight, designed to maintain and/or preserve those identified critical attributes of systems / equipment needed to support risk significant functions. They consider economical and efficient business practices while maintaining compliance with the basic HSE requirements. They do not reflect strict controls and potentially warrant a test / inspection interval extension or change in PM interval, however their reliability must be assured via a compensatory measure which can be a related testing or monitoring (condition / performance monitoring) function that serves to prevent the hidden failure. The test/inspection intervals and methodology may be modified based on the evaluation of design, service condition, performance history and compensatory actions.

'Category-C' is for business-driven program controls that can be applied to systems / equipment determined to have 'Low Risk' or 'Non-Risk' significance after balancing the risks. These are lower levels of control and oversight, limited to good engineering and business practices while maintaining the productivity targets and economics. They potentially warrant test / inspection interval extensions/deferrals and modifications in PM tasks/methodology, based on the service condition, failure modes and consequences, performance history and economics / production objectives.

The Assessed Asset Integrity Level approach provides a comprehensive risk management perspective throughout the operation and maintenance phase of the plant / process and covers the complete asset management activities starting from criticality/risk assessment, performance monitoring, equipment PM plans to in-service inspections/tests, surveillance tasks, corrective actions, recommendation/change management and long-term planning.

3.0 Conclusion

The Assessed Asset Integrity Level approach maps all the outcomes of the risk assessment methodologies to the corresponding assets / equipment to determine the overall asset criticality. This determination of asset criticality will help to develop the comprehensive equipment / asset strategies covering key aspects like maintenance tasks (PM / PdM/CM), PM templates, maintenance scheduling, inspection priorities and plans, testing requirements, life-cycle management and long-term planning. The Assessed Asset Integrity Level process assigns classifications by focusing on the risk relative to process safety risk and high economic /

production impact (tangible). When integrated to CMMS / ERP systems, this approach provides maximum benefits (high return on assets) as the maintenance history and test / inspection records can be used to validate the actual failure rates of the equipment and to verify whether they are in line with the estimated failure rates that were used during the design/analysis phase.

In an integrated Asset Integrity Management process (refer to Figure 8), the risk results of all methods can be easily presented and evaluated for overall balance. If a risk from one or more of the analysis methods is not within expected limits a recommendation for reassessment or system modification can be made. The expected benefits of the implementation of this process include identification and ranking of components and thereby strengthening the focus on PM activities, inspection plans, component reliability trending, performance monitoring and station budgeting. It also helps to integrate, focus and streamline existing risk management processes to establish long-range system reliability plans to achieve a goal of increased plant reliability and availability.

Figure 8. Integrated Asset Integrity Management Framework

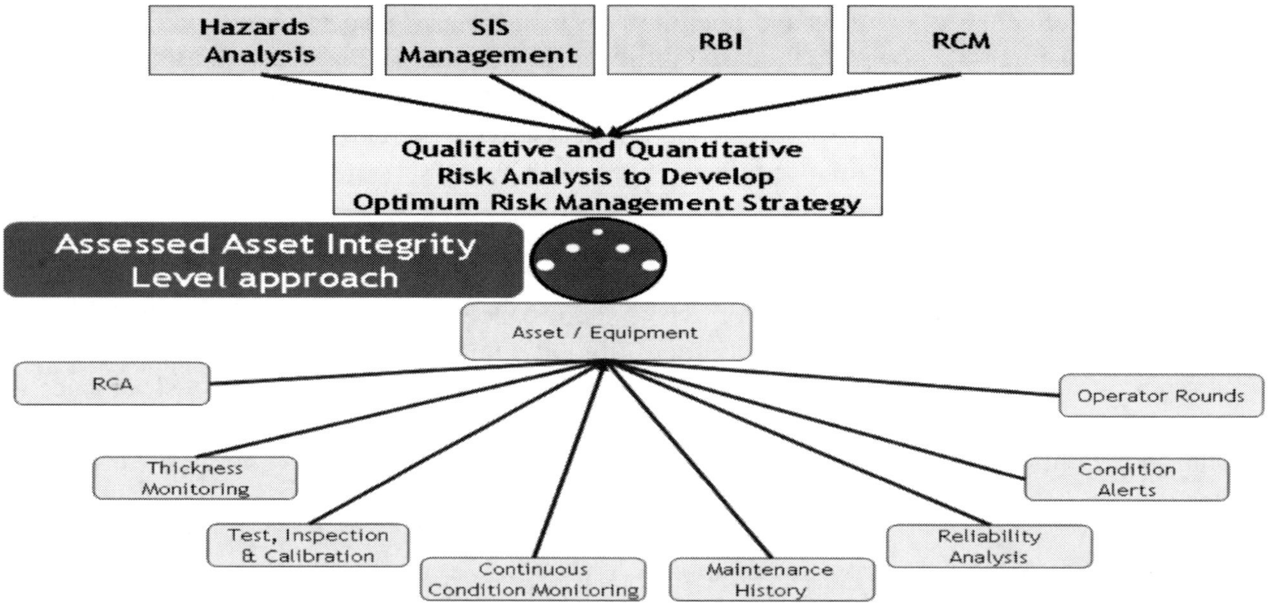

3.1 References

[1] SAE JA 1011, Evaluation Criteria for Reliability-Centered Maintenance (RCM) Processes, issued in January 2002

[2] Occupational Safety and Health Administration, 29 Code of Federal Regulations (CFR) 1910.119 Process Safety Management of Highly Hazardous Chemicals, 1992

[3] ANSI/ISA-84.00.01, Functional Safety: Safety Instrumented Systems for the Process Industry Sector, approved 2 September 2004

[4] API RP580 – Risk Based Inspection, American Petroleum Institute, May 2002

Successful Process Safety Management Requires a Barrier Risk Assessment Tool

Ronald G. Hallmark, CSP
Det Norske Veritas, USA
1400 Ravello Drive
Katy, Texas 77449
Ronald.Hallmark@DNV.com

Prepared for Presentation at
American Institute of Chemical Engineers
2010 Spring Meeting
6th Global Congress on Process Safety
San Antonio, Texas
March 22-24, 2010

UNPUBLISHED

Successful Process Safety Management Requires a Barrier Risk Assessment Tool

Ronald G. Hallmark, CSP
Det Norske Veritas, USA
1400 Ravello Drive
Katy, Texas 77449
Ronald.Hallmark@DNV.com

Keywords: bowtie, barrier, threats, controls, major event

Abstract

The expected outcome of Process Safety Management (PSM) is to minimize the likelihood of a major accident occurring and to ensure that the necessary mitigation and emergency preparedness mechanisms are in place in the event that an accident does happen. PSM programs are designed to establish comprehensive, robust and sustainable systems, practices and competencies for managing process safety and to ensure that a focus on process safety (low frequency-high consequence) issues is maintained. This paper describes in detail how a Barrier/BowTie approach to PSM is established within the framework of an established PSM program and provides the missing link in assurance of process safety. The barrier approach not only enhances process safety but ensures that the processes in place are functioning as expected. A case study example is used to illustrate how continuous improvement is achieved and maintained.

1 Introduction

Over the past 15-20 years, the occupational safety trend has been steadily improving. However, insurance loss claims for process safety incidents have not declined and are, in fact, slightly rising. Many companies focus on impressive occupational safety records as a sign of a good if not great safety performance and overlook low-frequency process vulnerabilities.

1.1 Case Study: Occupational Safety Focus, Process Safety Problem

Misplaced safety focus was evidenced in 1989, when the chemical plant of an unnamed major international Oil company exploded killing 23 and injuring over 100. In response, the plant manager stated that he couldn't understand how this disaster happened and cited a record number of years of operation without a lost time accident.

OSHA's investigation found several major contributors:

- lack of process hazard analysis
- inadequate standard operating procedures (SOPs)
- use of non-fail-safe block valve
- inadequate maintenance permitting system
- inadequate lockout/tagout procedures
- lack of combustible gas detection and alarm system
- presence of ignition sources
- inadequate ventilation systems for nearby buildings
- failure to maintain fire protection system in an adequate state of readiness
- proximity of high-occupancy structures (control rooms) to hazardous operations;
- inadequate separation between buildings; crowded process equipment;
- insufficient separation between the reactors and the control room for emergency shutdown procedures

1.2 Shifting Safety Focus

Is there a direct linkage with these failures and a good occupational safety program? Innovations in the past few decades have attempted to shift focus from narrowly defined occupational safety statistics to more systemic causes and have resulted in many enhancements to occupational safety and, to a lesser extent, process safety. Starting in the late 1980's Safety Management Systems (SMS) were introduced to the oil/gas majors who, in turn, mandated widespread industry use. In the early 1990's with the advent of tools such as QRAs, HAZOPs, and HAZIDs, risk management techniques began being applied. During the mid-1990's companies started experimenting with behavioral-based safety programs to influence "Safety Culture".

Still, recent studies have shown that while occupational health incidents have declined steadily, in many cases by as much as 80%, process safety incidents have not declined. Many of the investigation findings from incidents in 1989 are addressed directly by OSHA PSM regulations and requirements, but process safety management (PSM) measures have noticeable impacts on a company's facilities only when fully developed programs ensure that PSM controls/barriers are robust and healthy. OSHA PSM regulations do allow alternative suitable methods for PHAs. During the design stage, it is appropriate to use a line by line approach, e.g., HAZOP or What-If etc. During later operational phases, a risk-based approach like BowTie is a more powerful tool. The Barrier/BowTie approach is designed to ensure the health of controls/barriers in a way that is simply and clearly communicated.

1.3 BowTies: A Proven Tool Reapplied

The barrier concept is not new; it was conceived in the nuclear power industry of the 1960's. In the 1990's, the Netherlands government in cooperation with the oil and petrochemical industry started putting together the BowTie methodology, in essence a graphical combination of fault tree and event tree ideas. Shell put BowTies into practice in the 2000's and eventually made it a part of their world-wide risk assessment program.

Regulators in many countries including the UK, Abu Dhabi, and Belgium encourage use of Barrier/BowTie analysis. The IADC (International Association of Drilling Contractors) has endorsed this methodology for use in Safety Cases, the offshore industry's method to manage process safety.

Bowties have illustrated the ability to

- Focus on operations stage – key for long term safety
- Address risk in simpler terms than QRA
- Involve a cross section of operational staff who "own" the barriers, increasing buy-in and accountability amongst the operational staff
- Illustrate to senior managers the most important actions to be done to keep things safe
- Simplify audits by ensuring inspection and assessment focus on barriers
- Communicate risk simply and personally
- Continuously update and access controls – upon integration with Sharepoint web sites

1.4 Urgent Need for BowTie Integration into PSM Program

OSHA is becoming very aggressive and is scrutinizing details of PSM implementation in its National Emphasis Program. Furthermore, OSHA seems intent to look for both willful and negligence citations. A PSM program that clearly illustrates how the site "controls and mitigates" risks will greatly reduce the anxiety over OSHA inspections and the potential for citations and fines.

2 PSM Program Challenges

The objective of a PSM program is to establish comprehensive, robust and sustainable systems, practices and competencies for managing Process Safety and to ensure focus on Process Safety (low frequency-high consequence) issues is maintained. The expected outcome of a successful PSM program is to minimize the likelihood of a major accident from occurring and ensure that the necessary mitigation and emergency preparedness mechanisms are in place to minimize the effect in the event that an accident does happen.

2.1 Pitfalls to a Successful PSM Implementation

Seven major deterrents to a successful PSM program include:

- No push from shareholders to have a strong, effective PSM program – "Deafening Silence"
- No active participation in PSM by senior managers –"Invisible Senior Management"
- PSM gets lost in middle management/line management, "Middle Management Black Hole"
- Nobody likes change
- PSM goals and objectives are not measured
- It's somebody else's job – specific performance indicators are not clear

- Too many cooks in the kitchen, too many people deciding how PSM is implemented- Not clear
- Actively focusing on the wrong things, i.e., too many actions from PHA studies with low risk

2.2 Regulatory Compliance

The US OSHA PSM regulations were promulgated in the late 1980's and finalized in 1992 under OSHA 29 CFR 1910.119. Interestingly there have been no modifications of the standard since then, even though global standards have been modified to meet the latest technology. In the late 1980's and early 1990's, the barrier approach was not well developed except in parts of Europe.

In 2007, the Baker Panel concluded that current PSM programs are generally deficient at many US process facilities. The Panel felt that PSM programs have been allowed to degrade and are not as effective as they should be. OSHA has launched a National Emphasis Program focused on the revitalization of PSM programs.

3 Current Tools: Not an Operational Solution

Those familiar with PSM plants need no introduction to risk assessment and risk management. Most simply, the process requires identification of the hazards, determination of the risk associated with each identified hazard and reduction or elimination of risk.

3.1 Concept and Design Stage

Building or modifying a refinery, a chemical facility, or an upstream gas plant starts at the Concept Stage. This stage is the best place to incorporate the safety measures you need without a lot of cost. Many companies will use What-If teams with risk matrices and risk tools such as qualitative risk assessment to select between concept alternatives.

In the Design Stage, layout and technical design are confirmed. A combination of industry design standards and quantitative risk assessment are used to ensure you have proper separation distances, blast walls, firewater requirements, etc.

A Quantitative Risk Analysis (QRA) estimates numerical risk for a facility and supplies critical input to the Concept and Design Stages. Since QRA is conducted during the design, modification or enlargement of a facility, the result can represent stagnant risk picture that underestimates the effects of the management system and operational initiatives. Generally, the biggest problem with QRA is the difficulty of communicating the results (see Figure 1), and especially in answering the question "What does this graph or chart mean to me?"

Figure 1. Typical IR and FN Curve Report for QRA

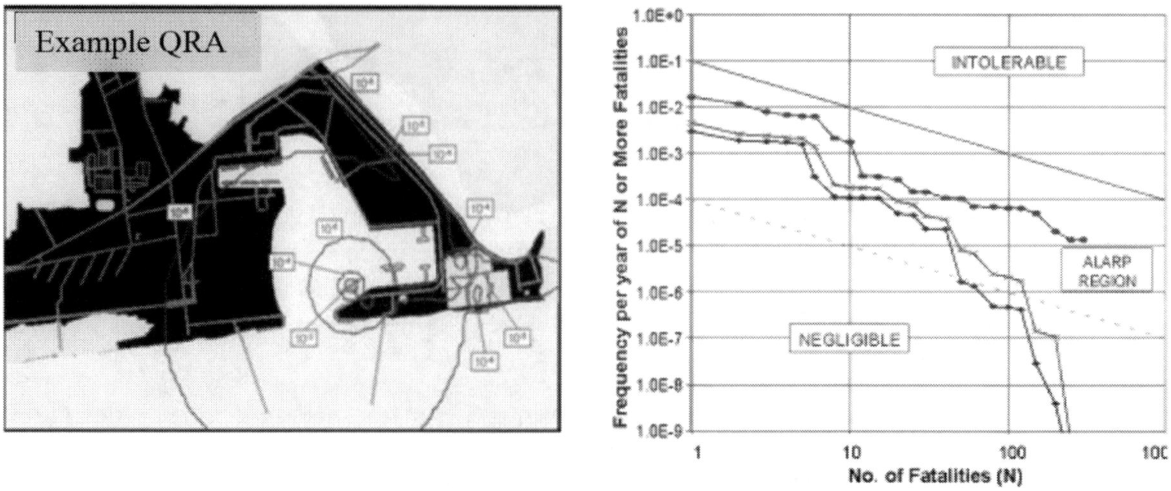

3.2 Operational Stage

Risk assessment in the Operational Stage is trickier and geared to ensure that ongoing operations achieve and maintain design risk levels. A wide range of risk tools need continual application.

OSHA PSM mandates that Process Hazard Analyses (PHA) be performed. PHA type and method are not specified. Regardless of the type of PHA conducted, the resulting list of hazards which have been identified and categorized by risk is combined with results from incident investigations, audits, inspections, and lessons learned to form the basis for the hazard and risk register. A hazard and risk register, shown in Figure 2, is a composite of all risks on the site, prioritized in a database for action and involves hundreds of lines and dozens of pages. Relevant information is difficult to find, contextualize and communicate.

Once identified, risks have to be communicated and managed. Process safety inspections, integrity maintenance, training and competency matrices, safe work procedures, maintenance of safety critical equipment and systems, emergency response plans, fire fighting, etc. require consistent management and measurement to ensure that the "intended controls and barriers" are in place, working and healthy.

3.3 Comparison of Stages

Design and operations stages are truly different on a number of counts. The design stage only needs a few specialists to conduct risk assessment who will decide on a risk reduction on a cost benefit approach. A variety of complex tools and approaches are available to these specialists to aid decision making. However, in the operations stage, it is **critical that everyone knows what keeps them safe and the role they have in maintaining their own safety**.

Figure 2. Sample Hazard and Risk Register

ID	FUNCTION	ISSUES	FAILURE MECHANISM	FAILURE EFFECT	SAFEGUARD	Consequence	Likelihood	Criticality	COMMENTS	ACTION ITEMS	Test	Analysis	Experience	Procedure
1.06.04		Welding quality	Undermatch in weld could overstress the weld due to differential expansion effects.	Potential fracture of the pipeline and leak of LNG to the environment	Filler material will be chosen to achieve overmatch (except INVAR use) Weld qualification (partial for the JIP) NDT	4	3	M	Procedure will be written for the JIP	3		7		3
1.07.00	*End Bulkheads - Stainless Steel*													
1.07.01		Corrosion (external)			Coating will be applied, CP design	3	2	L						
1.07.02		Transitions between different material				4	3	M	see item 1.06.01					
1.07.03		Stresses loads	Over stress due to additional load from no centric pipes.	Possible yielding of the pipe	Possible stress effects should be accounted for in the design.	4	2	M	The pipes shall not be installed with centralizers.		6	6		
1.08.00	*End Bulkheads - Carbon Steel*													
1.08.01		Transitions between				4	3	M	see item 1.06.01					

Operational management wants to be assured that the controls that are specified are actually in place and working as intended. BowTie/Barrier approach provides an up-to-date graphical illustration of the health of the safety controls for communicating to all personnel. BowTies are not too simple, and not too complex.

4 Integrating Barrier/BowTie Approach

The BowTie, illustrated in Figure 3, is a graphical combination of a fault tree and an event tree. The left side lists threats to a major event occurring and preventive "controls/barriers" that reduce the frequency of the major event. The right side lists potential consequences of the major event and mitigation "controls/barriers" that reduce the consequence of the major event. Controls can be process related, people related, plant related or a combination.

Some PSM/SMS requirements are controls. Some are actions linked to the health of specific controls. Others are fulfilled by the process of creating, maintaining and communicating the BowTies. The Barrier/BowTie approach incorporates PSM/SMS requirements such as:

- Active and visible leadership in PSM by senior and line managers
- Process safety and mechanical integrity inspection
- Record and method for PHA of all processes
- Use of software and hardware controls, including behavioral-based safety audits
- Use of risk assessment and risk management tools
- Verification of onsite PSM documentation and records
- Design engineering which includes an assessment of layout and escalation/ emergency controls

Figure 3. Illustration of a BowTie Diagram

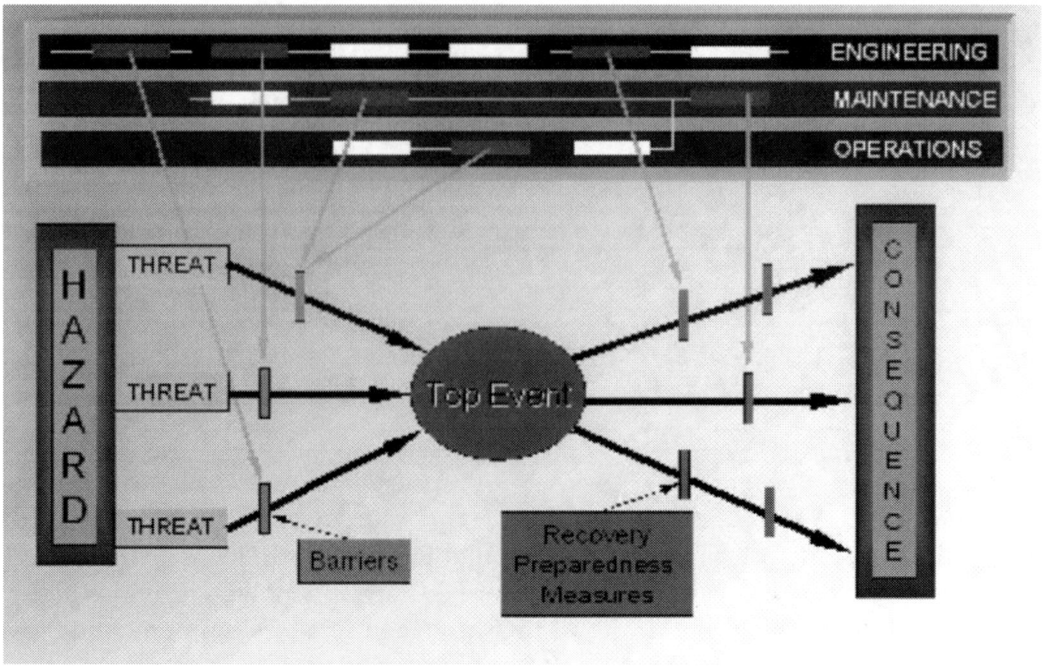

4.1 Steps to Develop a Barrier/BowTie Program

1. Utilize existing facility risk assessment such PHA and/or QRA studies with risk ranking.
2. Select higher risks for barrier analysis Typically 10 – 20 are sufficient.
3. Carry out Bow Tie sessions which are similar to PHA studies. Include a Facilitator, a scribe, operations team member, maintenance team member, safety team member, etc. The studies take approximately a half day per bow tie, for experienced teams. The first session may take a full day to get the team familiar with the methodology. The results of the studies can be documented in BowTie software.
4. Review the BowTies for consistency as team members may change during multiple sessions.
5. Score the barrier controls for adequacy. Experience has shown that 3 barriers are usually required for each threat and a minimum score of 3.0 where 1.0 is deemed a working acceptable control.
6. Assign control owners. If this is the introductory stage of barrier approach, a short workshop for control owners is recommended.
7. A briefing or feed-back presentation is recommended for appropriate staff and affected personnel.

8. Barrier study recommendations are entered into a Hazard and Risk register for actioning.
9. Monitor barriers by Management and Lead Operator site tours. Record results of monitoring, inspection, and maintenance programs which evaluate the health of the "control".
10. Incident investigations – determine which barriers failed and why. Feed these results back into your PSM program.
11. Continuous improvement through annual reviews and feedback sessions with Control Owners. Set Key Performance Indicators (KPIs) for managers and control owners to ensure that proactive actions are taken to keep controls working and in good shape.

4.2 Example of a Barrier/BowTie Diagram

Portions of a complete BowTie Diagram for a gasoline storage tank (hazard) overflow (event) are shown. Figure 4 and Figure 5 illustrate threats contributing to a storage tank overflow. In essence, each threat is an "or gate" in a fault tree which means under the right circumstance (failures), this threat alone could lead to the major event. A BowTie diagram could include ten or more threats. Each threat requires an evaluation of controls (barriers) needed to prevent the threat from becoming an event. Figure 4 illustrates the left side of the BowTie and illustrated mechanical failures which could lead to a tank spillage. Figure 5 illustrates how human factors/failures can lead to the major event. Figure 6 illustrates one of many consequences of the major event actually happening.

Figure 4. Mechanical Failure Threat to Tank Overflow

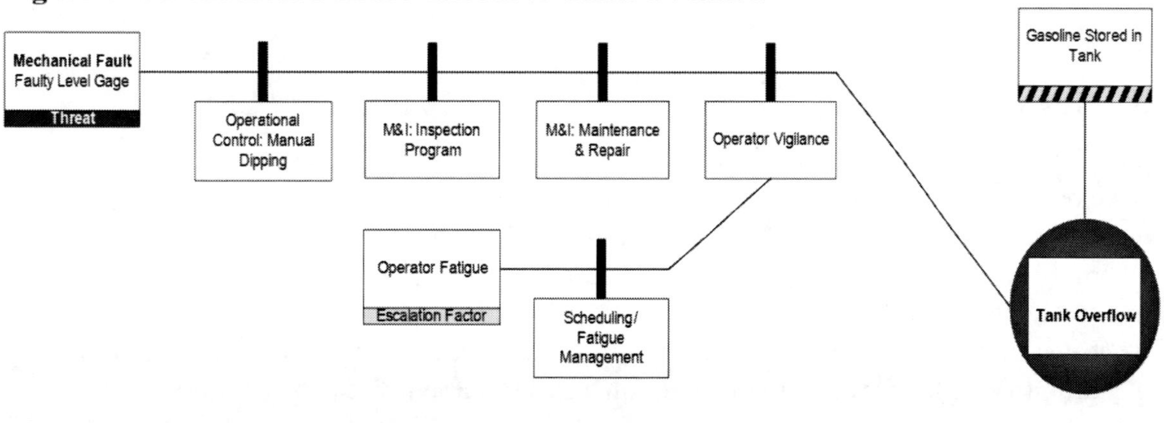

Figure 5. Human Error Threat to Tank Overflow

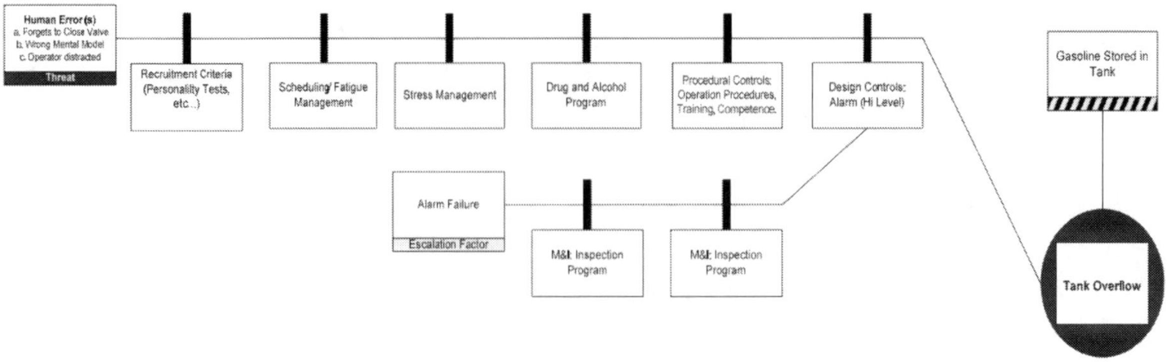

Figure 6. Consequence of Tank Overflow (right side of BowTie)

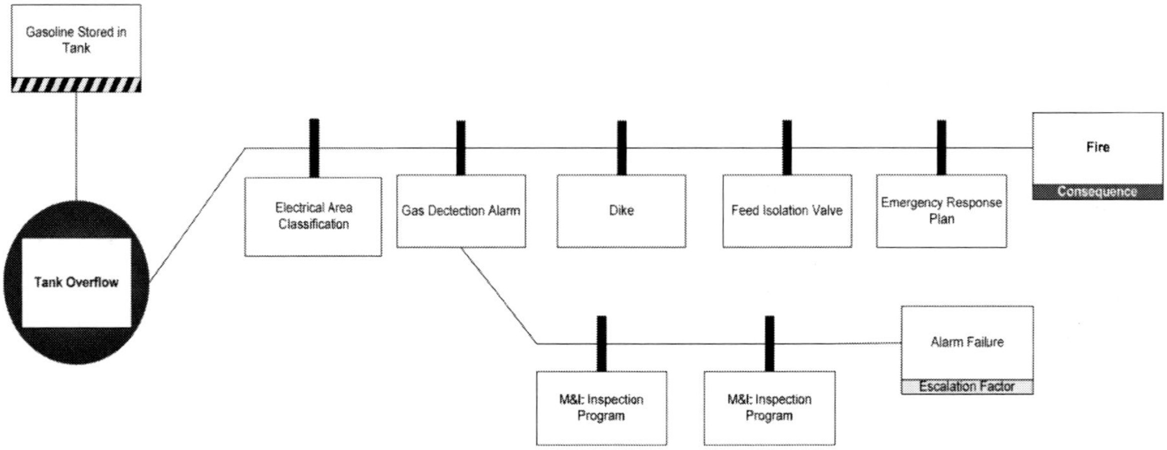

5 Key Benefits of the Barrier/BowTie Approach to a PSM Site

BowTies have illustrated the ability to:

- Focus on the Operations Stage – key for long term safety
- Address risk in simpler terms than QRA
- Involve a cross section of operational staff who "own" the barriers increasing buy-in and accountability amongst the operational staff.
- Illustrate to senior managers the most important actions to be done to keep things safe
- Simplify audits by ensuring inspection and assessment focus on barriers
- Communicate risk simply and personally
- Continuously update and access controls – upon integration with Sharepoint web sites

These advantages result in benefits to both the PSM Site and to the PSM Site manager.

Benefits to the PSM Site include:

- On-hand, actionable list of ranked risks in a Hazard and Risk Register
- Working knowledge of the controls and barriers necessary to prevent a major event

- Accountable control owners and measurements to ensure the health of the controls
- Ability to clearly communicate barrier/control management to all staff and workers

Benefits to PSM Site Management include:

- Direct evaluation method of the PSM health by the health of controls – through monitoring, audit, incidents, reporting of control health metrics
- Ability to manage change more easily by reviewing existing controls vs. new controls required.
- Direct graphical links between the action list, Bow Ties and the site PSM program
- Ability to clearly and easily communicate risks to staff and regulators
- Ability to evaluate risks almost continuously

6 Conclusions

- Current PSM programs have many deficiencies
- OSHA is now focusing on all covered PSM sites
- Key barriers are weak (e.g. operating procedures are deficient, relief systems poorly documented)
- A barrier program would enhance process safety by providing
  - Clear documentation of most important barriers
  - Metrics and expectations for barrier performance
  - Owners to barriers who maintain them through their lifetime
  - Standards for continuous improvement while focusing on highest risks

Looking at the history of oil/gas offshore activities shows that the barrier approach works well. After Piper Alpha, the major oil companies in the North Sea implemented a risk assessment and barrier system. The assessment identifies top risks and barrier management is applied to the most important controls. The results speak for themselves. Since 1990 when most offshore facilities adopted barrier approach, no major catastrophic events have occurred. It seems that the success of the barrier approach could be an effective means to rejuvenate PSM programs.

7 References

[1] Fisher, Mark., "DNV Operational excellence White Paper, BowTie barrier Analysis" Stockport, England, UK, 2009

[2] "Pitblado, Robin, PhD, BowTie Barrier Analysis Workshop, 09 October, 2009, Katy Texas. Presentation and lecture.

[3] Pitblado, Robin, "Safety Culture in Developing Countries", Presented AIChE Spring Conference, Tampa, FL, April 2009.

Optimizing Facility Siting and Layout through Mapping Risk Estimates on Plant Area and Monetizing

Seungho Jung
Chemical Engineering, Texas A&M University
College Station, Texas, 77843-3122, USA
Seungho.Jung@chemail.tamu.edu

Dedy Ng, Carl Laird, Sam M. Mannan
Chemical Engineering, Texas A&M University
College Station, Texas, 77843-3122, USA
Dedy.Ng@chemail.tamu.edu,
Carl.Laird@chemail.tamu.edu
Mannan@ tamu.edu

Prepared for Presentation at
American Institute of Chemical Engineers
2010 Spring Meeting
6th Global Congress on Process Safety
San Antonio, Texas
March 22-24, 2010

UNPUBLISHED

Optimizing Facility Siting and Layout through Mapping Risk Estimates on Plant Area and Monetizing

Seungho Jung
Chemical Engineering, Texas A&M University
College Station, Texas, 77843-3122, USA
Seungho.Jung@chemail.tamu.edu

Dedy Ng, Carl Laird, Sam M. Mannan
Chemical Engineering, Texas A&M University
College Station, Texas, 77843-3122, USA

Keywords: Mapping Risks, MILP, PHAST, Facility, Siting, Layout

Abstract

This research is focused on developing a methodology to find optimal placement for hazardous facilities and other process plant buildings using optimization theory and mapping of risks. Incidents can have possible consequences from flammable gas releases which can be evaluated using PHAST, and the probability of each can be derived from initial leak hole size estimation through event tree analysis. The available plot is divided in a square grid and risk scores are calculated for each square. In order to get financial risk, these risk-scores are combined with weighting factors depending on the type of structure and cost of potential buildings. The overall cost is a function of piping cost, management cost, land cost, and financial risk. The proposed approach uses a mixed integer linear optimization problem (MILP) that identifies attractive locations by minimizing the overall cost. Case studies are carried out for different production plants planned by taking account meteorological data for the area in order to show the applicability of the proposed method. The study will be useful in assisting the selection of locations for process plant buildings and for risk management.

1. Introduction

A site is a collection of process units that have their own support facilities or ones interdependent with another site. The support facilities in this framework include offices, maintenance shops, warehousing, a firehouse, and medical, transportation, parking, and security facilities. Site location and layout are key factors in determining the operational cost, design complexity, and safety of process operations and maintenance. Siting and layout are among the earliest steps in process design, and thus require careful planning to avoid frequent modifications once the site has been constructed. Well optimized siting and layout will not only minimize management costs, but also reduce the risk of losses throughout the site life-cycle. Table 1 shows a selected list of some serious incidents involving buildings in process plants.

Table 1. Accidents Involving Buildings in Process Plants

Date	Location	Fatalities	Description
2005	Texas City, Texas	15 (most in buildings)	Deaths mostly occurred in and around temporary office trailers that had been sited near a blowdown drum.
1992	La Mede, France	6 (most in buildings)	LPG leak resulted in an explosion that destroyed the unit and demolished the adjacent control room.
1992	Castleford, England	5 (3 in buildings)	A runaway reaction caused a jet flame that destroyed a wooden control room.
1989	Pasadena, Texas	23 (unknown number in buildings)	A release in a polyethylene plant resulted in an explosion that destroyed the facility, including the control room and administration building.
1988	Norco, Louisiana	7 (6 in buildings)	A corrosion-induced propane leak resulted in an explosion that destroyed the control room.
1987	Antwerp, Belgium	0	There was extensive damage to the unit and blast damage (glass breakage) as far as 10 km away.
1987	Grangemouth, Scotland	1 (0 in buildings)	A low-pressure separator was ruptured. Pieces of vessel were found more than 340 m away.

Accidents still may occur even though industry strives to avoid them. However, as indicated by some accidents in the table, proper design and siting of buildings can substantially reduce the risks of fatality. Many companies, as well as industry insurers, trade associations, and standards organizations, have developed specific criteria for spacing between process units, buildings, equipment, and property lines. These criteria are meant to reduce the impact of explosions or fires on major equipment and facilities, including adjacent units and buildings. Also, facility siting and layout must be addressed in Process Hazard Analysis (PHA). Hazard identification must assess the possible impacts of fires and explosions on equipment, structures, and life safety for existing facilities or equipment. In this paper a mathematical program combines a QRA approach for facility layout and addresses trade-offs between risk and the capital costs, as illustrated in a case study. The outcome of this work is to provide guidance for facility layout to address the explosion and fire impacts to process plant buildings and occupants. Prevention must be the first consideration when identifying options for reducing risks. Further, it is noted that constructing process plant buildings for increased blast or fire resistance may not always be the most effective means for reducing risk. Some situations may exist where, because of the potential risks to both indoor and outdoor personnel, risk reduction is best achieved by preventing the event from occurring; specifically, by reducing the probability of an incident.

2. Mathematical Formulation

2.1 Risk Scores and Consequence Modeling

BLEVEs are possible at the location of vessels containing substances. Using the amount of released material, temperature, and pressure, BLEVE overpressure calculations are feasible. Overpressures from BLEVEs usually yield a circular impact zone around the explosion point. Thus, calculating overpressures for each grid at a site depends on distances from the explosion point. PHAST (version 6.1), known as a conservative model simulator, is used for overpressure calculations. When vapor cloud explosions (VCEs) occur, the vapor is dispersed throughout the plant site while mixing with air. Often flammable vapors are initially heavier than air and are transported with the wind until the vapor cloud meets an ignition source. Thus, the explosion center cannot be assumed to be at the point of release, rather it is more rational to use a wind rose to account for directional effects. A challenge is that a large number of scenarios are required because the problem includes a very large number of uncertain parameters such as wind speed, wind direction, air stability, temperature, pressure, humidity of the area, and time delay to ignition. The latter is fixed at 180 seconds in this work for simplification of the calculations. We incorporated Monte Carlo sampling in a stochastic computer program to reduce the number of scenarios for an average real world application. After a few thousand samplings, an average size of vapor cloud at ignition can be derived and the VCE overpressure approximated in the pressure range of interest using the TNT equivalency method for each receptor. Eventually we can acquire averaged overpressures at each grid caused by VCE. Using the TNT equivalent model the equivalent mass of TNT explosion strength is:

$$W = \eta M \frac{E_c}{E_{cTNT}} \qquad \text{[Eq. 1]}$$

where
W = equivalent mass of TNT (kg)
M = actual mass of hydrocarbon
η = empirical explosion yield (assumed to be 0.1)
Ec = heat of combustion of hydrocarbon
E_{cTNT} = heat of combustion of TNT (4.6x10^6 J/kg)

Overpressure values are translated into the probability of structural damage using a probit function:

$$\Pr = -23.8 + 2.92 \ln(p^o)$$

where p^o is the peak overpressure (N/m^2). Using this method, the probability of structural damage is calculated for every grid in the plant and is converted into a risk score/index.

2.2 *Constraints*

2.2.1 Non-Overlapping Constraints

This mathematical problem is basically an assignment problem, assigning n numbers of facilities to k numbers of grids so as to minimize the cost, including the relationship cost between each other and the risk cost. In order to avoid overlapping between facilities, the binary variable B_{ik} is defined as:

$$\sum_{k=1}^{K} B_{ik} = 1 \qquad [\text{Eq. } 2]$$

$$\forall i \in \text{Facilities}, \forall k \in \text{all grids on the plane}$$

$$\sum_{i=1}^{n} B_{ik} \leq 1 \qquad [\text{Eq. } 3]$$

$$B_{ik} = 0 \text{ or } 1 \begin{cases} k = 1, 2, 3, \dots, K \\ i = 1, 2, , n \end{cases}$$

$$B_{ik} = \begin{cases} 1 & \text{if unit i is allocated to site area} \\ 0 & \text{otherwise} \end{cases}$$

2.2.2 Separation Distance Constraints

Besides the process units proper, there may be some facilities such as storage facilities and occupied buildings which should not be too close together. There are several recommendations about the separation distance between storage tanks and control rooms. For this, we need separation distance constraints which will have the form:

$$|x_i - x_j| + |y_i - y_j| \geq D_{ij} \qquad [\text{Eq. } 4]$$

$$\{xf(i) - xf(j)\} + \{yf(i) - yf(j)\} \geq D_{ij} \times sepB1_{ij} - M(1 - sepB1_{ij}) \qquad [Eq.5]$$

$$\{xf(i) - xf(j)\} - \{yf(i) - yf(j)\} \geq D_{ij} \times sepB2_{ij} - M(1 - sepB2_{ij}) \qquad [Eq.6]$$

$$-\{xf(i) - xf(j)\} + \{yf(i) - yf(j)\} \geq D_{ij} \times sepB3_{ij} - M(1 - sepB3_{ij}) \qquad [Eq.7]$$

$$-\{xf(i) - xf(j)\} - \{yf(i) - yf(j)\} \geq D_{ij} \times sepB4_{ij} - M(1 - sepB4_{ij}) \qquad [Eq.8]$$

$$sepB1_{ij} + sepB2_{ij} + sepB3_{ij} + sepB4_{ij} \geq 1 \qquad [Eq.9]$$

D_{ij} *is the minimum separation distance between* i, j

$i \in$ *occupied buildings,* $j \in$ *hazardous facilities such as storage tanks*

M *is a big number*

2.3 Objective Function

The objective function to be minimized includes the piping cost, the probable property damage cost, and the cost of protection devices. In order to simplify the problem, the most hazardous facility, usually a process unit, is located in the center of the given plot which is called the center grid. This has another purpose, namely to locate the most hazardous facility as far as possible from the plant site fenceline which may be close to residential areas or roads outside of the chemical complex.

$$\text{Min} \quad \sum_{i=1}^{n}\sum_{k=1}^{K}\{RS_k \times FC^i + RD_k \times UP^i\} \times B_{ik} \qquad \text{[Eq. 10]}$$

where
RS_k = Risk score of k grid caused from the center facility
RD_k = Rectilinear distance of k grid calculated from the center facility
FC^i = Facility building cost of i facility
UP^i = Unit piping (relationship) cost between i facility and the center facility
x_i: x coordinate of i facility
y_i: y coordinates of j facility
(Subject to non-overlapping, minimum–maximum separation distance constraints.)

3. Case Study – Hexane-Heptane Distillation Tower

The proposed approach is demonstrated with a case study that develops the facility layout around a hexane-heptane distillation tower. There are three major incident scenarios for this process unit such as catastrophic failure of a component, or a liquid or a vapor release from a hole in the process piping. The first incident scenario can be addressed through consideration of explosion impacts on buildings. Catastrophic failure can occur in case of failure of one of the vessels or of a full bore line rupture. In such cases it is assumed that the entire contents of the column, reboiler, condenser, and accumulator are lost instantaneously.

Figure 1. Event Tree Analysis (ETA)

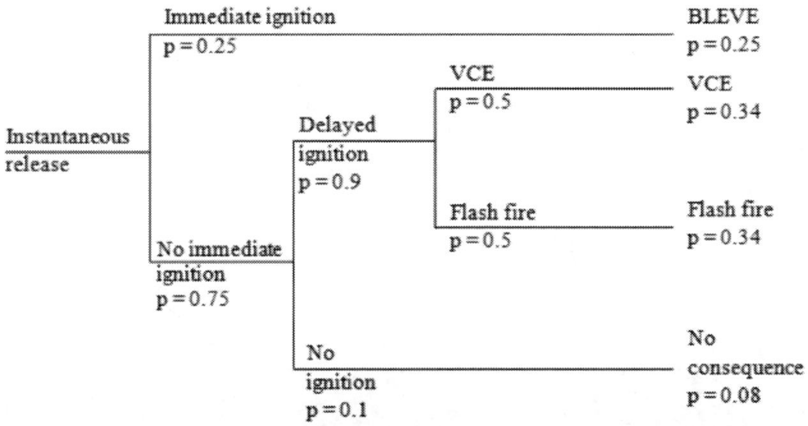

Catastrophic failure includes catastrophic rupture of any component in the fractionating system and catastrophic (full bore) rupture of any pipe work. There is approximately 25 m of 0.5-m-diameter piping and 55 m of 0.15-m-equivalent diameter piping included in this incident.

Hence, the frequency is calculated as follows:

- Catastrophic rupture of fractionating system 6.5×10^{-6} per year
- Full bore failure of -
  - 55 m of medium pipe $55 \times 2.6 \times 10^{-7}$ per m-year = 1.4×10^{-5} per year
  - 25 m of large pipe $25 \times 8.8 \times 10^{-8}$ per m-year = 2.2×10^{-6} per year

The total frequencyis 2.3×10^{-5} per year, and this is the value used for the instantaneous frequency in computations. Based on the ETA and release frequency calculations, the incident outcome frequency (yr^{-1}) can be obtained as given in Table 2.

Table 2 Incident Outcome Frequency

Incident Outcome	Incident Frequency (yr^{-1})	Incident Outcome Probability	Incident Outcome Frequency
BLEVE	2.3×10^{-5}	0.25	5.7×10^{-6}
VCE	2.3×10^{-5}	0.34	7.8×10^{-6}
Flash fire	2.3×10^{-5}	0.34	7.8×10^{-6}

As shown in Figure 1, in the case of a BLEVE immediate ignition of an instantaneous release is assumed. VCEs or flash fires can be due to delayed ignition of an instantaneous release. Various possible outcomes are identified for the risk analysis in the example, but BLEVE and VCE effects are expected to lead to building damages. Before calculating risk scores for all grids, the size of plant area is defined. In this case study, 100 grids are set into the given plot as shown in Figure 2. Thus, each grid will have a size of 10 m by 10 m. The process unit (distillation unit) is assumed to be in the center in the grids covered with black. The reason it is located in the center is to put the most hazardous unit as far as possible from the fenceline where residential areas might be located. Figure 2 shows the identification number for each grid square.

Figure 2. Grids in the Given Area

G01	G02	G03	G04	G05	G06	G07	G08	G09	G10
G11	G12	G13	G14	G15	G16	G17	G18	G19	G20
G21	G22	G23	G24	G25	G26	G27	G28	G29	G30
G31	G32	G33	G34	G35	G36	G37	G38	G39	G40
G41	G42	G43	G44			G47	G48	G49	G50
G51	G52	G53	G54			G57	G58	G59	G60
G61	G62	G63	G64	G65	G66	G67	G68	G69	G70
G71	G72	G73	G74	G75	G76	G77	G78	G79	G80
G81	G82	G83	G84	G85	G86	G87	G88	G89	G90
G91	G92	G93	G94	G95	G96	G97	G98	G99	G100

The overpressure in each grid is calculated by PHAST based on the distance from the center point of an explosion. Overpressures in each grid are converted into probabilities of structural damage via a probit function. For example, the distance between Grid 01 (G_{01}) and the center is 63.6 meters and if a BLEVE occurs in the center, the overpressure at this distance is 0.06 psi as calculated by PHAST. The overpressure is converted to the probability of structural damage via the probit function which gives 3.9%. The frequency of this BLEVE is 5.7x10^{-6} yr^{-1} from Table 2, and the lifetime of the plant is assumed to be 50 years. If a building is on the Grid 01, it means that it has a structural damage probability of 0.00001 for the entire lifetime due to a BLEVE. This number is multiplied by a weighting factor of 100, to give the risk score. Risk scores for all grids squares are calculated in this way, as shown in Figure 3.

Figure 3. Risk Scores from BLEVEs

1E-05	8E-05	4E-04	1E-03	0.002	0.002	1E-03	4E-04	8E-05	1E-05
8E-05	6E-04	0.003	0.007	0.011	0.011	0.007	0.003	6E-04	8E-05
4E-04	0.003	0.011	0.022	0.026	0.026	0.022	0.011	0.003	4E-04
1E-03	0.007	0.022	0.028	0.028	0.028	0.028	0.022	0.007	1E-03
0.002	0.011	0.026	0.028			0.028	0.026	0.011	0.002
0.002	0.011	0.026	0.028			0.028	0.026	0.011	0.002
1E-03	0.007	0.022	0.028	0.028	0.028	0.028	0.022	0.007	1E-03
4E-04	0.003	0.011	0.022	0.026	0.026	0.022	0.011	0.003	4E-04
8E-05	6E-04	0.003	0.007	0.011	0.011	0.007	0.003	6E-04	8E-05
1E-05	8E-05	4E-04	1E-03	0.002	0.002	1E-03	4E-04	8E-05	1E-05

Figure 4. Risk Scores from VCEs

0.026	0.027	0.028	0.029	0.027	0.026	0.025	0.022	0.02	0.018
0.026	0.029	0.03	0.031	0.03	0.029	0.026	0.023	0.021	0.018
0.027	0.03	0.032	0.033	0.033	0.032	0.028	0.024	0.021	0.018
0.028	0.031	0.034	0.036	0.036	0.035	0.029	0.024	0.021	0.018
0.027	0.03	0.034	0.037			0.029	0.024	0.02	0.018
0.027	0.03	0.033	0.037			0.028	0.023	0.02	0.017
0.026	0.028	0.031	0.033	0.032	0.031	0.027	0.023	0.02	0.017
0.025	0.027	0.029	0.029	0.029	0.028	0.025	0.022	0.02	0.017
0.023	0.025	0.026	0.027	0.026	0.025	0.023	0.022	0.02	0.017
0.022	0.023	0.024	0.025	0.024	0.023	0.022	0.02	0.019	0.017

Blast effects of VCEs are calculated with the TNT-equivalent model after assuming a release time of 180 seconds until meeting an ignition source. A C-program was used to conduct Monte Carlo simulation for a 10,000 set of meteorological data in Beaumont, Texas. Averaged overpressures for each grid are converted into probabilities of structural damage via the probit function. For example, the distance between Grid 01 (G_{01}) and the center is 63.6 meters and if a VCE occurs in the center, the calculated overpressure due to this distance and the direction from the center is 0.15 psi. The overpressure is converted to the probability of structural damage via the probit function, giving 22.2 %. The frequency of VCE is 7.8x10^{-6} / year and the assumed plant lifetime is 50 years. These numbers are multiplied by the weighting factor of 100, to yield a VCE risk score. Risk scores for all grids are calculated in this way and given in Figure 4. Calculating each Risk index from different incidents, we sum them directly and the result is the integrated risk score, as shown in Figure 5.

For this case study it is assumed there is one main control room, an office building, a building for maintenance, three storage tanks (one is larger than 38 m^3, two are smaller than 38 m^3), and a utility facility. It has been recommended what the separation distances need to be between certain types of facilities, as in Table 5, in order to be prepared for fire consequences.

Figure 5. Integrated Risk Scores – Process Unit Center Location Case Study

0.026	0.027	0.028	0.03	0.029	0.028	0.026	0.023	0.02	0.018
0.026	0.029	0.032	0.038	0.041	0.04	0.033	0.026	0.021	0.018
0.027	0.032	0.043	0.055	0.059	0.058	0.05	0.035	0.023	0.018
0.029	0.038	0.056	0.065	0.064	0.063	0.057	0.046	0.028	0.019
0.029	0.041	0.06	0.066			0.057	0.05	0.032	0.019
0.029	0.041	0.059	0.065			0.057	0.05	0.031	0.019
0.027	0.036	0.053	0.061	0.06	0.06	0.055	0.045	0.027	0.018
0.025	0.03	0.04	0.051	0.055	0.054	0.047	0.033	0.023	0.018
0.023	0.026	0.029	0.034	0.037	0.037	0.031	0.025	0.021	0.017
0.022	0.023	0.024	0.026	0.026	0.025	0.023	0.02	0.019	0.017

Table 5. Typical Spacing Requirements for On-Site Buildings

On-Site Building	Utilities	Atmospheric & Low Pressure Flammable & Combustible Storage Tanks (up to 1 atm) < 38 m^3	Atmospheric & Low Pressure Flammable & Combustible Storage Tanks (up to 1 atm) > 38 m^3	High Pressure Flammable Storage
Office, Lab, Maintenance, Warehouse	30	15	76	107
Substation, Motor Control-More than One Unit	30	30	76	76
Substation, Motor Control-One Unit	30	15	76	76
Control Room-Main	30	30	76	107
Control Room-More than One Unit	30	30	76	107
Control Room-One Unit	30	15	76	76

Some distances of interest are extracted for this case study in Table 6.

Table 6. Minimum Separation Distances between Facilities

	#4 Large Storage [m]	#5 Small Storage [m]	#6 Small Storage [m]	#7 Utility [m]
#1 Main control room	76	30	30	30
#2 Office	76	15	15	30
#3 Maintenance building	76	15	15	30

It is assumed that each facility has different building costs and relationship costs with the center process plant building as given in Table 7.

Table 7. Facility Cost and Unit Piping (UP) Cost of Each Facility

Num. (#)	Type	Facility Cost [$]	UP cost [$ / m]
1	Main control room	1,000,000	10
2	Office	300,000	0.1
3	Auxiliary building for maintenance	200,000	2
4	Large volume storage (> 38 m^3)	150,000	100
5	Small volume storage 1 (< 38 m^3)	100,000	100
6	Small volume storage 2 (< 38 m^3)	100,000	100
7	Utility	500,000	50

As discussed earlier, similar buildings such as #1,#2,#3 for occupied buildings and #4,#5,#6 for storage need to be located close, so the maximum separation distance is set at 30 m for each type. After including all separation distance constraints, the result is shown in Figure 6.

All of the occupied facilities have been located in the bottom-right area which would have less frequent impacts due to wind direction effects, given the location of the distillation unit. This result is shown in the VCE risk score. Due to minimum separation distances between storage and occupied buildings, storage areas are far from the control room, but they are close to the process unit because of high piping costs. Constraints for maximum separation distances allow each similar facility to be gathered within 30 m as shown. The utility area has been assigned to block G40 far from occupied areas and at a relatively low cost for risk. In this case study the process unit has been located in the center of the given plot in order to avoid getting close to potential properties outside of the plant site. However, if this plot was far removed from other properties, the process unit could be set in any grid for calculations.

If the process unit is located on the edge of the plot, such as in the bottom-left, risk scores from the same scenarios are as given in Figure 8, and the resultant layout is as shown in Figure 9.

Figure 6. Layout Results for the Process Unit Center Location Case Study

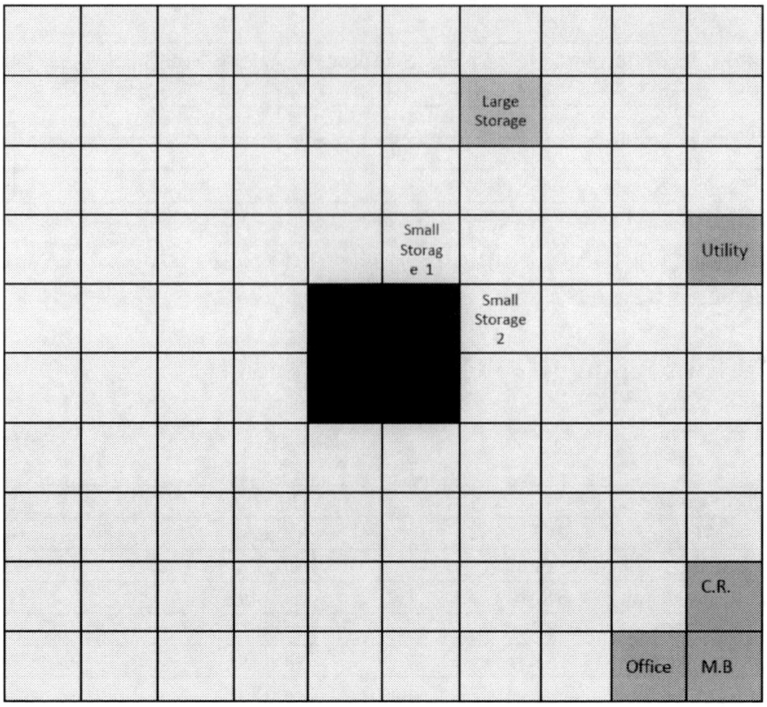

Figure 7. Integrated Risk Scores - Process Unit Edge Location Case Study

0.018	0.018	0.018	0.017	0.015	0.015	0.013	0.012	0.010	0.009
0.019	0.019	0.019	0.019	0.017	0.015	0.014	0.012	0.011	0.010
0.021	0.021	0.021	0.019	0.018	0.016	0.014	0.013	0.012	0.010
0.023	0.023	0.022	0.020	0.018	0.017	0.015	0.014	0.011	0.010
0.026	0.026	0.025	0.022	0.020	0.017	0.015	0.013	0.012	0.010
0.034	0.034	0.030	0.025	0.021	0.018	0.016	0.013	0.012	0.011
0.046	0.046	0.041	0.031	0.023	0.018	0.016	0.014	0.012	0.011
0.050	0.050	0.047	0.039	0.026	0.019	0.017	0.015	0.013	0.011
		0.047	0.043	0.029	0.020	0.017	0.015	0.013	0.011
		0.047	0.042	0.029	0.020	0.017	0.015	0.013	0.011

Figure 8. Layout Results for the Process Unit Edge Location Case Study

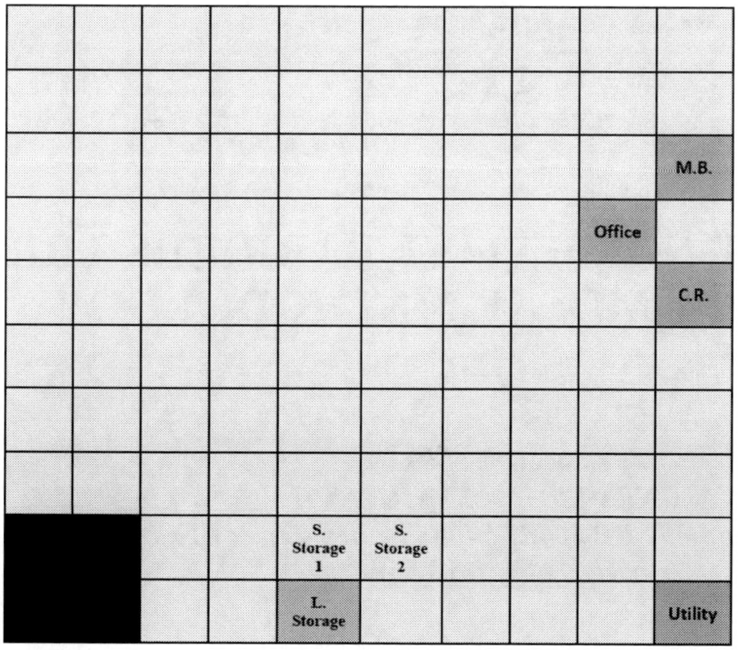

4. Conclusion

In this research, we present a new approach in integrating safety and economic decision making in the optimization of plant layout for fire and explosion scenarios. A new approach to obtaining a global optimum solution has been successfully achieved and shows a reasonable result. The linearized form converged very swiftly in the AMPL program. The optimization problem was formulated as a mixed integer linear programming problem (MILP). Future work will focus on the optimization of facilities with flammable gas scenarios in different plants such as for acrylic acid production, LPG storage, and for LNG terminals in order to expand the proposed optimization tool and develop safer layouts. Information from this work can be used to assist in risk assessment, and in providing advice for emergency preparedness and accident management at industrial facilities.

5. References

[1] "Guidelines for Evaluating Process Plant Buildings for External Explosions and Fires", AIChE / CCPS, New York (1996)

[2] "Guidelines for Facility Siting and Layout", AIChE / CCPS, New York (2003)

[3] "Guidelines for Consequence Analysis of Chemical Releases", AIChE / CCPS, New York (1999)

EVALUATION OF NORMATIVE BARRIERS THROUGH THE MADS/MOSAR METHODOLOGY

Laurent PERRIN
CNRS LRGP - ENSIC INPL
1, rue Grandville BP 20451 54001 Nancy Cedex - France
perrin@ensic.inpl-nancy.fr

Felipe MUÑOZ-GIRALDO
Universidad de los Andes
Carrera 1 N°18A - 10 Bogotá - Colombia
fmunoz@uniandes.edu.co

André LAURENT
CNRS LRGP - ENSIC INPL
1, rue Grandville BP 20451 54001 Nancy Cedex - France
alaurent@ensic.inpl-nancy.fr

Prepared for Presentation at
American Institute of Chemical Engineers
2010 Spring Meeting
6th Global Congress on Process Safety
San Antonio, Texas
March 22-24, 2010

UNPUBLISHED

EVALUATION OF NORMATIVE BARRIERS THROUGH THE MADS/MOSAR METHODOLOGY

Laurent PERRIN
CNRS LRGP - ENSIC INPL
1, rue Grandville BP 20451 54001 Nancy Cedex - France
perrin@ensic.inpl-nancy.fr

Felipe MUÑOZ-GIRALDO
Universidad de los Andes
Carrera 1 N°18A - 10 Bogotá - Colombia
fmunoz@uniandes.edu.co

André LAURENT
CNRS LRGP - ENSIC INPL
1, rue Grandville BP 20451 54001 Nancy Cedex - France
alaurent@ensic.inpl-nancy.fr

Keywords: safety barrier, legislation, MADS, MOSAR

Abstract

During the past 30 years, safety and environmental topics have moved up on the list of society's significant concerns. Administrations and regulatory bodies have to publish rules for the protection of the public, decide whether to authorize the use of a chemical product or the construction of a new plant, limit the environmental impact, regulate waste disposal, etc. All these rules i.e. the legislative, normative and procedural corpus of a country, a state, a city or a factory can be designated to a barrier function. These barriers are indispensable and can play an important role in industrial safety management. In developed countries, growth during the past few years in the number of laws directed at regulating environmental and safety aspects in industry is well known to everyone. And we consider that the definition, classification and performance of the legislative and normative barriers definitely allows the management of technological risks to perform risk reduction in order to enhance the acceptability in a regional context and permits, for a group of events, the identification of the flux of danger. We applied a systemic risk analysis methodology (MADS/MOSAR methodology) to Colombian industrial sectors and analyzed the current Colombian legislation link with three scenarios (a technological accident with an explosion, an occupational risk and a landscape alteration caused by a post-exploitation scenario). *In fine,* we can generate a global panorama of the different existing instruments allowing future management opportunities. Our analysis results propose a point of view, which could be useful for a local government or a whole national system to evaluate or improve the risk legislative and normative barriers. This modeling can, of course, be applied to other countries and other industrial sectors.

1. Introduction

As a part of the applied research conducted by the Reactions and Chemical engineering Laboratory's process safety group, studies concerning the evaluation of technological risks at an economical sector level are being carried out to assess the present barriers in legislation as an element of risk assessment.
This study is complex and requires modeling and a methodology of risk analysis to make sure that the suggested results are quite representative of the studied system. The model (MADS) and the methodology (MOSAR) are derived from the systemic approach. The MADS model allows the unfolding of complex systems and evaluation of the potential damage in specific targets. As shown in Figure 1, it allows identification and modeling of the mechanism of hazard between sources of hazard and targets. We can conduct a survey of the hazard process in which a source of hazard is linked to a target through the phenomena called "hazard flux". This is done in a very specific relation called fields of hazard that takes into account space and time dimensions.

Figure 1. The Generic Systemic MADS Model: Analysis Method for Malfunctions in Systems

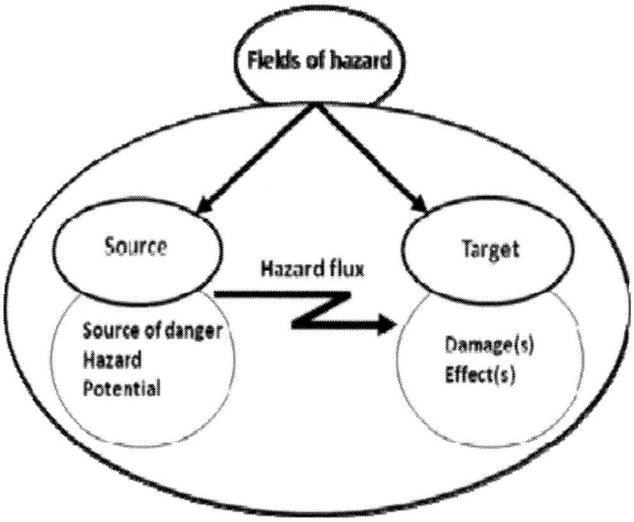

This model considers installation, hardware, software, human factors, management decision-making and organizational design as an integrated whole. The hazard analysis techniques based on this expanded model provide a means for obtaining the information necessary to design safety into the system and to determine the most critical parameters to monitor during the operations and how to respond to them. The reference model called "hazard process" puts into relation a danger source system and a target system by a hazard flux within an active environmental field.
We can illustrate a preventive deterministic approach by combining this systemic model with the different types of neutralization barriers (Figure 2); we can see that many barriers can be positioned in order to reduce the consequences of an undesired event.

MOSAR (Organized and Systemic Method of Risk Analysis) is based on the MADS model. Its aim is to find dysfunctions and to manage the risks in a complex system. The methodology proposes a structured scheme, exhaustive, progressive and quantitative if necessary. The MADS/MOSAR methodology is a generic approach providing a risk analysis of the system and at the same time, identifying the means for prevention, protection and mitigation necessary to minimize and manage risks. This combined method provides identification of the hazard sources distinguishing scenarios of undesired events, providing and evaluating risk analysis and then proposing safety fences. The MADS/MOSAR methodology enables us to highlight the main scenarios and to define the barriers necessary to set up, neutralize or reduce the occurrence of the undesirable event.

Figure 2. Safety Barriers Combined with the Generic Systemic MADS Model

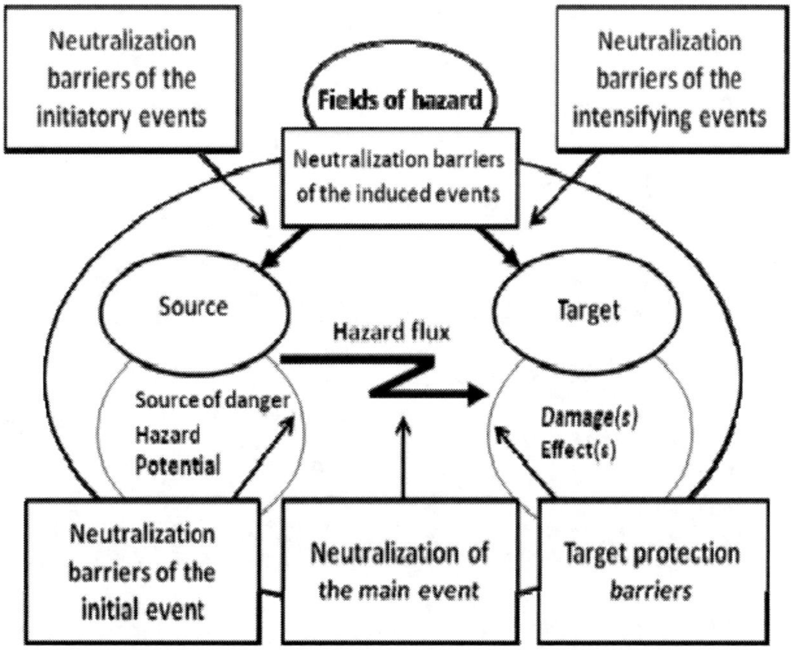

The MADS/MOSAR methodology may also be applied to the design of a new installation as well as to the diagnosis of an existing one. It works as a toolbox to assist the decision maker owing to the resulting choices that it provides. The first module (the A Module), which is the only one used in this paper, makes it possible to carry out a risk analysis of the system based on the process shown in Figure 3. This diagram shows the principal elements and the structure of the MADS/MOSAR methodology. After dividing the system into subsystems, the first step consists of the review of the most relevant release sources for each subsystem. One refers to a typology grid for the systems that are hazard sources and then uses the MOSAR model that connects them to the targets.

This methodology makes it possible to generate scenarios between the subsystems which are gathered for the same event. For each release scenario, a set of consequence chains are developed expressing the escalation of the events. The set of consequence chains is conveniently represented in the form of an event tree.

This risk analysis method also involves the estimation of the frequency and severity of a range of hazards. The risk of injury is then determined using injury relations, which give the probability as a function of the intensity of the physical effect. From a group of experts and the definition of the risk acceptability, we realize finally the scenario's hierarchy. The research of prevention means, i.e., technical barriers and operational barriers, is necessary to neutralize the hazardous scenarios and ensures the prevention of risks. The first module ends with the qualification of the identified barriers. The second module (the B module), allows a detailed analysis of a specific part of the installation to be carried out and in particular implements the reliability tools to search the technical dysfunctions (like a fault mode analysis). It also implements the tools of operational analysis to study the operational dysfunctions.

Figure 3. The MOSAR Methodology

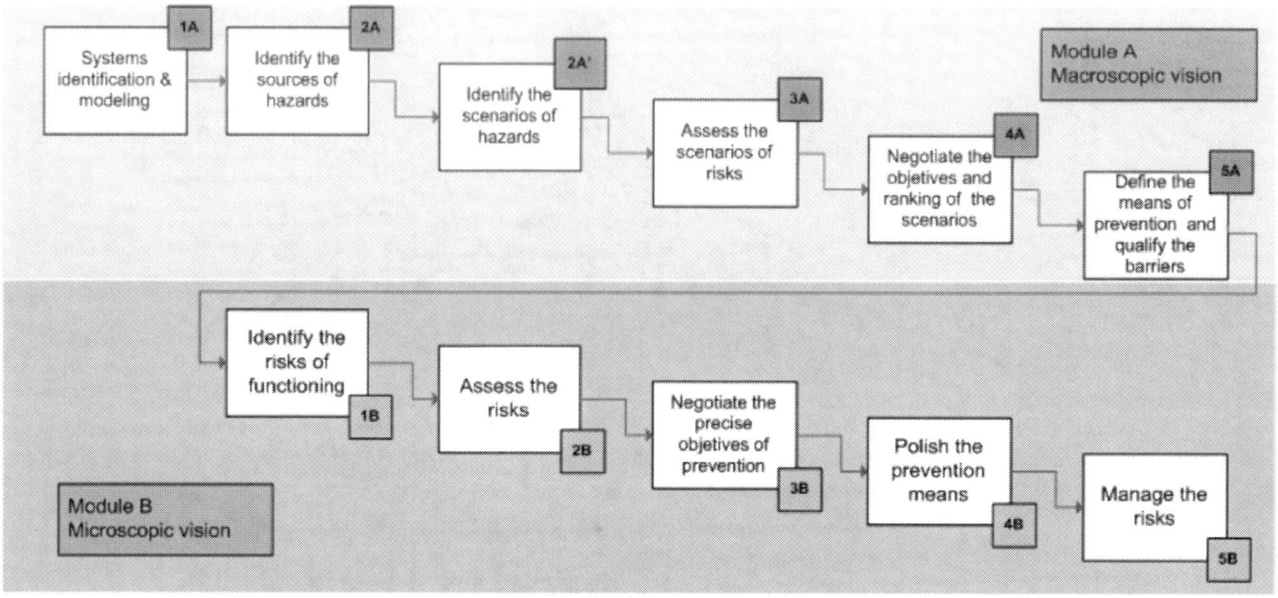

Using the MADS/MOSAR methodology combined with a classical typological classification (International Standard Industrial Classification of all Economical Activities - ISIC Rev.3), we can define the economic activities that will be the object of this analysis in a given context. In our case, we choose to apply this methodology to a South American emerging country: Colombia. We have focused on the mining exploitation, but we can transpose it in all the branches of industry.

The MADS/MOSAR methodology presents four main advantages for this kind of analysis: 1. the risk conception model is simple and pedagogic. 2. the methodology uses internationally well-known methods of risk analysis, frequently used in the process industries (i.e. checklists and Fault Tree Analysis). 3. it proposes a number of tools that make its application easier without needing a higher level of expertise. 4. MADS/MOSAR has a "*bottom-u*p" approach in its analysis procedure as well as in its workgroup conception.

This methodology allows a community to develop its own instruments according to its needs. This approach is more appropriate for local authorities than transferring directly a foreign solution to an emergent country. In the Colombian case, we have identified important opportunities for the development of instruments for the evaluation of major hazards, and the ways to harmonize existent ones through an impact evaluation.

2. What is a Barrier?

The word safety barrier is highly used since the early 1980´s. It defines the function or the object capable of eliminating or diminishing the evolution of an event and its propagation in a chain of non-desirable events. The proficiency of scientific documents, regulations, standards and practices has led to terms such as: defense in-depth, layer of protection analysis, safety critical function, safety critical element, and design for safety, among others to conceptualize and define. As a result of those efforts, we can identify the following: Guidelines for Safe Automation of Chemical Processes [1], Layer of Protection Analysis: Simplified Process Risk Assessment [2], Basic Safety Principles for Nuclear Power Plants [3], Part 1-7 Functional Safety of Electrical/Electronic/Programmable Electronic Safety-Related Systems [4], Function Safety-Safety Instrumented Systems for the Process Industry Sector [5], Petroleum and Natural Gas Industries - Offshore Production Installations - Guidance on Tools and Techniques for Hazard Identification and Risk Assessment [6] and the Petroleum and Natural Gas Industries - Control and Mitigation of Fires and Explosions on Offshore Production Installations - Requirements and Guidelines [7].

2.1 Defining a Safety Barrier

The use of the energy model for risk analysis in social-technical systems has been generalized, and it is in that one where the mechanism *Hazard-Barrier-Target* (HBT) is formalized [10].

It is not easy to define what a safety barrier is, because this one only makes sense within the HBT models. Normally, for its definition, we use function and system elements. That is why it is important for the authors to clarify the differences between a safety barrier, a safety function and a safety system.

2.1.1 Safety Barrier

A safety barrier is defined as: *It's a physical and or non-physical means planned to prevent, control or mitigate undesired events or accidents* [9].

2.1.2 Barrier Function

The barrier function has a direct relation with the sequence of the event or accident (normal condition - initial phase - concluding phase - injury phase). Haddon's initiative proposes three central strategies for the hazard control: elimination/reduction, energy control or mitigation. This first proposal has been evolved with the work of Kletz [11], Schupp [8], Duijm [12], Goossens [13], Johnson [14], Périlhon [16, 17], among others, and as a result of this evolution, we can define the barrier function as: *It's a function planned to avoid, prevent, control and/or mitigate undesired events or accidents.*

2.1.3 Barrier System

The system barrier is one or more functions within the action. There are various types of classifications to determine its structure, but generally they have a classical of process control (sensor-decision-action). Nevertheless, for our work, we can define a system barrier as: *It's the system designed and implemented to perform one or more barriers functions* [9].

Figure 4. Classification of Safety Barriers

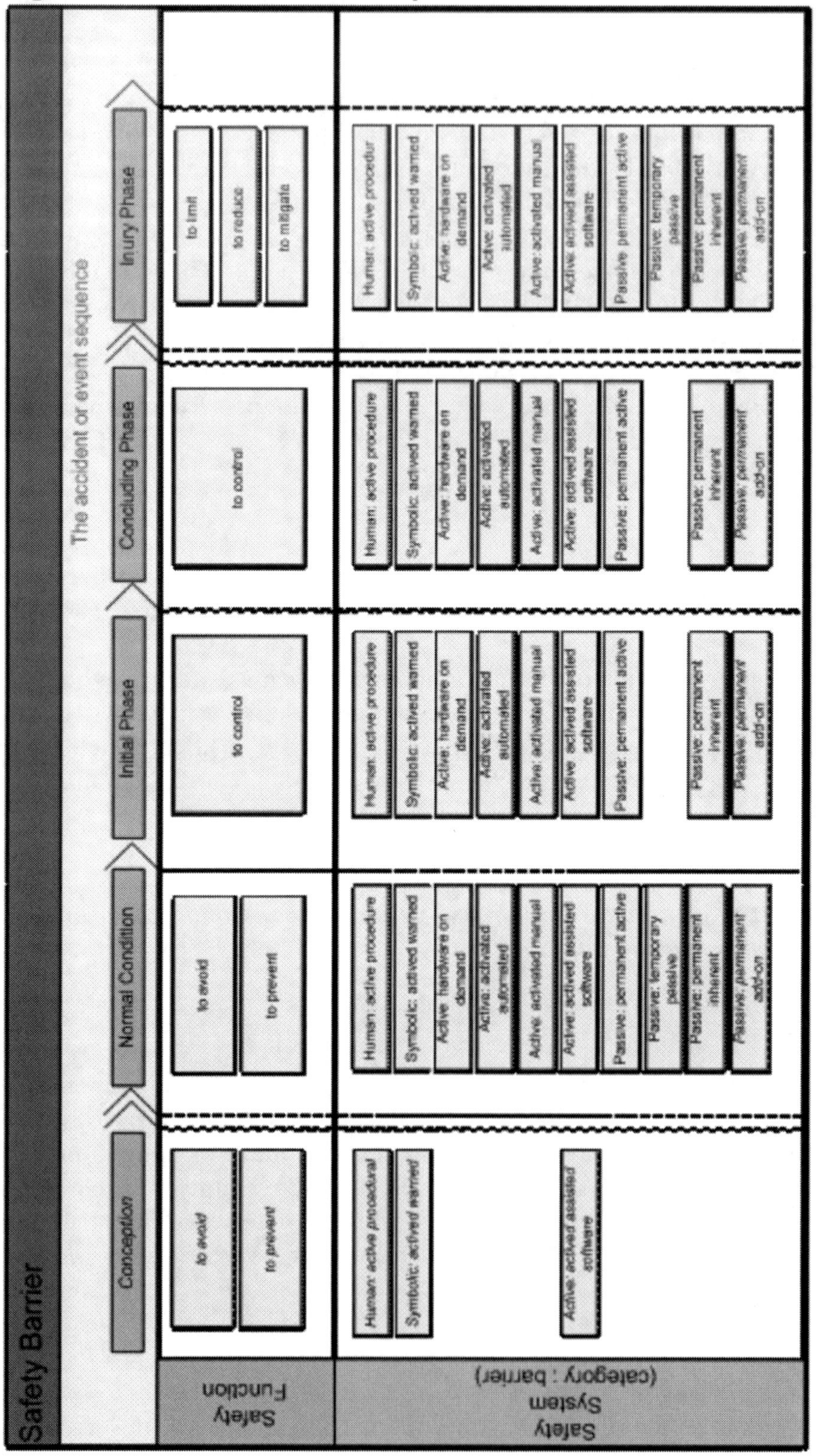

Based on those definitions, we propose a barrier classification that uses its functions and systems (Figure 4 and Table 1). This composition results from the concepts of the authors previously mentioned, and we have added some recommendations and a common structure, using an event or accident amplified sequence.

What can we understand through an accident amplified sequence? As mentioned above in the barrier definition, the action taken should be in direct relation with the event or accident sequence. Traditionally, this is defined by the sequence "standard, initial-condition, phase-concluding, phase-injury phase" [16], but it left behind events generated at the conception phase, and those are repeatedly pointed out in the post-accident analysis as beginners. The incorporation of this phase in an early basis gives some privilege to functions, such as avoid and prevent and induce systematically a reflection process about the possible sequences of events in the design process.

For every event or accident phase, there are one or multiple safety functions and a series of safety systems that can be used. We have to keep in mind that a barrier system can have one or more functions. This proposal should be seen as a toolbox that facilitates the identification and classification, and it generates unity in the criteria about what we understand as a safety barrier.

2.2 Normative Barriers

Can we classify a normative screen (e.g. national legislation, community, internationally known practices, standards, etc.) as a safety barrier? We think that the answer is YES.

How can we classify normative barriers with the definitions given above? A normative barrier is present in whole sequence of the event. Its functions are to avoid, to prevent, to control and to reduce, and its system is "human: active procedure". We should not forget that a normative barrier can trigger the implementation of others generating synergic effects over a productive system.

Standards for environmental protection, industrial safety and occupational health are the minimum range of a regional reference for the prevention, control and mitigation of non-desirable events and a function and system can be defined for each one of them.
In a globalization process, it is usual that social risk (class situation) coexists with an industrial risk. Ulrich Beck talks about this world phenomenon saying that: "…the proletariat of the World society of risk lives under the chimneys, near the refineries and the chemical industries inside the industrial centers of the third World." [17]. The transfer of process and industrial activities in the countries of the third world do not necessarily lead to a transposition of knowledge of its associated risk and is shown by the legislative remainder.

In this context, we develop our proposal that contains three hypotheses:

- It is possible through the MADS/MOSAR methodology to analyze the microsystems for the evaluation of sector risks and to suggest the implantation of barriers.

- It is possible to design normative barriers in an *a priori* way in the technological transfer process.
- An analysis through possible scenarios is a diagnosis/preventive tool for the improvement of a national legislative system that tends to the administration of industrial risk.

Table 1. Barrier Systems

Barrier System	Definition	Example
Human: active procedure	Barrier with human detection/diagnostic/action, and human or remote control to act	Legislation, procedures, operational actions, evacuation, actions coordination
Symbolic: active warned	Barrier with hardware detection and human diagnose/action/act	Safety alerts, safety signs, technical language in engineering
Active: hardware on demand	Barrier with detection, diagnose, action and act by hardware	Traditional sprinkler installation, relief systems
Active: activated automated	Barrier with detection an act by hardware, and diagnose/activate by software	Control system, programmable automated device
Active: activated manual	Barrier with hardware detection, human diagnose & action, and human or remote control to act	Shutdown adjustment in response or manual, detection and alarm systems with human action
Active: activated assisted software	Barrier with hardware detection, human or software diagnose & action, and human or remote control to act	Expert systems
Passive: permanent active	Barrier with act by hardware. Needs diagnose & activate by other process or human	Cathodic protection, defrosting systems, ventilation, explosion relief, inerting systems, lighting
Passive: temporary passive	Barrier with act by hardware, an putted on by human action	Blind flanges, insulation area barriers, personal protection equipment, reaction inhibitors
Passive: inherent	Barrier act by hardware. Inherent equipment property	Equipment thickness, equipment characteristic
Passive: add-on	Barrier act by hardware. Add-on or put-on a principal equipment	Secondary containment, drainage systems, rupture disc, earth connection

3. The Colombian Case

Colombia is ranked fifth worldwide in the production of thermal charcoals and has the biggest reserves in all of Latin America, reaching 59 millions of tons in the 2005 production [19].
The impact in the regional economy as well as in the social development caused by the mining activity is very significant

3.1 Structure of the Colombian Normative

We will understand the normative as the combination of the national legislation, technical standards and ministerial guides. The strength in the development of rules in Colombia for environmental protection, occupational health and industrial safety is given in international conventions such as Stockholm 1972, Basel 1987, ILO C.17, ILO C.144, ILO C.170, ILO C.174, among others (ILO : International Labour Organization).

Figure 4. Colombian Kelsen's Pyramid

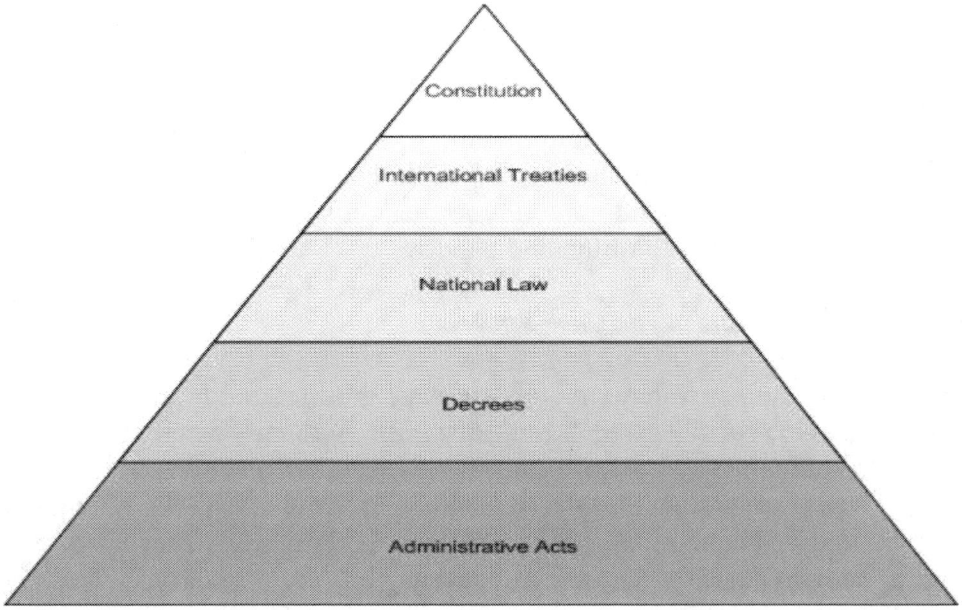

Colombia has determined a hierarchical legal order proposed by the Austrian philosopher Hans Kelsen, in his work *Pure Theory of Law*. The objective of this theory is to determine the order of the legal regulations, putting the national constitution as the maximum rule and from there the other instruments follow (Figure 4).

For this work, a Colombian legislative analysis was conducted from the point of view of safety barriers for the administration of risks in the mining sector. A complete analysis of the sector was done to determine the danger flux needed for each one of the non-desirable events and we determine also the legislation over them. This process reveals the development of the existing legislation in relation to the prevention, control and mitigation of a hazard flux. This cartography of events and barriers gives opportunities in the legislative development for the administration of risk associated with specific non-desirable events and determined as unacceptable.

The Colombian government in its initiative of stability in the judicial frame issued the Mines' Code, Law 685 of 2001, for the consolidation of national and foreign investment in Colombian mining and energetic resources. This initiative articulates the functions of the Mines and Energy Ministry as the manager of mining and energetic resources of the nation, and the environmental, lodging and territorial development Ministry as the manager of the territory with responsibility for the exploitation of resources.

Due to this relationship, technical guides had been developed for the mining sector; some of them have been listed below:

MINING ENVIROMENTAL GUIDES OF THE 18- 0861 RESOLUTION

- Mining Environmental Guide of Exploration [20]
- Mining Environmental Guide of Exploitation [21]
- Mining Environmental Guide of Benefit and Transformation [22]

OTHER MINING ENVIRONMETAL GUIDES

- Mining Environmental Guide to Underground Mining and Coal Parks [23]
- Mining Environmental Guide to Clinker Industry [24]

These documents became "the orientation and referent instrument of conceptual, methodological and procedure character for the mining planning, operation and pursuit" [23], for the owners of mining rights or mine holders in the Colombian territory. According to these guides, people who want mine titles or rights should make the proper request to the above-mentioned ministries in order to get the permits for exploiting and benefit.

3.2 Methodological Approach: Case Study

The analysis of complex systems requires a workgroup capable of understanding a variety of problems. To provide for this need, a multidisciplinary team was created and coordinated by the CNRS/LRGP (National Center for Scientific Research /Reactions and Chemical Engineering Laboratory - NB: The National Center for Scientific Research is a government-funded research organization, under the administrative authority of France's Ministry of Research) with the participation of GEODERIS. On specific occasions, the Ministries of Social Protection and Mining and Energy of Colombia also participated.

The work developed in France by the CNRS/LRGP and GEODERIS, was discussed in the field by representatives from the ministries above mentioned and some other Colombian organizations, such as the National Service of Learning -SENA-, in command of the mining teaching. The results presented below are a product of the interpretation made by the authors, taking into account the individual help of each of one of the participants, but this does not represent an official opinion of the Colombian government.

3.2.1 Systemic (1A)

To define what a system is and its unfolding complexity, we have used the concepts of Jean-Louis LE MOIGNE in his book "The Theory of the General System" [25]. Our case study is defined by:

- Unfolding complexity in four levels (Table 2): mining economical sector (mega-system), mining activity (supra-systems), unitary operations (systems) and available technologies (sub-systems).
- Basic unities for the analysis of the methodology: available Technologies (sub-systems).

Table 2. Complexity Unfolding Underground Mine

<table>
<tr><th>Mega-system</th><th colspan="3">Supra-system</th><th colspan="3">System</th><th colspan="3">Sub-system</th></tr>
<tr><td rowspan="19">Mining Sector</td><td rowspan="19">Underground mine</td><td rowspan="19">Environment (artificial – natural)</td><td rowspan="19">Man</td><td rowspan="19">Environment (artificial – natural)</td><td rowspan="19">Man</td><td rowspan="6">Breaking</td><td rowspan="19">Environment (artificial – natural)</td><td rowspan="19">Man</td><td>Mechanical pick</td></tr>
<tr><td>Bossing systems</td></tr>
<tr><td>Haulage systems</td></tr>
<tr><td>Drilling systems</td></tr>
<tr><td>Explosives</td></tr>
<tr><td>Others techniques</td></tr>
<tr><td rowspan="3">Mucking</td><td>Armored conveyor</td></tr>
<tr><td>Mechanical loader</td></tr>
<tr><td>Manual</td></tr>
<tr><td rowspan="5">Haul</td><td>Wagon</td></tr>
<tr><td>Conveyor</td></tr>
<tr><td>Truck</td></tr>
<tr><td>Shafts</td></tr>
<tr><td>Manual</td></tr>
<tr><td rowspan="5">Annexed operations</td><td>Ventilation</td></tr>
<tr><td>Drainage</td></tr>
<tr><td>Lighting</td></tr>
<tr><td>Support</td></tr>
<tr><td>Detection & Alarms</td></tr>
</table>

This communication does not deal with the procedure follows by the definition of the mega-system, supra-systems, systems and sub-systems, but each one was decreed with the strict sense of systemic.

3.2.1 Risk Assessment

MADS looks for a complete description of the danger mechanism. This mechanism is defined by the relation between: field danger - danger source and target. The combination of these three factors was evaluated for the stages: exploiting, maintenance and post operation.

The field danger will be defined by the relation between time and space that can be probable where a flux of danger can be generated between a source and a target. This is built by the natural environment (e.g. geology) and the artificial environment (e.g. other mining operations).

For the mining sector, the sources of danger are: available technologies (sub-systems), man itself, and the environment artificial or natural.

The targets are determined by the interests of the evaluation or protection. For this study, we determine as interest: the worker, the riparian, the industrial site and natural environment.

3.2.1.1 Dangers Typology (2A)

The first tool of MOSAR is what we call list 1. This is a list of dangers by categories defined by the feedback experience (Table 3) [15]. This semi-exhaustive list should be applied to every possible source of danger without taking into account the target. It then explores the intrinsic sources of hazards.

3.2.1.2 Construction of Scenarios (2A')

Once the risks are determined by source and the exploitation phase, the danger mechanism emerges. This then leaves the principal flux danger as well as the events: beginner, initial and reinforced. MOSAR systematizes this process in Table A.

The principal flux danger is expressed as material, energy and information, and it determines the relation between the elements exposed to this one in a specific space of danger.

To reduce the variety generated by Table A, the methodology suggests the use of the concept of black box. Here is where a change of scale of the analysis takes place. Until now, the work has been done with the last level of unfolding complexity (available technologies), but in this new step, we build a black box per system (unitary operation). It is very interesting how when we group the danger and its flux in each unitary operation the variety due by the typology coincidence gets reduced. We have to remember that the reduction is only possible in a same exploitation phase because it is the relation coincidence in time and space.

It is also important to maintain such a type of analysis until the construction of scenarios. This is the way we are going to analyze the whole compound of unitary actions inside a mining activity (from systems to supra-systems). This is our second change of scale and variety reduction. This phase will put in evidence the chain of events between the operation and the domino effects in each activity. The role of the experts group is to extract from the possible chains the non-desirable events and therefore, to define the scenarios to be studied in detail.

Finally, we just have to formalize its development and for this purpose, we have used the Fault Tree Analysis (FTA) method [18]. This construction is meant to symbolize, in a logical way, the evolution of the non-desirable event, making explicit its happening sequence. For this study, we have determined three non-desirable scenarios: an explosion, mine collapse, and respiratory (lung) diseases.

3.2.3 Placement of Existing Barriers (3A & 4A)

Once the FTA structure is determined, we propose to transpose the safety barrier grade defined at the beginning of this article. We will relate then the barrier function with the sequence of non-desirable events (Figure 5). For each determined flux downstream, we will use the strategies to avoid, prevent and control. Opposite to this, upstream, the barrier functions will be: limit, reduce and mitigate.

Table 3. General Hazards Typology: LIST 1

LIST 1: General Hazards Typology to Storage, Production and Transport

A. Sources of mechanical dangers

- Sources under pressure
  - gas or vapor
  - hydraulics
- Sources under mechanical constraints
- Sources in movement
  - solids
  - liquids
  - gas
- Sources of requiring a handling
  - manual
  - mechanics
- Sources of physical-explosions others that under pressure
  - implosion
  - BLEVE (Boiling Liquid Expansion Vapor Explosion)
  - mixture of liquids at very different temperatures
  - warming
- Sources of height fall
- Sources of level fall
- Sources of wounds (objects cutting, prickles, blunt...)
- Sources of noise and vibrations

B. Sources of chemical danger

- Chemical reaction sources
- Explosion sources
  - in condensed medium
  - in gas phase
- Sources of toxicity and aggressiveness
- Sources of pollution and odors
- Sources of lack of oxygen

C. Systems sources of electric danger of origin

- Sources AC/DC
- Sources of static electricity
- Condensing
- Generating of high frequencies

D. Sources of danger of development of fire

- Sources of lighting
- Bulk-headings sources (non-attendance, standard, fire resistance, distances between under systems)
- Materials sources (nature, implemented , fire performance, thermal load)
  - solar irradiation
  - ecological imbalances
- Sources related to ventilation
- Sources related to the extinction

E. Radiation sources

- Ionic-radiations
  - radiation sources
  - contamination sources
  - nuclear sources
  - sources to the evacuation of radioactive energy
- UV - IR - VISIBLE (lighting)
- Laser sources
- Microwave sources
- Magnetic fields

F. Systems sources of biological danger of nature

- Micro-organisms
- Animal and vegetable.
- GMO
- Animals
- Plants
- Human behavior
  - in normal situation
  - in a scenario
  - terrorism

H. Systems sources of dangers related to the active environment

- Artificial or active environment
  - transports
  - industrial facilities
- Natural origin
  - geological:
    - seismic
    - landslides
  - climatic:
    - Storms, cyclones, tornadoes, hurricanes...
    - fog, dryness, floods
    - fires of forests, lightning, freezing...

I. Social-economical sources

- finances - migrations - conflicts - criminality - violence
- large gatherings

Figure 5. Function Barriers on Event Sequences

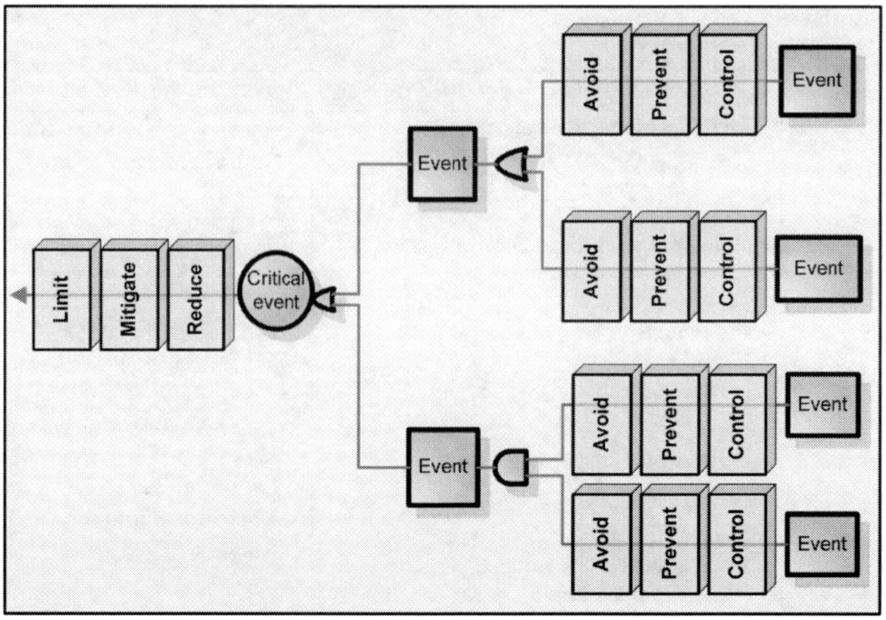

We have realized an analysis of the Colombian case applicable to these types of events. As a product of this revision becomes evident the assigned functions to each legislative instrument. We can then find in the appropriate FTA, the legislative instruments associated with the non-desirable event and the functions of the barrier that goes with it (Figure 6, e.g. explosion). From this exercise, we can visualize the distribution, localization and strategies defined by the Colombian legislative framework for the administration of critical events.

It became possible to identify the sequences of non-desirable events that already have normative barriers and can possibly generate a critical event. This identification is fundamental because it opens action possibilities for the development of future legislation. It generates a legislative cartography and proposes barrier functions that orientate system implantation.

It is not possible to realize a direct classification of the existent instruments, because Colombia does not have the information resources that can compile the appropriate index and indicators. A qualitative classification of the barriers can be done based on international feedback experience and thus determine the effectiveness of the legislative instrument in the danger control. This void in the national system makes our work a useful tool for the construction of a future index and indicators. It is important to say that this point of view explains the danger mechanism, the existent barriers and generates a common language for the risk managers.

4. Summary

This summary presents the main points as a result of the application of the MADS/MOSAR methodology for the evaluation of legislation related to the management of risk associated with the mining sector. We can highlight three dimensions of the evaluation: 1. the evaluation of the robustness of MADS/MOSAR for macro-system analysis, 2. the evaluation of the methodological ensemble used and 3. results obtained for the Colombian case.

Figure 6. Colombia's Function Legislation Barriers in Explosion Tree

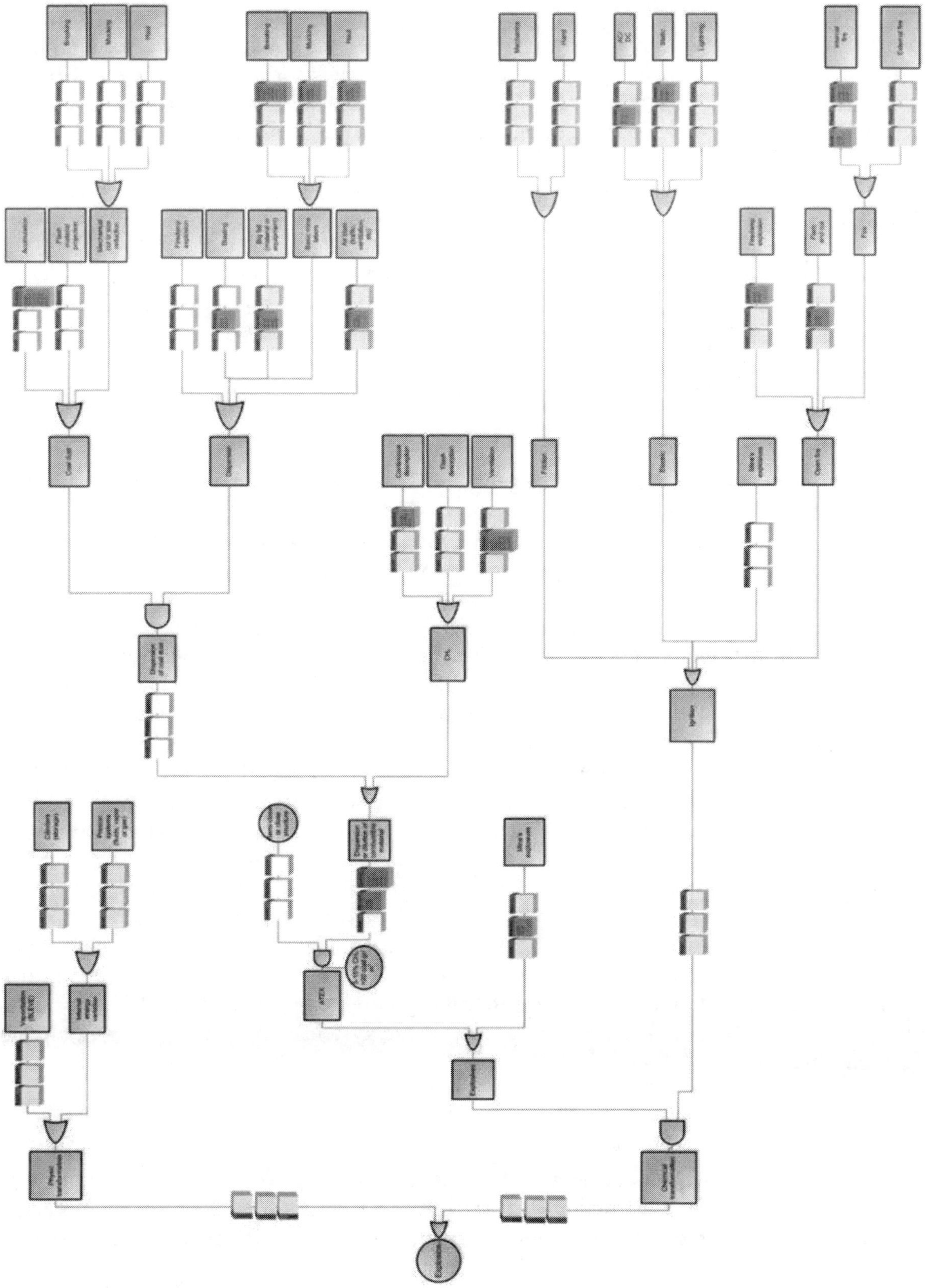

4.1 Robustness of MADS/MOSAR for Macro-System Analysis

The systemic basis of MOSAR made easy the analysis of a problem that appears diffuse. The constitution of a multidisciplinary group with complementary abilities made easy the construction of models of the studied system from different points of view. (e.g. public sector, industry sector). MOSAR is a bottom-up tool, where spaces for discussion are possible for the construction of a minimum frame of a reference between the actors. This frame has a shared language, different agreements that define the system of analysis, selection of the methods to use, methodological articulation, definition of possible scenarios, final construction of the mechanism of critical and non desirable events, localization of the existing barriers, and finally a shared diagnosis of the panorama.
MADS shows its versatility for the description of danger mechanisms and builds a shared model. It makes an integrated disciplinary pedagogical model, and it shows the complementarily in its strategies for the control of danger flux. In this case, the participants easily can find the events to propose control strategies from an action domain and in this way generate an ensemble of complementary barriers.

4.2 Evaluation of the Used Methodological Ensemble

MOSAR is conceived as like a toolbox and this is why one of the first steps of analysis is to define the modules to apply and its sequence. We decided to apply the whole Module A originally proposed by Périlhon (Figure 3). As a method for the visualization and formalization of the language in the definition of critical events we determine the use of FTAs.
Not all the different steps of Module A were easy to apply. Correspondent module for the source identification of sources of "hazards" has LIST 1 and this is an excellent tool for the systemic identification. In the related models for the identification of scenarios and the determination of risks, we have to be careful with the use of the language because Table A is a tool for the systematization of declarations. In this part, the workgroup declares the mechanism of danger for each one of the identified danger sources.
The negotiation stage is, without any doubt, the most enriching of the proposal. The common point of view of the different actors, having as a base a shared analysis, makes easier the comprehension of the problem and the search of agreements about united actions. It was interesting for the analysis group to present this proposal to the different Colombian actors. They know very well the problems, but they have trouble making them explicit.
Finally, the step correspondent to the definition of objectives prevention and localization of barriers was not the most appropriate. This one does not have clear definitions or matching mechanisms. We have proposed the classification barrier in Table 1 and we have unified the barrier function with the FTA. It is clear that the proposal does not allow the classification of proposed barriers but gives a general panorama of the system and its strategy for the control of the danger flux.

4.3 Results from the Colombian case

For the Colombian case, we considered three different scenarios of explosion, collapsing and pulmonary diseases. The construction of trees and the position of existent barriers generated a clear panorama about the existent national legislation for the administration of non-desirable

events. The opportunities in legislative development related with danger analysis were clear. None of the three scenarios had preventive barriers, elimination control or mitigation for danger flux that can generate major accidents.
In the field of industrial safety, Colombia has a decree that establishes the Basic measures of prevention (Decree 1335 of 1994). They are technical recommendations and procedures for the prevention of mining risks, and they do not study the sequence of events that can go beyond the limits of the exploitation system. The decree is clear about the a posteriori sanctions, but it does not promote a reflection about mining security during the stages of the mining life. For the underground mining work, we have developed complementary possible standards orientated to the design of:

- Procedures of maintenance for underground excavations.
- Transportation equipment used in mining
- Cables and electric installations underground
- Ventilation and dust control in the mining exploitations
- Prevention and extinction of fires in mining
- Drain of underground waters in mining
- Machinery and tools used in underground work

These complementary standards do not correspond to the functions of the barrier determined in its origin. That is the reason why today the occupational health committee for the mining sector works in the update of the decrees 1335 of 1987 and 2222 of 1993. The update of those decrees will lead to a revision of the complementary standards that we hope can be published in the next years. The design of this kind of standards is being made based on the concept from experts and it will surely target for a strategic control technique. This strategy has shown in the past the generation of management for parallel systems, and not an integration of protection concepts inside the whole business system.
In the field of environment, significant advances had been made in the ministry guides for the impact evaluation. This one has been developed bason the process life (e.g. exploration, exploitation), but the time horizon is limited to the closure of the operation and does not contemplate the study of major accident scenarios. The mining activity defers considerable impacts that had not been included inside the instruments as the radon emissions, mining effluents and impacts in the territory for total or partial collapsing. It gets clear the integration between the concepts of clean production and mining guides, but it is not very clear for the integration of the occupational health concepts - environmental protection - industrial security.
The national system has generated a total decentralization for the particular management of the extraction processes related to occupational health and extraction techniques. Amazingly, the same system wants to distribute the mining diversity into categories generating blindness about the emerging risks' product of the relation geology-technique-territory-culture in every exploitation. The legislation should be a frame of action and an institutional technical strength for the risk management and not a normative barrier manageable for its inapplicability.

5. Conclusion

The challenges for Colombia are as big as its mining potential. The promotion of the exploitation of mining resource under the concept of sustainability opens a horizon of teamwork for the integral measure of risk between the different national actors. The national institutions in charge

of the resources regulation (environmental, economical, social, etc), are designated to accelerate this process proposing a matching legislative framework.
The increase in the scale of exploiting generates a modification in mining national culture for which the national instruments of protection are not prepared.
The articulation of the legislative structure concerning risk becomes a challenge for Colombia. The manager of the system has to make choices and has to give national resources to public or private operators. Making choices using risk concepts is useful and should be reflected in levels such as:

- Exploitation alternatives or mining resource reserves.
- Environmental resources.
- Civilian protection.
- Territorial order.
- Workers protection
- Mining exploiting culture, ethnic and cultural minorities

According to the above-mentioned ideas, it can be concluded that: Related to other elements of the choice making we can conclude that:

i. The legislative development of danger studies integrated with impact studies becomes necessary. This process can be coordinated by the Environment, Lodging and Territorial development Ministry if the direction of sustainable development transforms itself technically and integrates the notions of integral risk.
ii. The incorporation of concepts such as danger mechanisms, risk and barriers will generate a change in national politics. No longer will the specific technical legislation for the measure of risk be privileged and that type of responsibility will be given to the mining titular during all the phases of the project. A reflection process appears equivalent to the one generated by environmental protection but with personal and collective protection and infrastructure as targets.
iii. When we incorporate the concept of danger flux in HBT or MADS models, a barrier classification becomes necessary. It is fundamental that the control organizations fixate clear barrier functions for each exploitation phase. It is the responsibility of the mining titular the design of systems safety with determined functions given by the control organizations and that they make tolerable limit risks pointed by them.
iv. We have to generate a reflection about the tolerable risks associated to the industrial activity. This decision should be made by the regulatory organizations and should be built in a dynamic way so it can be unique to every project or sub-type project. This decision should be communicated to the influence communities. This communication process of risk does not give responsibility to the communities or to the local managers.
v. Regarding the explosion scenario, the existing barriers of the Colombian legislation limit themselves to monitoring and mitigation of non desirable events. The classification of underground mining based on charcoal dust characteristics does not allow an integrated reflection of the danger mechanism. It could be useful to talk about this subject from the explosive atmosphere (ATEX) point of view and to include it in a hierarchical way in the barrier functions: elimination - prevention - control - mitigation. This strategy integrate the concept of collective and individual protection, turning the ATEX studies into an instrument of control-verification for the regulator.

6. References

[1] CCPS, *Guidelines for safe automation of chemical processes*. Center for Chemical Process Safety of the American Institute of Chemical Engineers ed. 1993, New York.

[2] CCPS, *Layer of protection analysis simplified process risk assessment* Center for Chemical Process Safety of the American Institute of Chemical Engineers ed. 2001, New York.

[3] IAEA, *Basic safety principles for nuclear power plants*. International Atomic Energy Agency ed. 1999, Vienna.

[4] IEC, *IEC:61508. Part 1-7 Functional safety of electrical/electronic/programmable electronic safety-related systems*. 1998, Geneva: International Electrotechnical Commission.

[5] IEC, *IEC: 61511. Function safety-safety instrumented systems for the process industry sector*. 2002, Geneva: International Electrotechnical Commission.

[6] ISO, *Petroleum and natural gas industries. Offshore production installations. Guidance on tools and techniques for hazard identification and risk assessment*. 2002, Geneva: International Organization for Standardization.

[7] ISO, *Petroleum and natural gas industries -- Control and mitigation of fires and explosions on offshore production installations -- Requirements and guidelines*. 1999, Geneva: International Organization for Standardization.

[8] Schupp, B., et al., *Design support for the systematic integration of risk reduction into early chemical process design*. Safety Science, 2006. 44(1): p. 37-54.

[9] Sklet, S., *Safety barriers: Definition, classification, and performance*. Journal of Loss Prevention in the Process Industries, 2006. 19(5): p. 494-506.

[10] Haddon, J.W., *Energy damage and the ten counter-measure strategies* Human Factors Journal, 1973.

[11] Kletz, T.A., *Inherently safer design-its scope and future*. Trans IChemE, 2003. 81(B).

[12] Nijs Jan Duijm, et al. *ARAMIS project : Assessing the effect of safety management efficiency on industrial risk*. in *PSAM 7—ESREL '04*. 2004. Berlin, Germany.

[13] Goossens, L., & Hourtolou, D., *What is a barrier?* 2003, ARAMIS-working document.

[14] Johnson, W.G., *Mort Safety Assurance Systems (Occupational safety and health)*. 4 ed. 1980: Marcel Dekker Inc 525.

[15] Périlhon, P., *MOSAR présentation de la méthode*. Techniques de l'Ingénieur, 2003. Traité sécurité et gestion des risques (SE4 060).

[16] C. Delvosalle, C. Fiévez, and A. Pipart, *Deliverable D.1.C.*, in *WP 1*, Accidental risk assessment methodology for industries in the context of the Seveso II directive, Editor. 2004, Faculté Polytechnique de Mons, Major Risk Research Centre: Mons

[17] Beck, U., *Risk Society: Towards a New Modernity*. Paidós Surcos ed. 2006, Barcelona. 393.

[18] Laurent, A., *Sécurité des procédés chimiques: connaissances de base et méthodes d'analyse de risques*. Génie des procédés de l'Ecole de Nancy, ed. E. TEC&DCO. 2003, Paris: Lavoisier.

[19] Unidad de Planeación Minero Energética UPME, *Boletín Estadístico de Minas y Energía 1999 - 2005*, S.d. Información, Editor. 2006. p. 132.

[20] Ministerio de Minas y Energía, -.M.d.A., Vivienda y Desarrollo Territorial, *Guía minera ambiental, exploración*, C.T.I.C.-C.-C.L.R. LIMITED, Editor. 2002, Camilo E. Peña Vargas, Diseño, Armada Electrónica y Producción. p. 112.

[21] Ministerio de Minas y Energía, -.M.d.A., Vivienda y Desarrollo Territorial, *Guía minera ambiental, explotación*, C.T.I.C.-C.-C.L.R. LIMITED, Editor. 2002, Camilo E. Peña Vargas, Diseño, Armada Electrónica y Producción. p. 154.

[22] Ministerio de Minas y Energía, -.M.d.A., Vivienda y Desarrollo Territorial, *Guía minera ambiental, beneficio y transformación*, C.T.I.C.-C.-C.L.R. LIMITED, Editor. 2002, Camilo E. Peña Vargas, Diseño, Armada Electrónica y Producción. p. 109.

[23] Ministerio de Minas y Energía, -.M.d.A., Vivienda y Desarrollo Territorial, *Guía minero ambiental de minería subterránea y patios de acopio de carbon*. 2004. p. 184.

[24] Ministerio de Minas y Energía, -.M.d.A., Vivienda y Desarrollo Territorial, *Guía minero ambiental de la industria del cemento*, V.y.D.T.-I.-. Ministerio de Minas y Energía - Ministerio de Ambiente, Editor. 2005. p. 387.

[25] Le Moigne, J.-L., *La théorie du système général : théorie de la modélisation*. 4e édition mise à jour ed. Systèmes-décisions. Section Systèmes de gestion. 1994, Paris: Presses Universitaires de France. xii, 330 p.

OSHA Refinery and Chemical NEP Findings

James Lay, P.E.
Lisa A. Long (Presenter)
Michael Marshall, P.E. (Presenter)

U.S. DOL – OSHA
200 Constitution Ave, NW
Washington, DC 20210
202-693-2409
Long.lisa@dol.gov

Prepared for Presentation at
American Institute of Chemical Engineers
2010 Spring Meeting
6th Global Congress on Process Safety
San Antonio, Texas
March 22-24, 2010

UNPUBLISHED

OSHA Refinery and Chemical NEP Findings

Lisa A. Long (Presenter)
USDOL-OSHA
200 Constitution Ave., NW
Room N3119
Washington, DC 20210
long.lisa@dol.gov

Michael L. Marshall, P.E. (Presenter)
USDOL-OSHA
200 Constitution Ave., NW
Room N3119
Washington, DC 20210
marshall.mike@dol.gov
U.S. DOL – OSHA

James Lay, P.E.
USDOL-OSHA
200 Constitution Ave., NW
Room N3119
Washington, DC 20210
lay.jim@dol.gov

Keywords: Refinery NEP, chemical NEP, chem NEP, OSHA, national emphasis program

Abstract

Currently there are two on-going U.S. DOL - OSHA (OSHA) enforcement initiatives involving PSM covered processes in the chemical and refining industries. On June 7, 2007, OSHA initiated a Petroleum Refinery Process Safety Management (PSM) National Emphasis Program (Refinery NEP) in response to a large number of fatal or catastrophic incidents in the petroleum refining industry. Knowing that similar catastrophic hazards exist in chemical facilities outside the refining industry, OSHA initiated its PSM Covered Chemical Facilities NEP (Chem NEP) on July 27, 2009. The Chem NEP is a pilot program targeted at chemical facilities with PSM covered processes. The paper describes OSHA's NEPs and associated findings to date, including the elements of the PSM standards that were most frequently cited.

1. Introduction

An NEP is a type of OSHA Special Emphasis Program. These programs provide for programmed (i.e., pre-planned) inspections of establishments. OSHA develops NEPs to focus outreach efforts and inspections on specific hazards in a workplace. The Refinery and Chem

NEPs provide OSHA inspection tools for evaluating catastrophic hazards (i.e., fire, explosion, and toxic) as they relate to employers' compliance with PSM standards.

Refinery NEP:
OSHA evaluated data from 1992 and 2005 and determined that serious PSM incidents occurred most often in the petroleum refining industry (SIC 2911/NAICS 32411). This evaluation defined serious PSM incidents as those involving an employee fatality, or the hospitalization of three or more workers, where the incident involved a PSM covered process. As a result of the incident history in petroleum refining, including the major catastrophe at the BP Texas City Refinery in 2005, OSHA developed the Refinery NEP, which became effective on June 7, 2007.

The Refinery NEP outlined policies and procedures for OSHA Compliance Officers to use in conducting inspections of petroleum refineries, and mandated that OSHA inspect all petroleum refineries, excluding those in the Voluntary Protection Program (VPP), within federal jurisdiction, over two years. Although VPP sites are excluded from NEP inspections, their PSM programs are evaluated as part of the VPP process. OSHA strongly encouraged State Plan States to adopt the NEP inspection strategy, and all states with refineries have either implemented the Refinery NEP or a similar program.

This NEP has since been extended in several regions; all Federal inspections are expected to be completed in 2011. Except for 15 inspections remaining to be conducted in OSHA's Region 6, all other OSHA Refinery NEP inspections have been either completed (citations issued) or are on-going.

The Refinery NEP emphasizes the *implementation* of PSM programs. Compliance officers use a list of static questions, which is published in the Refinery NEP directive, and a smaller list of dynamic questions, which is unpublished and periodically changed. The questions focus on issues identified as contributing to previous incidents and contain compliance guidance to assist Compliance Officers in evaluating PSM implementation in a unit selected by the inspection team. Contractors working in or adjacent to the selected unit are also inspected.

Due to the nature and large number of static and dynamic questions addressed during each Refinery NEP inspection, these inspections are both extensive and wide-ranging. As a result, the time to complete them is relatively long and the OSHA resources committed relatively large compared to the typical General Industry workplace inspections.

Chem NEP:
The Chem NEP, which became effective on July 27, 2009, is an OSHA enforcement initiative for inspecting facilities with PSM covered processes, i.e., those which contain highly hazardous chemicals (HHCs) in amounts at or greater than the threshold quantities listed in 29 CFR 1910.119. The Chem NEP is a pilot program and is scheduled to run for one year. It provides for programmed inspections in OSHA Regions 1, 7, and 10, and unprogrammed inspections in all regions. Unprogrammed inspections are not scheduled; they are inspections initiated as a result of an incident, employee complaint, media referral, etc.. VPP sites are exempt from programmed inspections, but are subject to unprogrammed Chem NEP inspections in the event of a incident, complaint, or referral related to a PSM covered processes.

There are roughly 150 petroleum refineries in the U.S. This number is small compared to the number of other PSM covered chemical facilities in the country. U.S. Environmental Protection Agency's (EPA) Risk Management Program (RMP) database lists 6,802 Program 3 processes, which are, in most cases, OSHA PSM covered processes. OSHA committed to inspecting each of the roughly 70 non-VPP refineries in Federal jurisdiction. However, due to the much larger number of chemical facilities with PSM covered processes, OSHA needed to employ a different inspection strategy. The Chem NEP was developed as a targeted inspection program. The target inspection pool includes facilities with PSM covered processes including:

- Facilities listed in EPA's RMP database as Program 3;
- Explosives manufacturing facilities;
- Facilities listed in OSHA's IMIS database as having been previously cited for PSM; and
- Facilities included through OSHA Area Offices' local knowledge.

OSHA has acknowledged that the current PSM compliance directive strategy - Program Quality Verifications (PQVs) inspection - was resource intensive and, therefore, OSHA could perform only a limited number each year. Consequently, very few PQV inspections have been conducted under the current PSM directive. The Chem NEP's new approach for inspecting PSM covered facilities better allocates OSHA's resources to perform a larger number of less resource-intensive inspections. These are intended to encourage facility compliance by increasing the probability that a facility will be inspected, and to collect accurate data on compliance with the PSM standard in the chemical process industries.

Besides its targeting strategy, the Chem NEP's new approach includes a focus on specific items to address during an inspection. To focus Compliance Officers on specific items to evaluate, a dynamic list of questions is provided. The dynamic list is not in the public domain; consequently, employers will not know which areas of PSM will be the focus of their inspection. The dynamic list is periodically updated to assure the integrity of the inspection process. Dynamic lists have been developed to address the following categories:

- PSM General;
- Ammonia Refrigeration;
- Chlorine for Water and/or Wastewater Treatment; and
- Chemical Processing

NEP Comparisons:
There are differences between the two OSHA PSM NEPs and other OSHA compliance directives (CPLs). The primary difference is that the PSM NEPs prescribe specifics items that OSHA Compliance Officers are required to evaluate.

Major differences between the two PSM NEPs are the lists of questions used in each, their respective targeting schemes, and the resources required to complete inspections.

<u>Lists</u>:

The Refinery NEP contains two types of list:

- The first list is the Static List. This list contains 95 questions that remain the same or static throughout all inspections conducted under this NEP. These questions are publically available through accessing this NEP on OSHA's website (www.osha.gov).
- The Refinery NEP also contains a list of questions that are changed periodically. This list is called the Dynamic List. Compliance Officers are to further evaluate PSM compliance using 8 to 15 additional questions using the Dynamic List. This list is not publically available.

The Chem NEP contains only a Dynamic List. Depending on the type of facility a CSHO inspects, they will use a series of 15 questions taken from 4 different categories (see above). Again, these lists are not available to the public and are changed periodically.

Targeting Scheme: The way facilities are chosen to be inspected is quite different for the two PSM NEPs.

- The Refinery NEP requires that all U.S. petroleum refineries in Federal jurisdiction be inspected under the program. All refineries are being targeted because of the sector's incident history and because the total number (roughly 100 petroleum refineries in Federal jurisdiction) is manageable given OSHA's resources.
- The number of other chemical PSM-covered processes is significantly larger than the number of petroleum refineries. Consequently, OSHA has employed a targeting system to identify those chemical facilities with PSM covered processes to be inspected. The sites which form the potential inspection pool are determined from four different areas (see discussion above). After determining the potential sites, each site is put on a local OSHA Area Office master list of potential sites to be inspected under this NEP. The sites to be inspected are then randomly selected from the Master List. Each Region in the pilot NEP is instructed to conduct 5 – 10 inspections, on average, for each Area Office in the regions.
- VPP sites will not be inspected under the Refinery NEP.
- The Chemical NEP does not include VPP sites for programmed inspections. However, VPP sites may be inspected under the Chemical NEP for unprogrammed, PSM related, inspections.

Resources: Early data from the Chem NEP shows that OSHA's assumptions about difference in resource requirements between the two PSM NEPs appear to be correct. The Refinery NEP requires a significantly larger commitment of Agency resources than the Chem NEP. Correspondingly, requirements for employers to resource these inspections (e.g., staff participation, document generation, legal assistance) appear to be proportional to the OSHA commitments, hence much less for the Chem NEP inspections.

2. Petroleum Refinery NEP

2.1 Inspection Statistics

Inspection data (from inspections conducted by Federal OSHA, not including inspections conducted by State-Plan States) demonstrates that from June 7, 2007 through January 13, 2010 OSHA has opened 211 inspections under the Refinery NEP. Of these inspections, 53 have involved refinery employers (i.e., separate refinery sites) and 158 inspections involved contractor employers working on or near PSM covered processes at these refinery sites. OSHA has averaged nearly 3 contractor inspections per refinery employer inspection. Currently, OSHA has completed the field inspection phase and issued citations in 47 of the refinery employer inspections. Nine other refinery employer inspections have been opened and are at some stage of completion. OSHA projects to have completed 69 refinery employer inspections by the end of September 2011.

2.2 Frequency of Citations by Standard Element

Inspection Numbers and Fines

OSHA has found a significant number of PSM violations in the refining industry. Table 1 illustrates the number and proposed penalties associated with the violations, broken down by classification.

Table 1: Refinery NEP Penalties (June 7, 2007 to January 13, 2010)

	Refineries		Contractors	
Violation Classification	**Number of Violations**	**Total Penalties**	**Number of Violations**	**Total Penalties**
Willful	10	$561,500	0	$0
Repeat	34	$877,500	0	$0
Serious	863	$3,137,708	18	$44,475
Unclassified	9	$429,000	0	$0
Other	63	$57,287	13	$15,412
Total	**979**	**$5,062,995**	**31**	**$59,887**

On average, Refinery NEP inspections for refinery employers have resulted in 20.8 violations with penalties of $107,723.

Violations broken down by Element and Sub-element of the PSM Standard

OSHA inspection data was analyzed to determine which element and sub-element of the PSM standard were most frequently violated. The following tables show the top 10 elements and sub-elements.

Table 2 – Refinery NEP Most Frequently Cited PSM Elements

Element	Description	Percentage of Total Violations
F	Operating Procedures (OP)	18%
J	Mechanical Integrity (MI)	17%
E	Process Hazard Analysis (PHA)	17%
D	Process Safety Information (PSI)	15%
L	Management of Change (MOC)	9%
M	Incident Investigation (II)	7%
H	Contractors	4%
O	Compliance Audits (CA)	4%
G	Operator Training	3%
N	Emergency Response (ER)	2%

Table 3 - Refinery NEP Most Frequently Specific PSM Standard

Sub-element	Description
e(5)	PHA – system to promptly address recommendations
d(3)(i)(B)	PSI – information pertaining to the equipment in the process – P&IDs
(d)(3)(ii)	PSI – equipment complies with RAGAGEP
(l)(1)	MOC – establish & implement MOC
j(2)	MI – written procedures
j(5)	MI – equipment deficiencies
e(3)(v)	PHA – facility siting
(f)(4)	(OP) – safe work practices
j(4)(i)	MI – inspection and testing not conducted
(d)(3)(i)(D)	PSI – relief system design and design basis

2.3 Refinery NEP Challenges and Learnings

The major challenge presented by the Refinery NEP is the significant resource commitment needed to complete NEP inspections. This resource issue impacts both OSHA and refinery employers.

Given petroleum industry incident history and the large number of processes, chemicals, equipment, and management systems involved, OSHA determined it needed to implement a comprehensive NEP for refineries. The PSM standard is a performance oriented standard that requires an extensive time commitment to both evaluate hazards and legally sustain resulting citations.

OSHA utilizes an inspection team concept for this NEP, with the teams consisting of 3 to 4 members. Many of the Refinery NEP inspections require teams to be engaged for the full 6 months that OSHA has to meet its statutorily mandated deadline for issuing citations. Also, many of these inspections have been contested. This results in additional resource requirements until any legal challenges are resolved.

Employers are also required to expend additional resources in the form of staff participation, document generation, legal assistance, etc. The resources needed by employers during these inspections have been roughly proportional to the resource commitments made by OSHA.

From a macro perspective, OSHA inspection resource commitment to Refinery NEP inspections means that while inspection resources are used in the petroleum refining industry, they are not being used in other industry sector workplaces. This can be a challenge to OSHA area offices and regions, who must balance the use of inspection resources in industries with traditionally high injury and illness rates with their use in preventing incidents resulting from high consequence – lower frequency hazards more typical in the Refining and Chemical sectors.

3. Chemical NEP

3.1 Inspection Statistics

The Chemical NEP has been effective since July 27, 2009. At this time of this writing, the data is preliminary; consequently, it is difficult to draw conclusions. As of January 20, 2010, OSHA opened approximately 26 Chemical NEP inspections. These inspections are equally split between programmed (planned inspections) and unprogrammed (complaints, referrals, and accidents).

In two of these inspections, Compliance Officers opened inspections, but then determined that workplaces did not have any PSM covered processes, and, therefore, closed the inspections. The remaining 24 inspections are still open, and citations have been issued in seven of them.

For the purposes of the Chemical NEP, OSHA divided PSM covered facilities into three categories: ammonia refrigeration, chlorine for water/wastewater treatment, and all other. The Chemical NEP instructed OSHA Regions, where possible, to conduct approximately 25% of programmed inspections in workplaces that contain PSM covered ammonia refrigeration process, 25% in workplaces that use chlorine for water/wastewater treatment, and the remaining 50% in all other facilities. This was done to ensure that OSHA inspectors visit a variety of covered processes. OSHA recognized that without this guidance, pilot Regions with large numbers of food processing facilities, might conduct more inspections in ammonia refrigeration and water treatment facilities, rather than facilities with other types of PSM covered processes.

Preliminary inspection data in the pilot Regions shows that, at this time, the majority of programmed inspections have taken place in PSM covered ammonia refrigeration processes. Preliminary information also suggests that many water/wastewater treatment facilities may have switched from using chlorine to non-PSM covered chemicals, and at this point, OSHA has not conducted any Chemical NEP inspections at water/wastewater treatment facilities. Finally, as expected, the majority of unprogrammed Chemical NEP inspections (all OSHA Regions) have been in facilities that have covered processes that contain chemicals other than ammonia for refrigeration or chlorine for water/wastewater treatment.

3.2 Frequency of Citations by Standard Element

Given the small number of citation issued to date for Chemical NEP inspections, it is difficult to draw conclusions on frequency of standard element violation. However, preliminary data shows OSHA has found multiple violations of the mechanical integrity, process safety information, and process hazard analysis elements of the PSM standard. This is generally consistent with the frequency ranking of PSM element citations found in the Refinery NEP.

Preliminary Chemical NEP citation data also suggests that numerous citations are being issued for standards other than the PSM standard. This should not be unexpected because, especially in the cases of unprogrammed inspections, OSHA may be looking at a wide variety of hazards.

3.3 Chemical NEP Inspections – What to expect

Employers will discover that their workplace has been selected for a Chemical NEP inspection when an OSHA Compliance Officer arrives at the site. Facilities cannot be notified in advance of an inspection.

Upon arrival, the Compliance Officer will schedule an opening meeting. During the opening meeting, Compliance Officers will attempt to understand the nature of the processes, establish PSM applicability, understand site hazards and emergency procedures, and gain a basic understanding of the PSM management system.

Compliance Officers will likely request some initial documents, and schedule a walkaround. During the walkaround, the CSHO will:

- Look for differences between what was presented in the opening meeting and actual conditions;
- Gather information to aid in the selection of the process unit(s) to be inspected;
- Obtain a basic overview of the facility's operations;
- Observe potential hazards including, but not limited to, pipework at risk of impact, corroded or leaking equipment, unit or control room siting and trailer location, relief devices and atmospheric vents that discharge to atmosphere, and ongoing construction and maintenance activities; and
- Solicit input from employees and their representatives, and contact employees concerning potential PSM program deficiencies.

Based on the initial information the Compliance Officer has gathered, they will select a unit or units to evaluate for compliance with the PSM standard. Compliance Officers may select more than one unit if they feel it is necessary to get a representative sample of the facility's covered processes based on the size and complexity of the facility.

Once the unit(s) is selected, the Compliance Officer will characterize the process type (ammonia refrigeration, chlorine for water/wastewater treatment, general PSM or storage only, or other processing) and evaluate compliance using the appropriate dynamic list of questions.

If the Compliance Officer determines during the evaluation that deficiencies may exist outside of the selected unit, the inspection may be expanded. Additionally, citations may be issued for violations of all standards regardless of whether they are specifically addressed in the Chemical NEP. Inspection time is expected to be brief by comparison with refinery inspections, but will likely be highly variable depending on the complexity of the process, the availability of documentation from the employer, and the quality of compliance at the site.

3.4 Chemical NEP - Challenges and Learnings

As the Chemical NEP progresses, OSHA will identify challenges and learning. OSHA's history with PSM inspections has demonstrated that they can be much more time and resource intensive than other OSHA inspections. A cursory review of Refinery NEP Inspection data reveals that the average Refinery NEP inspection required on the order of 1000 hours vs. less than 100 hours for all other inspections. Additionally, the duration of a Refinery NEP inspection has been approximately four times as long as an average OSHA inspection. Chemical NEP inspections are expected to be significantly less resource intensive than Refinery NEP inspections, but will still likely require more time and resources than an average OSHA inspection.

OSHA has invested a significant amount of time and effort into training compliance officers for Refinery and Chemical NEP inspections. This effort has been beneficial, and training will continue as the program progresses. We expect to make adjustments to our training materials and requirements based on field experience with this innovative enforcement program.

4. What is the Future of PSM NEPs?

The Refinery NEP is likely OSHA's largest PSM enforcement effort since the standard became effective in 1992. Based on preliminary results, OSHA feels that this effort has been very beneficial to worker safety. Refinery NEP inspections will continue until they are completed in all Regions, which is expected by the end of fiscal year 2011.

The pilot Chemical NEP expires after one year. After its expiration, the Chemical NEP will be evaluated using data from both programmed and unprogrammed inspections. After the evaluation, OSHA will consider necessary revisions and expansion of the program into all OSHA regions. The scope and duration of any follow-on NEP will not be determined until the results of the pilot NEP have been evaluated.

Experience and learning from both the Refinery and Chemical NEP will determine what inspection tools OSHA will use in the future to enforce the PSM standard in all workplaces with covered processes.

5. Disclaimer

This paper represents the views of the authors. Although it is based on OSHA programs, it is not an OSHA policy document.

6. References

(1) Occupational Safety and Health Administration, *Directive Number: CPL 03-00-010, Petroleum Refinery National Emphasis Program,* August 18, 2009, http://www.osha.gov/pls/oshaweb/owadisp.show_document?p_table=DIRECTIVES&p_id=4014, Accessed January 2010.

(2) Occupational Safety and Health Administration, *Directive Number 09-06 (CPL 02), PSM Covered Chemical Facilities National Emphasis Program,* July 27, 2009, http://www.osha.gov/pls/oshaweb/owadisp.show_document?p_table=DIRECTIVES&p_id=4008, Accessed January 2010.

(108b) Preparing for the Chemical NEP

Mark S. Dreux, Arent Fox PLLC, Washington, DC

On July 27, 2009, OSHA issued a National Emphasis Program for the chemical industry ("the Chemical NEP"). Pursuant to the Chemical NEP, OSHA will conduct PSM inspections of numerous chemical facilities.

This presentation will review lessons learned from OSHA's Refining NEP, particularly the OSHA citation patterns, its interpretations of the PSM Standard and its aggressive use of industry standards. This information will assist you in preparing for a Chemical NEP inspection and in making your facility safer.

See more of this Session: National Emphasis Programs - Preparation, Reviews, Outcomes & Challenges
See more of this Group/Topical: Topical 1: Global Congress on Process Safety

Complying with OSHA's RNEP, CNEP, and VPP RAGAGEP

Chad Patschke
Senior Risk/Reliability Engineer
ABSG Consulting Inc.
16855 Northchase Drive
Houston, TX 77060
(281) 673-2968
cpatschke@absconsulting.com

Prepared for Presentation at
American Institute of Chemical Engineers
2010 Spring Meeting
6th Global Congress on Process Safety
San Antonio, Texas
March 22-24, 2010

UNPUBLISHED

Complying with OSHA's RNEP, CNEP, and VPP RAGAGEP

Chad Patschke
Senior Risk/Reliability Engineer
ABSG Consulting Inc.
16855 Northchase Drive
Houston, TX 77060

Keywords: RNEP, VPP, CNEP, RAGAGEP, PSM, OSHA

Abstract

The OSHA Refinery National Emphasis Program (RNEP) and OSHA Voluntary Protection Program (VPP) incorporate specific questions relative to many recognized and generally accepted good engineering practices (RAGAGEP). Many of the same RAGAGEP can be expected to be used in the OSHA Chemicals NEP (CNEP) as well. In order to comply with OSHA's process safety management (PSM) regulation and improve process safety at their sites, chemical industry personnel should be aware of what RAGAGEP OSHA is recognizing and inspecting against and their key requirements.

This paper will discuss: (1) OSHA's expectations for compliance with RAGAGEP, including the RAGAGEP OSHA references in compliance directives, instructions, and citations, (2) some citations issued by OSHA involving RAGAGEP, and (3) common RAGAGEP compliance issues ABSG Consulting Inc. (ABS Consulting) observes when conducting PSM audits and assessments throughout the petrochemical and related industries.

1. Background and Introduction

On June 7, 2007, the U.S. Department of Labor, Occupational Safety and Health Administration (OSHA) launched a Refinery National Emphasis Program (RNEP). This program includes inspections of all refineries (SIC 2911/NAICS 324110) for compliance with OSHA's PSM regulation 29 CFR 1910.119. The complete OSHA directive is contained in CPL 03-00-004, which was replaced and extended by CPL 03-00-010 on August 18, 2009, to extend the time for completing inspections for Regions VI and VIII.

On April 18, 2008, OSHA replaced VPP directive CSP 03-01-002 with CSP 03-01-003. The new OSHA instruction includes changes to the approval application procedures, onsite evaluations, and the self-evaluations for applicants/participants subject to 29 CFR 1910.1119. For those companies covered by the PSM regulation and applying for VPP status, a PSM Application Supplement must be completed. For companies already participating in the VPP, responses to a supplemental PSM questionnaire (VPP Supplement B) must be included in the annual self-assessment. This questionnaire is also being used in the OSHA VPP "Star" certification and recertification audits. For the last 2 years, these supplemental questions have included numbers that refer to specific RAGAGEP.

On July 27, 2009, OSHA launched an NEP for chemical facilities (CNEP). This directive announces a 1-year CNEP for programmed inspections of facilities covered by the PSM/risk management program (RMP) regulations in Regions I, VII, and X known to OSHA to have a risk of catastrophic releases of a highly hazardous chemical (HHC), along with unprogrammed inspections in all regions. After 1 year, this pilot CNEP will be evaluated and considered for renewal and expansion to other regions. This CNEP could eventually impact over 7,000 facilities in the United States. According to OSHA, the CNEP is needed because:

- Very few programmed inspections have been conducted since Instruction CPL-02-02-045 was issued in 1994
- This NEP outlines a new approach for inspecting PSM-covered facilities that allows for a greater number of inspections through better allocation of OSHA's resources

All three of the aforementioned programs (i.e., RNEP, CNEP, and VPP) hold PSM-covered facilities accountable for following many good engineering practices.

2. Lessons Learned from the RNEP

What have we learned thus far from the RNEP inspections? A review of the information available from 46 RNEP inspections (as of August 2009) has shown a strong scrutiny of RAGAGEP-related PSM elements. Of the four PSM elements with the most instances of citations, three are heavily weighted toward RAGAGEP compliance (as indicated in bold):

(j) Mechanical Integrity (MI)	**27%**
(d) Process Safety Information (PSI)	**18%**
(f) Operating Procedures (OP)	15%
(e) Process Hazard Analysis (PHA)	**13%**

As you can see, the four elements above are involved in approximately 75% of all citations.

Breaking down the data even further (i.e., to the subelement), almost half of all citations are from subelements that may heavily involve RAGAGEP:

(d)(3) PSI – information pertaining to equipment in the process	17%
(j)(4) MI – inspection and testing	13%
(e)(3) PHA – requirements of PHA	7%
(j)(5) MI – equipment deficiencies	6%
(j)(2) MI – written procedures	5%

So, what is industry being held accountable to? OSHA has many tools ranging from codes and standards to publications to incident investigations, case studies, and safety bulletins, referred to or considered as RAGAGEP. RAGAGEP provide guidance on engineering, operating, and maintenance practices developed by industry and technical experts. They are usually based on, and/or revised to reflect, lessons learned from industry. Therefore, RAGAGEP are a valuable tool for any PSM program. Figure 1 provides an example of some RAGAGEP that apply to the petrochemical industry.

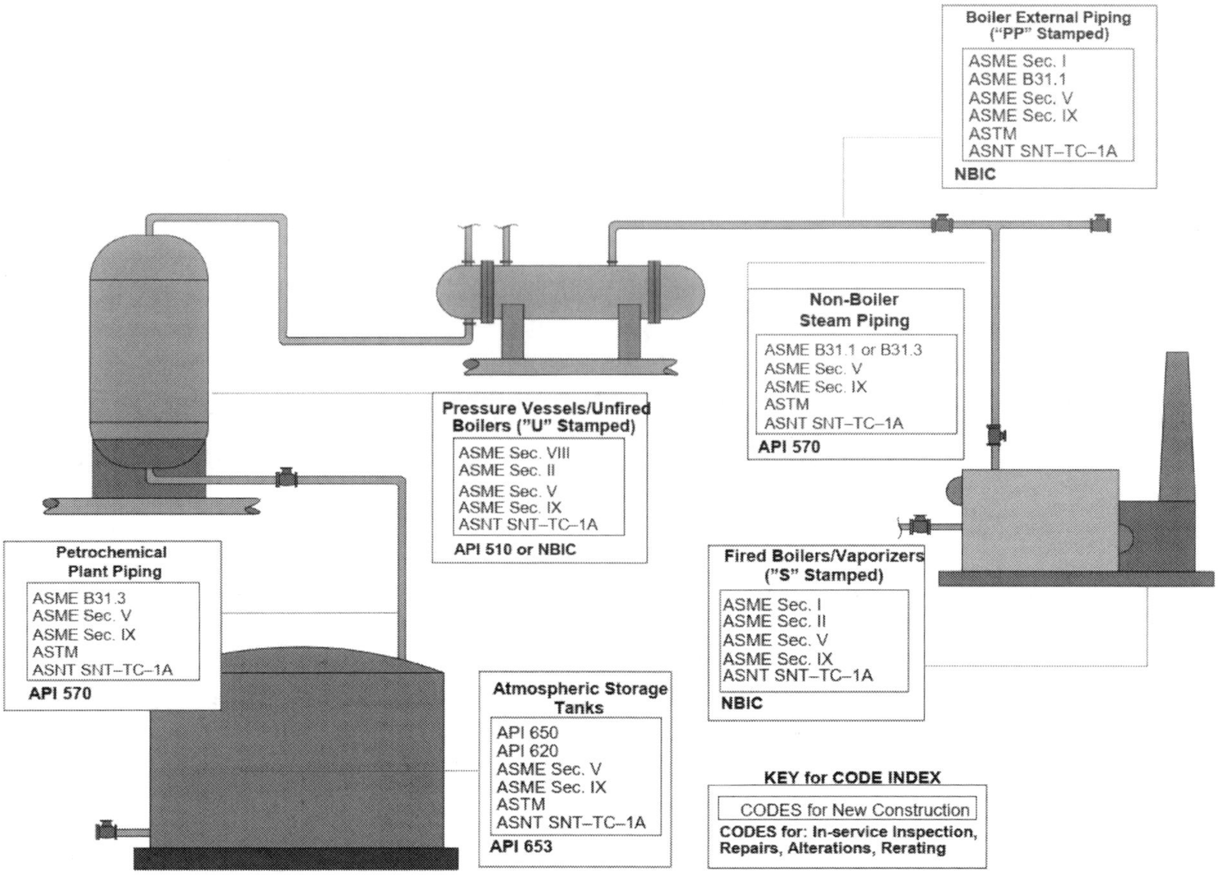

Figure 1: Examples of RAGAGEP for Petrochemical Process Equipment

3. RAGAGEP Resources

The most common RAGAGEP referenced by OSHA in RNEP inspections and VPP assessments are codes and standards from the American Society of Mechanical Engineers (ASME), the American Petroleum Institute (API), and publications from the Center for Chemical Process Safety (CCPS).

API has developed several standards, recommended practices, and publications to help fabricate, install, service, and maintain equipment in the petrochemical industry. The most prominent standards and recommended practices referenced in RNEP inspections are those dealing with inspection, repair, and alteration of pressure vessels and piping and design and inspection of relief systems. Other commonly referenced API documents address building and temporary building siting, positive material identification (PMI), and fitness-for-service.

Other codes and standards referenced by OSHA include the American National Standards Institute (ANSI) and the National Fire Protection Association (NFPA). ANSI provides guidance on identifying and maintaining safety instrumented systems and emergency shutdown systems. The NFPA standards referenced include the National Electric Code; Life Safety Code; standards on the inspection, testing, and maintenance of water-based fire protection systems; and standards for purged and pressurized electrical equipment.

OSHA's resources for RAGAGEP also include several publications from the CCPS, even though these were not written as or expected to be RAGAGEP. The most commonly referenced CCPS publication is the *Guidelines for Mechanical Integrity Systems*. Some of the OSHA references to this CCPS guideline include guidelines for quality assurance, temporary installations and repairs, equipment deficiency management, developing sampling criteria, corrosion under wet thermal insulation, example MI procedures, and widely accepted MI certifications. Other CCPS publications referenced as RAGAGEP by OSHA are shown in Table 1.

Table 1: Some CCPS Publications Referenced as RAGAGEP

- *Guidelines for Hazard Evaluation*
- *Guidelines for Engineering Design for Process Safety*
- *Guidelines for Safety Process Operations and Maintenance*
- *Guidelines for Safe Automation of Chemical Processes*
- *Guidelines for Facility Siting and Layout*
- *Guidelines for Evaluating Process Plant Buildings for External Explosions and Fires*
- *Safe Design and Operation of Process Vents and Emission Control Systems*
- *Layer of Protection Analysis: Simplified Process Risk Assessment*

Documents issued by the Chemical Safety Board (CSB) have also been used as a resource for RAGAGEP. The BP Texas City Investigation report, the fire at Formosa case study, and the PMI bulletin have been referenced in inspection documentation or instances for citation during inspections. (All of these documents are available on the Web site www.csb.gov.)

Since we have now reviewed codes and standards referenced by OSHA regarding RAGAGEP, the next sections will examine some common OSHA citations and audit findings, discuss what is expected, and what can be done to ensure compliance.

4. Equipment Process Safety Information RAGAGEP

There have been many instances where OSHA inspections or ABS Consulting audits revealed inaccuracies or incompleteness of process safety information with respect to PSM-covered equipment. One of the most common citations/findings addresses inaccurate process and instrument diagrams (P&IDs), which are easily audited. OSHA's apparent stance is that if the drawings do not match the actual installed equipment, the owner/user has not properly documented that the equipment complied with RAGAGEP. This also holds true for piping fabrication isometric drawings and as-builts. Another example of not documenting that equipment complies with RAGAGEP involves failure to document the code or standard the piece of equipment was designed to (e.g., ASME VIII for pressure vessels, API 610 for centrifugal pumps, API 520 for pressure relieving devices). This is usually documented on the equipment datasheet or drawing. Finally, fabrication related RAGAGEP deficiencies are a common audit finding. Examples include: (1) lack of documentation verifying that pressure vessels were fabricated to ASME Section VIII, and process piping to ASME B31.1, (2) failure of owner-users to ensure equipment is fabricated per RAGAGEP, and (3) failure to follow and/or adequately document the specific requirements of ASME Section X to ensure welders are qualified to weld on pressure vessels and process piping.

5. Maintenance Procedures RAGAGEP

A question often asked is whether a site-specific procedure is required or can the owner/user simply state that they adopted RAGAGEP. The answer is that it depends. If the RAGAGEP are specific and do not leave room for interpretation, then the RAGAGEP can be incorporated as-is in the PSM program. However, this is often not the case. Typically, the RAGAGEP are written as general guidance with the assumption that qualified individuals are involved and will make educated decisions based on the specific situation(s) at hand. An example of RAGAGEP requirements that are very specific and would not need additional details provided by a site-specific procedure is determining whether on-stream inspections can be conducted on pressure vessels instead of conducting an internal inspection. API 510 provides specific requirements that must be met to conduct an on-stream inspection in lieu of an internal inspection. On the other hand, an example of RAGAGEP that are not specific and would require a more detailed site-specific procedure is the determination of the quantity and location of thickness measurement locations (TMLs) on pressure vessels and piping. The API standards and recommended practices call for a representative number of TMLs to be identified, but it would be up to the owner/user to establish detailed guidelines on the quantity and location of the TMLs for their particular situation(s). However, be aware that if an owner/user procedure is more stringent than RAGAGEP requirements, the owner/user will be held accountable to following their own established requirements during an inspection or audit, under OSHA's general duty clause.

6. Relief Systems RAGAGEP

As a result of the BP Texas City investigation, relief systems are usually assessed in detail during an RNEP inspection. First, the review of the process safety information for relief systems can lead to several citations if all the information suggested by the RAGAGEP (API 521) is not available. A typical relief device datasheet, when completed properly, will provide much of the information required. The datasheet should include the (1) basic information (i.e., model and serial number, size, relief pressure) and (2) worst-case scenario used to size the relief device. RAGAGEP also suggest that the state of the material (e.g., liquid, vapor, liquid-vapor) and physical properties of the relieving material (e.g., vapor weight, molecular weight) be documented. The design calculations, including the device capacity, required capacity, and inlet and outlet pressure drops should be available as well. If the relief device is connected directly to a vessel, there needs to be documented proof that the inlet pressure loss does not exceed 3%.

One common audit finding is that not all sources for overpressure considered when sizing the relief valve were documented. This is usually accomplished by including a table listing all scenarios considered (e.g., fire, blocked discharge, exothermic reaction). Also, OSHA expects to see documentation that each relief discharge location was evaluated for risks associated with the release of the material from the discharge (e.g., toxic/flammable dispersion, radiant heat if the vent lights off).

Other relief system issues identified during OSHA RNEP inspections include:

- The set pressure of the relief device was higher than the design pressure of the vessel it was protecting
- The knockout drum retention time was inadequate for the relief system
- Inadequate program to control isolation valves in relief systems (e.g., car-seal program)

If the facility has a relief system with a flare, other things to assess include:

- Flare and blowdown sizing calculations should coincide with the material and energy balance throughput for the unit(s) the relief header(s) protects
- A radiant heat study for the flare should be available
- Flare inspection records should include inspection of the stack and flare tips

Citations have also been issued for failure to comply with RAGAGEP when inspecting, testing, and maintaining relief devices. A common instance cited is the failure to inspect a relief device within the established frequency or within the maximum allowed time per API 510. RAGAGEP also require the owner/user to adjust the frequency of testing if the relief device is found to be heavily fouled. It is important to conduct an initial pop test when the device is removed to better understand how the relief device would have performed under operating conditions. The owner/user should have a procedure explaining how to adjust the frequency of inspection if the device is found to be fouled, and when it is acceptable to extend the frequency between inspections if the relief device is in a clean, nonfouling service. A common example approach is to reduce the frequency by 50% if the device is found to be fouled at the current inspection and to extend the frequency after two consecutive inspections demonstrate that the device was found to be in good working condition (not to exceed 10 years). If a procedure does not exist and the facility elects not to develop one, the reduction and extension of frequencies could be addressed through the management of change (MOC) process.

7. Pressure Vessels and Piping RAGAGEP

Many of the instances resulting in citations during RNEP inspections involve the inspection and repair of pressure vessels and piping. Several citations have been issued regarding TML management. As mentioned above, the RAGAGEP do not provide specific guidance on the quantity and location placement of TMLs. It is up to the owner/user to develop the guidance to specify how many TMLs and the placement of these TMLs would constitute a representative sample. Also, the owner/user should establish a means to ensure that readings are taken at the same location each time to ensure quality of the corrosion monitoring data. Another common citation regarding thickness readings is the mismanagement of anomalous data, sometimes referred to as growth. Should the current thickness reading be greater than the previous thickness reading, assuming the piping or vessel section was not replaced, the owner/user should have a procedure or plan to address how the data are managed. It is common during audits to find growth when reviewing thickness data records and nothing has been done to address the issue of a negative value being calculated for the short-term corrosion rate. Another common problem is the lack of a management system when a TML falls below a minimum thickness requirement. Typically, corrosion monitoring programs will have two flags, the calculation minimum thickness for pressure and an established minimum thickness for structural support of piping between supports. In some cases, the actual piping thickness will be below the structural minimum thickness without any record of further analysis being done by a qualified engineer. The pressure and structural minimum thickness values are considered safe control limits and to operate outside those limits would require equipment deficiency management under subelement (j)(6). Another common problem is the development of baseline data. If the corrosion rate is not known and cannot be reasonably predicted from experience, literature, or equipment in similar service, a corrosion rate should be established after approximately 1,000 hours of service. An inspector or auditor can easily check this by asking for inspection records for the newest process at a facility.

Specific to piping, a common audit finding is the failure to classify piping in accordance with API 570, which provides guidance on the not-to-exceed intervals for external visual and UT inspections. Another issue specific to piping inspection programs not in alignment with RAGAGEP is the failure to identify injection points, deadlegs, soil to air interfaces, and areas where corrosion under insulation may be a problem. All of these have unique inspection requirements.

Specific to pressure vessels, having procedures addressing the inspection of liners (e.g., bonded, nonmetallic) is something that OSHA is assessing during the RNEP inspections. Another common audit finding is the lack of a procedure addressing the inspection of nonmetallic process vessels and piping, as API 510 and 570 do not address nonmetallic process equipment. Therefore, facilities must provide guidance based on other external/internal RAGAGEP in these cases.

8. Instrumentation, Electrical, and Controls RAGAGEP

Much of the process safety information for instrumentation and controls can be obtained by referencing a vendor-supplied manual for the equipment and cross referencing the model number of the equipment to determine the materials of construction, installation requirements, etc. However, documenting the design codes and standards employed can sometimes be tricky. Take electrical classification of areas for example. When reviewing the drawing indicating electrically classified areas or reviewing the document that describes the electrically classified areas, often the design code or standard used to develop the drawing/document is not listed. A simple note on the drawing stating that the drawing is based on API RP 500 or NFPA 70 would satisfy this requirement. Another common issue with electrical classification drawings is the imaginary barrier at process unit boundaries and roadways. It is frequently observed that the hatching indicating classified areas on the drawings stops at the unit boundaries and roadways, regardless of whether the classified area surrounding a source extends beyond the unit boundary or into a roadway. If a classified area surrounding a potential source of flammable vapors extends beyond the unit boundary or into a roadway, it should be represented on the drawing.

OSHA is also auditing safety instrumented systems and asking if safety integrity level (SIL) assessments have been conducted and testing plans established in accordance with ANSI/ISA 84. The details of these RAGAGEP are complex and will not be discussed in this paper, but can be found in this standard as well as the CCPS *Guidelines for Safe Automation* book.

Pressurized enclosures, including electrical cabinets and buildings, are another area of focus during RNEP inspections. Insufficient documentation regarding the design basis for the purging and pressurization systems for these enclosures, as well as the HVAC systems for control rooms, is a common audit finding. The placement and testing of alarms associated with these systems have been cited as well. In one instance, the air intake vapor monitor alarmed at the HVAC control panel, which was in a remote hallway of an operations control room. The facility was cited because the system did not alarm in a continuously occupied location.

9. PHA RAGAGEP (Including Human Factors and Facility Siting)

Not only do RAGAGEP apply to design and maintenance of equipment, they are also used when conducting process hazard analyses (PHAs) and facility siting studies. There are several CCPS guideline publications addressing safe operations, hazard evaluations, and facility layout. Citations have been issued to companies for not including human factors and facility siting assessments in PHAs. A related citation has been a lack of completeness when conducting such assessments. "Yes/no" answers or not answering some questions on an assessment questionnaire could be grounds for a citation in OSHA's view because the short answer, or lack thereof, may not adequately address the assessment of the hazard.

Obviously, after the BP Texas City event, temporary building siting became another area of concern for facility siting. Citations have been issued for placing temporary buildings in overpressure zones as determined by API RP 753. In addition, upgrades or replacements of permanent buildings that have not been implemented in a timely manner (in OSHA's view) have been cited.

10. Conclusion

There are many resources that OSHA uses to assess petrochemical facilities against RAGAGEP and many opportunities for companies to improve their mechanical integrity, process safety information, process hazard analysis, and PSM programs in general by fully incorporating and properly implementing applicable RAGAGEP. The key is to know what resources are available and how to apply them to the facility appropriately. And finally, RAGAGEP regarding facility siting considerations and methodologies have evolved even further since the BP Texas City event, and the latest guidance on this topic was issued in December 2009 in API RP 752.

11. Resources

[1] ABSG Consulting Inc., *Mechanical Integrity, Course 130*, Process Safety Institute, Houston, TX, 2009.

[2] American Institute of Chemical Engineers, *Guidelines for Hazard Evaluation Procedures, Second Edition with Worked Examples*, Center for Chemical Process Safety, New York, NY, 2008.

[3] American Institute of Chemical Engineers, *Guidelines for Mechanical Integrity Systems*, Center for Chemical Process Safety, New York, NY, 2006.

[4] American Institute of Chemical Engineers, *Safe Design and Operation of Process Vents and Emission Control System*, Center for Chemical Process Safety, New York, NY, 2006.

[5] American Institute of Chemical Engineers, *Guidelines for Facility Siting and Layout*, Center for Chemical Process Safety, New York, NY, 2003.

[6] American Institute of Chemical Engineers, *Layer of Protection Analysis: Simplified Process Risk Assessment*, Center for Chemical Process Safety, New York, NY, 2001.

[7] American Institute of Chemical Engineers, *Guidelines for Evaluating Process Buildings for External Explosions and Fires*, Center for Chemical Process Safety, New York, NY, 1996.

[8] American Institute of Chemical Engineers, *Guidelines for Safe Process Operations and Maintenance*, Center for Chemical Process Safety, New York, NY, 1995.

[9] American Institute of Chemical Engineers, *Guidelines for Engineering Design for Process Safety*, Center for Chemical Process Safety, New York, NY, 1993.

[10] American Institute of Chemical Engineers, *Guidelines for Safe Automation of Chemical Process*, Center for Chemical Process Safety, New York, NY, 1993.

[11] American Petroleum Institute, *Management of Hazards Associated with Location of Process Plant Permanent Buildings,* API 752, 3rd Edition, Washington, DC, December 2009.

[12] American Petroleum Institute, *Piping Inspection Code: Inspection, Repair, Alteration, and Rerating of In-service Piping*, API 570, 3rd Edition, Washington, DC, November, 2009.

[13] American Petroleum Institute, *Welded Storage Tanks for Oil Storage*, API 650, 11th Edition, Washington, DC, November, 2009.

[14] American Petroleum Institute, *Tank Inspection, Repair, Alteration, and Reconstruction*, API 653, 4th Edition, Washington, DC, April 2009.

[15] American Petroleum Institute, *Design and Construction of Large, Welded, Low-Pressure Storage Tanks*, API 620, 11th Edition, Washington, DC, March, 2009.

[16] American Petroleum Institute, *Management of Hazards Associated with Location of Process Plant Portable Buildings*, API Recommended Practice (RP) 753, Washington, DC, November, 2007.

[17] American Petroleum Institute, *Pressure Vessel Inspection Code: Maintenance Inspection, Rating, Repair and Alteration*, API 510, 9th Edition, Washington, DC, June, 2006.

[18] American Petroleum Institute, *Centrifugal Pumps for Petroleum, Petrochemical and Natural Gas Industries*, API 610/ISO 13709, Washington, DC, 2004.

[19] American Petroleum Institute, *Classification of Locations for Electrical Installation at Petroleum Facilities Classified as Class I, Division 1 and Division 2*, API Recommended Practice (RP) 500, Washington, DC, November, 2000.

[20] American Petroleum Institute, *Design Installation of Pressure-Relieving Systems*, API 520, Washington, DC, November, 2000.

[21] American Petroleum Institute, *Pressure-Relieving and Depressuring Systems*, API 521, Washington, DC, November, 2000.

[22] American Petroleum Institute, *Material Verification Program for New and Existing Alloy Piping Systems*, API Recommended Practice (RP) 578, Washington, DC, 1999.

[23] American Society of Mechanical Engineers, *Process Piping*, ASME B31.3, New York, NY, 2008.

[24] American Society of Mechanical Engineers, *International Boiler and Pressure Vessel Code*, New York, NY, 2007.

[25] American Society of Mechanical Engineers, *Power Piping*, ASME B31.1, New York, NY, 2007.

[26] American Society of Nondestructive Testing, *Recommended Practice for Personal Qualification and Certification in Nondestructive Testing,* ASNT-TC-1A, 2006 Edition, Columbus, OH, 2006.

[27] Environmental Protection Agency, *Accidental Release Prevention Requirements: Risk Management Programs*, Clean Air Act, Section 112 (r)(7), Washington, DC, 1996.

[28] International Society for Measurement and Control, *Functional Safety: Safety Instrumented Systems for the Process Industry Sector - Part 1: Framework, Definitions, System, Hardware and Software Requirements*, ANSI/ISA-84.00.01-2004 Part 1 (IEC 61511-1 Mod), Research Triangle Park, NC, 2004.

[29] National Board of Boiler and Pressure Vessel Inspectors, *National Board Inspection Code*, 2007 Edition, Columbus, OH, 2007.

[30] National Fire Protection Association, *Life Safety Code®*, NFPA 101, Quincy, MA, 2009.

[31] National Fire Protection Association, *National Electrical Code®*, NFPA 70, Quincy, MA, 2008.

[32] National Fire Protection Association, *Standard for Purged and Pressurized Enclosures for Electrical Equipment*, NFPA 496, Quincy, MA, 2008.

[33] Occupational Safety and Health Administration Instruction CPL 03-00-010, *Petroleum Refinery Process Safety Management National Emphasis Program*, August 18, 2009.

[34] Occupational Safety and Health Administration Instruction CPL 02-09-06, *PSM Covered Chemical Facilities National Emphasis Program*, July 27, 2009.

[35] Occupational Safety and Health Administration Instruction CPL 03-01-003, *Voluntary Protection Program Policies and Procedures Manual*, April 18, 2008.

[36] Occupational Safety and Health Administration Instruction CPL 02-02-045, *Process Safety Management of Highly Hazardous Chemicals -- Compliance Guidelines and Enforcement Procedures*, September 13, 1994.

[37] Occupational Safety and Health Administration, *Process Safety Management of Highly Hazardous Chemicals*, 29 CFR Part 1910, Section 119, Washington, DC, 1992.

6TH GLOBAL CONGRESS ON PROCESS SAFETY

Its Coming Right For Us!
What to Expect and How to Prepare for the Chemical NEP

Christie A. Arseneau
AcuTech Consulting
2500 City West Blvd, Suite 300, Houston, TX 77042
carseneau@acutech-consulting.com

Mark L. Farley
Pillsbury Winthrop Shaw Pittman LLP
909 Fannin, Suite 2000, Houston TX 77010
mark.farley@pillsburylaw.com

Prepared for Presentation at
American Institute of Chemical Engineers
2010 Spring Meeting
6th Global Congress on Process Safety
San Antonio, Texas
March 22-24, 2010

UNPUBLISHED

Its Coming Right For Us!
What to Expect and How to Prepare for the Chemical NEP

Christie A. Arseneau
AcuTech Consulting
2500 City West Blvd, Suite 300, Houston, TX 77042
carseneau@acutech-consulting.com

Mark L. Farley
Pillsbury Winthrop Shaw Pittman LLP
909 Fannin, Suite 2000, Houston TX 77010
mark.farley@pillsburylaw.com

Keywords: Chemical NEP, national emphasis program, OSHA, inspection, citations

Abstract

On July 27, 2009 OSHA published Directive 09-06 (CPL 02), *PSM Covered Chemical Facilities National Emphasis Program*. This instruction, which follows the agency's process safety emphasis program for petroleum refineries, details the agency's plans to conduct programmed inspections at chemical facilities covered by the PSM Standard. Although the first year of the Chemical NEP is only applicable to non-VPP facilities in Regions I, VII, and X, it is anticipated that OSHA will extend the NEP program nationwide to possibly 15,000 facilities total. Most importantly, the Chemical NEP will be used to conduct unprogrammed inspections of workplaces in response to PSM-related complaints, referrals, accidents or catastrophes even in regions not covered by the pilot. For these reasons, it is important for line managers, health and safety professionals, and in-house legal counsel to be familiar with this OSHA instruction. This paper presents both a technical and legal perspective on the Chemical NEP, makes the business case for preparing enterprise-wide for the Chemical NEP, and presents a practical strategy that corporations can utilize in anticipation of an OSHA Chemical NEP inspection.

1. Applicability of OSHA's Chemical National Emphasis Program

OSHA's published Directive 09-06 (CPL 02), *PSM Covered Chemical Facilities National Emphasis Program* extends an inspection program similar to the Refinery NEP to those facilities in the chemical industry that are covered by the PSM Standard and is effective through July 27, 2010. Until then, OSHA will be executing the programmed Chemical NEP on a pilot basis in three regions where the population of PSM-covered chemical facilities is light (as opposed to Regions II, V, and VI in particular). Refineries (NAICS code 32411) will not be subject to Chemical NEP inspections because the previous NEP program focused on the refinery sector. Unprogrammed NEP inspections in response to alleged hazardous working conditions, or actual incidents, can be conducted in all OSHA regions. As with the Refinery NEP, it is expected that OSHA will be arriving unannounced at chemical plants to initiate these inspections.

Chemical facilities that are Voluntary Protection Programs (VPP) sites will not be subject to the programmed Chemical NEP inspections. While all VPP participants covered by PSM, including those in the chemical industry, were required to complete a PSM self-assessment using a subset of the Refinery NEP questions for submission to OSHA as part of their 2007 annual safety and health management program self-evaluation required under the VPP, the recently published Chemical NEP (CPL-02-09-06) does not explicitly address this self-assessment.

Chemical facilities that are in state-plan states will not be subject to federal NEP inspections, but may be subject to state NEP or NEP-like inspections. Industry's experience with the Refinery NEP indicates that some states, such as California, Minnesota, New Mexico, Utah and Wyoming that implemented an in-state Refinery NEP, will choose to implement an in-state Chemical NEP program, and some will not. States are required by the new CPL to inform OSHA of their intent and OSHA will post the states' responses on its website when they are received.

Given the approximate number of PSM-covered facilities, excluding the state plans, VPP, and refineries, upwards of 15,000 facilities in the U.S. could be subject to federal OSHA NEP inspections. OSHA has previously estimated that the Chemical NEP could be an eight-year program.

OSHA has established three categories of facilities for this program:

1) Facilities having anhydrous ammonia as a refrigerant onsite.
2) Facilities having chlorine for water treatment onsite.
3) Facilities that have other highly hazardous chemicals on site, or are using anhydrous ammonia or chlorine for other purposes.

In an effort to identify facilities to inspect, OSHA will compile lists of facilities that fit within these categories by:

1) Using the list of approximately 8,000 facilities that have submitted Program 3 RMPs to EPA (note that notification of PSM coverage to OSHA has never been required).
2) Adding the list of facilities with the NAICS code for manufacturing explosives.
3) Adding facilities from the OSHA IMIS database that have been cited for PSM.
4) Local OSHA Area Office knowledge of facilities in their jurisdictions that are PSM covered.

2. Chemical National Emphasis Program Inspection Protocol and Process

Unlike the Refinery NEP, which featured a publically available static list of inspection questions in addition to a dynamic list of questions available only to OSHA personnel, the Chemical NEP will use only a dynamic list of questions that will be available solely to OSHA personnel via their Intranet. No two facilities can expect to be inspected with exactly the same set of questions since only a portion of the dynamic questions will be used. Given the timing of the Refinery and Chemical NEPs, it is anticipated that the dynamic Chemical NEP list of questions will include those questions from the static and dynamic refinery question lists, edited to remove or modify questions that were refinery-specific. Although the Chemical NEP dynamic question list is not currently available, the following inspector direction is quoted from Appendix A of Directive 09-06 (CPL 02), which summarizes the dynamic list:

> "This list based evaluation is a gap analysis formatted in a series of questions that have been developed to assess and verify the employer's PSM compliance with specific issues such as *design, fabrication, installation, startup, operation, maintenance, change, controls (engineering and administrative), safe work practices, contractor safety*, etc., at the facility…" (emphasis added)

This direction indicates that the Chemical NEP will, like the Refinery NEP, focus on the following PSM elements: Mechanical Integrity, Operating Procedures, Management of Change/Pre-startup Safety Review, and Contractor Safety. As evidenced by the citations issued by OSHA, the Refinery NEP also concentrated on PHAs. This is likely to also be true for the Chemical NEP.

The facility category, as described above, will determine the number of questions and which questions are used as follows from a table on page 18 of Directive 09-06 (CPL 02):

If the process is:	Then Use:
Ammonia Refrigeration Only	Ammonia Refrigeration dynamic list – the first 10 questions PSM General dynamic list – the first 5 questions
Water and/or Wastewater Treatment Only	Water and Wastewater treatment dynamic list – the first 10 questions PSM General dynamic list – the first 5 questions
Storage Only	PSM General dynamic list – all questions
Chemical Processing and all other categories not listed above	Chemical Process dynamic list – the first 10 questions PSM General dynamic list – the first 5 questions

This distribution and number of questions indicates a more focused and presumably a briefer inspection than was experienced in the Refinery NEP, and certainly briefer than what OSHA originally planned for the PSM program in their PQV inspection protocol published in the PSM Compliance document (CPL 02-02-045).

In addition to OSHA asking questions from the dynamic list, the Chemical NEP inspection will also include an examination of PSM applicability, which is a standard practice for some industry PSM audits. It is also anticipated that, like the Refinery NEP, the Chemical NEP inspections are to focus more on the implementation of the PSM program, or the lack of it, rather than just the mere existence of a documented program. In other words, OSHA will examine the program from a "Say what you do, and do what you say" viewpoint rather than just examine written policies and procedures for compliance.

Upon arrival at a facility, OSHA is likely to select a specific process unit to focus their NEP inspection, which is a typical government inspection and industry auditing practice. The inspectors may expand the NEP inspection to other units if the results indicate that the results seen in the focused unit may extend beyond to other units.

While OSHA is on-site, inspection of contractors and subcontractors that are performing work at the host site also will be included in the Chemical NEP.

The inspection will examine the PSM management systems that are in place, even though, for many PSM elements, these systems are not explicitly required by the PSM Standard. For example, the OSHA inspectors will likely review the mechanical integrity system for managing inspections and testing of equipment, including management of late and overdue inspections, selection on inspection frequencies, approval process for modifications to frequencies, etc., despite the standard not specifically requiring an MI management system to be in place. The following direction to the inspectors is contained in Directive 09-06 (CPL 02): "The requirements of the PSM standard are intended to eliminate or mitigate catastrophic releases of HHC. The provisions of the standard present closely interrelated requirements, emphasizing the application of management controls when addressing the risks associated with handling or working near HHC."

OSHA will examine the interrelationship between process safety elements and will audit the flow of information among these elements. For example, if a valve is changed, the management of change (MOC), mechanical integrity (MI), process safety information (PSI), and process hazard analysis (PHA) elements may all be affected, and that change should appear in the procedures and records associated with all affected elements. To assure that all the employer's process safety management systems/elements are being fully implemented, inspectors will consider citing all applicable violations in different PSM elements.

The interrelationship between elements can also result in the one hazard – several abatement concept. For example, for mechanical integrity, abatement requirements include: management system/program requirements – e.g., the employer must develop mechanical integrity program procedures that include piping inspection procedures, 1910.119(j)(2), and specific employer action/task abatement requirements – e.g., the employer must inspect the piping, 1910.119(j)(4). Therefore, to assure that all the employer's process safety management systems/elements are being fully implemented, inspectors will consider citing all applicable violations in different PSM elements. Grouping these violations may be appropriate.

As with the Refinery NEP, and as is typical with OSHA enforcement practices, inspectors can also generate citations for alleged violations of other general OSHA standards, such as lockout/tagout, HAZCOM, confined space, etc., if conditions supporting these violations are observed during the Chemical NEP inspections.

3. Preparing for the Chemical NEP

Because it is unlikely that a site's triennial compliance audit went deep enough to examine the same issues that are covered in the Chemical NEP inspections, we encourage PSM covered chemical facilities subject to the program to prepare for the Chemical NEP utilizing a multipronged strategy, as depicted in Figure 1.

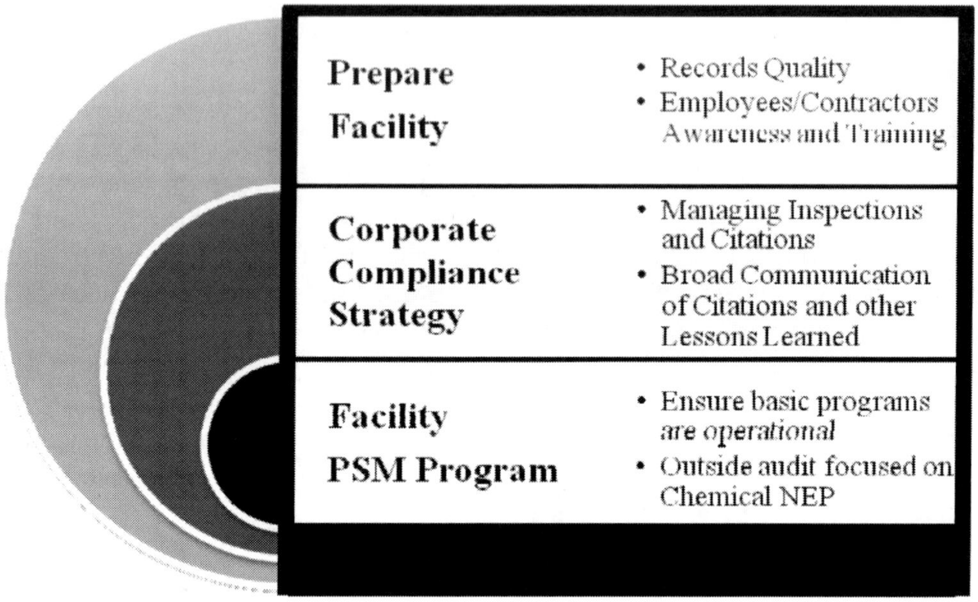

Figure 1 – Chemical NEP Preparation Strategy

The core of the approach is to ensure that the facility PSM program is operational and functioning as intended. If a facility is in Region I, VII, or X, they should take action and prepare for a Chemical NEP inspection. If a facility is outside the Regions where the pilot will be held, they should still take action and prepare for a possible unprogrammed inspection and/or an expansion of the pilot nationwide.

Facilities preparing for the Chemical NEP inspection should ensure that their covered processes have been accurately identified and that the boundaries of the covered process are known. While OSHA does not require a facility to document the covered processes or their boundaries, it is a good practice to document this information in an effort to facilitate the inspection.

Upon reviewing the facility's covered processes, the facility can further review their programs to ensure that they are being implemented as designed by conducting a self-assessment or another triennial compliance audit if it is due or soon to be due. The facility also should consider answering the non-refinery specific static questions from the Refinery NEP in an effort to anticipate some of the questions that OSHA may ask. Completing the self-assessment or compliance audit, along with reviewing the Refinery NEP static questions, will help the facility identify areas for improvement. Reviewing the Refinery NEP static questions will also help the facility anticipate responses to the Chemical NEP, assuming OSHA recycles some of those questions. In addition, the facility should focus on closing out open action items from incident investigations, MOCs, PHAs, and audits. The facility should also ensure that mechanical integrity inspections are current and the inspection backlog is minimized or eliminated. It may be helpful for the review of the facility's PSM program and focus on NEP aspects to be completed by someone external to the facility or external to the unit under review as part of an enterprise-wide preparation strategy. An outsider's perspective – colleague, corporate, consultant – can help a facility identify corrective actions in anticipation of a compliance inspection.

In any event, the employer should carefully consider whether to conduct any such self-assessments or other internal review pursuant to an applicable legal privilege (e.g., attorney-client privilege). By structuring a self-assessment in this manner, a company may be able to protect against the disclosure of the assessment and any underlying work papers in subsequent enforcement by the government.

Upon the completion of the self-assessment, it is imperative that the site develop and implement plans to close out any gaps that are identified. To the extent that the site cannot close a gap immediately, the site should develop a specific plan and schedule for abating the condition. Where the final abatement may take some time (e.g., when a capital expenditure is required or if the fix must await a maintenance turnaround), it is extremely important for the site to implement interim measures to mitigate the hazard to avoid the allegation that the site was aware of a hazardous condition and yet did nothing to address it.

The next layer of preparedness is to coordinate with internal affiliates and corporate offices to ensure that there is a common approach to preparing for Chemical NEP inspections. This may include preparing and utilizing standardized tools to complete Chemical NEP specific self-assessments using the Refinery NEP static questions or other self-assessment protocols. The corporation may request certain metrics from the facility level regarding percentage of open action items from the most recent audit or from all PHAs. These metrics provide a snapshot of each facility and may be helpful in allocating resources to prepare for an inspection. The corporation can also network with their counterparts in industry to keep apprised of the current status of the Chemical NEP and share learnings with others. These learnings can then be disseminated internally to better prepare facilities for the inspection.

The last layer is to prepare the facility. At a minimum, we recommend informing personnel that may be involved of the upcoming potential inspection and ensuring that they have the knowledge and resources necessary to effectively manage the inspection. This includes not only full-time employees, but contract personnel who may either be interviewed during an inspection or otherwise interface with the inspector. It is important, for example, that the site have in place a plan for (1) ensuring that employees and contractors understand their rights with respect to government interviews and witness statements; (2) managing information requests and the production of documents; (3) safeguarding company trade secrets; and (4) managing the day-to-day presence of the inspection team. As part of this effort, the facility should ensure that the facility records related to PSM are in order and easily accessible during an inspection.

4. Conclusion

After the Refinery NEP was issued in 2007 and inspections were underway, the chemical industry began anticipating their turn with the OSHA inspector for their own NEP. The Chemical National Emphasis Program issued in 2009 is being rolled out as a pilot program initially for Regions I, VII, and X, but is expected to be implemented nationally. Whether or not a facility is in one of the pilot regions, they can begin preparing for a Chemical NEP inspection at both the facility and corporate level. Preparation for the Chemical NEP is key to not only possibly reducing regulatory liability, but is also key to helping ensure a safe and productive workplace for all employees.

5. References

29 CFR 1910.119 - Process Safety Management of Highly Hazardous Chemicals;
http://www.osha.gov/pls/oshaweb/owadisp.show_document?p_table=STANDARDS&p_id=9760

Preamble to 29 CFR 1910.119 – Process Safety Management of Highly Hazardous Chemicals;
http://www.osha.gov/pls/oshaweb/owadisp.show_document?p_table=PREAMBLES&p_id=1041

CPL 02-02-045– OSHA PSM Compliance Directive;
http://www.osha.gov/pls/oshaweb/owadisp.show_document?p_table=DIRECTIVES&p_id=1558

CPL 00-03-004 - Refinery NEP Compliance Directive;
http://www.osha.gov/pls/oshaweb/owadisp.show_document?p_table=DIRECTIVES&p_id=3589

Directive 09-06 (CPL 02) - Chemical NEP Compliance Directive;
http://www.osha.gov/pls/oshaweb/owadisp.show_document?p_table=DIRECTIVES&p_id=4008

6TH GLOBAL CONGRESS ON PROCESS SAFETY

Challenges in Hosting an NEP Inspection

Ken Hanchey
ABSG Consulting Inc.
16855 Northchase Drive
Houston, TX 77060
khanchey@absconsulting.com

James R. Thompson
ABSG Consulting Inc.
16855 Northchase Drive
Houston, TX 77060
jthompson@absconsulting.com

Prepared for Presentation at
American Institute of Chemical Engineers
2010 Spring Meeting
6th Global Congress on Process Safety
San Antonio, Texas
March 22-24, 2010

UNPUBLISHED

Challenges in Hosting an NEP Inspection

Ken Hanchey
ABSG Consulting Inc.
16855 Northchase Drive
Houston, TX 77060
khanchey@absconsulting.com

James R. Thompson
ABSG Consulting Inc.
16855 Northchase Drive
Houston, TX 77060
jthompson@absconsulting.com

Keywords: OSHA NEP, Inspections, Teams, Strategy, Procedures

Abstract

A National Emphasis Program (NEP) inspection can be very resource intensive for a plant. For example, OSHA requested and received well over 10,000 pages of documentation from each of many refineries during NEP inspections. Developing a strategy that aligns with your corporate governance and protocols could be the key to a successful outcome. The strategy should include the predetermined tactical plan for the overall response to each inquiry from OSHA. The tactical plan will govern your response to the inspection process, but must align with the overall strategic intent.

To ensure that OSHA receives the proper information about the state of PSM compliance and that your resources are used efficiently, you need to be ready to welcome, escort, and work with OSHA inspectors from the moment they arrive at your facility until they finish the inspection. Your inspection management team will need procedures to (1) check credentials, (2) escort the inspectors, (3) provide documents, (4) take photographs, (5) plan interviews, (6) respond to findings, (7) track your actions, and (8) communicate inspection progress to the workforce, management, and maybe the public.

You will need to train all the personnel who may interface with the OSHA inspectors on their roles and their rights under OSHA.

OSHA has a plan (NEP) to inspect your facility. You need a plan to manage that inspection!

1. Challenges in Hosting an NEP Inspection (Introduction)

OSHA has said "it's a new day at the Department of Labor" [1] and OSHA's inspections have changed. That was after the Refinery National Emphasis Program (NEP) [2] had been in place for 2 years. Because an NEP inspection can be very time consuming (e.g., providing thousands of pages of evidence requested, interviewing hundreds of plant personnel, spending months hosting the inspection), you need to be prepared to effectively manage an inspection of that magnitude.

2. Strategy

Every company that has facilities subject to OSHA's NEP should have a corporate strategy to manage these inspections. The purpose of this strategy should include compliance with the law, public image, uniformity, efficiency, and above all else, to ensure the safety and health of its employees and neighbors. The goals are to (1) limit exposure and expense by supplying the agency with requested information, (2) have inspectors leave the site as soon as possible to minimize site disruptions, and (3) make compliance improvements as efficiently as possible.

The strategy should outline the structure of your preparation for and response to OSHA inspections, including:

- Will you perform a gap assessment of your facilities against the requirements and the bias of each directive?
- Will you mount an effort to close the identified gaps in your process safety management system?
- Will you make an effort to develop a common inspection logistics procedure to be used at your sites?
- At what level will you cooperate with OSHA?
- When will you dispute their claims – at the inspection, at the citation, or in court?
- Will you require a subpoena to provide some or all of the information?
- Will you involve outside resources before, during, or after the inspection, or in court?
- Will you dictate tactics such as how to communicate citations or how to track citations? Remember, similar citations at multiple sites may be considered repeat violations.

The strategy also needs to consider communication to the public because citations must be posted to the workforce and thus will become public. Your side of the story may be better expressed and received early on rather than heard publicly through the citations.

3. Tactics

Your tactics need to conform to your corporate strategy. Be ready to welcome, escort, and work with OSHA inspectors from the moment they arrive at your facility until they finish the inspection. Be ready to ensure OSHA gets the proper PSM compliance information the first time and your resources are used efficiently Use (1) teams of trained personnel to ensure the work load can be managed 24/7/365 and (2) procedures and forms so your performance meets the demands of the inspection.

Quickly providing the requested information demonstrates that you are knowledgeable and in control of compliance with the regulation. In addition, providing only the information requested limits your exposure.

3.1 Teams

3.1.1 Security

Your security team will be the first contact with an agency. It represents your company and must confirm that these are authorized inspectors. The team will then locate the right company contacts and will "escort" the inspectors until the company contacts arrive.

3.1.2 Leadership

You will need a leadership team to manage inspections. These team members should be the most knowledgeable individuals regarding the inspection process and the requirements applicable to the inspection being performed. The team should be comprised of leaders such as Health & Safety Manager, Environmental Manager, Site Leader, Legal Department, etc.

3.1.3 Escorts

Escorts accompany the inspector(s) within the facility and are selected for this role based upon their knowledge of the area and/or process being visited. Team members must be trained in interfacing with inspectors, possess good communication and interpersonal skills, make sound judgments, and accurately interpret inspectors' requests. They see that the inspectors follow the safety rules and agreed-upon procedures.

3.1.4 Subject Matter Experts

Subject matter experts (SMEs) are the individuals most knowledgeable of a particular aspect related to the inspection and are used as a resource for that area of focus. They consist of roles such as Training Manager/Coordinators, Operations Supervisors, Maintenance Supervisors, Industrial Hygienist, Process Engineers, EH&S staff, etc.

3.1.5 Support Staff

Support resources are the remaining roles that may provide support to the inspection, such as office professionals and interviewees.

3.1.6 Document Team

The document team obtains and controls the documents provided to the inspecting agency. Document team members must be knowledgeable of the various filing systems within the facility, such as equipment files, training records, operating procedures, maintenance inspection records, etc.

3.2 Procedures

3.2.1 Day 1

On the first day of the inspection, ask for the reason for the inspection and the nature or scope of the inspection. Confirm, if possible, the need for legal assistance (this need should have already been discussed internally and possible outside counsel resources identified). Determine if the agency has a subpoena. Decide if you will require a subpoena.

If you can present an overview of your manufacturing process and the process you use to comply with the covered regulation it will help the inspection. Lists of covered plants, permits, pertinent data, [3] [4] and personnel may be useful.

You need to agree on the ground rules for the inspection. These should cover personal protective equipment (PPE), accessing the facility, procuring documents, taking pictures, interviewing personnel, and providing daily procedures. Make plans known to office agency personnel.

3.2.2 Daily Procedures

The daily procedure should start with a brief meeting, including a safety review and plans for the day.

Daily wrap-up meetings should be scheduled with the inspectors to assess potential concerns, respond to potential violations, know the target areas, and begin addressing areas of concern. The wrap-up meetings should encompass the activities of the day, any findings pertaining to those activities, any pending documents or information requested by the inspectors and the status, collection of documents no longer needed by the inspectors, and planned activities for the next day.

3.2.3 Document Procedures

Ask for document requests to be in writing. You need a process to:

- Identify the documents that meet the scope of the request
- Stamp “Business Confidential” or “Trade Secrets” on appropriate documents to limit the potential release of information through Freedom of Information Act requests
- Label each page with a Bates number
- Make appropriate number of copies. (Consider scanning all documents by initial Bates number in the group. Remember there may be 10,000 pages or more)
- Create a master index of the documents provided, including the request, the Bates numbers, a description of what was provided, and when it was provided

- Retain copies per your document retention schedule
- Ensure "Attorney-Client Privileged" documents are not submitted to an inspector
- Ensure the agency is not allowed access to a range of information greater than that requested. This can be done by not providing access to a company computer, file cabinet, etc.

3.2.4 Picture and Video Procedures

Try to work out an agreement with OSHA about sharing collected evidence (e.g., videos, photographs). To avoid duplicating photographs for example, arrange to have someone take photos and then supply OSHA and the facility. Hot work permits may be required to take pictures. Take pictures right after the inspector and at the same perspective. Notify the inspector if "Business Confidential" or "Trade Secrets" are in the picture. Make a note of why the picture was taken and create a picture log.

3.2.5 Interviews

Ask to schedule interviews in advance to enable job coverage and other interviews. (If the facility is union, a union representative may be present at the interview – stipulate how this is to be addressed for your facility.) The escort should attempt to attend the interview but is not to participate. However, the employee has the right to be interviewed in private if he/she wishes.

Interviews should be conducted in a private area, such as a conference room, spare office, or other similar location.

The date, time, name of interviewer, and name of interviewee are to be documented and retained. Key topics discussed should be documented if possible.

Train all personnel who might talk to the agency on how to be interviewed. For example:

- Understand the question before responding. Ask for clarification when necessary
- There are no "off-the-record" topics
- It is acceptable to talk to or have an attorney during the interview
- Always tell the truth; provide facts and actual knowledge
- Answer only the questions asked
- Avoid speculation or opinion. If the interviewee does not know the answer to a question, it is better to say so
- The employee has the right to refuse to be interviewed
- Company will not take action with respect to the employee's choice

The inspector will take notes and ask the interviewee to sign. The interviewee does not have to sign. However, if the interviewee chooses to sign, it is best that the interviewee read the inspector's notes carefully and make any necessary corrections before signing.

3.2.6 Sampling

Try to schedule sampling a day in advance to ensure you have the right PPE, procedures, equipment, and forms. Will you share samples, collect duplicate samples, share results?

3.2.7 Training

Assign all the roles, including company spokesperson, interviewing, writing minutes, tracking document requests, and tracking responses, and train appropriately. Train everyone on "Worker Rights Under the Occupational Safety and Health Act of 1970." [5]

3.2.8 Communication

You need a plan to communicate the results of the inspection to all the team members, to management, and to the public.

4. Conclusion

It can take a lot of effort to prepare for a large inspection. However, if you do not prepare in advance of a large inspection, you will not make a good impression and will invite an expanded inspection.

5. References

[1] Remarks as Prepared for Jordan Barab, Acting Assistant Secretary of Labor for Occupational Safety and Health, American Society of Safety Engineers, Safety 2009 Conference, San Antonio, Texas, June 29, 2009 *http://www.osha.gov/pls/oshaweb/owadisp.show_document?p_table=SPEECHES&p_id=2002*

[2] CPL 03-00-10 - Petroleum Refinery Process Safety Management National Emphasis Program *http://www.osha.gov/pls/oshaweb/owadisp.show_document?p_table=DIRECTIVES&p_id=4014*

[3] CPL 02-09-08 - Injury and Illness Recordkeeping National Emphasis Program (RK NEP) *http://www.osha.gov/pls/oshaweb/owadisp.show_document?p_table=DIRECTIVES&p_id=4074*

[4] CPL 02-09-06 - PSM Covered Chemical Facilities National Emphasis Program *http://www.osha.gov/pls/oshaweb/owadisp.show_document?p_table=DIRECTIVES&p_id=4008*

[5] "Worker Rights Under the Occupational Safety and Health Act of 1970" *http://www.osha.gov/as/opa/worker/rights.html*

End of Manuscript

(108f) OSHA Chemical NEP Panel Discussion

Mark S. Dreux, Arent Fox PLLC, Washington, DC

Lisa Long, Office of General Industry Enforcement, U. S. Department of Labor - OSHA, Washington, DC

David A. Moore, AcuTech Group, Inc., McLean, VA

Mark L. Farley, Pillsbury Winthrop Shaw Pittman LLP, Houston, TX

Ken Hanchey, ABS Consulting, Houston, TX

Chad Patschke, ABS Consulting, Houston, TX

This presentation will provide additional time for the session speakers to discuss additional aspects of the OSHA Chemical National Emphasis Program (NEP), as well as discuss additional questions and comments from the session attendees.

See more of this Session: National Emphasis Programs - Preparation, Reviews, Outcomes & Challenges
See more of this Group/Topical: Topical 1: Global Congress on Process Safety

AUTHOR INDEX